Nanotechnology: Applications in Energy, Drug and Food

Shafiquzzaman Siddiquee • Gan Jet Hong Melvin
Md. Mizanur Rahman
Editors

Nanotechnology: Applications in Energy, Drug and Food

Springer

Editors
Shafiquzzaman Siddiquee
Biotechnology Research Institute
Universiti Malaysia Sabah
Kota Kinabalu, Sabah, Malaysia

Gan Jet Hong Melvin
Material and Mineral Research Unit (MMRU)
Faculty of Engineering
Universiti Malaysia Sabah, Jalan UMS
Kota Kinabalu, Sabah, Malaysia

Md. Mizanur Rahman
Faculty of Engineering
Energy Research Unit (ERU)
Universiti Malaysia
Kota Kinabalu, Sabah, Malaysia

ISBN 978-3-319-99601-1 ISBN 978-3-319-99602-8 (eBook)
https://doi.org/10.1007/978-3-319-99602-8

Library of Congress Control Number: 2018960259

This Springer imprint is published by the registered company Springer Nature Switzerland AG
The registered company address is: Gewerbestrasse 11, 6330 Cham, Switzerland

Contents

1 **Carbon Nanomaterials for Energy Storage Devices** 1
Zhipeng Wang and Gan Jet Hong Melvin

2 **Zinc Based Spinel Oxides for Energy Conversion and Storage Applications** 31
Faheem K. Butt and Sajid Ur Rehman

3 **Nanotechnology in Renewable Energy: Critical Reviews for Wind Energy** 49
W. K. Muzammil, Md. Mizanur Rahman, A. Fazlizan, M. A. Ismail, H. K. Phang, and M. A. Elias

4 **Nanomaterials: Electromagnetic Wave Energy Loss** 73
Gan Jet Hong Melvin, Yaofeng Zhu, and Qing-Qing Ni

5 **Emerging Nanotechnology for Third Generation Photovoltaic Cells** 99
Biju Mani Rajbongshi and Anil Verma

6 **Nanotechnology: Emerging Opportunities for Fuel Cell Applications** 135
Wai Yin Wong and Nabila A. Karim

7 **Nanotechnology-Based Drug Delivery Systems: Past, Present and Future** 175
Riana Awang Saman and Mohammad Iqbal

8 **Targeted Therapeutic Nanoparticles for Cancer and Other Human Diseases** 187
Rabiatul Basria S. M. N. Mydin, Wan Nordiana Rahman, Rosmazihana Mat Lazim, Amirah Mohd Gazzali, Nur Hazirah Mohd Azlan, and Said Moshawih

9 **Solid Lipid Nanoparticles: A Modern Approach for the Treatment of Neurodegenerative Diseases** 209
Anisha A. D'Souza

10 **Nanoparticles in Nanomedicine Application: Lipid-Based Nanoparticles and Their Safety Concerns** 227
Rabiatul Basria S. M. N. Mydin and Said Moshawih

11 **Nanomaterials in Drug Delivery System** 233
Nur Izzati Mohd Razali, Noor Syazwani Mohd Saufi, Raha Ahmad Raus, Wan Mohd Fazli Wan Nawawi, and Dayang Fredalina Basri

12 **Drug Discovery: A Biodiversity Perspective** 249
Kholis A. Audah

13 **Nano TiO_2 for Biomedical Applications** 267
Khairul Arifah Saharudin, Srimala Sreekantan, Rabiatul Basria S. M. N. Mydin, Siti Nor Qurratu Aini Abd Aziz, and G. Ambarasan Govindasamy

14 **Nanotechnology: Recent Trends in Food Safety, Quality and Market Analysis** 283
Zamri Nurfatihah and Shafiquzzaman Siddiquee

15 **Nanotechnology Applications in Food: Opportunities and Challenges in Food Industry** 295
Afiqah Silon Ummi and Shafiquzzaman Siddiquee

16 **Improvement of Food Packaging Based on Functional Nanomaterial** 309
Bambang Kuswandi and Mehran Moradi

17 **Applications of Polymeric Nanoparticles in Food Sector** 345
Norizah Abdul Rahman

18 **Acetylcholinesterase (AChE) Biosensors for Determination of Carbamate Pesticides** 361
Anwar Samsidar and Shafiquzzaman Siddiquee

19 **Nanosensors Based Detection of Foodborne Pathogens** 377
Mohd Hazani Mat Zaid, Jerro Saidykhan, and Jaafar Abdullah

20 **Electrochemical Methods to Characterize Nanomaterial-Based Transducers for the Development of Noninvasive Glucose Sensors** 423
Nur Alya Batrisya Ismail, Firdaus Abd-Wahab, Nurul Izzati Ramli, Mamoun M. Bader, and Wan Wardatul Amani Wan Salim

Contributors

Jaafar Abdullah Institute of Advanced Technology, Universiti Putra Malaysia, Serdang, Selangor, Malaysia

Department of Chemistry, Faculty of Science, Universiti Putra Malaysia, Serdang, Selangor, Malaysia

Norizah Abdul Rahman Department of Chemistry, Faculty of Science, Universiti Putra Malaysia, Serdang, Selangor Darul Ehsan, Malaysia

Firdaus Abd-Wahab Department of Biotechnology Engineering, Faculty of Engineering, International Islamic University Malaysia, Selangor, Malaysia

Riana Awang Saman Biotechnology Research Institute, Universiti Malaysia Sabah, Kota Kinabalu, Sabah, Malaysia

Afiqah Silon Ummi Biotechnology Research Institute, Universiti Malaysia Sabah, Kota Kinabalu, Sabah, Malaysia

Kholis A. Audah Department of Biomedical Engineering and Directorate of Academic Research and Community Services, Swiss German University, Tangerang, Indonesia

Siti Nor Qurratu Aini Abd Aziz School of Materials & Mineral Resources Engineering, Engineering Campus, Universiti Sains Malaysia, Nibong Tebal, Pulau Pinang, Malaysia

Nur Hazirah Mohd Azlan Oncological and Radiological Sciences Cluster, Advanced Medical and Dental Institute, Universiti Sains Malaysia, Kepala Batas, Pulau Pinang, Malaysia

Mamoun M. Bader Department of Chemistry, College of Science and General Studies, Alfaisal University, Riyadh, KSA

Dayang Fredalina Basri Biomedical Science Programme, School of Diagnostic & Applied Health Sciences, Faculty of Health Sciences, Universiti Kebangsaan Malaysia, Kuala Lumpur, Malaysia

Faheem K. Butt Division of Science and Technology, Department of Physics, University of Education, Lahore, Pakistan

Physik-Department, ECS, Technische Universität München, Garching, Germany

Anisha A. D'Souza Formulation Development Laboratory, Pharmaceutical R&D, Piramal Enterprises Limited, Mumbai, India

M. A. Elias Solar Energy Research Institute (SERI), Universiti Kebangsaan Malaysia, Bangi, Malaysia

A. Fazlizan Solar Energy Research Institute (SERI), Universiti Kebangsaan Malaysia, Bangi, Malaysia

G. Ambarasan Govindasamy School of Materials & Mineral Resources Engineering, Engineering Campus, Universiti Sains Malaysia, Nibong Tebal, Pulau Pinang, Malaysia

Mohammad Iqbal Biotechnology Research Institute, Universiti Malaysia Sabah, Kota Kinabalu, Sabah, Malaysia

M. A. Ismail Faculty of Engineering, Material and Mineral Research Unit (MMRU), Universiti Malaysia, Kota Kinabalu, Sabah, Malaysia

Nur Alya Batrisya Ismail Department of Biotechnology Engineering, Faculty of Engineering, International Islamic University Malaysia, Selangor, Malaysia

Nabila A. Karim Fuel Cell Institute, Universiti Kebangsaan Malaysia, Bangi, Selangor, Malaysia

Bambang Kuswandi Chemo and Biosensors Group, Faculty of Pharmacy, University of Jember, Jember, East Java, Indonesia

Rosmazihana Mat Lazim Medical Radiation Programme, School of Health Sciences, Universiti Sains Malaysia (Health Campus), Kubang Kerian, Kelantan, Malaysia

Gan Jet Hong Melvin Material and Mineral Research Unit (MMRU), Faculty of Engineering, Universiti Malaysia Sabah, Jalan UMS, Kota Kinabalu, Sabah, Malaysia

Amirah Mohd Gazzali School of Pharmaceutical Sciences, Universiti Sains Malaysia, Minden, Pulau Pinang, Malaysia

Mehran Moradi Department of Food Hygiene and Quality Control, Faculty of Veterinary Medicine, Urmia University, Urmia, West Azarbaijan, Iran

Said Moshawih Jordan Center for Pharmaceutical Research, Amman, Jordan

W. K. Muzammil Faculty of Engineering, Material and Mineral Research Unit (MMRU), Universiti Malaysia, Kota Kinabalu, Sabah, Malaysia

Rabiatul Basria S. M. N. Mydin Oncological and Radiological Sciences Cluster, Advanced Medical and Dental Institute, Universiti Sains Malaysia, Kepala Batas, Pulau Pinang, Malaysia

Wan Mohd Fazli Wan Nawawi Department of Biotechnology Engineering, Faculty of Engineering, International Islamic University of Malaysia, Jalan Gombak, Selangor, Malaysia

Qing-Qing Ni Department of Functional Machinery and Mechanics, Shinshu University, Ueda, Tokida, Japan

Zamri Nurfatihah Biotechnology Research Institute, Universiti Malaysia Sabah, Kota Kinabalu, Sabah, Malaysia

H. K. Phang Faculty of Engineering, Energy Research Unit (ERU), Universiti Malaysia, Kota Kinabalu, Sabah, Malaysia

Md. Mizanur Rahman Faculty of Engineering, Energy Research Unit (ERU), Universiti Malaysia, Kota Kinabalu, Sabah, Malaysia

Wan Nordiana Rahman Medical Radiation Programme, School of Health Sciences, Universiti Sains Malaysia (Health Campus), Kubang Kerian, Kelantan, Malaysia

Biju Mani Rajbongshi Sustainable Environergy Research Lab (SERL), Department of Chemical Engineering, Indian Institute of Technology Delhi, Delhi, India

Nurul Izzati Ramli Department of Biotechnology Engineering, Faculty of Engineering, International Islamic University Malaysia, Selangor, Malaysia

Raha Ahmad Raus Department of Biotechnology Engineering, Faculty of Engineering, International Islamic University of Malaysia, Jalan Gombak, Selangor, Malaysia

Nur Izzati Mohd Razali Department of Biotechnology Engineering, Faculty of Engineering, International Islamic University of Malaysia, Jalan Gombak, Selangor, Malaysia

Sajid Ur Rehman State Key Laboratory on Integrated Optoelectronics, Institute of Semiconductors, Chinese Academy of Sciences, Beijing, China

Khairul Arifah Saharudin School of Materials & Mineral Resources Engineering, Engineering Campus, Universiti Sains Malaysia, Nibong Tebal, Pulau Pinang, Malaysia

Jerro Saidykhan Department of Chemistry, Faculty of Science, Universiti Putra Malaysia, Serdang, Selangor, Malaysia

Anwar Samsidar Biotechnology Research Institute, Universiti Malaysia Sabah, Kota Kinabalu, Sabah, Malaysia

Noor Syazwani Mohd Saufi Department of Biotechnology Engineering, Faculty of Engineering, International Islamic University of Malaysia, Jalan Gombak, Selangor, Malaysia

Shafiquzzaman Siddiquee Biotechnology Research Institute, Universiti Malaysia Sabah, Kota Kinabalu, Sabah, Malaysia

Srimala Sreekantan School of Materials & Mineral Resources Engineering, Engineering Campus, Universiti Sains Malaysia, Nibong Tebal, Pulau Pinang, Malaysia

Anil Verma Sustainable Environergy Research Lab (SERL), Department of Chemical Engineering, Indian Institute of Technology Delhi, Delhi, India

Wan Wardatul Amani Wan Salim Department of Biotechnology Engineering, Faculty of Engineering, International Islamic University Malaysia, Selangor, Malaysia

Zhipeng Wang Institute of Advanced Materials, Jiangxi Normal University, Nanchang, Jiangxi Province, China

Wai Yin Wong Fuel Cell Institute, Universiti Kebangsaan Malaysia, Bangi, Selangor, Malaysia

Mohd Hazani Mat Zaid Institute of Advanced Technology, Universiti Putra Malaysia, Serdang, Selangor, Malaysia

Yaofeng Zhu Key Laboratory of Advanced Textile Materials and Manufacturing Technology, Ministry of Education, Zhejiang Sci-Tech University, Hangzhou, People's Republic of China

About the Editors

Shafiquzzaman Siddiquee is an associate professor in the Biotechnology Research Institute of Universiti Malaysia Sabah (UMS), Malaysia. Dr. Siddiquee obtained his MSc (Molecular Microbiology) and PhD (Bioelectrochemistry) degrees in 2008 and 2010, respectively, from University Putra Malaysia (UPM), Malaysia. In 2011, he received a *Best Post Graduate Award* from UPM for being the most outstanding postgraduate student of the year. In 2014, he was the winner of the *ProSPER.net-Scopus Young Scientist Award* in the category of Sustainable Agriculture. He served as a guest professor at the Institute of Analytical Chemistry, Chemo- and Biosensors of the Regensburg University, Germany, in 2016. Currently he is a head of the UMS Biosensor Research Group. He has involved in varying types of research areas such as chemosensors- or biosensors-based research, management of food wastes, and *Trichoderma* biodiversity and its applications. He has published over 85 refereed journal articles, 4 books, 12 book chapters, and 38 proceeding papers. His findings yielded, so far, 4 patents, 5 invited speaking engagements, and 40 research awards, including notable award of *MOHE entrepreneurial Project Award 2015* and International Conference and Exposition on Inventions by Institutions of Higher Learning 2015 *(PECIPTA 2015)*. Dr. Siddiquee has established relevant research group networks both nationally and internationally. In pop culture, Dr. Siddiquee's findings have been broadcast in national (Malaysia) TV stations (TV1, TV2, and TV7) as well as highlighted in various print media.

Gan Jet Hong Melvin received his Ph.D. from Shinshu University, Japan, in 2015. Then, he joined the Institute of Carbon Science and Technology, Japan, for 1 year as a researcher. Now, he is a senior lecturer in Mechanical Engineering Program and a research fellow of Material & Mineral Research Unit at the Faculty of Engineering, Universiti Malaysia Sabah, Malaysia.

His major field is materials science related journals, which includes nanoparticles, hybrid nanomaterials, carbon-based nanomaterials, and composite materials. Some of his previous and current active projects and works are related to polymer-based electroactive actuators, electromagnetic wave absorber, and energy storage devices, which are published in peer-reviewed journals, conference proceedings, and abstracts. He is an active reviewer for several material-related journals and a member of the Board of Engineers, Malaysia.

Md. Mizanur Rahman is working as a senior lecturer in the Mechanical Engineering Program, Faculty of Engineering at Universiti Malaysia Sabah. He has interests in both fundamental and applied aspects of renewable energy technologies, especially in new technology to harvest electricity from solar power and hydro power. Firstly he has started his carrier at the Renewable Energy Technologies in Asia (RETs in Asia) Phase—II Project in 1999 as a Research Engineer under the Department of Mechanical Engineering at Khulna University of Engineering and Technology (KUET) and Asian Institute of Technology (AIT), Bangkok, Thailand, before joining as a Program Support Specialist in 2005 at BRAC, Bangladesh. Dr. Rahman was appointed as an Assistant Manager Technical in 2006 at Rural Power Company Ltd and as a lecturer in 2009 at TAS Institute of Oil and Gas. Finally, he moved to Universiti Malaysia Sabah as senior lecturer in the year 2012. Dr. Rahman has obtained BSc in Engineering (Mechanical), MSc in Environmental Management, and PhD in Mechanical Engineering from BIT Khulna, Bangladesh: jointly from the University of San Francisco, USA, Mahidol University, Bangkok, Thailand, and Universiti Malaysia Sabah in 1998, 2004, and 2012, respectively. He has published research papers in various journals and national and international conference proceeding. He is a chartered engineer, member of the Institution of Mechanical Engineers (IMechE), UK; fellow of the Institute of Engineers, Bangladesh (IEB); and member of the Bangladesh Society of Mechanical Engineers (BSME), Institute of Material Malaysia (IMM), and the American Society of Mechanical Engineering (ASME).

Chapter 1
Carbon Nanomaterials for Energy Storage Devices

Zhipeng Wang and Gan Jet Hong Melvin

1.1 Introduction

Extreme climate changes and the decreasing availability of fossil fuels, such as coal, natural gas, oil, and so on, require the users, providers, and society to gear towards sustainable and renewable resources (Simon and Gogotsi 2008; Yang et al. 2011). This fact is strongly supported and apparent as the global energy consumption has been speeding up at an alerting rate due to the fast economic expansion worldwide, increase in world population, and the rapidly growing of energy-based appliances including increasing human reliance on them (Dai et al. 2012). For instance, it was estimated that the world need to double its energy supply by 2050 to fulfill the demands (Dai et al. 2012; Chen and Dai 2013).

Accordingly, we can observe the boost in renewable energy production and utilization, for example from sun and wind. However, sun does not shine during the night and wind does not blow on request, which means that power generation does not necessarily correspond to demand. Furthermore, it is worth to concern that the renewable resources are localized and often away from the load centers (Yang et al. 2011). Moreover, it is undeniable that fossil fuels, nuclear heat, renewable energies are available to be converted to electrical energy. However, this conversion is performed with large energy losses, for instance only about 30% of the nuclear heat is converted to electrical energy (Fauvarque and Simon 2010). Thus, energy storage systems play a significant role and essential in our lives, to prevent

Z. Wang
Institute of Advanced Materials, Jiangxi Normal University, Nanchang, Jiangxi Province, China

G. J. H. Melvin (✉)
Material and Mineral Research Unit (MMRU), Faculty of Engineering, Universiti Malaysia Sabah, Jalan UMS, Kota Kinabalu, Sabah, Malaysia
e-mail: melvin.gan@ums.edu.my

S. Siddiquee et al. (eds.), *Nanotechnology: Applications in Energy, Drug and Food*,
https://doi.org/10.1007/978-3-319-99602-8_1

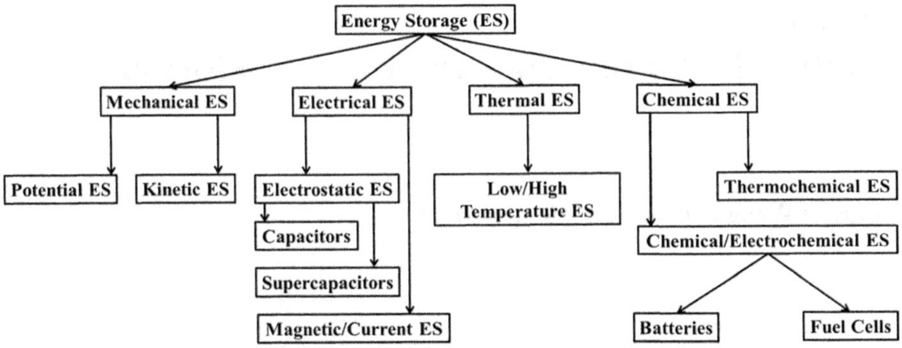

Fig. 1.1 Various types of energy storage technology

unnecessary losses and supplying necessary energy which satisfies the demands. From this point of view, electrical energy storage systems such as batteries and electrochemical capacitors (ECs) have been studied extensively and becoming the center of highlight for this particular reason. Majorly, energy storage can be divided into chemically, electrochemically, and electrically stored (Pumera 2011). The classification of major energy storage technologies is presented in Fig. 1.1.

In our daily life, batteries are one of the most common electrical energy storage devices owing to their capability to store large amount of energy in a relatively small volume and weight, and in the same time are able to provide appropriate and suitable levels of power for wide broad applications. However for batteries, even though high energy densities can be obtained, slow power delivery has agonized them (Zhai et al. 2011). This is one of the reasons which limited or restricted their utilization, especially when fast storage coupled with high power is the demands. Alternatively, to tackle the constraints, ECs or usually described as supercapacitors or ultracapacitors have been given the attentions. Some of the benefits of ECs are, they can provide high specific power, a long cycle life, fast charge/discharge process, and so on (Zhai et al. 2011). However, within seconds charge/discharge process of ECs caused their energy density is lower than the batteries, but faster power delivery can be accomplished in short time (Simon and Gogotsi 2008, 2013). ECs play an essential role in complementing or substituting batteries in the energy storage field, for example they can act as back-up power supply utilized to protect against power disruption, load levelling, and so on (Simon and Gogotsi 2008; Zhai et al. 2011).

Generally, electric capacitor is a sandwich structure of several materials, where two conductive plates neighboring a dielectric/insulator, as illustrated in Fig. 1.2. The charge accumulation will occur during charging process and the charge separation will occur during discharging process (Yu et al. 2013). They can be divided into electrostatic and electrolytic (utilization of electrolytes as dielectrics) capacitors which are considered as the first and second generation capacitors. These capacitors function mainly as the components in electrical circuit to store micro- to picofarad charges of direct current or to filter the frequencies for alternating current circuits. Rapid progress in materials science, advancement of technologies, and high

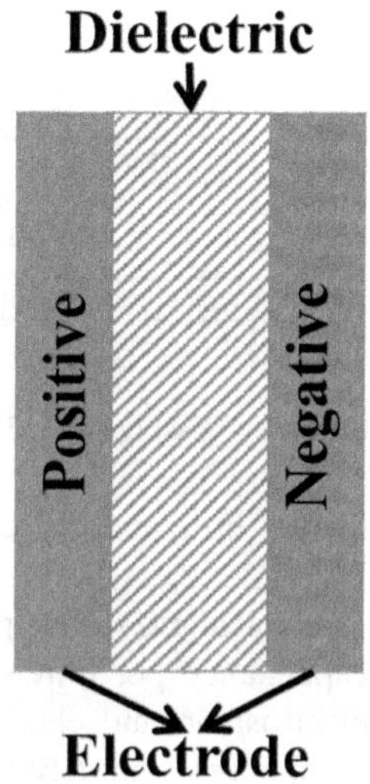

Fig. 1.2 Conventional electric capacitor

demands are some of the factors that lead to the invention of third generation, supercapacitors.

Capacitive behavior of ECs can be majorly categorized into two types, depending on their charge storage mechanism and active materials utilized. First type is the electrochemical double layer capacitors (EDLCs), where active materials with high surface area are ideal and dependent on electrostatic attraction between ions and the charged electrode's surface or electrostatic charge accumulation at the electrode/electrolyte interface (Simon and Gogotsi 2008; Chen and Dai 2013; Zhai et al. 2011). Carbon materials are one of the prominent candidates to be selected as the active materials (He et al. 2017). Second type is redox supercapacitors or also known as pseudo-capacitors, which is highly associated with fast redox or Faradaic charge transfer reactions of the electro-active materials, such as metal oxides (Wallar et al. 2017) or conducting polymers (Jo et al. 2017), on the surface of the electrodes, for charge storage (Simon and Gogotsi 2008; Zhai et al. 2011; Yu et al. 2013). It is also worth to notice the existence of hybrid capacitors that combine the capacitive or pseudo-capacitive electrode with a battery electrode, which integrate the properties of both, capacitor and battery. The classification of various supercapacitors is presented in Fig. 1.3.

The earliest EC with extinguish high capacitance was described and patented in 1957 by Becker (Becker 1957). Carbon with a high specific surface area coated on a metallic current collector in a sulphuric acid solution was utilized. In 1966, Standard Oil of Ohio (SOHIO) developed another version and patented a device that stored energy in double layer interface (Rightmire 1966). Advancing with further adjustments, Boos invented the first practical supercapacitor and patented in 1970 (Boos 1970). Then in 1971, Nippon Electronic Company (NEC), Japan, licensed the technology from SOHIO and developed aqueous-electrolyte capacitors for powersaving units in electronics, and this application can be reflected as the pioneer

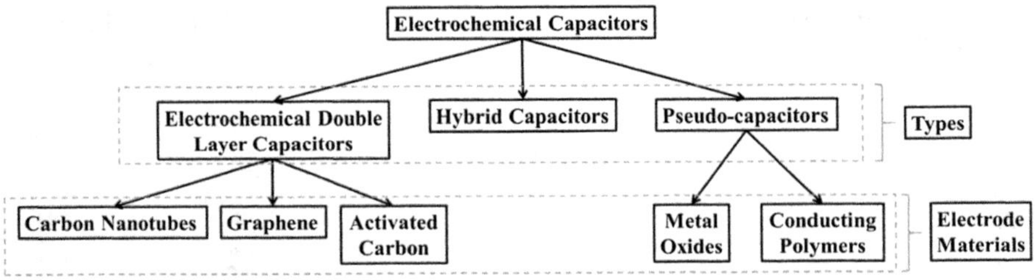

Fig. 1.3 Classification of various supercapacitors

for EC utilization in commercial products (Simon and Gogotsi 2008; Sharma and Bhatti 2010).

Typical ECs or supercapacitors use electrolyte solutions but have even greater capacitance per unit volume due to their porous electrode structure compared to electrostatic and electrolytic capacitors. They are constructed with three essential components, namely the electrodes, the electrolyte, and the separator. The overall performance of supercapacitors is determined by the physical properties of both the electrode and the electrolyte materials. Nevertheless, the electrode is one of the most important components for charge storage/delivery, and plays a vital role in determining the energy and power densities of a supercapacitor. The electrochemical performance of a supercapacitor can be characterized by cyclic voltammetry and galvanostatic charge-discharge measurements. Furthermore, ECs are categorized into two major groups, which are symmetric and asymmetric (Sharma and Bhatti 2010). Symmetric ECs (SECs) use the similar electrode material (usually carbon) for both the positive and negative electrodes. Asymmetric ECs (AECs) use two different materials for the positive and negative electrodes. SECs get their electrostatic charge from the accumulation and separation of ions at the interface between the electrolyte and electrodes. Aqueous or organic electrolyte solutions can be utilized for SECs. The electrolyte solution comprises aqueous substances (such as potassium hydroxide or sulfuric acid) or organic substances (such as acetonitrile or propylene carbonate). An SEC using aqueous electrolyte is also known as a Type I SEC and an SEC using organic electrolyte is known as a Type II SEC. Similarly, an AEC using an aqueous electrolyte is known as a Type III AEC and one using organic electrolyte is known as a Type IV AEC. A typical charge storage mechanism of electrochemical double-layer capacitor (EDLC) is depicted in Fig. 1.4.

It is undeniable that nanomaterials, especially carbon nanomaterials play a great role in the development of ECs. This is due to their remarkable structure, electrical properties, electrochemical stability, and so on. High performance and functionality can be expected from carbon nanomaterial-based ECs. In this chapter, the influence factors of EC performance focusing on graphene and carbon nanotubes (CNTs) will be discussed further.

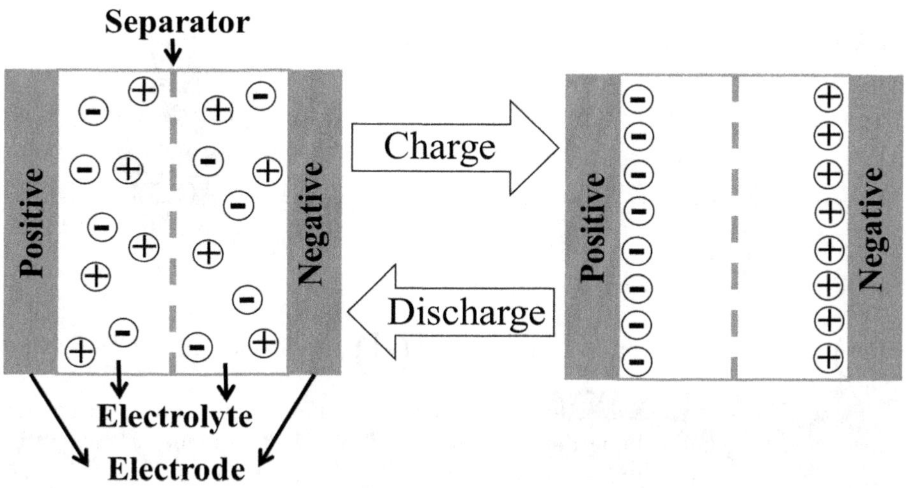

Fig. 1.4 Charge/discharge process of electrochemical double-layer capacitor

1.2 Carbon Nanomaterials

1.2.1 Carbon Nanotubes

There are great numbers of reviews and publications related to the history, synthesis, excellent properties, and development as ECs of CNTs. Since the discovery of fullerene (C_{60}, buckyball) by Kroto et al. (1985), among the first synthesis of CNTs by Oberlin et al. (1976), observation of both multi- and single-walled CNTs soon afterwards by Iijima in 1991 (Iijima 1991) and 1993 (Iijima and Ichihashi 1993), respectively, the interest in CNTs has rapidly developed. Furthermore, CNTs are famous for their outstanding mechanical, electrical, chemical, thermal, etc. properties that allow them to be manipulated for various applications and research studies.

Generally, CNTs can be grouped into single-walled CNT (SWCNT) and multi-walled CNT (MWCNT). A SWCNT can be thought of as a rolled-up sheet of a structure called graphene, which is a single layer of an allotrope of carbon called graphite, and the edges of the sheet are joined together to form a seamless tube (Young and Lovell 2011; Hierold et al. 2008). Several tubes of different diameters can be fitted into each other to make a MWCNT. Figure 1.5 shows the images of CNT and graphene.

There are mainly three methods to produce CNTs: arc discharge, laser ablation, and chemical vapor deposition (CVD). Each of these techniques had its advantages and disadvantages and briefly discussed below.

Arc discharge and laser ablation depend on the evaporation of a graphite target to create gas phase carbon fragments that recombine to form CNTs. The temperature

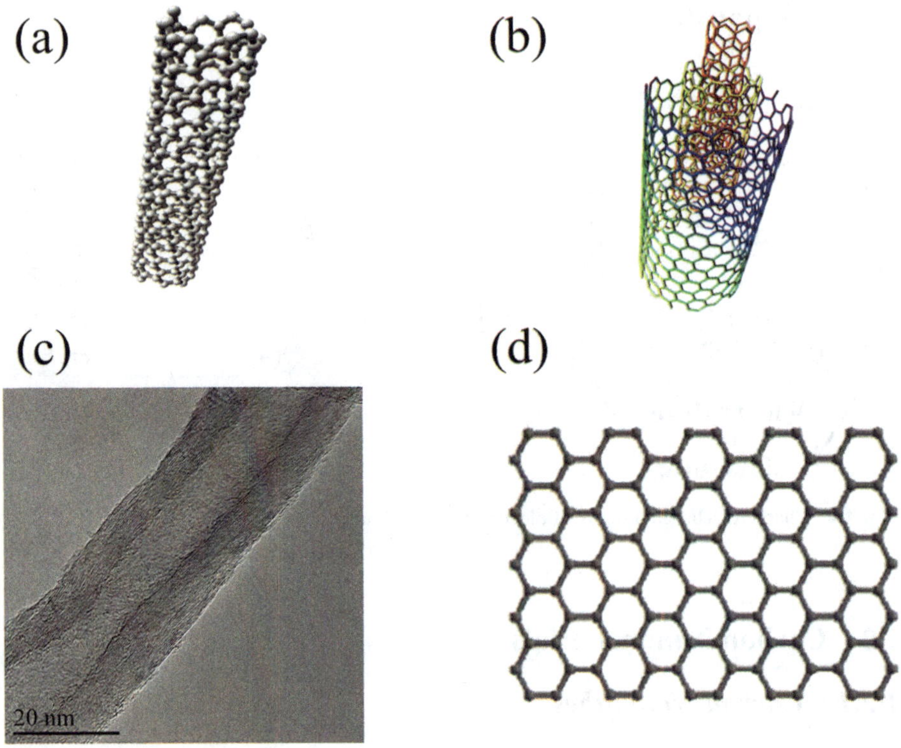

Fig. 1.5 (**a**) SWCNT, (**b**) MWCNT, (**c**) TEM image of MWCNT, (**d**) graphene

reached in these processes is in the range 2000–3000 °C, more than sufficient for the carbon atoms to rearrange into the tube structure. In order to promote the yield of CNTs, several different metals in concentrations about 1% are incorporated into the target materials that is evaporated (Hierold et al. 2008). For application of CNTs for composite materials where large quantities of CNTs are required, these methods would make the cost of CNTs prohibitive (Thostenson et al. 2001). However, a large amount of non-tubular graphitic and amorphous carbon is also produced during the process (Hierold et al. 2008; Thostenson et al. 2001). Thus, purification steps are essential before the usage of them.

CVD is the most widely used method for the production of CNTs. Generally, the CVD process includes catalyst-assisted decomposition of hydrocarbons, usually ethylene or acetylene, in a tube reactor at 400–1100 °C and growth of CNTs over the catalyst upon cooling the system (Hierold et al. 2008; Hu et al. 2010; Popov 2004). The growth temperature depends on the type of CNTs to be grown and the catalyst composition (Hierold et al. 2008). The advantages of this method are the ability to fabricate aligned arrays of CNTs with controlled diameter and length, and under the right condition only nanotubes are produced and no unwanted graphitic material (Hierold et al. 2008; Thostenson et al. 2001).

1.2.2 Graphene

Since the extraction of graphene from bulk graphite in 2004 (Novoselov et al. 2004), which this discovery was recognized by Nobel Prize in Physics (2010), the interest in graphene has rapidly increased. In that study, graphene sheets were obtained by using Scotch tape to repeatedly split graphite crystals into increasingly thinner pieces until individual atomic planes (monolayer of graphite) were reached. Graphene is a one-atom-thick planar sheet of two-dimensional (2D) sheet sp^2 bonded carbon atoms that are densely packed in a honeycomb crystal lattice. As the mother and components of all graphitic forms, graphene is a building block for carbon materials of all other dimensionalities, for instance 0D buckyballs, 1D nanotubes, and 3D graphite. Graphene, in fact, including multi-layered graphene, possesses high aspect ratio, large surface area, excellent electrical, thermal, mechanical properties, and so on (Dai et al. 2012; Pumera 2010).

Generally, preparation of graphene can be categorized into top-down and bottom-up approaches. Several factors are to be considered in graphene preparation, such as cost effectiveness, scaled-up production, high electrochemical activity, conductivity, and so on (Lv et al. 2016).

Top-down approach usually utilizes mechanical force or chemical intercalation to overcome the van der Waals forces between the graphene layers to achieve separation of graphene from bulk graphite, such as micromechanical cleavage (mechanical exfoliation), oxidation-exfoliation-reduction, intercalation exfoliation, solid exfoliation, and so on (Lv et al. 2016; Dong et al. 2017). Particularly for EC applications, mechanical exfoliation by Scotch tape is inappropriate because of the low yield, even though the graphene produced possesses high quality (without abundant defects).

Bottom-up approach usually utilizes a small molecule precursor to grow into graphene by chemical vapor deposition (CVD) or chemical synthesis. For graphene prepared by CVD, they show some excellent properties, as a result of their large crystal domains, monolayer structure and less defects in the graphene sheets, which are beneficial for boosting carrier mobility in electronic applications (Ke and Wang 2016). Furthermore, the layers and defects of graphene can be controlled by adjusting growth parameters such as temperature, time, catalyst, and so on.

By utilizing physical or chemical method appropriately, preparation of porous graphene and doped graphene sheets can be realized, and high performance/functionality graphene can be prepared with introduction of organic/inorganic materials. In the same time, the production of graphene with low cost, high yield, and high quality is also a crucial factor. Generally, the chemical exfoliation of graphite into graphene oxides (GOs), followed by controllable reduction of GOs (with reduction agent such as hydrazine hydrate) into graphene can be considered as an efficient and low-cost method (Dong et al. 2017).

1.2.3 Carbon Nanomaterials Derived from Waste Materials

Carbon nanomaterials such as CNTs and graphene have been utilized widely for the development of ECs due to their exceptional properties. Not limited to chemical process or reaction to obtain carbon nanomaterials, carbonization of waste materials, which is comparatively low cost and environment friendly, can be considered as a beneficial method (Melvin et al. 2017a, b). Since most of the waste materials are available in large scale, greener technology to produce bulk quantity of valuable carbon nanomaterials from waste materials can be promoted. Waste materials can be in the form of agricultural wastes, synthetic compound wastes, and so on.

Agricultural waste materials, such as rice husks, have been exploited to produce carbon nanomaterials (Muramatsu et al. 2014; Wang et al. 2015). A good quality of carbon nanomaterials (graphene, derivatives of graphene) can be obtained. They offer high performance and promising applications in carbon-based energy storage and conversion devices.

Furthermore, through chemical or physical activation of carbon materials derived from waste materials, activated carbon (AC) can be produced. ACs are usually used due to their porous structure that can lead to high capacitance, which is favorable in ECs field. Some of the recent reports, activated carbon are produced from corn straws (Lu et al. 2017), oil palm shells (Abioye et al. 2017), sawdusts (Huang et al. 2017), which are dedicated to the development of high performance ECs.

1.3 Influence Factors of Energy Storage

1.3.1 Porosity Effect

A simplest model of an EC is composed of two pieces of conductive electrodes (mainly porous carbon materials) with high-surface-area porous structures, which were soaked into electrolyte and separated by an ion-conducting but electron-insulating separator membrane (Conway 1999). The two electrodes are identical for a symmetric supercapacitor, but different for an asymmetric supercapacitor (ASC). By applying voltage to the two electrodes, the electrolyte ions with the opposite signs accumulate on the surface of each electrode, and the concentration of ions is commonly proportional to the applied voltage. For EDLC, the capacitance originates from the pure electrostatic charge adsorption on the interface between electrode and electrolyte. Therefore, the capacitance is strongly dependent on the surface area of the electrode materials which is accessible to the electrolyte ions. The porous carbon materials with large surface area enable EDLC to possess high capacitance. As for pseudo-capacitor, it involves chemical and electrochemical interactions with the electrolyte. These interactions with fast and reversible Faradaic processes provide additional charge storage, resulting in higher energy densities for the ECs. When an EC stores charges by a capacitive carbon electrode with a pseudocapacitive electrode, it is then called a hybrid supercapacitor. Due to their excellent physiochemical properties as well as controllable structures and relatively low costs, porous carbon materials have been widely used as the electrode materials in all types of ECs.

The energy density of ECs is determined by the capacitance of their electrodes and the operating maximum voltage. The EC energy could be calculated according to following equation (Gu and Yushin 2014):

$$E = \left(\frac{C_{-} \cdot C_{+}}{C_{-} + C_{+}}\right) \cdot V_{\max}^{2} \tag{1.1}$$

where E is the energy, $V_{\max}$ is the maximum voltage between two electrodes, C_{-} and C_{+} are the capacitances of the negative and positive electrodes, respectively. When the capacitances of the both electrodes are the same, the maximum energy is achieved:

$$E = \frac{C \cdot V_{\max}^{2}}{2} \tag{1.2}$$

In a symmetric EDLC, the specific capacitance of each electrode could be identified by a galvanostatic charge-discharge measurement, where the specific capacitance is determined by the following equation:

$$C = Idt/dV \tag{1.3}$$

The power density is another crucial parameter for evaluating the EC performance, and can be obtained using equation:

$$P = \frac{V_{\max}^{2}}{4R} \tag{1.4}$$

where P is the power density, R is the equivalent series resistance including intrinsic resistance of electrode materials, contact resistance between the electrode materials and current collectors, diffusion resistance of ions in electrode materials and through the separator, and ionic resistance of electrolytes.

One strategy to improve the energy density of EDLC is to maximize the capacitance, which is affected by not only specific surface area (SSA), but also pore size and distribution, pore volume, and electrical conductivity. Initially, it was expected that the larger SSA, the higher capacitance. Thus, most efforts have been focusing increasing the specific capacitance of EDLCs by increasing the SSA of carbon materials. However, it was found that the specific capacitance of carbons increases with increasing SSA at relatively low values, but it rapidly tends to saturation when the SSA is greater than 1200–2000 $m^2\ g^{-1}$ (Barbieri et al. 2005). It means that there is no linear relationship between the SSA and the capacitance. For example, ACs with SSA as high as 2500–3000 m^2 g^{-1} possessed a relatively small capacitance of less than 10 μF cm^{-2}, which is smaller than the theoretical capacitance of EDLCs (15–25 μF cm^{-2}) (Raymundo-Piñero et al. 2006). On the contrary, activated carbon fibers with a SSA of nearly 1000 m^2 g^{-1} (i.e., 730–1274 m^2 g^{-1}) show extremely high capacitances corresponding to those for conventional AC materials with a SSA of 3000 m^2 g^{-1} (Kim et al. 2012). Obviously, the wrong impression was almost given when the SSA was employed to discuss the capacitance of EDLC electrodes which were made of activated carbon materials. Recently, many attempts have been made to investigate the relationship between the porosity and the capacitance of carbon materials. The results showed that an appropriate pore is more important than a high SSA to achieve high capacitance performance. As we know, based on the International Union of Pure and Applied Chemistry (IUPAC) classification, pore sizes can be classified into three types: micropores (<2 nm), mesopores (2–50 nm), and macropores (>50 nm). Previous studies showed most of the surface area of AC electrode materials locates in the scale of micropores (Frackowiak and Béguin 2001). These pores are often poorly or non-accessible for electrolyte ions, especially for organic electrolytes, and are not suitable for forming an electrical double layer on the pore walls. However, Chmiola et al. (2006a) reported that the ion salvation shell will be highly distorted and can further enter the micropores, resulting into an anomalous increase in the capacitance for carbon materials with pore sizes less than 1 nm, as shown in Fig. 1.6. In the related research, carbide-derived carbon with pore sizes of 0.8–1.0 nm also showed high specific surface capacitance in 1 M H_2SO_4 due to the desolvation of the electrolyte ions entering subnanometer pores, which cause a sieving effect at pore sizes less than the solvated ion sizes (Chmiola et al. 2006b). The same group further investigated the relation between the ion size and pore size for the EDLC application using solvent-free electrolyte, e.g. ionic liquid (IL), and demonstrated a maximal normalized capacitance at the pore size of about 0.7 nm, which is nearly close to the electrolyte ion sizes (Largeot et al. 2008). The similar results were also concluded in the organic electrolytes (Segalini et al. 2012). These interesting finds challenged the long-term ideals that only pores with size bigger than that of solvated electrolyte ions can contribute to the capacitance of EDLC. In fact, it has been reported that mesopores contribute the most to the capacitance in the EDLCs (Tanahashi et al. 1990). Very recently, Lin et al. obtained the high specific capacitance from nitrogen-doped mesoporous carbon with a biomodal pore size distribution centered around 1.8 nm and 3.5–4 nm in all three mesporous structures (Lin et al. 2015). However, the straight mesoporous channels are only effective for a fast transportation of electrolyte ions to the small micropores where they are stored (Vix-Guterl et al. 2005). Therefore, the relation between the pore size and capacitance should be future investigated both theoretically and experimentally, which is propitious to design the carbon materials for EC electrodes with high performance.

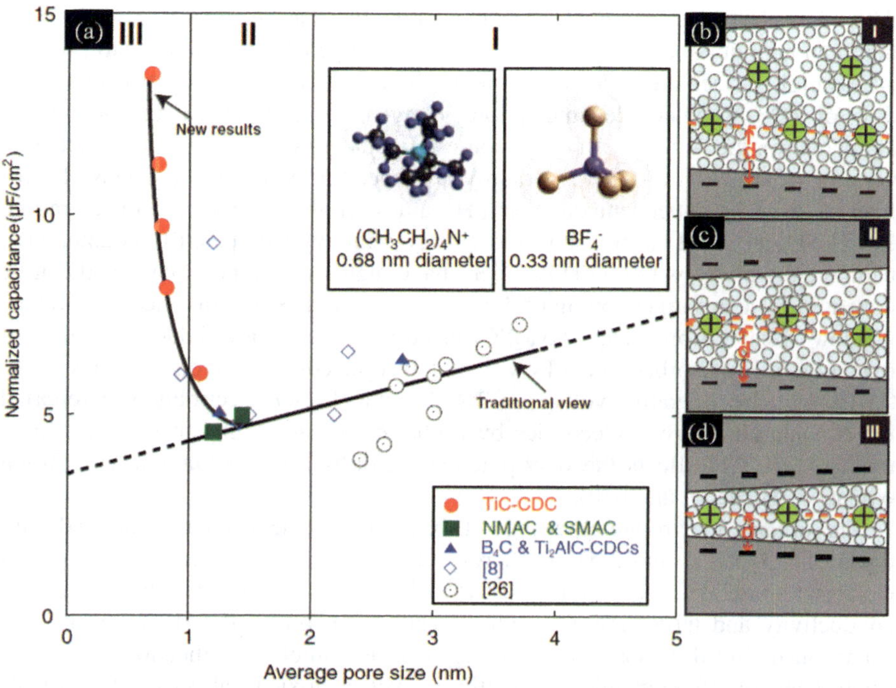

Fig. 1.6 (**a**) Specific surface capacitance normalized by BET SSA for carbide-derived carbon vs average pore size shows an increasing trend with decreasing pore size below 1 nm. Diagrams of solvated ions residing in pores with distance between adjacent pore walls: (**b**) greater than 2 nm, (**c**) between 1 and 2 nm, and (**d**) less than 1 nm demonstrate this behavior schematically (Chmiola et al. 2006a)

1.3.2 Electrolyte Effect

According to Eq. (1.2), besides improvement of capacitance of the electrode materials, an alternative strategy to increase the energy density is enlarging the applied voltage, which is strongly dependent on the selected electrolyte. As an ideal electrolyte for ECs, it should be satisfied with some features as follows: high ionic conductivity, low viscosity to access small pores, chemical inertness (no reaction with the electrode materials), well-matched with the electrolyte materials, a low volatility and flammability, and low cost. Currently employed electrolytes in ECs include aqueous, mostly H_2SO_4, KOH, and Na_2SO_4, organic, mostly acetonitrile (ACN), and propylene carbon (PC), and ionic liquid electrolytes. The typical characteristics of electrolyte, e.g. ion type and size, the ion concentration and solvent, and the interaction between the ion and the solvent, play important roles in the EDLC capacitance and pseudocapacitance. The electrochemical stability of electrolyte affects directly the maximal operation voltage of an EC, and further determines the energy and power densities of the EC. The aqueous electrolytes have high conductivity and small ion size, which easily penetrates inside small pores to access high surface area. However, due to the aqueous electrolyte's potential window of 1.23 V (water decomposition voltage is 1.23 V at 1.0 atm and room temperature), the aqueous electrolyte-based ECs have usually an operating potential window of around 1.0–1.3 V. Certainly, the operation voltage of aqueous electrolytes is strongly dependent on their pH values. High H^+ or OH^- concentration, e.g. H_2SO_4 or KOH, would limit the voltage window. Compared to counterparts, neutral electrolytes with low H^+ or OH^- have higher operating voltage and can be increased by appropriate doping of the electrode materials. For instance, the voltage window of 2.0 V from aqueous Na_2SO_4 electrolyte was achieved due to the electro-oxidation of H_2 absorbed in CNTs with high oxygen content and disorder (Hsu et al. 2012). A higher operating voltage of 2.4 V in the Na_2SO_4 electrolyte was reported for N-containing carbon electrodes by Bichat et al., suggesting that surface functionality strongly affects the over-potential of di-hydrogen evolution and carbon oxidation (Bichat et al. 2010).

The organic electrolyte-based and IL-based ECs generally have potential windows of 2.5–2.7 V and 3.5–4.0 V, respectively, which are higher than aqueous electrolyte-based ECs (Zhong et al. 2015). However, organic electrolytes have low conductivity and larger ion size, which results into power deterioration and low capacitance. Besides those, other issues such as complex purification procedure, safety concerns (flammability), volatility as well as toxicity also exist for organic electrolytes. ILs, known as room temperature molten salts, are nonflammable, which is important for many mobile electronic devices and hybrid vehicles. Besides this, ILs have other advantages including high electrochemical stability, high electrochemical stability over a wide potential window, non-toxicity, and various combination choices of cations and anions. But, ILs have typically high current cost, high viscosity liquids, and low ionic conductivity at room temperature, which limits the charge/discharge rate of IL-based EDLCs.

As for the development of EC electrolytes, widening the potential window of the electrolyte materials can effectively improve the energy density as seen from the Eq. (1.2). It is worth to notice that increasing the applied voltage of ECs would be more efficient than increasing the electrode capacitance in terms of energy density improvement, ascribing to the energy density is proportional to the square of the applied voltage. Thus, developing electrolytes with wide potential windows should be given even higher priority efforts than the development of electrode materials with high capacitance.

Besides the potential window of electrolytes, the interaction between the electrolyte and the electrode materials should be attended because it plays a critical role in the EC performance. As mentioned above, the electrolyte ion size and the pore size of porous carbon materials should be matched for increasing the specific capacitance of EDLCs. The pseudo-capacitors which were composted of carbon-based materials and metal oxides/hydroxides are also affected by the properties of the electrolytes. The internal resistance of ECs is strongly dependent on the ionic conductivity of the electrolytes especially for organic and IL electrolytes. In addition, the operating temperature and life time of the ECs are heavily influenced by thermal stability of electrolytes, e.g. viscosity, boiling point, and freeze point, and electrochemical decomposition of the electrolytes.

1.4 Carbon Nanomaterials for Supercapacitor Applications

Until now, carbon family has many members including from traditional diamond, graphite, activated carbons (ACs) to novel nanocarbons, which contain fullerene, CNTs, graphene, and their derivatives. Previously, AC, a form of disordered carbon (amorphous carbon), has been widely used electrode materials in ECs due to their high surface area, relatively excellent electric properties and low cost. ACs are generally produced by thermal decomposition and physical or chemical activation of carbonaceous sources, including biomass, petroleum coke, and phenolic resins. However, due to wide and random distribution of pore sizes produced by the activation process, ACs only exhibited the limited performance in ECs.

With the successive discovery of fullerene C_{60}, CNTs, and graphene; carbon nanomaterials have attracted great interest from basic theoretical studies to practical applications. Fullerene C_{60}, a soccer-like structure, is a perfect electron acceptor, and fullerene and its derivatives have been widely used in solar cells for charge separation. Compared to fullerene, CNTs, graphene, and their composites have been extensively studied as the electrode materials in the energy storage because of their excellent physiochemical properties, e.g. high conductivity, high surface area, and electrochemical activity.

Currently, the hottest topics of research and development on energy storage mainly focus on lithium ion batteries and ECs. Compared to the counterparts, ECs have some advantages of higher power density, longer cyclic stability, higher Coulombic efficiency as well as faster full charge-discharge cycles.

1.4.1 Carbon Nanotubes in Electrochemical Capacitors

Due to their unique properties such as good mechanical, thermal stability, high specific surface area, high conductivity, high porosity, high charge transport capability, and high electrolyte accessibility, CNTs have attracted great interest in electrode materials for developing high-performance ECs. According to the published papers, the specific surface area of pure CNTs has been achieved in the range of 120–500 $m^2\ g^{-1}$ with the specific capacitance ranging from 2 to 200 $F\ g^{-1}$ (Yang et al. 2015). Initially, Niu et al. reported that MWCNT-based supercapacitor electrode, in which MWCNTs are about to be ~20 μm in length and 8 nm in diameter, exhibited a high specific capacitance of 102 $F\ g^{-1}$ with surface area of 430 $m^2\ g^{-1}$ and a power density of 8 $kW\ kg^{-1}$ in concentrated (38 wt%) H_2SO_4 aqueous electrolyte (Niu et al. 1997). Frackowiak et al. employed the MWCNTs with 10–20 nm outer diameters and 2–5 nm inner diameters as supercapacitor electrodes in 6 M KOH electrolyte to achieve BET SSA in the range of 128–411 m ^{2} g^{-1} with the specific capacitances of 4–80 F g^{-1}, which are remarkably dependent of the CNT diameters (Frackowiak et al. 2000). Obviously, the diameter of the CNTs plays an important role in affecting the intrinsic surface area.

Generally, SWCNTs have higher SSA than MWCNTs, and will provide higher specific capacitance although they always tend to bundle. An et al. utilized the SWCNT bundles with a diameter of 10–20 nm as the electrode materials in a solution of 7.5 N KOH, and obtained a specific capacitance of 180 F g^{-1}, power density of 20 kW kg^{-1}, and energy density of 7 Wh kg^{-1}, respectively (An et al. 2001). Aligned CNTs may also offer higher SSA than misaligned CNTs. For example, a SWCNT forest was synthesized using the zipping effect of water, which allowed the bulk materials to remain their intrinsic properties, and preserves the high BET SSA of 1000 m^2 g^{-1}. The specific capacitance of these aligned SWCNT electrode reached 80 F g^{-1} with an energy density of 35 Wh kg^{-1} in TEATFB-based organic electrolyte (Futaba et al. 2006).

Besides the structure modulation of CNTs, many efforts have been devoted to improve the specific capacitance of CNTs by increasing their specific surface area via chemical activation (KOH) or plasma treatment. A chemically-activated MWCNTs showed that the BET SSA reach 1050 m^2 g^{-1} with the specific capacitances of 90 F g^{-1} and 65 F g^{-1} in the solution of 6 M KOH electrolyte and 1.4 M TEATFB-based organic electrolyte, respectively (Frackowiak et al. 2002). Another KOH-activation of MWCNTs present the positive results that the BET SSA increased from 194 to 510 m^2 g^{-1} with a corresponding increasing of the specific capacitance from 25 to 50 F g^{-1} in a $LiClO_4$-based organic electrolyte (Jiang et al. 2002).

Plasma treatment has some advantages on CNTs for energy storage such as defect production and functionalization, which makes the modified CNTs more hydrophilic because of functional groups, and results in the improvement of specific capacitance (Yoon et al. 2004). Yoon et al. grew vertically aligned MWCNTs on Ni foil substrate using hot filament plasma enhanced chemical vapor deposition, and modified them by ammonia plasma with the substrate temperature of 650 °C (Yoon et al. 2004). The results demonstrated that the plasma etching causes the MWCNT supercapacitor electrodes in 6 M KOH with increasing BET SSA from 9.6 to 86.5 m^2 g^{-1}, and the specific capacitance from 37 to 207 F g^{-1}, respectively (Yoon et al. 2004). The similar high performance of supercapacitor electrodes, which were composed of O_2 plasma-activated MWCNTs, has been achieved by Dai group (Lu et al. 2009). Figure 1.7 shows that the vertically aligned MWCNTs were synthesized on Si/SiO_2 by thermal chemical vapor deposition process, and were etched by O_2 plasma to result in opening of the CNT tips, increasing the SSA to 400 m^2 g^{-1}, and high specific capacitance of 400 F g^{-1} with a high energy density of 148 Wh kg $^{-1}$ in the electrolyte of ionic liquid [EMIM][Tf_2N] (Lu et al. 2009). Recently, the MWCNT powders were treated by microwave and O_2 plasma induced by a radio frequency of 13.56 MHz for enhancing the capacitance of MWCNT electrode (Dulyaseree et al. 2016). The contact angle of droplet of 1 M Na_2SO_4 aqueous

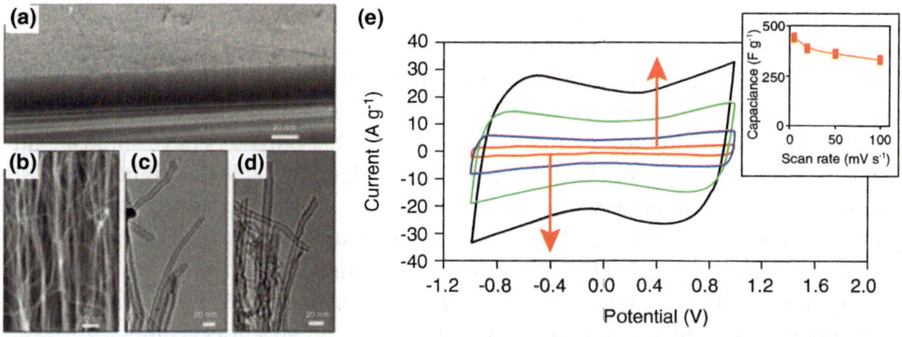

Fig. 1.7 (**a**) Low- (**b**) high-magnification SEM images of vertically-aligned CNTs etched by O_2 plasma. TEM images of the CNTs before (**c**) and after (**d**) plasma etching. (**e**) CV curves of plasma-etched vertically-aligned CNTs as elctrode in [EMIM][Tf_2N] electrolyte at different scan rates. Inset exhibits capacitances as a function of scan rate (Lu et al. 2009)

solution decreased from 113.84° for original MWCNT powders to 36.48° and 19.87° for microwave- and O_2 plasma-treated MWCNTs, respectively (Dulyaseree et al. 2016). This suggests that treated MWCNTs are hydrophilic. The corresponding specific capacitances in 1 M Na_2SO_4 electrolyte increase from 61.5 F g^{-1} for original MWCNT powders to 214 and 238 F g^{-1} for microwave- and O_2 plasma-treated MWCNTs, respectively (Dulyaseree et al. 2016). The capacitance improvement is ascribed to the increase in the number of oxygen-containing functional groups. However, plasma activated carbon electrodes decorated with functional groups may potentially have some issues with low cycle stability and high leakage current. Therefore, such research involving plasma activation should be further investigated in the coming days.

Apart from improving the specific surface area, there is another way to increase the specific capacitance by doping CNTs with heteroatoms, which leads to increasing the electrical conductivity and active sites on CNTs. For example, Gueon et al. synthesized N-doped CNT spherical particles by emulsion-assisted evaporation of hexadecane, followed by N-doping using melamine (Gueon and Moon 2015). A specific capacitance of 215 F g^{-1} has been obtained at a current density of 0.2 A g^{-1}, which is 3.1 times higher than that of the untreated CNTs. The improved performance may be attributed to more active sites and higher electrical conductivity that stem from the N-doping. Xu et al. prepared porous N-doped CNTs from tubular polypyrrole in a N_2 atmosphere, followed by activation of KOH solution. The resulting porous N-doped CNTs provide a high SSA of 1765 m^2 g^{-1} with a specific capacitance of 210 F g^{-1} at a current density of 0.5 A g^{-1}, and have higher specific capacitance (174 A g^{-1}) than the self-assembled graphene hydrogen under the same conditions (160 F g^{-1} at 1 A g^{-1}) (Xu et al. 2013). Moreover, 99% of the initial capacitance was maintained after 5000 cycles. The excellent performance of porous N-doped CNTs can be attributed to high electrical conductivity, large surface area, and unique pore-size distribution (Xu et al. 2013).

Compared to EDLCs, pseudo-capacitors exhibit higher capacitance due to their charge storage mechanism (fast and reversible redox reactions). In order to achieve higher specific capacitance, CNTs have been extensively utilized as pseudo-capacitors in combination with other active components, e.g., conductive polymers, metal oxides and hydroxides. Among them, conducting polymers have some advantages including mass production, environmental friendliness, low weight as well as low-cost. The conducting polymers providing good electrical conductivity, intrinsic porosity, and plenty of redox moieties have demonstrated high specific capacitance when composited with CNTs. At present, most studied conducting polymers are polypyrrole (PPy), plythiophene (PT), polyaniline (PANI), and poly (3,4-ethylenedioxythiophene) (PEDOT) for pseudo-capacitors. Jurewicz et al. deposited a PPy layer of 5 nm on MWCNTs by electrochemical polymerization of pyrrole, and increased the specific capacitance of 163 F g^{-1} from the pristine MWCNTs of 50 F g^{-1} in a 1 M H_2SO_4 electrolyte (Jurewicz et al. 2001). Higher specific capacitance of 249 F g^{-1} for PPy/CNT composite electrodes has been realized with CNT content of 81.8 wt% (Li et al. 2014). Xu et al. reported the PPy/CNT sponges have an increase in specific capacitance from 224 to 350 F g^{-1} with increasing CNT content from 30 wt% to 49% (Xu et al. 2015). However, with further increasing CNT content to 66 wt%, the capacitance decreased to 153 F g^{-1}. This may ascribe to the lower intrinsic capacitance of CNT and smaller contribution from PPy for this pseudo-capacitor (Paul et al. 2010). In comparison with other conducting polymers, PANI has a higher theoretical specific capacitance (Li et al. 2009). Bavio et al. synthesized PANI/CNT composites through a chemical method of self-organization, and investigated the PANI/CNT composites in a 0.5 M H_2SO_4 electrolyte (Bavio et al. 2014). The specific capacitance increased to 838 F g^{-1} at a current density of 2 Ag^{-1} from pristine PANI of 314 F g^{-1}, and further increased to 1744 F g^{-1} when the embedded CNTs were previously functionalized in 2.2 M HNO_3.

Metal oxides and hydroxides including Co_3O_4, Fe_3O_4, NiO, MnO_2, Mn_3O_4, RuO_2, SnO_2, TiO_2, and $Ni(OH)_2$ have also attracted great attention in the field of pseudo-capacitors, in which fast and reversible redox reactions take place on the electrode surfaces. The redox phenomenon of all these oxides involves into the polyvalent nature of the transition metals in the oxidation states. The oxides interact with protons and/or hydroxide anions on changing their oxidation states and the corresponding redox behavior not only exists on the surface of the electrodes but also influences the oxide bulk. Despite their excellent specific capacitance, most of the oxides encountered some issues, e.g. low electrical conductivity (except RuO_2), poor stability, and low rate capability. The carbon materials can not only overcome these disadvantages mentioned above for these oxides, but also restrict the volumetric change of the oxides during the charge-discharge processes. The combination of metal oxides with carbon materials has already demonstrated high performances in pseudo-capacitors.

MnO_2 have a high theoretical specific capacitance of 1370 F g^{-1} (Toupin et al. 2004). The MnO_2/CNT composites have shown the high specific capacitance between 115 and 950 F g^{-1} in pseudo-capacitors (Yan et al. 2009a). As shown in

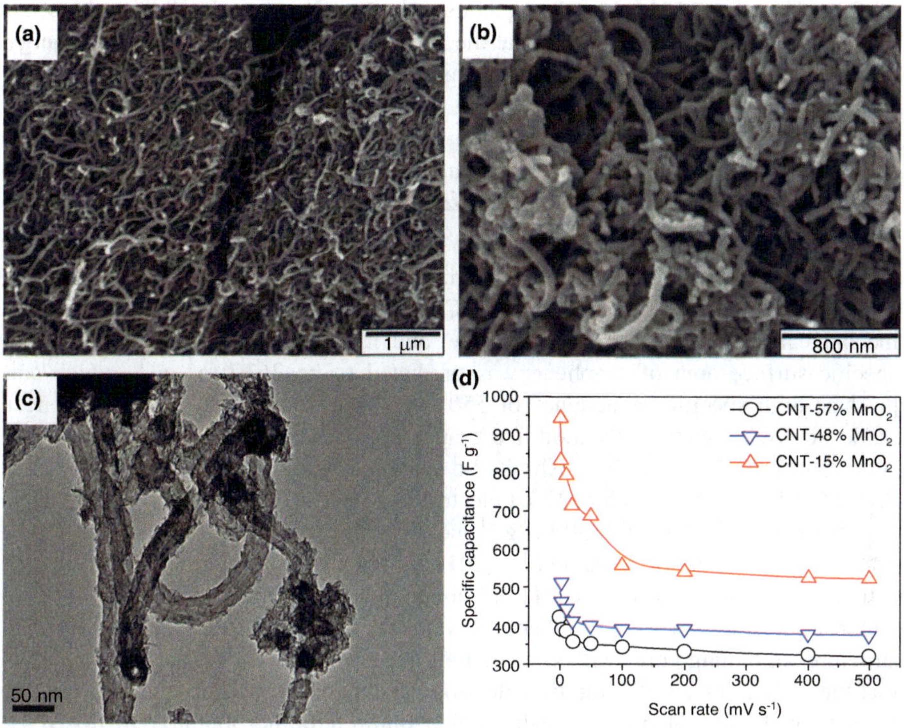

Fig. 1.8 (**a**) SEM image of purified CNTs. (**b**) SEM and (**c**) TEM images of CNT-57%MnO_2 composite. (**d**) Specific capacitance of CNT/MnO_2 composites based on MnO_2 at different rates (Yan et al. 2009a)

Fig. 1.8, Yan et al. reported that MnO_2/CNT composites were synthesized by reduction of $KMnO_4$ under microwave irradiation, and showed a specific capacitance of 944 F g^{-1} at the scan rate of 1 mV s^{-1} in 1 M Na_2SO_4 aqueous solution for the MnO_2/CNT composites (containing 15 wt% MnO_2). These supercapacitor electrodes have the maximum power density of 45.4 kW kg^{-1} and the energy density of 25.2 Wh kg^{-1} when the MnO_2 content is 57 wt%.

As RuO_2 have relatively high conductivity, highly reversible redox reactions, three oxidation states, and wide operation potential window (1.2 V in acidic solutions), RuO_2/CNT composites as supercapacitor electrodes have been widely studied for pseudo-capacitors with high specific capacitance (Borenstein et al. 2017). Initially, the amorphous or nanoscale crystalline $RuO_2{\cdot}xH_2O$ powders were be mixed mechanically with the HNO_3-treated MWCNTs to fabricate the RuO_2/CNT composites for supercapacitor electrodes (Ma et al. 2000). Depending on the weight content of $RuO_2{\cdot}xH_2O$, the composites have reached the specific capacitance of 145–560 F g^{-1} in a 38% H_2SO_4 solution. Yan et al. also synthesized RuO_2 nanoparticles on MWCNTs by microwave-assisted irradiation for supercapacitor electrodes, and displayed high specific capacitance up to 493.9 F g^{-1} at the scan rate of 50 mV s^{-1} in 1 M H_2SO_4 aqueous solution (Yan et al. 2009b). Most of investigations have demonstrated that the RuO_2/CNT composites have high energy densities at high power densities, and exhibit excellent cycling stability.

1.4.2 Graphene in Electrochemical Capacitors

Graphene, consisting of carbon atoms arranged in a hexagonal network, has been considered an electrode candidate for ECs due to high carrier mobility, excellent mechanical properties, chemical stability, and high surface area. The theoretical specific surface area of graphene was predicted to be 2630 m^2 g^{-1}, which can produce high specific capacitance of 550 F g^{-1} (El-Kady et al. 2012). A pioneer work employing chemically-modified graphene (reduced graphene oxide, rGO) as supercapacitor electrodes had exhibited a specific capacitance of 137 F g^{-1} with a SSA of 705 m^2 g^{-1} in a 5.5 M KOH electrolyte (Stoller et al. 2008), which is lower than the theoretic SSA of 2630 m^2 g^{-1}, possibly due to the aggregation of rGO. In order to improve the SSA, Zhu et al. (2011) utilized chemical activation process to activate exfoliated GO with KOH, and obtained a very high SSA of 3100 m^2 g^{-1}, which is even higher than the theoretic value and ascribed to the presence of 3D structure containing pores with sizes of 1–10 nm (Fig. 1.9a, b). Figure 1.9c, d shows that the authors assembled the two-electrode symmetrical supercapacitors based on the activated GO in 1-butyl-3-methyl-imidazolium tetrafluoroborate (BMIM BF4)/ AN electrolyte, and obtained the specific capacitance from the galvanostatic charge/

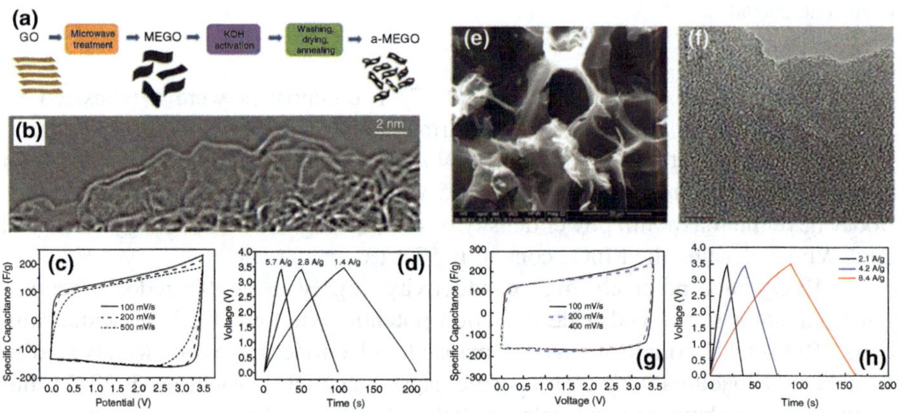

Fig. 1.9 (**a**) Schematic diagram displaying the microwave exfoliation/reduction of GO and the following chemical activation of MEGO with KOH. (**b**) High-resolution TEM image from the edge of a-MEGO. (**c**) CV curves of the a-MEGO-based supercapacitor at different scan rates. (**d**) The corresponding Galvanostatic charge/discharge under different constant currents. (**e**) SEM and (**f**) TEM images of activated hollow spheres of microwave-expanded GO (asMEGO). (**g**) CV curves of asMEGO at different scan rates. (**h**) The corresponding Galvanostatic charge/discharge under different constant currents (Zhu et al. 2011; Kim et al. 2013)

discharge curve with values of 165, 166, and 166 F g^{-1} at the current densities of 1.4, 2.8, and 5.7 A g^{-1}, respectively (Zhu et al. 2011). In another work, as shown in Fig. 1.9e–h, the same group further improved the SSA to 3290 m^2 g^{-1} for highly porous graphene, which derived from carbons with hierarchical pore structures consisting of mesopores integrated with macroporous scaffolds, and exhibited a specific capacitance of 174 F g^{-1} with energy density and power density of 74 Wh kg^{-1} and 338 kW kg^{-1}, respectively (Kim et al. 2013).

Like doped CNTs, heteratom-doped graphene has better performance in the application of energy storage compared to pristine graphene, ascribing to the improved electrical and electrochemical properties. Jeong et al. synthesized nitrogen-doped graphene by a nitrogen plasma process from GO, and obtained a specific capacitance of 282 F g^{-1} for N-doped graphene, which is four times than that of pristine graphene (Jeong et al. 2011). This is attributed to the introduction of charge-transferring sites stemming from N-doping and improvement of electrical conductivity of graphene. Moreover, a power density of 8×10^5 W kg^{-1} and an energy density of 48 Wh kg^{-1} were achieved in the organic electrolyte of TEA BF4. Besides N-doping, other elements including boron, sulfur, phosphorous and their co-doping have been employed to improve the specific capacitance of graphene (Han et al. 2012; Chen et al. 2014; Karthika et al. 2013; Wang et al. 2014b).

Graphene has also been investigated with other active materials, e.g., conducting polymers, metal oxides and hydroxides, to form composites as electrodes for pseudo-capacitors. Firstly, freestanding and flexible GO/PANI composite paper was prepared by in-situ anodic electropolymerization of PANI film on GO paper, and exhibited the gravimetric and volumetric capacitances of 233 F g^{-1} and 135 F cm^{-3} in the electrolyte of 1 M H_2SO_4, respectively, much higher than those of the GO paper (Wang et al. 2009). Zhang et al. employed in-situ polymerization of aniline monomer in the GO surface to fabricate graphene/PANI nanofiber composites under acid conditions, and achieved a specific capacitance of 480 F g^{-1} at a current density of 0.1 A g^{-1} in a 2 M H_2SO_4 for a PANI-doped graphene composite (Zhang et al. 2010). The results showed that the specific capacitance of PANI/GO electrodes is very strongly dependent of the mass ratio of PANI to GO. Besides PANI, PPy has also been composited with graphene to synthesize PPy/GO films by electrooxidation of pyrrole in the GO aqueous solution, which were further reduced electrochemically to form PPy/rGO composites (Chang et al. 2012), as shown in Fig. 1.10. These PPy/rGO composite films exhibited a specific capacitance of 424 F g^{-1} in 1 M H_2SO_4 at a current density of 1 A g^{-1}, which was higher than those of a surfactant-intercalated GO (194 F g^{-1}) (Zhang et al. 2011), resorcinol-formaldehyde resin/GO composite (316 F g^{-1}) (Rightmire 1966), and a PANI nanofiber/GO composite (210 F g^{-1}) (Zhang et al. 2010). The high specific capacitance for the PPy/rGO composite ascribed to the fast insertion/extraction of doping ions in PPy/rGO matrix, high conductivity and high effective SSA of the composite films.

Similar to the case for CNTs, metal oxides or hydroxides have also been investigated in the graphene-based composites to improve the pseudocapacitance. Graphene was used for loading MnO_2 to form composite electrodes with high

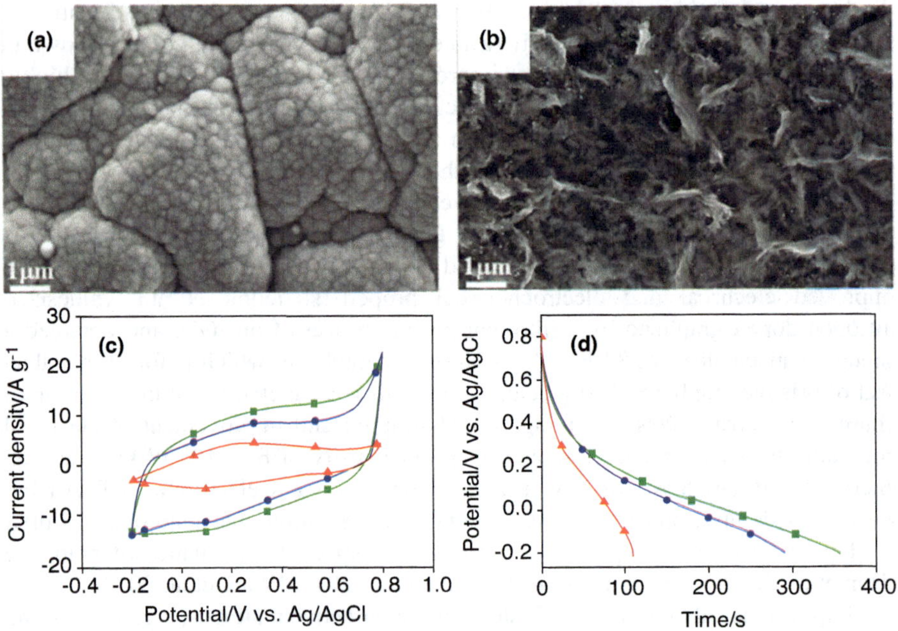

Fig. 1.10 SEM images of (**a**) PPy film and (**b**) PPy/GO composite film coated on gold electrode. The films were synthesized by passing 130 mC cm^{-2} of charge at +0.8 V vs. Ag/AgCl. (**c**) CV curves of PPy (red triangle), PPy/GO (blue dot), and PPy/rGO (green square) modified gold electrodes in 1 M H_2SO_4 at 0.1 V s^{-1}. (**d**) Galvanostatic discharge/charge curves of PPy (red triangle), PPy/GO (blue dot), and PPy/rGO (green square) modified gold electrodes in 1 M H_2SO_4 at a current density of 1 A g^{-1} (Chang et al. 2012)

conductivity, large surface area, and excellent stability. The MnO_2/graphene composites have demonstrated high specific capacitance in the range of 300–520 F g^{-1} (Yan et al. 2010). Yan et al. synthesized MnO_2/graphene composites by the self-limiting deposition of nanoscale MnO_2 on the GO surface under microwave irradiation, and presented a specific capacitance as high as 310 F g^{-1} and 228 F g^{-1} in 1 M Na_2SO_4 electrolyte at the scan rates of 2 mV s^{-1} and 500 mV s^{-1}, respectively, which are higher than these of pure graphene (104 F g^{-1}) and birnessite-type MnO_2 (103 F g^{-1}). Moreover, these composites displayed excellent cycling stability and a decrease of 4.6% in the initial capacity after 15,000 cycles. These excellent electrochemical performances can be attributed to the increased conductivity and surface area of graphene network in the MnO_2/graphene composites. Alternatively, Cheng et al. electrodeposited gamma-type MnO_2 on the graphene surface to increase the specific capacitance of 328 F g^{-1} of the MnO_2/graphene composites from 245 F g^{-1} of pure graphene electrode in 1 M KCl at a charging current of 1 mA (Cheng et al. 2011). The MnO_2/graphene composites also demonstrated an energy density of 11.4 Wh kg^{-1} and a power density of 25.8 kW kg^{-1}, respectively.

Besides flat graphene related to substrate, vertical graphene (VG) is another type of graphene, consists of a large amount of edges with open structures, and has some advantages compared to flat graphene as follows (Miller et al. 2010): (1) VGs can provide direct pathways for ion faster diffusion; (2) edge planes have higher capacitance of 50–70 μF cm^{-1} in comparison to basal planes of about 3 μF cm^{-1}; (3) open structures lead to minimal porosity effects, and reduce ionic resistance; (4) VG (using plasma technique) can grow on the conducting substrates, and be used as electrode without binder to minimize electronic resistances. The VG, which grew on carbon fibers by radio frequency plasma-enhanced chemical vapor deposition in H_2-CH_4 system, have the specific capacitance of 0.076 F cm^{-2} and a value of 14,900 F at maximum working voltage of 0.6 V in 6 M H_2SO_4 electrolyte (Zhao et al. 2009). The EDLCs made of VG, which deposited directly on the Ni foil from the decomposition of CH_4, have demonstrated fast response with efficient line-filtering at 120 Hz (Miller et al. 2010). It ascribes to the open morphology of VG that provides high conductance channels for ingress and egress of the electrolyte ions between the vertical sheets, which minimize porous electrode behavior and allow fast response. The specific capacitance of the EDLCs can be improved using the feedstock of C_2H_2 instead of CH_4 or increasing the VG growth temperatures (Cai et al. 2014), as shown in Fig. 1.11.

For graphene-based electrode materials, N-doping is one of the popular methods for enhancing the performance of supercapacitors. For example, Yen et al. synthesized N-doped VG on flexible carbon cloths by microwave PECVD, where followed by introducing in situ NH_3 plasma (Yen et al. 2015). The resulting N-doped VG supercapacitor electrodes exhibited a specific capacitance of 991.6 F g^{-1} at a current density of 14.8 A g^{-1}, obtained the highest specific energy density of 275.4 Wh kg^{-1} a power density of 14.8 kW kg^{-1}, and maintained near 98% Coulombic efficiency after 5000 cycles at in 1 M H_2SO_4 aqueous solution (Yen et al. 2015). These excellent performances attributed to the ultrahigh surface area of VG and unique specificity to the pyridinic N-doping configuration (Yen et al. 2015). The VG can also provide a support for loading metal oxides or hydroxides on its surface in the applications of pseudo-capacitors (Xiong et al. 2013; Hassan et al. 2014a, b; Wang et al. 2014a; Dinh et al. 2014; Han et al. 2017). Xiong et al. reported that VG was prepared by microwave PECVD on commercial CNT substrates and subsequently coated with a thin layer of MnO_2, and investigated the electrochemical performances of the MnO_2/VG composites (Xiong et al. 2013). A specific capacitance of 580 F g^{-1} at a scan rate of 2 mV s^{-1}, and an energy density of 28 Wh kg^{-1} and a power density of 25 kW kg^{-1} at a high current density of 50 A g^{-1} have been achieved in 1 M Na_2SO_4 aqueous electrolyte (Xiong et al. 2013). This good performance could be obtained due to the VG/CNT architecture without binder and the metallic nature of MnO_2/VG composites with a facile conduction path for electron transport in the charge/discharge process (Xiong et al. 2013).

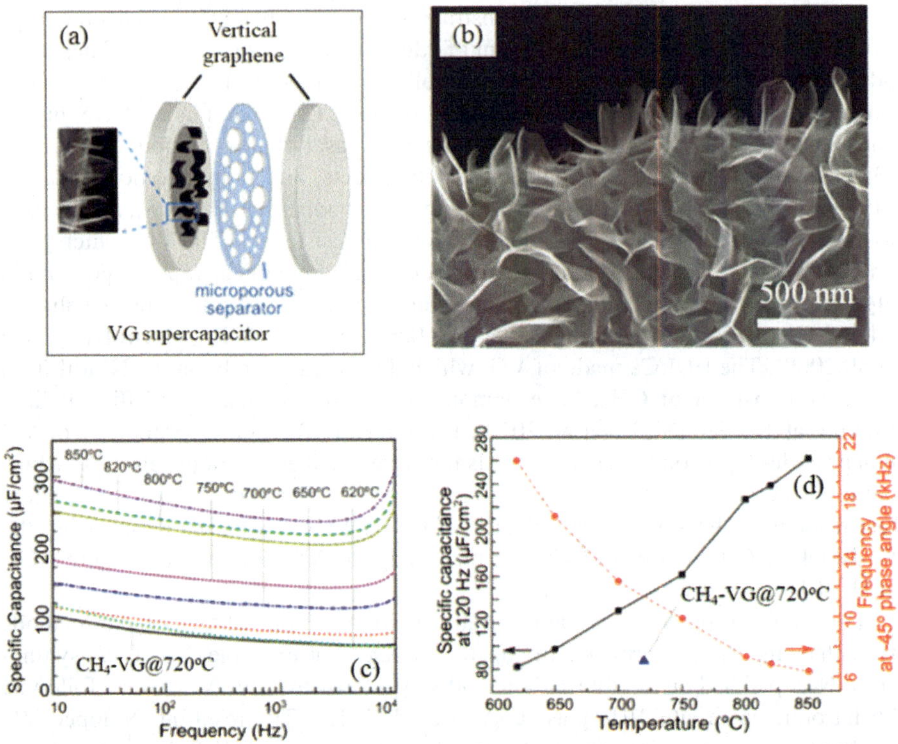

Fig. 1.11 (**a**) Schematic diagram of VG supercapacitor which consists of two same VG sheets. (**b**) SEM images of VG sheets. (**c**) Specific capacitance of EDLCs, which consist of C_2H_2-VG at different temperatures, as a function of frequency. The higher the growth temperature, the higher the capacitance over the entire frequency range for all samples. (**d**) The specific capacitance of EDLCs at 120 Hz (black square blocks) for 10 min growth and the characteristic frequency f_0 (red circular dots) as a function of growth temperature. The EDLC made of CH_4-VG@720 °C as a comparison (Cai et al. 2014)

1.5 Future Perspectives, Opportunities, and Challenges

In comparison with batteries, it is undeniable that even though supercapacitors offer higher power, relatively good cycle life, and higher reliability, they exhibited much lower energy density and higher self-discharge (Weinstein and Dash 2013). Nevertheless, some supercapacitors have been applied for emergency doors, memory backup, and energy recovery, despite the fact that limited energy density still remain as the big challenge.

Carbon materials derived from waste materials such as agricultural wastes, especially coconut shell ACs can be considered as the popular option for active material (Weinstein and Dash 2013). Kuraray (Japan) is one of the major suppliers for most of this product. One of the main reasons is that coconut shell ACs offers far more lower cost, in contrast to another types of carbon materials.

Carbon nanomaterials such as CNTs and graphene have received much attention, to be developed as high performance supercapacitor electrode material. For the time being, although their costs are higher than AC, they are able to close the performance gap, which is essential so that supercapacitor can be applied for wide broad applications in various fields. In addition to electrode materials, electrolyte also plays a crucial role in determining the performance outcome of supercapacitors. It is known that energy storage density is proportional to the square of the operating voltage, thus using alternative electrolyte systems with significantly wider operating voltage stability windows can significantly improve the energy storage capability of supercapacitors (Yu et al. 2013). Future advancement of supercapacitors can be shifted to develop appropriate morphology and porous structure of carbon nanomaterials that have high compatibility with the improved electrolyte, development of novel nanostructured metal oxides/conductive polymers to improve energy storage capabilities, development of computational tools to investigate the mechanisms during charge/discharge process, and so on.

1.6 Conclusions

In this chapter, energy storage performance of electrochemical capacitors or supercapacitors was discussed focusing on carbon nanomaterials, specifically CNTs, graphene, and their derivatives. The affecting factors of two main elements, namely electrode and electrolyte materials were elaborated, which is highly associated with the overall supercapacitor's performance. Various types of materials utilized together with carbon nanomaterials, such as metal oxides, conductive polymers, doped materials, and so on, were also included. The development of safe, low cost, high energy storage, fast charge/discharge process, long cycle life supercapacitors is essential in order to produce efficient energy storage device.

References

Abioye AM, Noorden ZA, Ani FN (2017) Synthesis and characterizations of electroless oil palm shell based-activated carbon/nickel oxide nanocomposite electrodes for supercapacitor applications. Electrochim Acta 225:493–502. https://doi.org/10.1016/j.electacta.2016.12.101

An KH, Kim WS, Park YS, Moon J-M, Bae DJ, Lim SC, Lee YS, Lee YH (2001) Electrochemical properties of high-power supercapacitors using single-walled carbon nanotube electrodes. Adv Funct Mater 11(5):387–392. https://doi.org/10.1002/1616-301X/01/0510-0387

Barbieri O, Hahn M, Herzog A, Kotz R (2005) Capacitance limits of high surface area activated carbons for double layer capacitors. Carbon 43:1303–1310. https://doi.org/10.1016/j.carbon.2005.01.001

Bavio MA, Acosta GG, Kessler T (2014) Synthesis and characterization of polyaniline and polyaniline-carbon nanotubes nanostructures for electrochemical supercapacitors. J Power Sources 245:475–481. https://doi.org/10.1016/j.jpowsour.2013.06.119

Becker H I (1957) Low voltage electrolytic capacitor. US Patent 2800616

Bichat MP, Raymundo-Piñero E, Béguin F (2010) High voltage supercapacitor built with seaweed carbons in neutral aqueous electrolyte. Carbon 48(15):4351–4361. https://doi.org/10.1016/j.carbon.2010.07.049

Boos D L (1970) Electrolytic capacitor having carbon paste electrodes. US Patent 3536963

Borenstein A, Hanna O, Attias R, Luski S, Brousse T, Aurbach D (2017) Carbon-based composite materials for supercapacitor electrode: a review. J Mater Chem A 5(25):12653–12672. https://doi.org/10.1039/c7ta00863e

Cai M, Outlaw RA, Quinlan RA, Premathilake D, Butler SM, Miller JR (2014) Fast response, vertically oriented graphene nanosheet electric double layer capacitors synthesized from C_2H_2. ACS Nano 8(6):5873–5882. https://doi.org/10.1021/nn5009319

Chang H-H, Chang C-K, Tsai Y-C, Liao C-S (2012) Electrochemically synthesized graphene/polypyrrole composites and their use in supercapacitor. Carbon 50(6):2331–2336. https://doi.org/10.1016/j.carbon.2012.01.056

Chen T, Dai L (2013) Carbon nanomaterials for high-performance supercapacitors. Mater Today 16 (7-8):272–280. https://doi.org/10.1016/j.mattod.2013.07.002

Chen X, Chen X, Xu X, Yang Z, Liu Z, Zhang L, Xu X, Chen Y, Huang S (2014) Sulfur-doped porous reduced graphene oxide hollow nanosphere frameworks as metal-free electrocatalysts for oxygen reduction reaction and as supercapacitor electrode materials. Nanoscale 6 (22):13740–13747. https://doi.org/10.1039/C4NR04783D

Cheng Q, Tang J, Ma J, Zhang H, Shinya N, Qin L-C (2011) Graphene and nanostructured MnO_2 composite electrodes for supercapacitors. Carbon 49(9):2917–2925. https://doi.org/10.1016/j.carbon.2011.02.068

Chmiola J, Yushin G, Dash R, Gogotsi Y (2006a) Effect of pore size and surface area of carbide derived carbons on specific capacitance. J Power Sources 158(1):765–772. https://doi.org/10.1016/j.jpowsour.2005.09.008

Chmiola J, Yushin G, Gogotsi Y, Portet C, Simon P, Taberna PL (2006b) Anomalous increase in carbon capacitance at pore sizes less than 1 nanometer. Science 313(5794):1760–1763. https://doi.org/10.1126/science.1132195

Conway BE (1999) Electrochemical supercapacitors: scientific fundamentals and technological applications. Kluwer Academic/Plenum Publishers, New York, NY

Dai L, Chang DW, Baek J-B, Lu W (2012) Carbon nanomaterials for advanced energy conversion and storage. Small 8(8):1130–1166. https://doi.org/10.1002/smll.201101594

Dinh TM, Achour A, Vizireanu S, Dinescu G, Nistor L, Armstrong K, Guay D, Pech D (2014) Hydrous RuO_2/carbon nanowalls hierarchical structures for all-solid-state ultrahigh-energy-density micro-supercapacitors. Nano Energy 10:288–294. https://doi.org/10.1016/j.nanoen.2014.10.003

Dong Y, Wu Z-S, Ren W, Cheng HM, Bao X (2017) Graphene: a promising 2D material for electrochemical energy storage. Sci Bull 62(10):724–740. https://doi.org/10.1016/j.scib.2017.04.010

Dulyaseree P, Yordsri V, Wongwiriyapan W (2016) Effects of microwave and oxygen plasma treatments on capacitive characteristics of supercapacitor based on multiwealled carbon nanoutbes. Jpn J Appl Phys 55:02BD05. https://doi.org/10.7567/JJAP.55.02BD05

El-Kady MF, Strong V, Dubin S, Kaner RB (2012) Laser scribing of high-performance and flexible graphene-based electrochemical capacitors. Science 335(6074):1326–1330. https://doi.org/10.1126/science.1216744

Fauvarque JF, Simon P (2010) Principles of electrochemistry and electrochemical methods. In: Béguin F, Frackowiak E (eds) Carbons for electrochemical energy storage and conversion systems. CRC Press, Boca Raton, pp 1–36

Frackowiak E, Béguin F (2001) Carbon materials for the electrochemical storage of energy in capacitors. Carbon 39(6):937–950. https://doi.org/10.1016/S0008-6223(00)00183-4

Frackowiak E, Metenier K, Bertagna V, Beguin F (2000) Supercapacitor electrodes from multiwalled carbon nanotubes. Appl Phys Lett 77:2421–2423. https://doi.org/10.1063/1.1290146

Frackowiak E, Delpeux S, Jurewicz K, Szostak K, Cazorla-Amoros D, Beguin F (2002) Enhanced capacitance of carbon nanotubes through chemical activation. Chem Phys Lett 361(1–2):35–41. https://doi.org/10.1016/S0009-2614(02)00684-X

Futaba DN, Hata K, Yamada T, Hiraoka T, Hayamizu Y, Kakudate Y, Tanaike O, Hatori H, Yumura M, Iijima S (2006) Shape-engineerable and highly densely packed single-walled carbon nanotubes and their application as super-capacitor electrodes. Nat Mater 5 (12):987–994. https://doi.org/10.1038/nmat1782

Gu W, Yushin G (2014) Review of nanostructured carbon materials for electrochemical capacitor applications: advantages and limitations of activated carbon, carbide-derived carbon, zeolite-templated carbon, carbon aerogels, carbon nanotubes, onion-like carbon and graphene. WIREs Energy Environ 3(5):424–473. https://doi.org/10.1002/wene.102

Gueon D, Moon JH (2015) Nitrogen-doped carbon nanotube spherical particles for supercapacitor applications: emulsion-assisted compact packing and capacitance enhancement. ACS Appl Mater Interfaces 7(36):20083–20089. https://doi.org/10.1021/acsami.5b05231

Han J, Zhang LL, Lee S, Oh J, Lee K-S, Potts JR, Ji J, Zhao X, Ruoff RS, Park S (2012) Generation of B-doped graphene nanoplatelets using a solution process and their supercapcitor applications. ACS Nano 7(1):19–26. https://doi.org/10.1021/nn3034309

Han ZJ, Pineda S, Murdock AT, Seo DH, Ostrikov K, Bendavid A (2017) RuO_2-coated vertical graphene hybrid electrodes for high-performance solid-state supercapacitors. J Mater Chem A 5 (33):17293–17301. https://doi.org/10.1039/C7TA03355A

Hassan S, Suzuki M, Mori S, El-Moneim AA (2014a) MnO_2/carbon nanowalls composite electrode for supercapacitor applications. J Power Sources 249:21–27. https://doi.org/10.1016/j.jpowsour.2013.10.097

Hassan S, Suzuki M, Mori S, El-Moneim AA (2014b) MnO_2/carbon nanowall electrode for future energy storage application: effect of carbon nanowall growth period and MnO_2 mass loading. RSC Adv 4(39):20479–20488. https://doi.org/10.1039/c4ra01132e

He N, Yildiz O, Pan Q, Zhu J, Zhang X, Bradford PD, Gao W (2017) Pyrolytic-carbon coating in carbon nanotube foams for better performance in supercapacitors. J Power Sources 343:492–501. https://doi.org/10.1016/j.jpowsour.2017.01.091

Hierold C, Brand O, Fedder GK, Korvink JG, Tabata O (2008) Carbon nanotube devices: properties, modeling, integration and applications, vol 8. John Wiley & Sons, Chichester

Hsu Y-K, Chen Y-C, Lin Y-G, Chen L-C, Chen K-H (2012) High-cell-voltage supercapacitor of carbon nanotube/carbon cloth operating in neutral aqueous solution. J Mater Chem 22 (8):3383–3387. https://doi.org/10.1039/C1JM14716A

Hu L, Hecht DS, Gruner G (2010) Carbon nanotube thin films: fabrication, properties, and applications. Chem Rev 110(10):5790–5844. https://doi.org/10.1021/cr9002962

Huang Y, Liu Y, Zhao G, Chen JY (2017) Sustainable activated carbon fiber from sawdust by reactivation for high-performance supercapacitors. J Mater Sci 52(1):478–488. https://doi.org/10.1007/s10853-016-0347-0

Iijima S (1991) Helical microtubules of graphitic carbon. Nature 354(6348):56–58. https://doi.org/10.1038/354056a0

Iijima S, Ichihashi T (1993) Single-shell carbon nanotubes of 1-nm diameter. Nature 363 (6430):603–605. https://doi.org/10.1038/363603a0

Jeong HM, Lee JW, Shin WH, Choi YJ, Shin HJ, Kang JK, Choi JW (2011) Nitrogen-doped graphene for high-performance ultracapacitors and the importance of nitrogen-doped sites at basal planes. Nano Lett 11(6):2472–2477. https://doi.org/10.1021/nl2009058

Jiang Q, Qu MZ, Zhou GM, Zhang BL, Yu ZL (2002) A study of activated carbon nanotubes as electrochemical super capacitors electrode materials. Mater Lett 57(4):988–991. https://doi.org/10.1016/S0167-577X(02)00911-4

Jo EH, Jang HD, Chang H, Kim SK, Choi J-H, Lee CM (2017) 3D network-structured crumpled graphene/carbon nanotube/polyaniline composites for supercapacitors. ChemSusChem 10 (10):2210–2217. https://doi.org/10.1002/cssc.201700212

Jurewicz K, Delpeux S, Bertagna V, Beguin F, Frackowiak E (2001) Supercapacitors from nanoubes/polypyrrole composites. Chem Phys Lett 347(1–3):36–40. https://doi.org/10.1016/S0009-2614(01)01037-5

Karthika P, Rajalakshmi N, Dhathathreyan KS (2013) Phosphorus-doped exfoliated graphene for supercapacitor electrodes. J Nanosci Nanotechnol 13(3):1746–1751. https://doi.org/10.1166/jnn.2013.7112

Ke Q, Wang J (2016) Graphene-based materials for supercapacitor electrodes—a review. J Mater 2 (1):37–54. https://doi.org/10.1016/j.jmat.2016.01.001

Kim YJ, Yang C-M, Park KC, Kaneko K, Kim YA, Noguchi M, Fujino T, Oyama S, Endo M (2012) Edge-enriched, porous carbon-based, high energy density supercapacitors for hybrid electric vehicles. ChemSusChem 5(3):535–541. https://doi.org/10.1002/cssc.201100511

Kim T, Jung G, Yoo S, Suh KS, Ruoff RS (2013) Activated graphene-based carbons as supercapacitor electrodes with macro- and mesopores. ACS Nano 7(8):6899–6905. https://doi.org/10.1021/nn402077v

Kroto HW, Heath JR, O'Brien SC, Curl RF, Smalley RE (1985) C60: Buckminsterfullerene. Nature 318(6042):162–163. https://doi.org/10.1038/318162a0

Largeot C, Portet C, Chmiola J, Taberna P-L, Gogotsi Y, Simon P (2008) Relation between the ion size and pore size for an electric double-layer capacitor. J Am Chem Soc 130(9):2730–2731. https://doi.org/10.1021/ja7106178

Li H, Wang J, Chu Q, Wang Z, Zhang F, Wang S (2009) Theoretical and experimental specific capacitance of polyaniline in sulfuric acid. J Power Sources 190(2):578–586. https://doi.org/10.1016/j.jpowsour.2009.01.052

Li P, Shi E, Yang Y, Shang Y, Peng Q, Wu S, Wei J, Wang K, Zhu H, Yuan Q (2014) Carbon nanotube-polypyrrole core-shell sponge and its application as highly compressible supercapacitor electrode. Nano Res 7(2):209–218. https://doi.org/10.1007/s12274-013-0388-5

Lin T, Chen I-W, Liu F, Yang C, Bi H, Xu F, Huang F (2015) Nitrogen-doped mesoporous carbon of extraordinary capacitance for electrochemical energy storage. Science 350(6267):1508–1513. https://doi.org/10.1126/science.aab3798

Lu W, Qu L, Henry K, Dai L (2009) High performance electrochemical capacitors from aligned carbon nanotube electrodes and ionic liquid electrolytes. J Power Sources 189(2):1270–1277. https://doi.org/10.1016/j.jpowsour.2009.01.009

Lu Y, Zhang S, Yin J, Bai C, Zhang J, Li Y, Yang Y, Ge Z, Zhang M, Wei L, Ma M, Ma Y, Chen Y (2017) Mesoporous activated carbon materials with ultrahigh mesopore volume and effective specific surface area for high performance supercapacitors. Carbon 124:64–71. https://doi.org/10.1016/j.carbon.2017.08.044

Lv W, Li Z, Deng Y, Yang Q-H, Kang F (2016) Graphene-based materials for electrochemical energy storage devices: opportunities and challenges. Energy Storage Mater 2:107–138. https://doi.org/10.1016/j.ensm.2015.10.002

Ma R, Wei B, Xu C, Liang J, Wu D (2000) The development of carbon nanotubes/ $RuO_2{\cdot}xH_2O$ electrodes for electrochemical capacitors. Bull Chem Soc Jpn 73(8):1813–1816. https://doi.org/10.1246/bcsj.73.1813

Melvin GJH, Wang Z, Siambun NJ, Rahman MM (2017a) Carbon materials derived from rice husks at low and high temperatures. IOP Conf Ser Mater Sci Eng 217:012017. https://doi.org/10.1088/1757-899X/217/1/012017

Melvin GJH, Wang Z, Ni QQ, Siambun NJ, Rahman MM (2017b) Fabrication and characterization of carbonized rice husk/barium titanate nanocomposites. IOP Conf Ser Mater Sci Eng 229:012024. https://doi.org/10.1088/1757-899X/229/1/012024

Miller JR, Outlaw RA, Holloway BC (2010) Graphene double-layer capacitor with ac-line filtering performance. Science 329(5999):1637–1639. https://doi.org/10.1126/science.1194372

Muramatsu H, Kim YA, Yang K-S, Cruz-Silva R, Toda I, Yamada T, Terrones M, Endo M, Hayashi T, Saitoh H (2014) Rice husk-derived graphene with nano-sized domains and clean edges. Small 10(14):2766–2770. https://doi.org/10.1002/smll.201400017

Niu C, Sichel EK, Hoch R, Moy D, Tennent H (1997) High power electrochemical capacitors based on carbon nanotube electrodes. Appl Phys Lett 70:1480–1482. https://doi.org/10.1063/1.118568

Novoselov KS, Geim AK, Morozov SV, Jiang D, Zhang Y, Dubonos SV, Grigorieva IV, Firsov AA (2004) Electric field effect in atomically thin carbon films. Science 306(5696):666–669. https://doi.org/10.1126/science.1102896

Oberlin A, Endo M, Koyama T (1976) Filamentous growth of carbon through benzene decomposition. J Cryst Growth 32(3):335–349. https://doi.org/10.1016/0022-0248(76)90115-9

Paul S, Lee Y-S, Choi J-A, Kang Y-C, Kim D-W (2010) Synthesis and electrochemical characterization of polypyrrole/multiwalled carbon nanotube composite electrodes for supercapacitor applications. Bull Kor Chem Soc 31(5):1228–1232. https://doi.org/10.5012/bkcs.2010.31.5.1228

Popov VN (2004) Carbon nanotubes: properties and application. Mater Sci Eng R 43(3):61–102. https://doi.org/10.1016/j.mser.2003.10.001

Pumera M (2010) Graphene-based nanomaterials and their electrochemistry. Chem Soc Rev 39 (11):4146–4157. https://doi.org/10.1039/c002690p

Pumera M (2011) Graphene-based nanomaterials for energy storage. Energy Environ Sci 4 (3):668–674. https://doi.org/10.1039/C0EE00295J

Raymundo-Piñero E, Leroux F, Béguin F (2006) A high-performance carbon for supercapacitors obtained by carbonization of a seaweed biopolymer. Adv Mater 18(14):1877–1882. https://doi.org/10.1002/adma.200501905

Rightmire R A (1966) Electrical energy storage apparatus. US Patent 3288641

Segalini J, Iwama E, Taberna P-L, Gogotsi Y, Simon P (2012) Steric effects in adsorption of ions from mixed electrolytes into microporous carbon. Electrochem Commun 15(1):63–65. https://doi.org/10.1016/j.elecom.2011.11.023

Sharma P, Bhatti TS (2010) A review on electrochemical double-layer capacitors. Energy Convers Manag 51(12):2901–2912. https://doi.org/10.1016/j.enconman.2010.06.031

Simon P, Gogotsi Y (2008) Materials for electrochemical capacitors. Nat Mater 7(11):845–854. https://doi.org/10.1038/nmat2297

Simon P, Gogotsi Y (2013) Capacitive energy storage in nanostructured carbon-electrolyte systems. Acc Chem Res 46(5):1094–1103. https://doi.org/10.1021/ar200306b

Stoller MD, Park S, Zhu Y, An J, Ruoff RS (2008) Graphene-based ultracapacitors. Nano Lett 8 (10):3498–3502. https://doi.org/10.1021/nl802558y

Tanahashi I, Yoshida A, Nishino A (1990) Electrochemical characterization of activated carbon-fiber cloth polarizable electrodes for electric double-layer capacitors. J Electrochem Soc 137 (10):3052–3057. https://doi.org/10.1149/1.2086158

Thostenson ET, Ren Z, Chou T-W (2001) Advances in the science and technology of carbon nanotubes and their composites: a review. Compos Sci Technol 61(13):1899–1912. https://doi.org/10.1016/S0266-3538(01)00094-X

Toupin M, Brousse T, Bélanger D (2004) Charge storage mechanism of MnO_2 electrode used in aqueous electrochemical capacitor. Chem Mater 16(16):3184–3190. https://doi.org/10.1021/cm049649j

Vix-Guterl C, Frackowiak E, Jurewicz K, Friebe M, Parmentier J, Béguin F (2005) Electrochemical energy storage in ordered porous carbon materials. Carbon 43(6):1293–1302. https://doi.org/10.1016/j.carbon.2004.12.028

Wallar C, Luo D, Poon R, Zhitomirsky I (2017) Manganese dioxide-carbon nanotube composite electrodes with high active mass loading for electrochemical supercapacitors. J Mater Sci 52 (7):3687–3696. https://doi.org/10.1007/s10853-016-0711-0

Wang D-W, Li F, Zhao J, Ren W, Chen Z-G, Tan J, Wu Z-S, Gentle L, Lu GQ, Cheng H-M (2009) Fabrication of graphene/polyaniline composite paper via in situ anode electropolymerization for high-performance flexible electrode. ACS Nano 3(7):1745–1752. https://doi.org/10.1021/nn900297m

Wang X, Liu J, Wang Y, Zhao C, Zheng W (2014a) $Ni(OH)_2$ nanoflakes electrodeposited on Ni foam-supported vertically oriented graphene nanosheets for applications in asymmetric supercapacitors. Mater Res Bull 52:89–95. https://doi.org/10.1016/j.materresbull.2013.12.051

Wang X, Sun G, Routh P, Kim D-H, Huang W, Chen P (2014b) Heteroatom-doped graphene materials: syntheses, properties and applications. Chem Soc Rev 43(20):7067–7098. https://doi.org/10.1039/C4CS00141A

Wang Z, Ogata H, Morimoto S, Ortiz-Medina J, Fujishige M, Takeuchi K, Muramatsu H, Hayashi T, Terrones M, Hashimoto Y, Endo M (2015) Nanocarbons from rice husk by microwave plasma irradiation: from graphene and carbon nanotubes to graphenated carbon nanotube hybrids. Carbon 94:479–484. https://doi.org/10.1016/j.carbon.2015.07.037

Weinstein L, Dash R (2013) Supercapacitor carbons. Mater Today 10(16):356–357. https://doi.org/10.1016/j.mattod.2013.09.005

Xiong G, Hembram KPSS, Reifenberger RG, Fisher TS (2013) MnO_2-coated graphitic petals for supercapacitor electrodes. J Power Sources 227:254–259. https://doi.org/10.1016/j.jpowsour.2012.11.040

Xu G, Ding B, Nie P, Shen L, Wang J, Zhang X (2013) Porous nitrogen-doped carbon nanotubes derived from tubular polypyrrole for energy-storage applications. Chem Eur J 19 (37):12306–12312. https://doi.org/10.1002/chem.201301352

Xu R, Wei J, Guo F, Cui X, Zhang T, Zhu H, Wang K, Wu D (2015) Highly conductive, twistable and bendable polypyrrole-carbon nanotube fiber for efficient supercapacitor electrodes. RSC Adv 2015(28):22015–22021. https://doi.org/10.1039/C5RA01917F

Yan J, Fan Z, Wei T, Cheng J, Shao B, Wang K, Song L, Zhang M (2009a) Carbon nanotube/MnO_2 composites synthesized by microwave-assisted method for supercapacitors with high power and energy densities. J Power Sources 194(2):1202–1207. https://doi.org/10.1016/j.jpowsour.2009.06.006

Yan S, Wang H, Qu P, Zhang Y, Xiao Z (2009b) RuO_2/carbon nanotubes composites synthesized by microwave-assisted method for electrochemical supercapacitor. Synth Met 159(1-2):158–161. https://doi.org/10.1016/j.synthmet.2008.07.024

Yan J, Fan Z, Wei T, Qian W, Zhang M, Wei F (2010) Fast and reversible surface redox reaction of graphene-MnO_2 composites as supercapacitor electrodes. Carbon 48(13):3825–3833. https://doi.org/10.1016/j.carbon.2010.06.047

Yang Z, Zhang J, Kintner-Meyer MCW, Lu X, Choi D, Lemmon JP, Liu J (2011) Electrochemical energy storage for green grid. Chem Rev 111(5):3577–3613. https://doi.org/10.1021/cr100290v

Yang Z, Ren J, Zhang Z, Chen X, Guan G, Qiu L, Zhang Y, Peng H (2015) Recent advancement of nanostructured carbon for energy applications. Chem Rev 115(11):5159–5223. https://doi.org/10.1021/cr5006217

Yen H-F, Horng Y-Y, Hu M-S, Yang W-H, Wen J-R, Ganguly A, Tai Y, Chen K-H, Chen L-C (2015) Vertically aligned epitaxial graphene nanowalls with dominated nitrogen doping for superior supercapcitors. Carbon 82:124–134. https://doi.org/10.1016/j.carbon.2014.10.042

Yoon B-J, Jeong S-H, Lee K-H, Kim HS, Park CG, Han JH (2004) Electrical properties of electrical double layer capacitors with integrated carbon nanotube electrodes. Chem Phys Lett 388 (1–3):170–174. https://doi.org/10.1016/j.cplett.2004.02.071

Young RJ, Lovell PA (2011) Introduction to polymers. CRC Press, Boca Raton

Yu A, Chabot V, Zhang J (2013) Electrochemical supercapacitors for energy storage and delivery: fundamentals and applications. CRC Press, Boca Raton

Zhai Y, Dou Y, Zhao D, Fulvio PF, Mayes RT, Dai S (2011) Carbon materials for chemical capacitive energy storage. Adv Mater 23(42):4828–4850. https://doi.org/10.1002/adma.201100984

Zhang K, Zhang LL, Zhao XS, Wu J (2010) Graphene/polyaniline nanofiber composites as supercapacitor electrodes. Chem Mater 22(4):1392–1401. https://doi.org/10.1021/cm902876u

Zhang K, Mao L, Zhang LL, On Chan HS, Zhao XS, Wu J (2011) Surfactant-intercalated, chemically reduced graphene oxide for high performance supercapacitor electrodes. J Mater Chem 21(20):7302–7307. https://doi.org/10.1039/C1JM00007A

Zhao X, Tian H, Zhu M, Tian K, Wang JJ, Kang F, Outlaw RA (2009) Carbon nanosheets as the electrode material in supercapacitors. J Power Sources 194(2):1208–1212. https://doi.org/10.1016/j.jpowsour.2009.06.004

Zhong C, Deng Y, Hu W, Qiao J, Zhang L, Zhang J (2015) A review of electrolyte materials and compositions for electrochemical supercapacitors. Chem Soc Rev 44(21):7484–7539. https://doi.org/10.1039/c5cs00303b

Zhu Y, Murali S, Stoller MD, Ganesh KJ, Cai W, Ferreira PJ, Pirkle A, Wallace RM, Cychosz KA, Thommes M, Su D, Stach EA, Ruoff RS (2011) Carbon-based supercapacitors produced by activation of graphene. Science 332(6037):1537–1541. https://doi.org/10.1126/science.1200770

Chapter 2
Zinc Based Spinel Oxides for Energy Conversion and Storage Applications

Faheem K. Butt and Sajid Ur Rehman

2.1 Spinel Oxides

The general formula of spinels is AB_2O_4 where “A” and “B” includes metal ions respectively (Zhao et al. 2017). The most precious stones are made up of spinels. Cr^{3+}, Fe^{2+} and Zn^{2+} are some of the metal ions used in the formation of AB_2O_4 spinel oxides. For example, Cr^{3+}, Fe^{2+} and Zn^{2+} are used in the formation of red and blue color spinel materials respectively (Sutherland et al. 2009). The electronic configuration and electronic structure are used to calculate optical, electrical, intrinsic magnetic and catalytic properties. Different magnetic properties like ferromagnetic and paramagnetic properties are governed by numerous spinel oxides predominantly containing Fe, Co, Cr or Ni species (Phanichphant 2012; Marco et al. 2001; Jeppson et al. 2006). The spinel oxides are always important for electronic industry (spintronic devices) (Bitla et al. 2015), biotechnology (magnetic sensing) (Marco et al. 2001; Vomir et al. 2016) and information technology (data storage) (Marco et al. 2001) for the potential applications. In laser devices and magneto-optical recording of transparent or semi-transparent spinel oxides have demonstrated excellent electrochemical luminescence, photoluminescence (Sonoyama et al. 2006) and parasitic light absorption characteristics (Goldstein et al. 2013). Supercapacitors (Wei et al. 2016; Shanmugavani and Selvan 2016; Zhang et al. 2016; Butt et al. 2014a) and Lithium/metal-ion batteries (Whittingham 2004, 2014) are the energy storage compact devices, which have extended the applications of allowed electrical

F. K. Butt (✉)
Division of Science and Technology, Department of Physics, University of Education, Lahore, Pakistan

Physik-Department, ECS, Technische Universität München, Garching, Germany

S. U. Rehman
State Key Laboratory on Integrated Optoelectronics, Institute of Semiconductors, Chinese Academy of Sciences, Beijing, China

S. Siddiquee et al. (eds.), *Nanotechnology: Applications in Energy, Drug and Food*,
https://doi.org/10.1007/978-3-319-99602-8_2

characteristics of spinel oxide materials. The redox capacitance of spinel oxides can be tuned owing to their intrinsic capability of alterable chemical valence states. In Li-ion batteries spinel compounds are used as an anode and cathode (Zhao et al. 2017).

AB_2X_4 is the common composition of normal spinel materials where "A" having elements like Li, Mn, Zn, Cu, Co, Mg, Ni, Cd Ge, Fe, Ca etc. takes place at the center of tetrahedral coordinated positions, "B" having elements like Al, Cr, Mn, Fe, Co, Ni, Ga, In, Mo etc. at the center of octahedral coordinated positions and "X" acts as anions having elements O, S, Se, Te, N etc. at the polyhedral vertices (Sickafus et al. 1999; Hill et al. 1979).

2.2 Zinc Vanadium Oxide (ZnV_2O_4)

Recently, zinc vanadium oxide (ZnV_2O_4) spinel oxide has been proposed as promising material for energy storage applications such as lithium-ion batteries, supercapacitors and hydrogen storage (Butt et al. 2014a, b, c; Xiao et al. 2009). Various nanostructures of ZnV_2O_4 have been reported such as clew like hollow structures, hollow spheres, nanosheets, and glomerulus nano/microspheres etc. (Butt et al. 2014c; Duan et al. 2011). However, a few reports discuss about the formation of hierarchical nanostructures of ZnV_2O_4 as well as their electrochemical applications. Figure 2.1 shows the possible applications of ZnV_2O_4 spinel oxide.

One of the most important spinel is Zinc vanadium oxide (ZnV_2O_4), has gained much attention of researchers, which has cubic spinel structure and crystal having mott-insulator crystalline structure (Butt et al. 2014a, b; Xiao et al. 2009). ZnV_2O_4 lattice parameters are a = b = c = 8.393 Å and have symmetry constraints $\alpha = \beta = \gamma = 90$ (Butt et al. 2014c). For ZnV_2O_4 in the general composition of spinel "AB_2O_4" is that the "B" is occupied by ions of V^{3+} and takes place in the spinel structure at tetrahedral position and the oxygen has six ions which takes place at octahedral position. At low temperature frustrated magnetism takes place due to magnetic ordering. Therefore, due to its interesting magnetic and electrochemical features ZnV_2O_4 has gained attention in the spinel oxides family. In ZnV_2O_4, ZnO_4 and VO_6 exist in tetrahedral and octahedral formation (Rechuis et al. 2003). ZnV_2O_4 has Fd3m symmetry group due to its FCC spinel structure. The ions of zinc and vanadium occur at 18a and 16d sides at tetrahedral and octahedral positions respectively, and 32e site is occupied by the ions of oxygen. Zinc, vanadium and oxygen atoms have been represented by the red, yellow and green spheres respectively as shown in Fig. 2.2. In this chapter we present the synthesis of important ZnV_2O_4 nano/micro structures and their uses in various electrochemical devices and applications including LIB, supercapacitors, Hydrogen storage and thermoelectric applications.

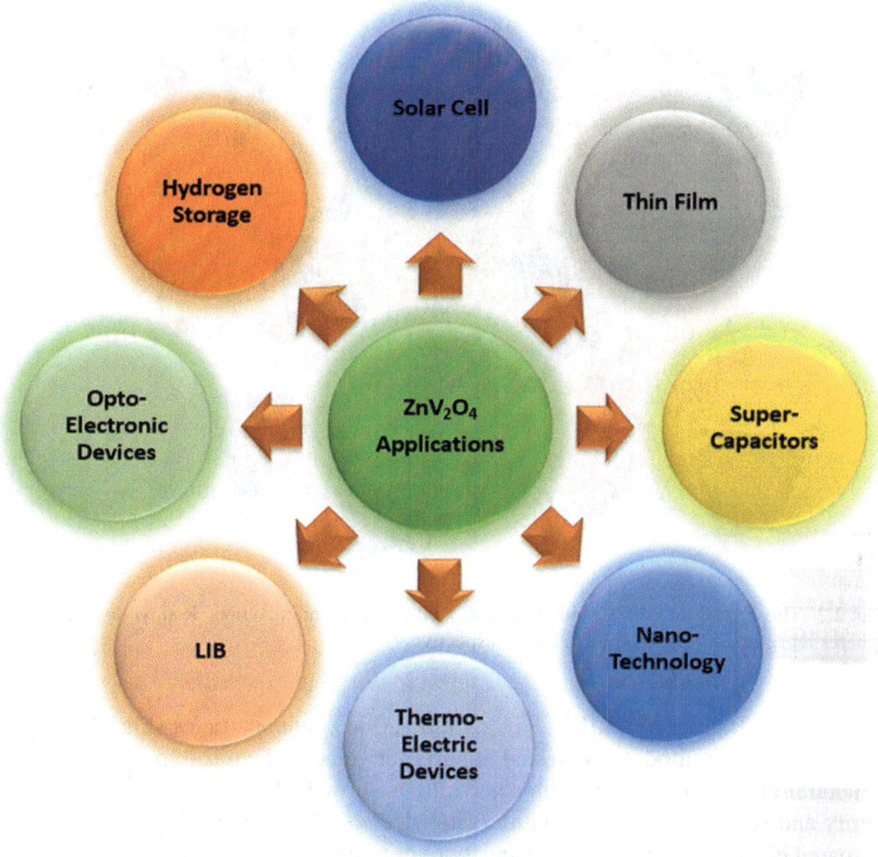

Fig. 2.1 A representation of possible applications of zinc vanadium oxide materials

2.3 Lithium Ion Batteries (LIBs)

Lithium is the lightest element among other elements which has high electrochemical potential. Lithium acts as anode/cathode in the rechargeable batteries which can give extraordinary high energy density. Lithium ion batteries (LIBs) replaced the lead acid batteries because LIBs have high energy density, tiny memory effect and low self-discharge in portable electronic devices. The light weight lithium ion batteries packs provide the same voltage as lead-acid batteries. The positive and negative electrodes in the electrochemical reactants provide the conductive medium for lithium ions (Etacheri et al. 2011; Fotouhi et al. 2016). The lithium-based batteries have high gravimetric and volumetric capacity. On earth's crust, the amount of lithium is enough to power the automobiles present in our society. Lithium ion batteries become the important source for portable electrochemical energy storage (Lu et al. 2013).

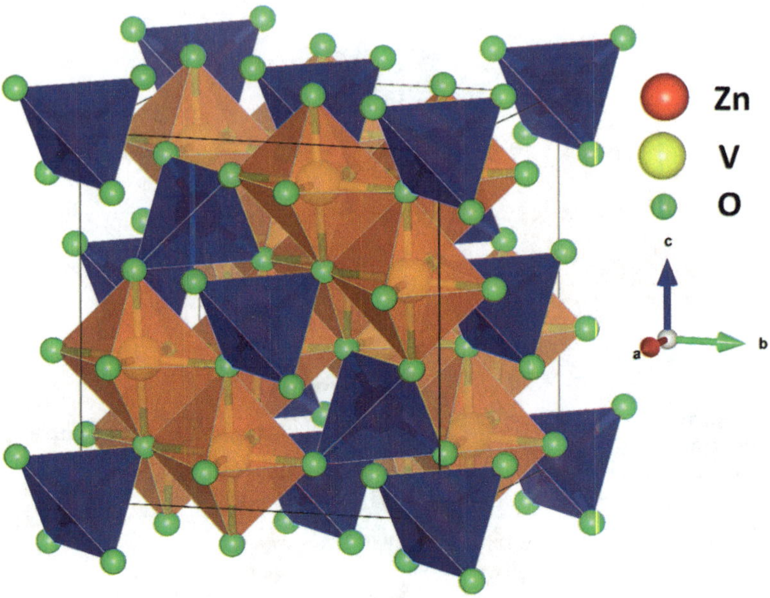

Fig. 2.2 The crystal structure of spinel ZnV_2O_4 with corresponding atoms. Red, yellow and green spheres represent the Zn, V and O atoms, respectively

Xiaoming Zhu et al. (2015) synthesized ZnV_2O_4 nanoparticles (Fig. 2.3a) by a hydrothermal method and tested them for electrochemical properties. The structural and morphological characterizations show that ZnV_2O_4 sample has high purity and well crystallization with crystal size less than 20 nm (Fig. 2.3b). The as prepared electrode shows stable capacity over 660 mAh g^{-1} in the voltage range of 0.01–3.0 V at 50 mAg^{-1} (Fig. 2.3c, d). It shows that the pristine ZnV_2O_4 is transformed to isostructural spinel $LixV_2O_4$ ($\times$ close to 7.6) and metal Zn phase during the first lithiation process. Then the spinel $LixV_2O_4$ seems to perform a topotactic intercalation reaction mechanism and the in-situ formed $LixV_2O_4$ can still keep its spinel matrix while allowing more than 5.7 lithium reversibly into/out over 50 cycles.

Lingxing Zeng et al. (2012) synthesized nanocomposite of ZnV_2O_4–CMK by using a low temperature carbothermal reduction route. Li-ion acts as anode material in nanocomposites which gives cycling stability, large reversible capacity and high rate performance. The pure ZnV_2O_4-CMK nanocomposites are formed which is confirmed by the peaks obtained by XRD. The crystalline size of ZnV_2O_4-CMK is calculated mathematically by Bragg peaks and Scherer formula. The ZnV_2O_4-CMK nanocomposites show rod-like structure and forming an irregular bulk (Fig. 2.4a) for pure ZnV_2O_4-CMK nanocomposites which is synthesized by hydrogen reduction route. The mesoporous channels are shown in Fig. 2.4b presented by the TEM image.

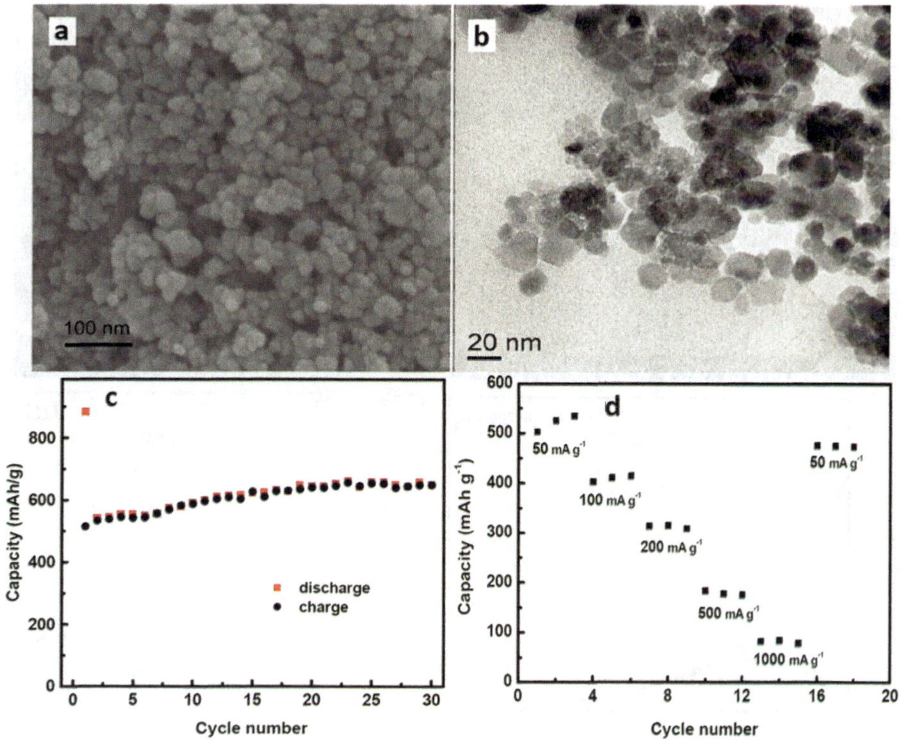

Fig. 2.3 (**a**) SEM, (**b**) TEM images of the as-prepared ZnV_2O_4 (**c**) discharge-charge capacity vs. cycling number plot of ZnV_2O_4 electrode at 50 mA g^{-1} and (**d**) rate capability of ZnV_2O_4 electrode at various discharge-charge. Reprinted with permission from ref. Zhu et al. (2015). Copyright 2015, Elsevier

The TEM image shows the size of ZnV_2O_4 particles are in range of 30 to 80 nm. To improve capacity and increase the rate capability is one of the possibilities to shorten path for Li-ion and electron transport which provide more sites to insert Li-ion with a large surface area of ZnV_2O_4 -CMK nanocomposites.

Cheng Zheng et al. (2014) synthesized hierarchical ZnV_2O_4 microspheres via ethanol thermal reduction route. N_2 adsorption-desorption measurements and thermogravimetric analysis (TG) characterization techniques are used to characterized ZnV_2O_4 microspheres. The formation of homogeneous solution need strong stirring of 0.178 g Zn $(NO_3)_2{\cdot}6H_2O$, 0.204 g NH_4VO_3 and 0.09 g CTAB which are added in 30 mL of ethanol to synthesize hierarchical ZnV_2O_4 microspheres. The volume changing takes place during the cycle of hierarchical nanostructure. Long-term cycling stability, high rate performance and high reversible capacity are exhibited by LIBs which acts as an anode material.

In Fig. 2.5a, b the images are taken by the characterization method of SEM. The reaction time plays an important role in the structural morphology and size of

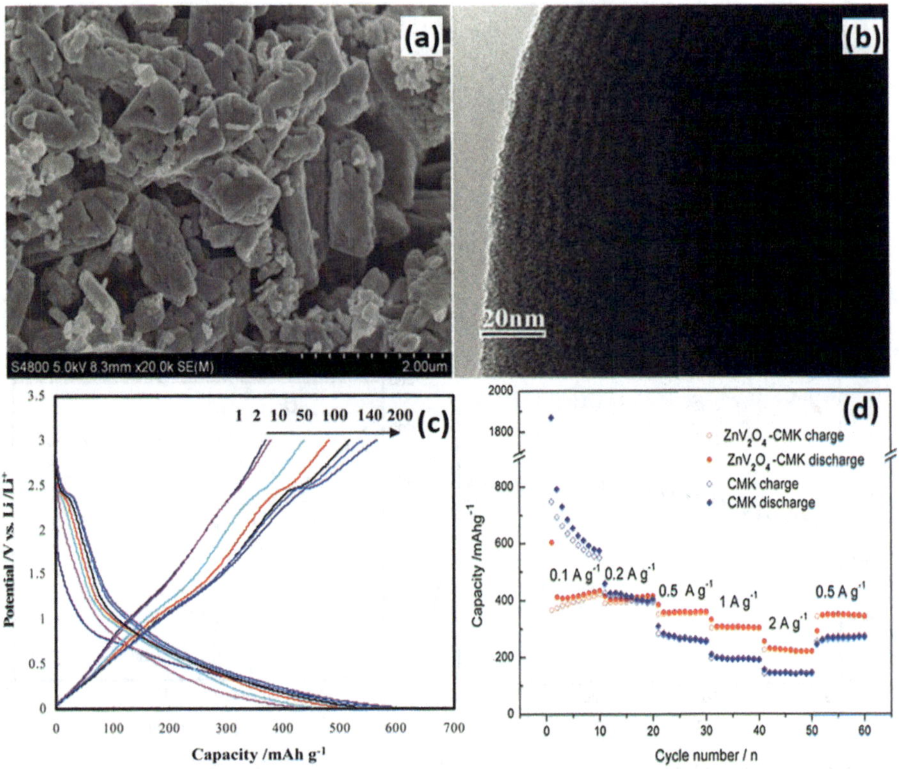

Fig. 2.4 (**a**) SEM and (**b**) TEM images of ZnV_2O_4-CMK. (**c**) Charge-discharge capacity and (**d**) rate capacity of ZnV_2O_4-CMK. Reprinted with permission from refs. Zhu et al. (2015) and Zeng et al. (2012). Copyright 2012, Royal Society of Chemistry

ZnV_2O_4. In Fig. 2.5, different reaction time gives the distinct morphological structure of ZnV_2O_4. In Fig. 2.5a, nanoflakes flowers of irregular particles are observed by the SEM characterization. With the increase in reaction time, morphological changes take place in the structure of ZnV_2O_4. At Fig. 2.5b, the structure of ZnV_2O_4 changes from nanoflakes microspheres to hierarchical microsphere. The change in structure takes place with the change in size from 0.5 to 2 μm due to the increase in reaction time. In Fig. 2.5c, d, shows the TEM and HRTEM images of ZnV_2O_4 nanostructures. In Fig. 2.5c, from nanoflakes microsphere structure a 3D hierarchical structure of ZnV_2O_4 is obtained as can be seen by TEM characteristics. In Fig. 2.5d, it is observed that ZnV_2O_4 is single crystal structure with plane (311) of lattice fringe spacing. It is clearly observed from results that ZnV_2O_4 microsphere acts as a good anode for LIBs batteries with high performance. Figure 2.5e, f shows the charge-discharge graphs of hierarchical ZnV_2O_4 microspheres. The ZnV_2O_4 displayed by a long slope and two short plateaus. The charge-discharge curve at various current densities Fig. 2.5f shows the cycling performance of hierarchical ZnV_2O_4 microspheres. The cycling process of charge and discharge slowly improves the

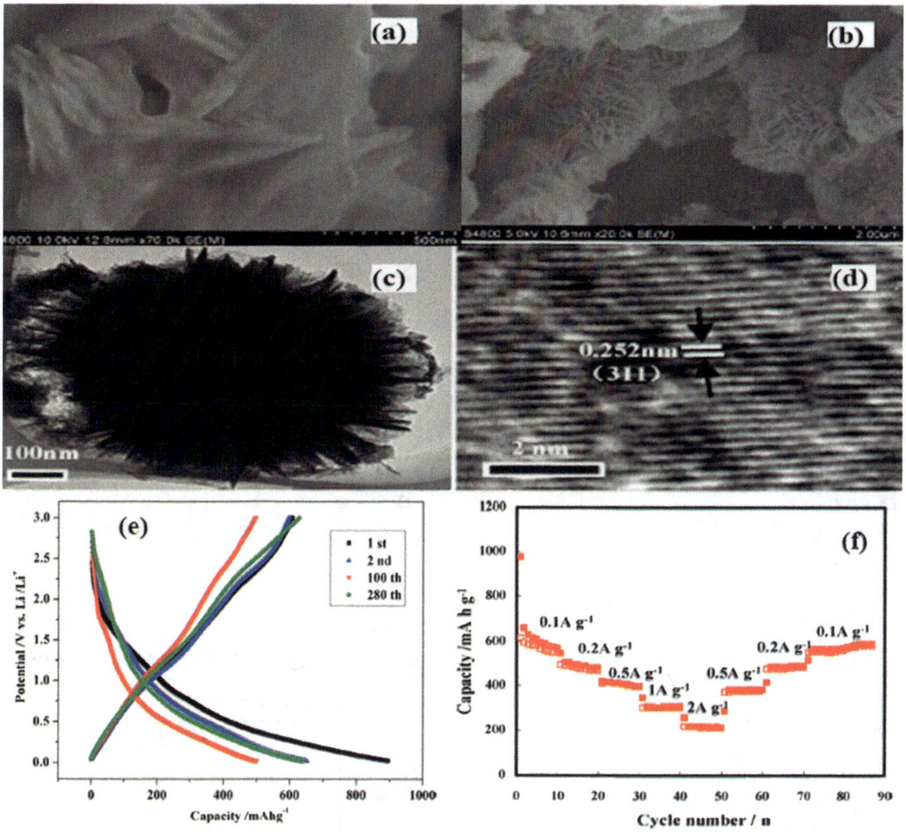

Fig. 2.5 (**a**, **b**) SEM, (**c**) TEM, (**d**) HRTEM images of ZnV_2O_4, (**e**) charge-discharge cycling capacity and (**f**) rate capacity at various current densities. Reprinted with permission from ref. Zheng et al. (2014). Copyright 2014, Royal Society of Chemistry

electrochemical activity of hierarchical microspheres structure of ZnV_2O_4. The cycling process due to increased capacity improved the Li+ diffusion kinetics by the process of stabilization and activation.

In Fig. 2.5f, at specific current density the reversible capacity of hierarchical microspheres structure of ZnV_2O_4 is obtained. The current density is reinstated to its original form when the cycling capacity attains the high rate performance. The Li ion insertion is considered to be improved by the stability of 3D hierarchical microspheres.

Lifen Xiao et al. (2009) synthesized hollow ZnV_2O_4 microspheres with a clew-like feature by mixing ammonium metavanadate and zinc nitrate hexahydrate in benzyl alcohol at 180 °C. Time-dependent experiments have shown that the growth mechanism of the clew-like ZnV_2O_4 are hollow microspheres (Fig. 2.6a) obtained due to a unique multistep pathway. The lithium storage ability of clew-like ZnV_2O_4

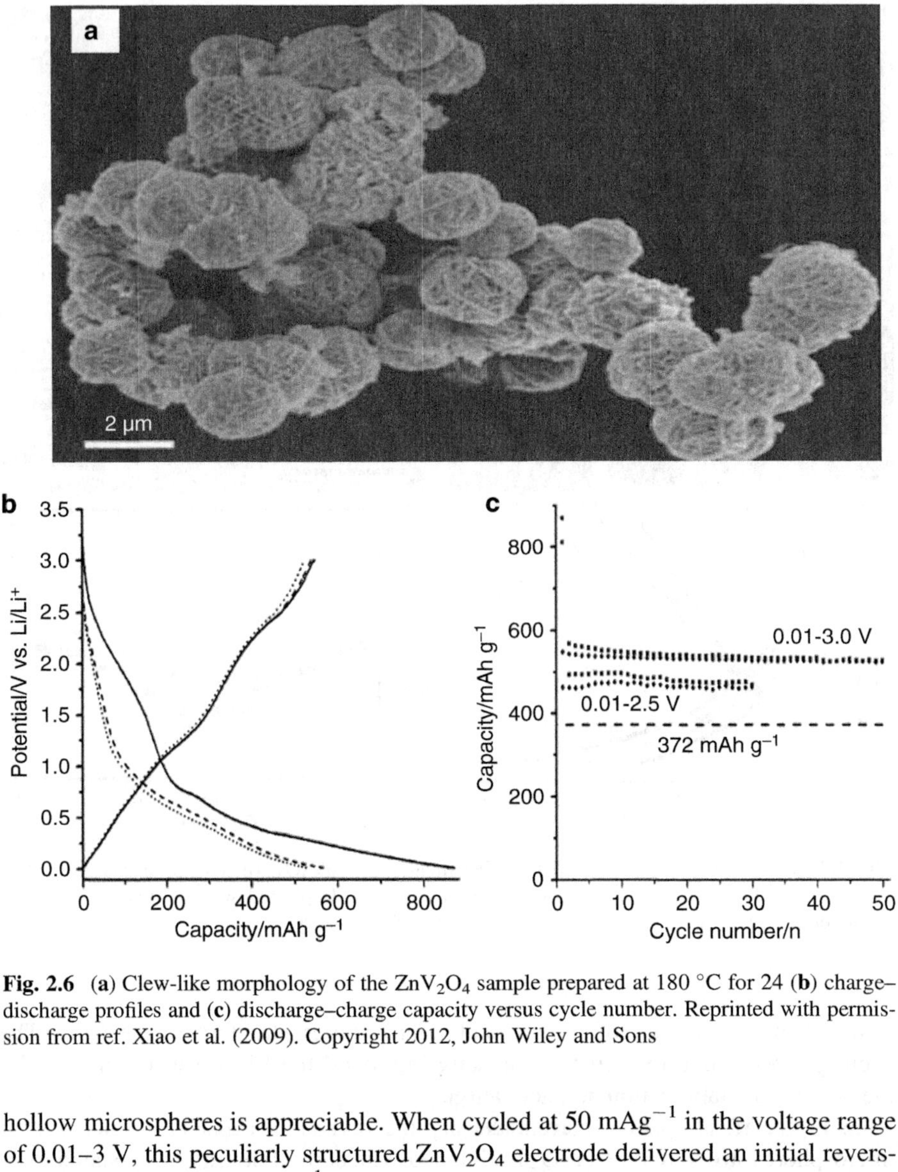

Fig. 2.6 (**a**) Clew-like morphology of the ZnV_2O_4 sample prepared at 180 °C for 24 (**b**) charge–discharge profiles and (**c**) discharge–charge capacity versus cycle number. Reprinted with permission from ref. Xiao et al. (2009). Copyright 2012, John Wiley and Sons

hollow microspheres is appreciable. When cycled at 50 mAg^{-1} in the voltage range of 0.01–3 V, this peculiarly structured ZnV_2O_4 electrode delivered an initial reversible capacity of 548 $mAhg^{-1}$ and exhibited almost stable cycling performance to maintain a capacity of 524 $mAhg^{-1}$ over 50 cycles. This attractive lithium storage performance suggests that the resulting clew-like ZnV_2O_4 hollow spheres are promising for lithium-ion batteries (Xiao et al. 2009).

It can be summarized that among many structures of ZnV_2O_4 the clew-like structure is very promising material for Li-ion batteries due to its striking storage ability for lithium. Their morphological and structural characterizations show that ZnV_2O_4 sample has high purity and good crystallinity.

2.4 Supercapacitors

In energy storage devices, supercapacitors become prominent new technology. Supercapacitors are also known as Ultra capacitors and electrochemical dual layer capacitor (EDLC) (Simon et al. 2017). As, compared to batteries, supercapacitors are used in power demand due to its low energy density and high-power density. The electrostatic energies are stored by the supercapacitors. Supercapacitor may have aqueous or organic electrolyte. The bank load on battery is distributed by using the supercapacitors. The limitations of lead acid batteries can be overcome by supercapacitors. Supercapacitors provide excellent power performance to energy storage devices (Zhang et al. 2017).

Faheem et al. (2014a) proposed a facile and template free method to synthesize novel hierarchical nanospheres (NHNs) of ZnV_2O_4 by a hydrothermal method. In Fig. 2.7a, b, it can be seen clearly, the main building blocks of nanospheres are made up of some major nanosheets-like structures. The thin nanosheets-like structure has maximum size 10–15 nm in thickness. In Fig. 2.7c, d, hierarchical nanospheres can be seen under the TEM. In Fig. 2.7d the sample is characterized by the technique of high resolution transmission electron microscopy (HRTEM) with an interplanar spacing of 0.48 nm which corresponds to the spinel structure of ZnV_2O_4 having plane (111).

Figure 2.8a, b shows the electrochemical supercapacitor measurements of hierarchical ZnV_2O_4 grown at 200 °C. Figure 2.8a shows the discharge curves of device at various current densities. Figure 2.8b shows the cycling specific capacitance. The measurements show that the supercapacitor device is stable and shows higher capacitance values as compared to other ternary oxides and binary counterparts of ZnV_2O_4 (Butt et al. 2014c). Figure 2.8d shows the hydrogen absorption properties of ZnV_2O_4 at three different temperatures. It was indicated that the storage capabilities of ZnV_2O_4 is higher than ZnO and V_2O_5 nanomaterials which point out the potential of this material as an efficient energy storage material. Reported study reveals that ZnV_2O_4 is stable and shows higher capacitance values as compared to its other ternary oxides and binary counterparts, which make it an efficient and reliable energy storage material for supercapacitors applications.

2.5 Hydrogen Storage

Hydrogen is light weight element and highly abundant element used as a synthetic fuel. To minimize the greenhouse gas emission, hydrogen is an efficient and sustainable energy source used to produce power. The hydrogen based energy has

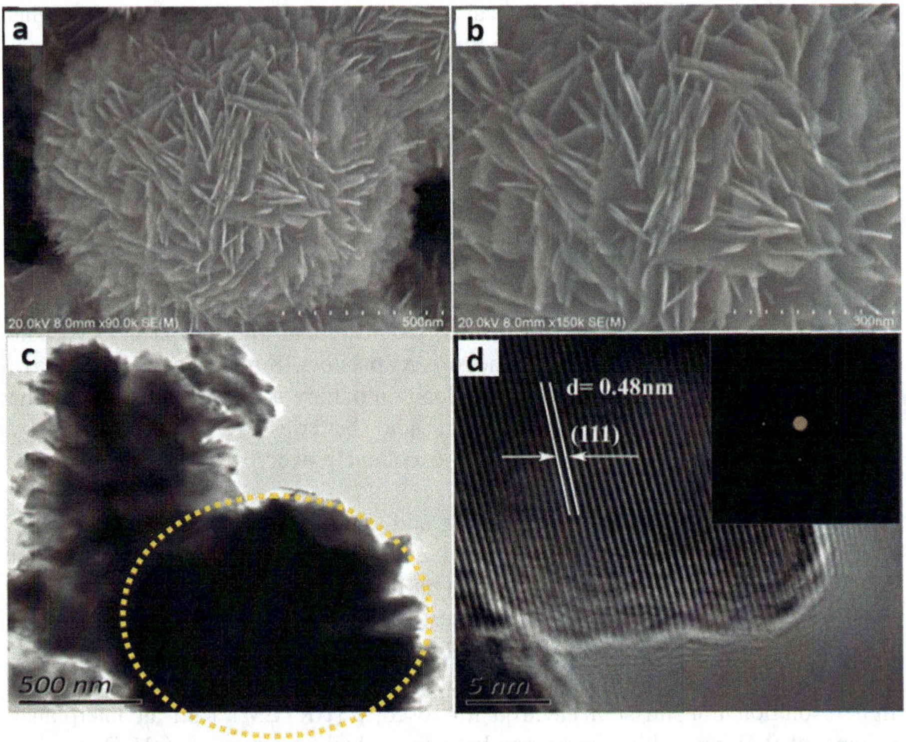

Fig. 2.7 (**a, b**) FESEM, (**c**) TEM and (**d**) HRTEM of ZnV_2O_4. Reprinted with permission from ref. Butt et al. (2014a). Copyright 2015, American Chemical Society

high volumetric density for hydrogen storage (Stetson et al. 2015). Hydrogen plays an important role for the production of energy carries. Hydrogen storage has high energy production rate when obtained from the renewable energy sources. The hydrogen has an advantage of emitting water as a waste product which in result powers the vehicles and other systems efficiently with less environmental concerns (Chilev and Lamari 2016).

Butt et al. (2014c) presented study on hydrogen storage properties of ZnV_2O_4 glomerulus nano/microspheres synthesized via template free route by utilizing common laboratory reagents. The presented method is facile, cost effective and can be used for mass production. An isovolumetric method is used to measure hydrogen absorption-desorption kinetics of ZnV_2O_4 spinel oxide. The process of XRD performed on the sample shows sharp peaks of ZnV_2O_4 spheres in a good quality where value of peak is (311) having lattice parameters a = b = c = 8.404 Å and position is 35.364°.

On the surface of sample glomerulus type structures can be seen having nanospheres of 450–500 nm diameter as shown in the Fig. 2.9a, b. The zinc, vanadium and oxygen are the elements which are analyzed by EDS having different ratios. By HRTEM and SAED characterization of the sample as shown in Fig. 2.9c, d, the value of interplanar distance was found. Due to SAED characterization, the nano

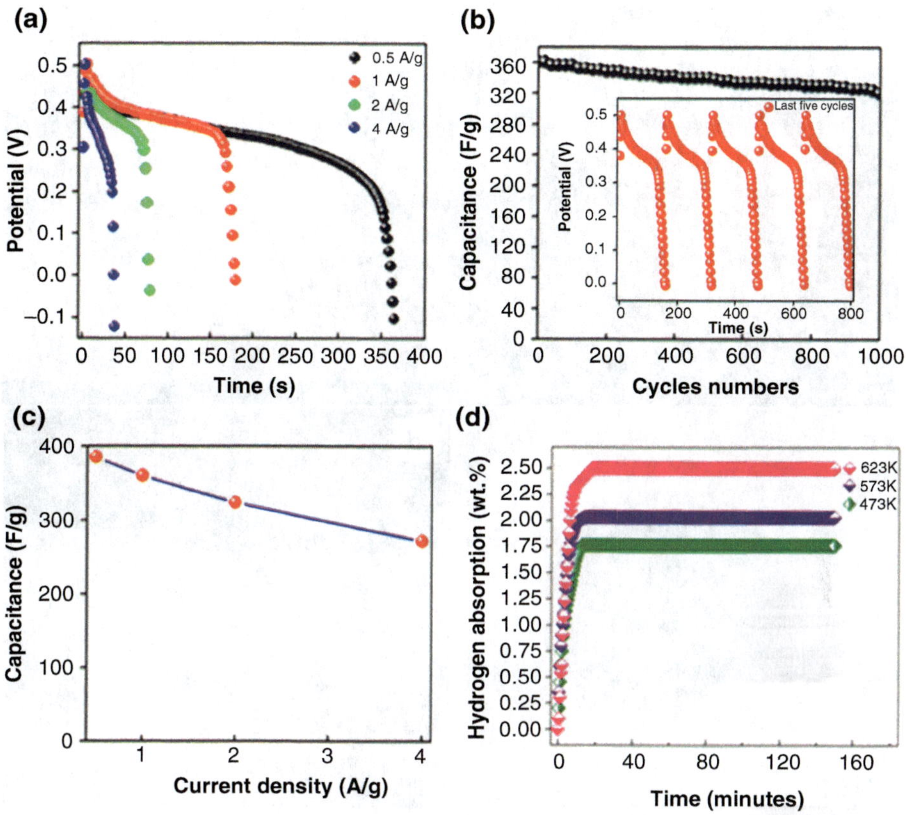

Fig. 2.8 (**a**) First charge-discharge curves of ZnV_2O_4 nanospheres at different current densities, (**b**) cycling capacitance measurements, (**c**) capacitance as a function of current density and (d) hydrogen absorption curve. Reprinted with permission from ref. Butt et al. (2014a). Copyright 2015, American Chemical Society

crystalline structure is given by concentric rings formed by the bright spot. In Fig. 2.9e, at different temperature nano/microspheres of ZnV_2O_4 were tested for hydrogen absorption.

Novel nanosheets of ZnV_2O_4 spinel oxide achieved by template free route for the first time were explored for its potential hydrogen storage properties (Butt et al. 2014b). ZnV_2O_4 is an excellent candidate in 2D layered nanostructures for storage applications. The maximum value in ZnV_2O_4 nanosheets for hydrogen absorption was 1.74 wt.% at 573 K, and 1.36 wt.% at 473 K respectively (see in Fig. 2.10c, d). Faheem et al. (2014b) measurements for hydrogen storage along ZnV_2O_4 shown superiority as compared to previous reports on the value of hydrogen absorption concerning nitrides, oxides, and chalcogenides. To understand in depth authors also applied the various kinetics models and rate-limiting mechanism. The calculations reveal that kinetics model (with constant interface velocity) is governed by 3D

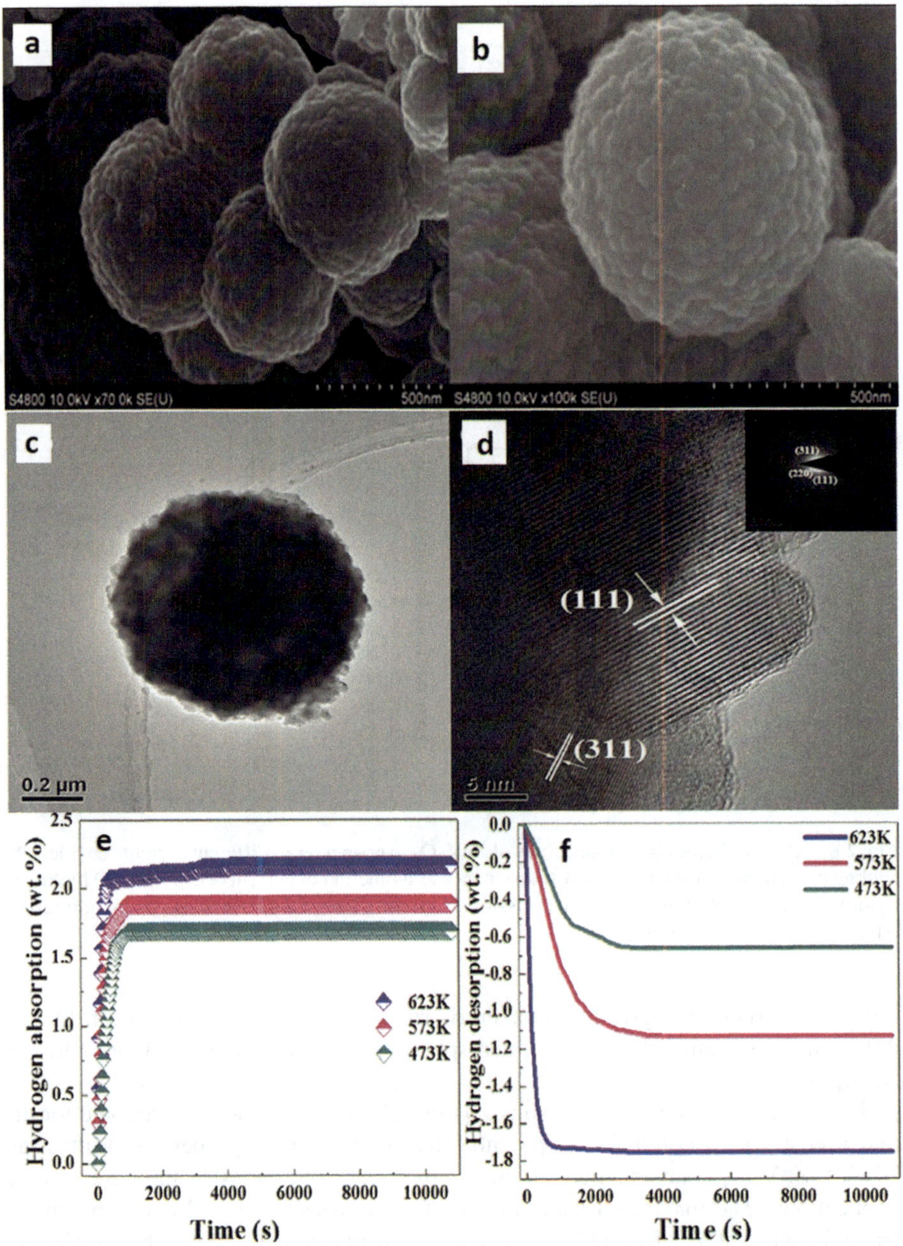

Fig. 2.9 (**a**, **b**) SEM, (**c**) TEM, (**d**) HRTEM images and hydrogen absorption and desorption of ZnV_2O_4 spinel oxide. (**e**, **f**) Hydrogen absorption and desorption curves at various temperatures. Reprinted with permission from ref. Butt et al. (2014c). Copyright 2015, Elsevier

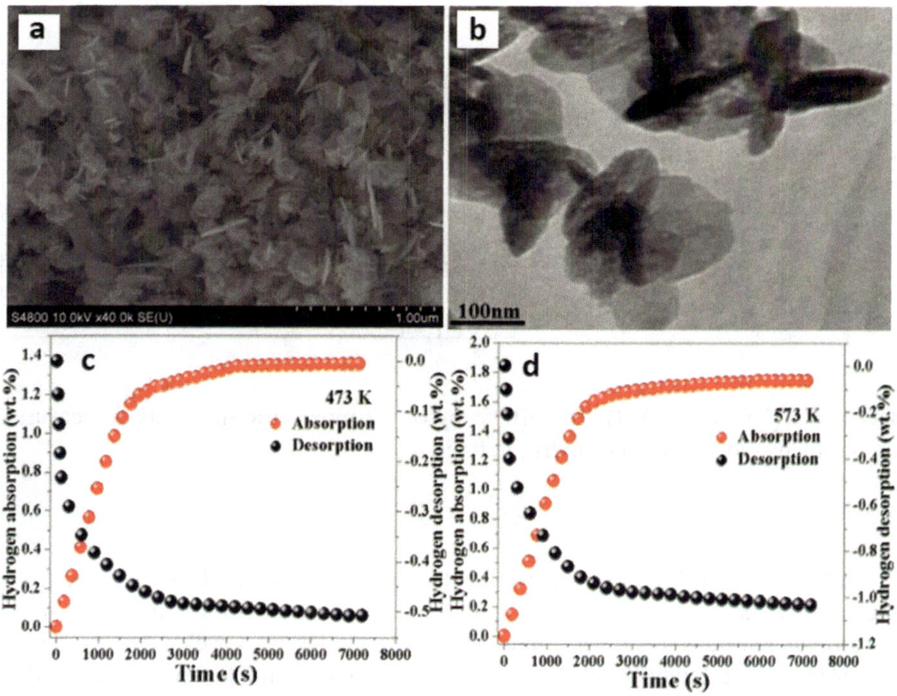

Fig. 2.10 (**a**) FESEM and (**b**) TEM images of ZnV_2O_4 nanosheets respectively (**c**) Hydrogen storage of ZnV_2O_4 nanosheets at 473 K and (**d**) 573 K. Reprinted with permission from ref. Butt et al. (2014b). Copyright 2014, American Chemical Society

growth. The results show that ZnV_2O_4 spinel oxide as a promising hydrogen storage material.

The study of ZnV_2O_4 as a hydrogen storage shows superiority as compared to previous reports on the value of hydrogen absorption concerning nitrides, oxides, and chalcogenides etc. Various applied kinetic models endorse that ZnV_2O_4 is a potential material for the storage of hydrogen.

2.6 Thermoelectric Properties

To fulfil the energy demands in a current situation, the attention of researchers has been diverted towards the use of alternative energy sources and green technologies to effectively replace the fossil fuels. So in current scenario, the electrical power can be generated through thermoelectric devices, which are used as a recovery waste heat or heat pumps (Zhang and Zhao 2015; Elsheikh et al. 2014). Thermoelectric (TE) materials can directly convert waste heat into electricity and

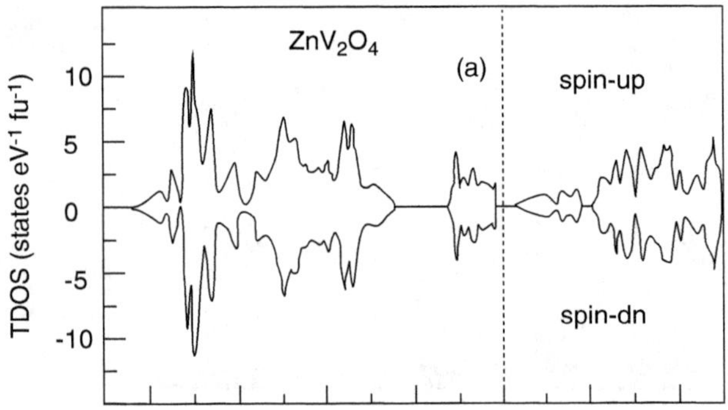

Fig. 2.11 Total and partial density of states plots for ZnV_2O_4. Reprinted with permission from ref. Singh et al. (2016). Copyright 2016, IOP

vice-versa (Pei et al. 2012). The efficiency of thermoelectric materials is measured by a dimensionless figure of merit, defined as

$$ZT = S^2 \sigma T / \kappa$$

where *S, σ, κ,* and *T* are the Seebeck coefficient, electrical conductivity, electronic thermal conductivity and temperature, respectively. $S^2\sigma$ is also called power factor. By Boltzmann transport theory, first principles calculations for electronic structure are used to investigate the thermoelectric properties. Saurabh Singh et al. (2016) reported the investigation on thermoelectric properties of ZnV_2O_4. The experimental results and Density functional theory (DFT) calculations shows that ZnV_2O_4 is thermally stable at high temperature and have good thermoelectric efficiency. The effective mass calculations revealed that the holes in valence band (VB) are approximately four times that of electrons in the conduction band (CB). The large effective mass values for holes are primarily responsible for the observed positive and large S values of ZnV_2O_4. The ab-initio calculation shows that ZnV_2O_4 has antiferromagnetic structure. To represent antiferromagnetic structure of the vanadium ions (V^{3+}) spin up and spin down sequence is shown in Fig. 2.11.

The determined power factor (PF) shows that thermoelectric properties in high temperature range can be increased by fine-tuning of ZnV_2O_4 synthesis conditions and appropriate doping. In the temperature range 900–1400 K, the figure-of-merit (ZT) for p-type doped ZnV_2O_4 is 0.3. Saurabh Singh et al. (2016) suggested that by suitable quantity of p-type doping, ZnV_2O_4 could be a good thermoelectric candidate at high temperatures (Fig. 2.12).

The ZnV_2O_4 is frequently utilized for electrochemical supercapacitor and hydrogen storage applications with hierarchical nanospheres and nanosheets morphologies. These nanostructures also have very good performance when they are used for energy storage devices like in lithium ion batteries and supercapacitors. ZnV_2O_4 nanostructured photo-luminescence (PL) measurements demonstrate the potential for violet/blue optoelectronic devices. Therefore, the "nano-effect" or nanostructures of ZnV_2O_4 can lead to higher performance as compared to bulk and micro-structure.

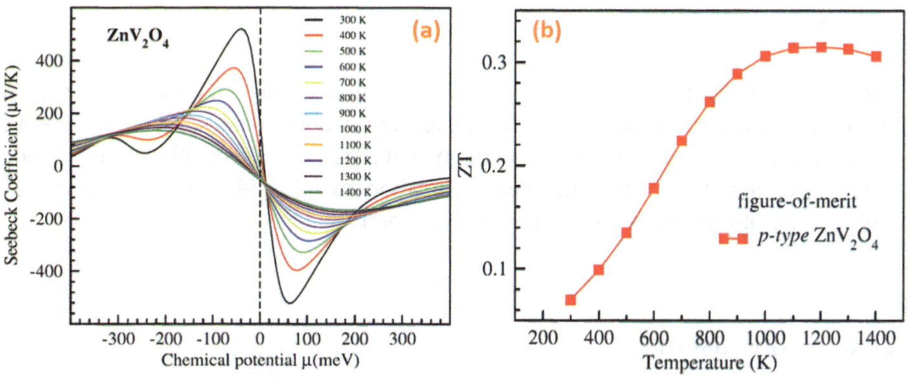

Fig. 2.12 The Seebeck coefficient (α) and dimensionless *figure-of-merit* (*ZT*) for *p-type* ZnV_2O_4 compound at various temperature. Reprinted with permission from ref. Singh et al. (2016). Copyright 2016, IOP

Thus, the novel hierarchical nanospheres (NHNs), glomerulus nano/microspheres and spinel oxide nanosheets of ZnV_2O_4 can be potential applicants in different energy storage applications in near future.

2.7 Future Perspectives

Zinc Vanadium Oxide (ZnV_2O_4) offer a great perspective for the most efficient and cost-effective material for a wide variety of conversion and energy storage applications. It is also possible to add different species at zinc (Zn) site and check their stability and performance for different energy storage and conversion devices. Further, the universal route for the synthesis of ZnV_2O_4 can be used for fabrication of vanadium based transition metal oxides with higher conductivity which will enable a higher device performance for various optoelectronics and energy storage systems.

2.8 Opportunities

One of the big opportunities in zinc vanadium oxide (ZnV_2O_4) is to dope transition metal at zinc site and develop a transition metal library (TML) and test this TML as energy storage and energy conversion devices.

2.9 Challenges

There is a great challenge to synthesize zinc vanadium oxide (ZnV_2O_4) in pure form at low temperature because, vanadium has four valence states. For this reason, vanadium is known as the goddess of beauty and fertility. The versatile compositions can be attained from vanadium based transition metal oxides but the control of morphology and phase is a great challenge to perform.

2.10 Conclusions

In this chapter we present the review on the synthesis of template free, efficient and facile techniques for the fabrication of novel hierarchical nanospheres (NHNs), glomerulus nano/microspheres and spinel oxide nanosheets of ZnV_2O_4. The advanced technological potential applications in the field of hydrogen storage, lithium ion batteries and in supercapacitors for energy storage are presented. XRD and TEM results along with its study for thermoelectric properties has endorsed that ZnV_2O_4 is a very promising and efficient material to be used for energy storage and thermoelectric devices. All in all, we believe this comprehensive review will kindle broader interests in ZnV_2O_4 based energy storage system within the new generation materials that lead it as one of the promising solutions for energy crisis.

Acknowledgements Financial support from Alexander von Humboldt Foundation and Federal Ministry for Education and Research (BMBF) is gratefully acknowledged.

References

Bitla Y et al (2015) Origin of metallic behavior in NiCo2O4 ferrimagnet. Sci Rep 5:15201

Butt FK et al (2014a) Synthesis of novel ZnV_2O_4 hierarchical nanospheres and their applications as electrochemical supercapacitor and hydrogen storage material. ACS Appl Mater Interfaces 6 (16):13635–13641

Butt FK et al (2014b) Synthesis of novel ZnV_2O_4 spinel oxide nanosheets and their hydrogen storage properties. CrystEngComm 16(5):894–899

Butt FK et al (2014c) Synthesis, evolution and hydrogen storage properties of ZnV_2O_4 glomerulus nano/microspheres: a prospective material for energy storage. Int J Hydrogen Energy 39 (15):7842–7851

Chilev C, Lamari FD (2016) Hydrogen storage at low temperature and high pressure for application in automobile manufacturing. Int J Hydrog Energy 41(3):1744–1758

Duan F et al (2011) Template-free synthesis of ZnV_2O_4 hollow spheres and their application for organic dye removal. Appl Surf Sci 258(1):189–195

Elsheikh MH et al (2014) A review on thermoelectric renewable energy: principle parameters that affect their performance. Renew Sust Energ Rev 30:337–355

Etacheri V et al (2011) Challenges in the development of advanced Li-ion batteries: a review. Energy Environ Sci 4(9):3243–3262

Fotouhi A et al (2016) A review on electric vehicle battery modelling: from Lithium-ion toward Lithium–Sulphur. Renew Sust Energ Rev 56:1008–1021

Goldstein A et al (2013) Parasitic light absorption processes in transparent polycrystalline $MgAl_2O_4$ and YAG. J Am Ceram Soc 96(11):3523–3529

Hill RJ, Craig JR, Gibbs G (1979) Systematics of the spinel structure type. Phys Chem Miner 4 (4):317–339

Jeppson P et al (2006) Cobalt ferrite nanoparticles: achieving the superparamagnetic limit by chemical reduction. J Appl Phys 100(11):114324

Lu L et al (2013) A review on the key issues for lithium-ion battery management in electric vehicles. J Power Sources 226:272–288

Marco JF et al (2001) Cation distribution and magnetic structure of the ferrimagnetic spinel NiCo2O4. J Mater Chem 11(12):3087–3093

Pei Y, Wang H, Snyder GJ (2012) Band engineering of thermoelectric materials. Adv Mater 24 (46):6125–6135

Phanichphant S (2012) Cellulose-precursor synthesis of nanocrystalline Co 0.5 Cu 0.5 Fe 2 O 4 spinel ferrites. Mater Res Bull 47(2):473–477

Rechuis M et al (2003) Crystallographic and magnetic structure of ZnV_2O_4. Eur Phys J B 35:311–316

Shanmugavani A, Selvan RK (2016) Improved electrochemical performances of $CuCo_2O_4/CuO$ nanocomposites for asymmetric supercapacitors. Electrochim Acta 188:852–862

Sickafus KE, Wills JM, Grimes NW (1999) Structure of spinel. J Am Ceram Soc 82(12):3279–3292

Simon P, Brousse T, Favier F (2017) Electrochemical double-layer capacitors (EDLC). In: Simon P, Brousse T, Favier F (eds) Supercapacitors based on carbon or pseudocapacitive materials. ISTE Ltd; Wiley, London; Hoboken, NJ, pp 1–25

Singh S, Maurya R, Pandey SK (2016) Investigation of thermoelectric properties of ZnV_2O_4 compound at high temperatures. J Phys D Appl Phys 49(42):425601

Sonoyama N et al (2006) Electrochemical luminescence of rare earth metal ion doped $MgIn_2O_4$ electrodes. J Electrochem Soc 153(3):H45–H50

Stetson NT, McWhorter S, Ahn CC (2015) 1 - Introduction to hydrogen storage. In: Gupta Ram B, Basile A, Veziroğluin TN (eds) Compendium of hydrogen energy, vol 2. Woodhead Publishing, Cambridge, pp 3–25

Sutherland FL et al (2009) Gem-corundum megacrysts from east Australian basalt fields: trace elements, oxygen isotopes and origins∗. Aust J Earth Sci 56(7):1003–1022

Vomir M et al (2016) Dynamical torque in $Co_xFe_{3-x}O_4$ nanocube thin films characterized by femtosecond magneto-optics: a π-shift control of the magnetization precession. Nano Lett 16 (8):5291–5297

Wei C et al (2016) Valence change ability and geometrical occupation of substitution cations determine the pseudocapacitance of spinel ferrite XFe_2O_4 (X= Mn, Co, Ni, Fe). Chem Mater 28 (12):4129–4133

Whittingham MS (2004) Lithium batteries and cathode materials. Chem Rev 104(10):4271–4302

Whittingham MS (2014) Ultimate limits to intercalation reactions for lithium batteries. Chem Rev 114(23):11414–11443

Xiao L et al (2009) Clewlike ZnV_2O_4 hollow spheres: nonaqueous sol–gel synthesis, formation mechanism, and lithium storage properties. Chem Eur J 15(37):9442–9450

Zeng L et al (2012) ZnV_2O_4–CMK nanocomposite as an anode material for rechargeable lithium-ion batteries. J Mater Chem 22(28):14284–14288

Zhang C et al (2016) Hierarchically porous Co_3O_4/C nanowire arrays derived from a metal–organic framework for high performance supercapacitors and the oxygen evolution reaction. J Mater Chem A 4(42):16516–16523

Zhang L et al (2018) A review of supercapacitor modeling, estimation, and applications: a control/management perspective. Renew Sust Energ Rev 81:1868–1878

Zhang X, Zhao L-D (2015) Thermoelectric materials: energy conversion between heat and electricity. J Mater 1(2):92–105

Zhao Q et al (2017) Spinels: controlled preparation, oxygen reduction/evolution reaction application, and beyond. Chem Rev 117(15):10121–10211

Zheng C et al (2014) Synthesis of hierarchical ZnV_2O_4 microspheres and its electrochemical properties. CrystEngComm 16(44):10309–10313

Zhu X et al (2015) Nanophase ZnV_2O_4 as stable and high capacity Li insertion electrode for Li-ion battery. Curr Appl Phys 15(4):435–440

Chapter 3
Nanotechnology in Renewable Energy: Critical Reviews for Wind Energy

W. K. Muzammil, Md. Mizanur Rahman, A. Fazlizan, M. A. Ismail, H. K. Phang, and M. A. Elias

3.1 New Challenges for Wind Turbine Blades

According to Tan et al., until 2010, oil is the main contribution in the world energy demand which is about 34% followed by coal (26%), natural gas (22%), biomass (9%), nuclear (6%), and others which are mainly from renewable sources (1%) (Tan et al. 2013). Among all the renewable energy sources, wind energy is cleaner; require less maintenance with a sustainable energy source for electricity generation. The generation of electricity from wind was started in around 1888 in Denmark. During this time, the wind turbine blade was made from stainless steel, but the blades of the turbine failed within 100 h of operation. Since then, different types of materials such as aluminium, plastics, and wood are used for the development of wind turbine blades, but after the 1970s, composite materials were used for the wind turbine blades (Mishnaevsky et al. 2017). The evolution of wind turbine technology is shown in Fig. 3.1.

In general, the main components of the turbine are generator, tower, gearbox, and blades. Among all the components, the blade is considered as the costliest component in the wind turbine. In a large wind turbine, the lifetime of the turbine blades is expected to last up to 20 years whereas for the small wind turbine it is less than

W. K. Muzammil · M. A. Ismail
Faculty of Engineering, Material and Mineral Research Unit (MMRU), Universiti Malaysia, Kota Kinabalu, Sabah, Malaysia
e-mail: khairulm@ums.edu.my

M. M. Rahman (✉) · H. Phang
Faculty of Engineering, Energy Research Unit (ERU), Universiti Malaysia, Kota Kinabalu, Sabah, Malaysia
e-mail: mizanur@ums.edu.my

A. Fazlizan · M. A. Elias
Solar Energy Research Institute (SERI), Universiti Kebangsaan Malaysia, Bangi, Malaysia
e-mail: a.fazlizan@ukm.edu.my

S. Siddiquee et al. (eds.), *Nanotechnology: Applications in Energy, Drug and Food*,
https://doi.org/10.1007/978-3-319-99602-8_3

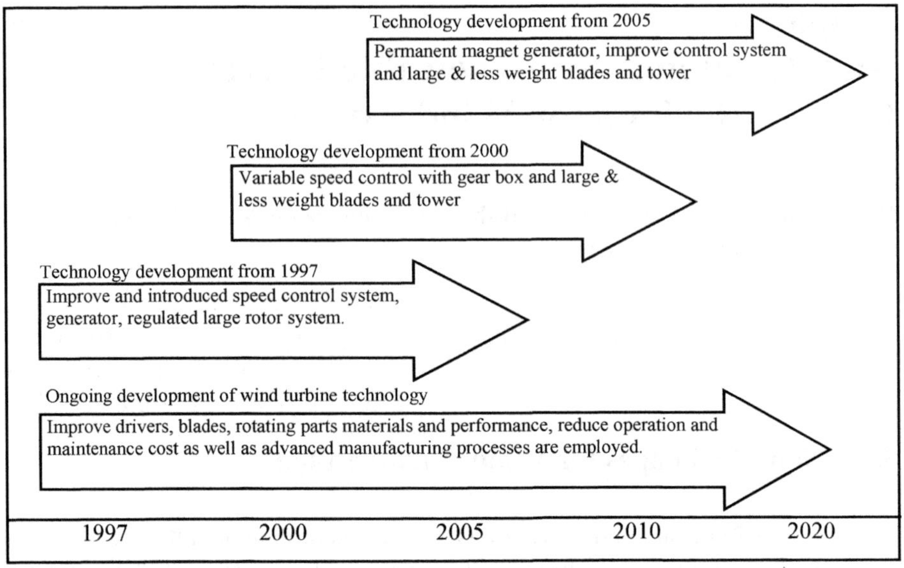

Fig. 3.1 Development of wind turbine technology (Ancona and McVeigh 2001)

20 years (Song 2012). The lifespan of wind turbine blades is mainly dependent on the type of materials used. Therefore, the materials should have high stiffness, low density and long fatigue cycles to guarantee aerodynamic performance and less weight (Ancona and McVeigh 2001; Song 2012; Martinez et al. 2009).

There are many types of research concentrated on the development of specific components of the wind turbine. The general development trends for a wind turbine are mainly focused on using lighter materials with improved life cycle and minimal cost. In case of wind turbine blades, glass fibre reinforced plastic (GRP) material is introduced in modern wind turbines. Other potential materials involving steel, composite materials and carbon filament reinforced plastic (CFRP) have also been explored and introduced as wind turbine blade materials due to their high strength with good fatigue resistance characteristics (Ancona and McVeigh 2001; Griffith et al. 2012).

3.1.1 Wind Turbine and the Energy in the Wind

A wind turbine is a device that is used to convert the kinetic energy in the wind into electrical energy. Wind turbines capture the power from wind through blades, which are built on the aerodynamic principles to harness energy. The geometry and the position of the airfoil are determined to quantify the drag and lift force in the wind turbine to generate tangential force. Figure 3.2 shows an example of the aerodynamics of wind turbine blade. When the airflow passes over (and under) an airfoil, the lift and drag forces are generated due to the difference in air pressure (Fig. 3.2a).

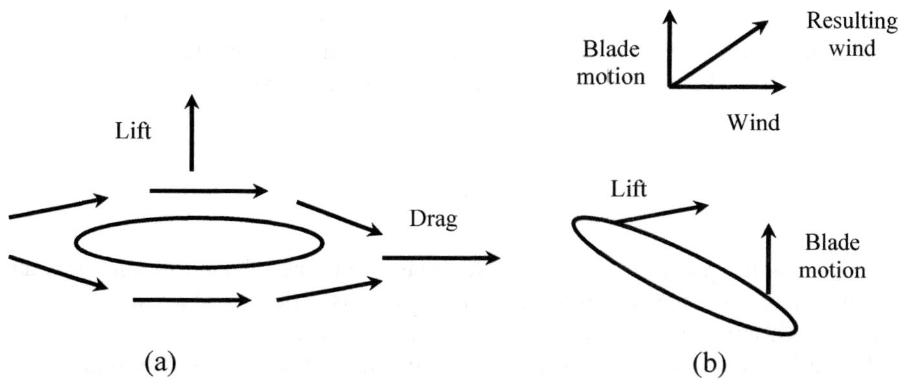

Fig. 3.2 The combined effect of wind and the relative wind for blade motion causes the turbine blade to move and creates a lift up force (Zobaa and Bansal 2011), (**a**) Lift and drag forces generated by wind, (**b**) Relative blade motion and oncoming wind

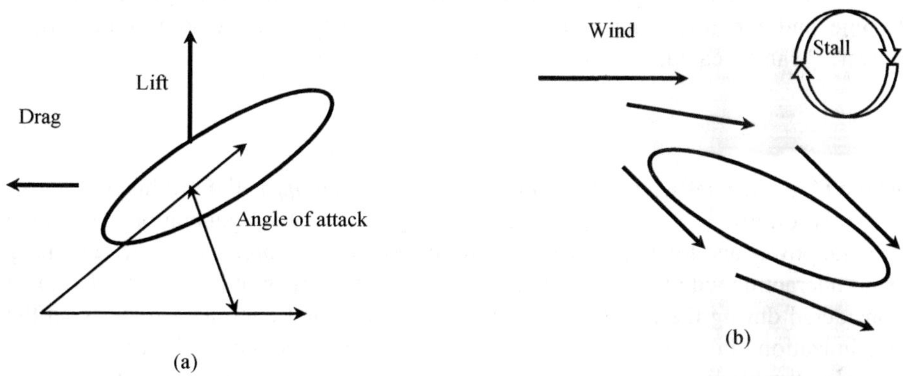

Fig. 3.3 An increase in the angle of attack causes the wing to stall (Zobaa and Bansal 2011), (**a**) A typical angle of attack between blade motion and oncoming wind, (**b**) High angle of attack could lead to stall

Most of today's wind turbines rely on the generated lift force for rotation. Figure 3.2b shows the relative motion of the blade during the operation of the wind turbine. The resultant oncoming wind that interacts with the blade at a certain angle of attack creates the blade motion (Ragheb 2012; Kim et al. 2014; Schubel and Crossley 2012; Barakati 2011).

The angle between the airfoil and the inflow wind is known as the angle of attack as shown in Fig. 3.3. The lift force at the expense of increased drag force can be improved by optimising the angle of attack. However, if the angle of attack is increased too much, it would result in stalling of the airfoil blades with reduced performance of the wind turbine (Elliot et al. 1986).

Consequently, the total wind power and the coefficient of power can be found through the conversion of wind into mechanical power. The total wind power extracted by moving air energy is expressed as a sum of kinetic energy. By assuming the constant area or ducted flow, the continuity equation states that the mass flow rate, $m = \rho A v$. Thus, the total wind power can be calculated from the following equation:

$$P_w = \frac{1}{2} m v^2 = \frac{1}{2} \rho A v^3 \quad (3.1)$$

where, P_w is the power of wind, m is the mass flow rate of wind in kg/s, v is air velocity in m/s, A is the area of blades in m^2, and ρ is the density of air in kg/m^3 (ρ is a function of temperature, relative humidity and pressure).

It is also found that optimum energy can be extracted through wind turbine if the outlet velocity of the wind is approximately one-third of the inlet wind speed, which indicates that the theoretical efficiency of the wind turbine cannot be exceeded by 59.3%. This parameter is known as the Betz limit which helps to determine the maximum theoretical power that can be extracted in the wind system ($0.593P_w$). Therefore, the final output power depends on the coefficient of power of wind turbine and the mechanical and generator efficiency. The wind turbine output power, P can be calculated by using the following formula:

$$P = C_p \eta_m \eta_g P_w \quad (3.2)$$

where C_p is the coefficient of power for a wind turbine, η_m is the mechanical power transmission efficiency, and η_g is the efficiency of electricity generation. The value of wind power and the wind turbine power depend on the direction of the oncoming wind interacting with the turbine blades. There are two main objectives needed to be considered during the design of a wind turbine. The first parameter focuses on the maximization of the average power output from the wind source and the second is to optimize the blade design parameters and reduce the mass of the blade to enhance performance. Aerodynamic principle, potential power from wind, the efficiency of the blade, and blade load are some of the critical parameters to be taken into consideration to optimize the design and the efficiency of the wind turbine.

The blade design of a wind turbine is very important to maximise its potential of extracting wind energy. The actual velocity and the airflow at blade skin are different due to the boundary layer interaction between the blade and the wind. The airflow at the skin of the blade depends on the angle of attack. In turn, the value of lift and drag forces depend upon the critical angle of the airfoil in the wind turbine blade. Any changes in the angle of attack may create additional parasitic forces due to circulation effect on the wind turbine blade, which would reduce the performance of the wind turbine. There are many types of blade design proposed by researchers (Schubel and Crossley 2012), but the maximum theoretical efficiency has not yet to be achieved. Therefore, during the design of a wind turbine, parameters such as blade profile, blade number and tip speed ratio are some of the key parameters required for calculating the optimum angle of attack for optimal performance of wind turbine.

Also, most of the wind turbines on the market today could start to generate electricity when the wind speed is between 3 and 4 m/s where the turbine blades

Table 3.1 Power density per unit area for different wind classes at 50 m height (Elliot et al. 1986)

Wind power class	Wind power density (W/m^2)	Rayleigh average wind speed (m/s)
1	0–200	0–5.6
2	200–300	5.6–6.4
3	300–400	6.4–7.0
4	400–500	7.0–7.5
5	500–600	7.5–8.0
6	600–800	8.0–8.8
7	800–2000	8.8–11.9

had sufficient energy to rotate. The National Renewable Energy Lab (NREL) based on the potential wind power density (WPD) has recognised several classes of wind energy. The WPD signifies the region's potential of wind energy based on the mean wind speed at predetermined heights. Rayleigh distribution can be used to approximate the distribution of wind speeds as shown in Table 3.1. Areas of Class 7 have the highest power density with a range of 800–2000 W/m^2 (mean wind speed: 8.8–11.9 m/s at 50 m height). Whereas, areas of Class 1 have the lowest power density with a range of 0–200 W/m^2 (mean wind speed: 0–5.6 m/s). As logic dictates, the implementation of wind energy devices must be prioritised for areas with the higher average wind speed. Current commercial wind turbines are also limited to regions with Class 3 or higher WPDs. However, due to the on-site energy demand of future cities, the placement of wind energy devices would be nearer to or in the urban areas. Lower WPD class regions are usually found in the urbanised areas, and therefore the installed wind turbines in cities or towns are smaller. Recently, there is an increasing trend of small and medium scale wind turbines installed in the urban areas due to many factors, such as the wind conditions, noise pollution, aesthetic and structural safety issues. Therefore, small wind turbines in the context of integration in an urban environment are considered as one of the most promising technologies for efficient diffusion of renewable energy sources.

Wind turbine blade is usually designed by applying the design process of aircraft wings. Furthermore, during the design process, different operating parameters and mechanical loads of wind on the turbine blade must be taken into consideration. Moreover, the amount of potential wind power that can be extracted increases with the increase of blade length. However, with the increase of blade length, some structural enhancement must be made to reduce the risk of failures. In a modern wind turbine, the blade is made of glass fibre reinforced polyester or epoxy. Steel and aluminium alloys are also used in the fabrication of small wind turbines. However, the use of these materials may cause high load and fatigue when the size of the turbine blade is increased. The diameters of the rotor for wind turbine depend on the power rating of the wind turbine. The relation between power rating and rotor diameter is tabulated in Table 3.2. The diffusion of wind energy technology and the size of the turbine is normally based on the siting of the device. Small wind turbines having the diameter of fewer than 20 m with power production of less than 10 kW are usually found in urban areas, homes, farms, and remote areas. Typical

Table 3.2 Relation between power rating and wind turbine rotor blade diameter (Edelstein et al. 2003)

Power rating in kW	Rotor diameter in m
300	27–33
500	33–40
600	40–44
750	44–48
1000	48–54
1500	54–64
2000	64–72
2500	72–80

modern wind turbines for large wind farms have diameters of 40–90 m with power rating between 500 and 2500 kW.

3.1.2 Environmental Issues Affecting Wind Turbines

Apart from the maximum power input and optimal tip speed ratio of the turbine blade, environmental issues on turbine blade affect the energy production of wind turbines. The environment is one of the crucial factors to relate the effectiveness and reliability performance of wind turbine, particularly in the application of offshore wind power. Development of offshore wind power is challenging and has been mainly located in northern European countries such as Denmark, Ireland, The Netherlands, Sweden and the UK. The effects of environment on the turbine blades could pose significant problems to the operation of wind turbines. The erosion of blades and the build-up of ice on the unprotected skin of turbine blades could degrade the aerodynamic performance and power output of the wind turbine (Castorrini et al. 2016). Moreover, the accretion of ice on the blades can affect the structural integrity of the whole wind turbine, and even more so when ice sheds from a blade could damage other blades or the other components of the wind turbine. Blade erosion and icing could also have a significant impact on the annual energy production of a wind farm, which mainly depends on the weather, periods of raining and winter seasons, installation sites, and turbine designs.

Ice protection systems (IPS) include passive and active systems such as the ice-resistant coatings and electro-thermal systems have been proposed to alleviate environmental issues. Blade coatings, for example, is applied on the surface of turbine blades to minimise the build-up of ice and maximise heat absorption. This passive system is a cost-effective approach to prevent icing on the blades, however over the years; the coatings would require re-application. Active systems work by heating the turbine blades using thermal devices. These devices depend on the ice-detection method and may require the turbine to stop before they can be activated completely. In some research, ultrasonic de-icing using sandwich transducers on aluminium alloy plate (Zeng and Song 2017) and the application of surface mounted continuous carbon fibre sheets on wind turbine blades (Xu et al. 2018) have shown great potential for de-icing purposes. The verification of the performance of real wind turbines in actual conditions is essential for reliability. Additional research and development should be carried out in the future, particularly for offshore wind turbines where extreme environmental effects are common.

3.2 Wind Turbine Blades—The Story So Far

There are two general configurations of wind turbine technology, which are defined by the axis at which the turbine rotates; horizontal axis wind turbine (HAWT) and vertical axis wind turbine (VAWT). Beside this general configuration, wind turbines are also categorised into how the blades are dominantly propelled; through lift force or drag force. Despite the different types of blades, they all share the same role for wind turbines regardless of their configuration; convert kinetic energy from the wind into torque or rotary mechanical energy which further converted into electricity.

With the advancement of research and technology, new wind turbine blades are becoming lighter. These blades carry less weight than the conventional blades. Hence, inertial forces are reduced, and therefore improving the wind turbine efficiency. Complex shapes are also introduced, for example, large wind turbine blade design, especially for the HAWT, have aerodynamically improved through a complex algorithm to calculate the optimal twist angle. These long and large blades are made possible through the ingenious configuration setup for internal blade structure. Advances in materials and fabrication technology have also assisted these design improvements where complex blade shapes, as well as a tight constraint for strength-to-weight ratio, can be achieved.

3.2.1 Structural Elements of Wind Turbine Blades

Structurally, wind turbine blades are associated with the classical beam theory. This is true for all types of blades, HAWT, VAWT, lift type and drag type (although drag type blades may be closer to plate theory rather than beam). The essential load that is exerted on these beams is the flap-wise bending load. It is known that the longer the moment arm; in this case, the blade span, the larger the bending load is. The structural design, especially the internal structure, is highly dependent on this load. External structure and shape, on the other hand, are defined by aerodynamic loads.

Figure 3.4 depicts the typical cross section of a lift-type wind turbine blade. The essential parameters for structural consideration depicted in the figure are the lift force and drag force, which sum up to the total aerodynamic force that theoretically acts on the centre of a pressure point in the cross section. Also, notice the angle of attack which highly affects the lift and drag forces. While the leading edge and trailing edge hold the shape of the cross section that is necessary to achieve the pre-determined total aerodynamic force.

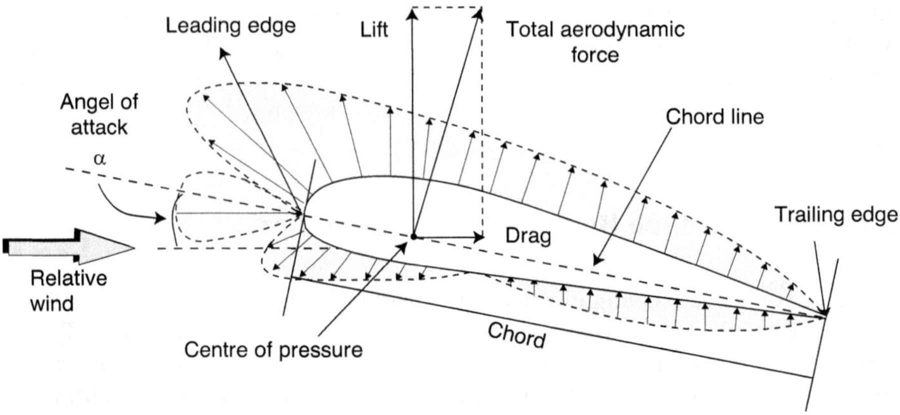

Fig. 3.4 Typical cross section of a lift type blade (Monroy Aceves et al. 2012)

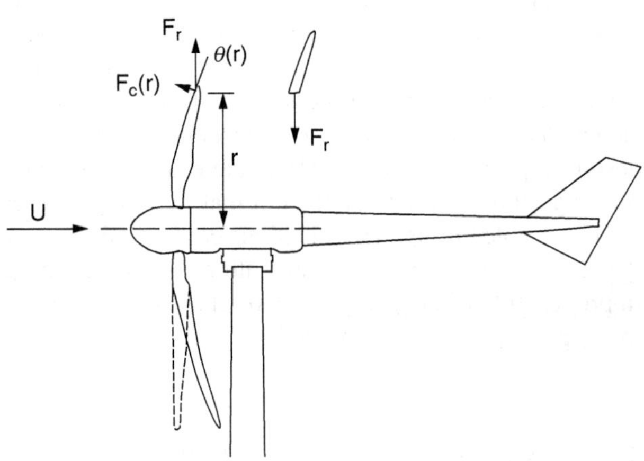

Fig. 3.5 Centrifugal force in deflected blades, $F_c(r)$ (Pourrajabian et al. 2016)

As discussed, the major load exerted onto blades is the flap-wise bending load. Also previously discussed is the aerodynamic loads. Other loads that need to be resisted by the blades are edge-wise bending and torsional loads. For small wind turbines, centrifugal force is dominant, and this force reduces the flap-wise bending (Pourrajabian et al. 2016), therefore, reduces blade deflection as depicted in Fig. 3.5. On the other hand, there is no dispute that flap-wise bending is the dominant load for large wind turbines that structural design for blades is mainly affected by this load.

There are several types of internal configuration or structure typically designed for wind turbine blades that can resist all loads. The semi-monocoque structure is a concept where there are shell structures stiffened by other internal structures. Figure 3.6 depicts a typical cross section of a semi-monocoque blade structure. In this case, the skin and the shear webs resist the edge-wise bending and the torque, while the spar flanges resist the flap-wise bending loads. Also, the skin holds the shape of the blade, and shear webs resist the shear force as well (Cox and Echtermeyer 2012).

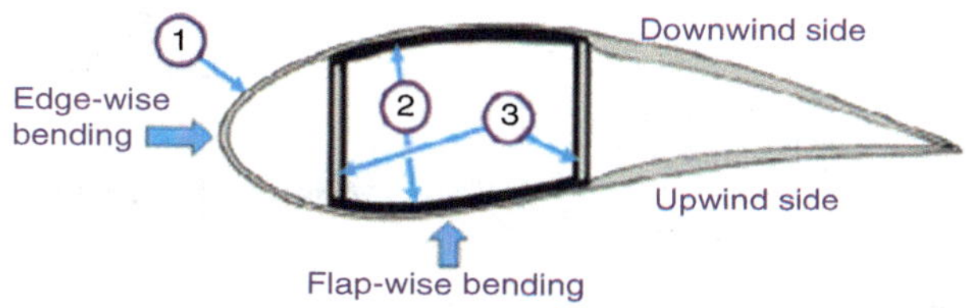

Fig. 3.6 Typical semi-monocoque structure for the blade. No. *1*: skin, no. *2*: spar flanges, no. *3*: shear webs (Cox and Echtermeyer 2012)

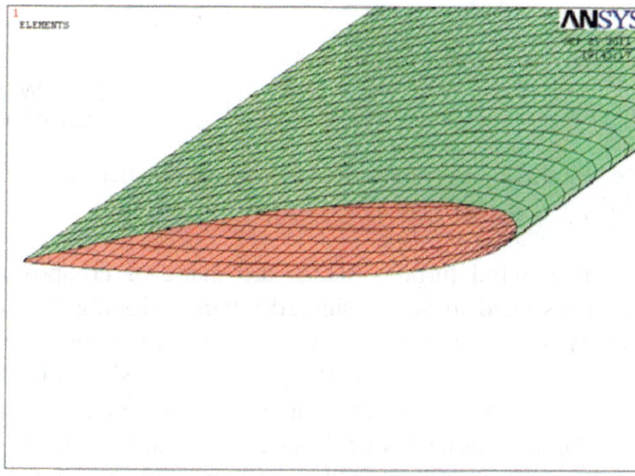

Fig. 3.7 Typical blade geometry (Hameed and Afaq 2015)

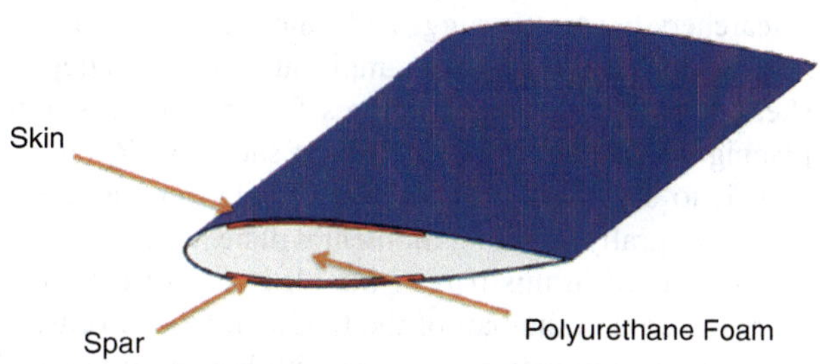

Fig. 3.8 Typical skin-spar-foam structure concept for the blade (Kong et al. 2011)

The monocoque structure, on the other hand, has stiffer skin structure and do not require any supporting or stiffening structure. In other words, all exerting loads discussed previously for blades are resisted by single structure; which is the skin of the blades that cover the whole span of the blade. Therefore, it is typical for this type of structure to have thick skin and high stiffness material. Figure 3.7 illustrates the typical cross section of a monocoque structure for the blade. This type of structure is usually utilised for small wind turbines.

Another type of internal structure concept is the skin-spar-foam as described in Fig. 3.8. This type is similar to the semi-monocoque structure; the only difference is this concept uses foam to assist in resisting the edge-wise bending, torque and shear loads. In fact, the foam also holds the shape of the blade and provide additional stiffness to the overall structure, hence will be able to reduce the material used for other subcomponents.

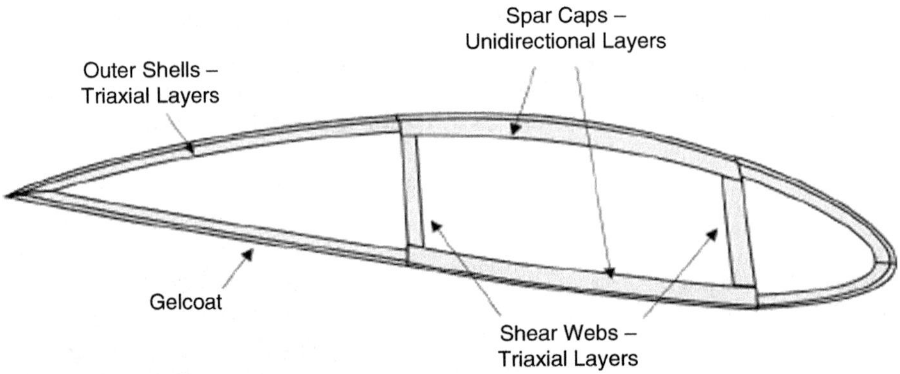

Fig. 3.9 Example of composite layups for three different subcomponents in blades (Fagan et al. 2017)

Many wind turbine blades are made of composite material. Huge structural elements need to be considered when designing the composites layups. Different layups and orientations of the composite fibres provide different strength, and they are designed based on specific purposes in resisting loads. For example, spar caps or spar flanges, which are resisting the flap-wise bending are suitable to have unidirectional fibre composites while shear webs and skins, which are resisting the twist and edge-wise bending are suited with triaxial direction fibre composites. Figure 3.9 illustrates the fibre composites direction and their location in a blade. Many researchers have also suggested other sets of fibre composites orientation to resist such loads. Yang et al. for example suggested that depending on the ratio of axial and shear stresses, placing the fibres 0° and 90° resist the flap-wise bending while placing the fibres 15° and 30° resist shearing (Yang et al. 2014).

It is to be noted that the fibre orientation may also vary across the span of the blade. Naturally, bending moment will be high at the root of the blade. Therefore, the materials used in this part of the blade are stiffer. This can be achieved by adding more material to the part of the blade, in other words, thicker skins, spars and shear webs, or even utilising materials with higher strength. Figure 3.10 illustrates different layups utilised for different locations along the blade. Notice the different layups are also true for leading edge and trailing edge. As described by Chen et al. (2013), the thickness of the blade shell can be varied according to the stress needs through certain optimisation method and therefore substantial amount of material can be saved. Figure 3.11 illustrates different types of materials with different spar cap thicknesses for the same blade sample. It is also to be noted that the blade design, especially for leading and trailing edges, need to be balanced with the manufacturing limit. For example, the sharp and small profile of the trailing edge may require certain minimum thickness that the material can achieve.

Types of resins used to bind the fibres together are also part of the structural elements that need to be considered when utilising composites for blade material. Other than binding the fibres, the resins also provide structural strength, and their stiffness is integrated into the overall blade stiffness in structural analysis. Examples of resin typically used as composites binders are polyester and vinyl ester. Special

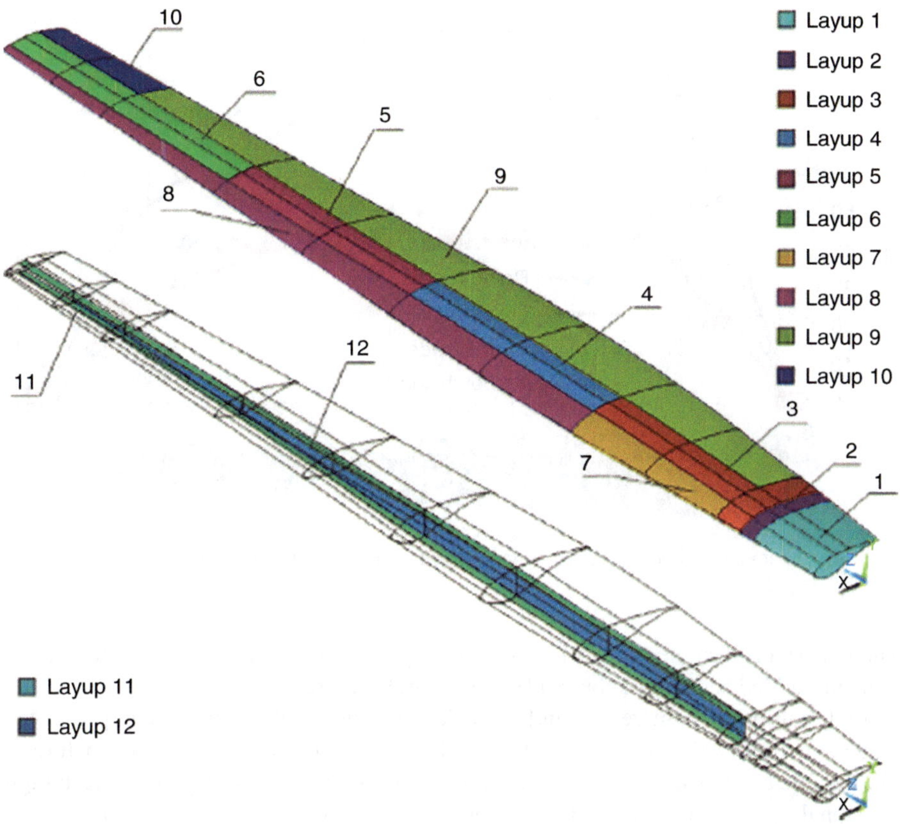

Fig. 3.10 Different sets of fibre composites layups for different locations along the blade (Dal Monte et al. 2017)

coating on the blade also needs to be considered. The two main purposes of special coatings are to improve blade durability and protection against environmental effects, for example, extreme weather condition and corrosion.

3.2.2 Material Requirements for Turbine Blades

Through recent technology, wind turbines are getting bigger in sizes and can produce a substantial amount of power. Large wind turbines require large blades. The challenge nowadays for the wind turbine industry is to get the gigantic blade to be as light as possible as to ease the transportation and installation while maintaining high strength as well as the efficiency of power generation. Classically, wind turbine blades are designed by employing the safe-life philosophy. It is similar to the conservatism concept since the blades are assumed to work at 100% performance.

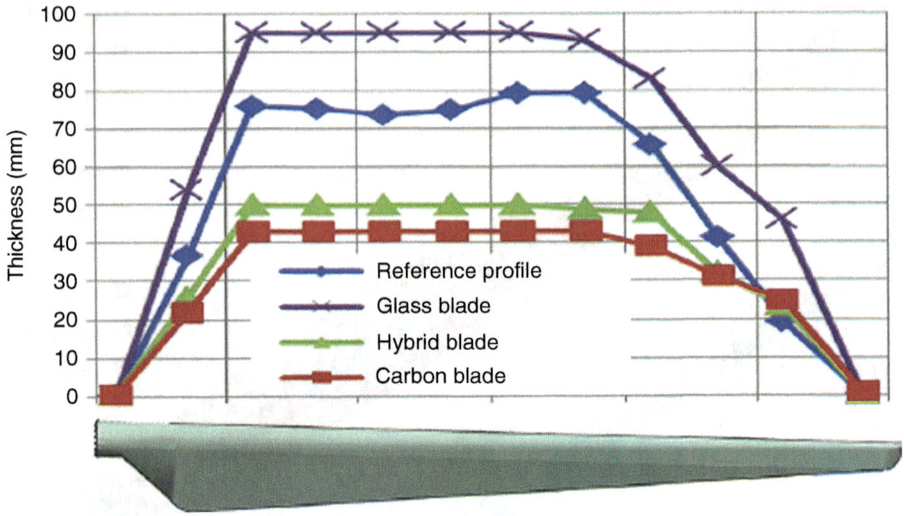

Fig. 3.11 Spar cap thickness across the blade span for different materials (Nijssen and de Winkel 2016)

The material is designed to work at a lower capability level. Therefore, more mass is typically added to achieve the blades' structural requirement.

Materials with lightweight and have low density will be able to improve the blade's inertial loads. In other words, the blade with this type of material is naturally more stable and has less momentum, which is good for such blades that rotate at high rotational speed. This is as Hameed et al. (2015) have suggested for a material selection for wind turbine blades, the following should be considered; high yield strength, high stiffness, and low density. These three criteria must be respectively followed to achieve longer blade operational life, improved aerodynamic performance and reduced gravitational loads on the blade. This applied to both small and large wind turbines. In fact, small wind turbine blades that experience a dominant centrifugal force, given a lower material density, can achieve a reduction in centrifugal loads.

It is important to note that wind turbine blades are designed to rotate at a certain speed and have dynamic loads. Therefore, careful calculations to identify the blade's natural frequencies are important, so that materials with the correct stiffness are selected to avoid catastrophic fracture.

The three main criteria discussed previously also need to be balanced with the cost of material and manufacturing processes. For example, carbon fibre composites have substantially higher stiffness than glass fibre composites. Nevertheless, carbon fibre composites are costlier, and in many of the cases, glass fibres are the preferred materials. An exception may be for the blades that require specific high magnitude of strength and where cost is not an issue.

Durability is also paramount when selecting the blade material. The typical operating life of wind turbines, large and small, is around 20–25 years. Such blade that operates well throughout that amount of time will be required to be durable, structurally as well as environmentally. Structurally is where fatigue loads need to be considered. High yield stress is not the key parameters for fatigue, in fact, repetitive low loads over a huge amount of cycles are the key parameters for fatigue. The material selected should be able to withstand these repetitive loads. The number of cycles and the variability of the cycles is also needed to be considered in performing the fatigue analysis (Nijssen and Brøndsted 2013). Environmentally is where the material should be able to withstand harsh and extreme weather conditions, including high wind, high or low humidity, extremely high or low temperatures and corrosion.

Future trend for blade material selection reveals that it will be essential for the wind turbine blades to be sustainable and are easily recyclable. One of the evidence that this trend is emerging is the utilisation of green composites for wind turbine blades. Natural fibres have lower density and lower cost compared to glass fibres. Furthermore, natural fibres are carbon dioxide neutral material. More importantly, the natural fibres are biodegradable and recyclable, which means natural fibre blades can be disposed of more cleanly compared to other synthetic fibres (Hamdan et al. 2014).

Another future trend is the utilisation of new advanced types of materials. An emerging type is the thermoplastics. Although the thermoplastic material is relatively costly, it is claimed to have unlimited shelf-life and greener in terms of its environmentally friendly attributes. Besides this, thermoplastic composites also have higher stiffness and higher fatigue (Brown and Brooks 2010). On the other hand, there is also a practice called structural health monitoring where the materials of the blade will be able to adapt to the instantaneous condition and predict wind loading. This can be achieved through the utilisation of "smart" materials. However, there is still a huge gap as further understanding of blade structural behaviour utilising the smart material is still required (Nijssen and Brøndsted 2013).

3.3 Application of Nanotechnology in Wind Turbines

Wind energy is one of the fastest growing renewable energy resources in the world today in which the cumulative installed wind energy capacity has been increased significantly since the year 1996. It was reported that at the end of 2015, the total wind capacity was approximately 433 GW and it is projected to reach 2000 GW by the year 2030 (Chong et al. 2017). This rapid expansion of wind energy industry requires advances in both the design and materials for wind energy technology. Rotor blades, in particular, need to be stronger and lighter to compensate for the ever-increasing demand for larger turbine blades. This is due to the larger swept area that can be produced by large wind turbines to produce more wind energy output. Moreover, as the lifetime for these wind turbines is between 20 to 25 years, the operational lifetime of the blades need to be extended to increase the cost-effectiveness of this technology. Advances in nanotechnology have been employed in developing cost-effective materials with the higher strength-to-mass ratio. Protection of wind turbine blades surface from ice formation, and many researchers have also explored the solutions for ice adhesion. This sub-section aims to review state-of-the-art nanotechnology applications in wind turbine technology, particularly the rotor blades.

3.3.1 Nano-Reinforced Materials for Rotor Blades

In recent years, large wind turbines with a rotor diameter of up to 100 m have been employed in wind farms around the world. These large rotor blades are mostly made from fibreglass reinforced epoxy resins with some components of carbon fibre and balsa wood. A higher percentage of carbon fibres in rotor blades is expected to increase as the design of wind turbines keep increasing in size despite its high costs. The main challenges of large wind turbines include high bending forces, aerodynamic loads, highly turbulent winds, dynamic thrust and torque, and high cyclic fatigue effects. In high wind velocity environment, cases of blade failures have been reported (Chou et al. 2013). In 2015, it was stated that annual blade failures are estimated at around 3800 a year, and it is the primary cause of insurance claims in the US onshore market ahead of gearbox and generator failures (Campbell 2015). Therefore, some requirements for high-performance blades must be met to ensure better overall performance and longer lifetime of wind turbines. These requirements are (Mishnaevsky 2012; Golfman 2012):

- High reliability and fatigue resistant materials for stable operations.
- High stiffness material for better stability due to the aerodynamic loads and orientation of blade during operation.
- High strength material to resist extreme wind conditions and blade gravity loads.
- Low-density material to reduce the overall load on the tower due to gravity and aerodynamic forces.
- Materials with properties to resist environmental effects such as icing, humidity and temperature.

The discovery of multi-walled carbon nanotubes (CNTs) by Iijima in 1991 (Iijima 1991) and single-walled CNTs (Bethune et al. 1993; Iijima and Ichihashi 1993) in 1993 have made possible for nanotechnology to be incorporated in wind energy technology. Multi-walled CNTs are made of single-walled CNTs that are arranged concentrically. Single-walled CNTs can be illustrated as single-layered carbon atoms rolled up into a shape of a cylinder. CNTs usually have a high aspect ratio (length-to-diameter ratio greater than 1000) due to the length of the nanotubes are measured in micrometre or millimetre. Some distinctive properties of CNT and its differences with other carbon-based materials are shown in Table 3.3. Researchers have been incorporating CNTs as a reinforcing agent to improve the strength of composite materials (Thostenson et al. 2001). Parts of wind turbine blades such as the shear web, root and spar are made of fibre reinforced polymers (FRPs) materials.

Table 3.3 Properties of various carbon-based materials (Ma and Kim 2011)

	Carbon-based materials				
Properties	CNT	Diamond	Carbon fibre	Graphite	Fullerene
Tensile strength (GPa)	>10	1.2	1.0–5.6	~10[a] <0.01[b]	N/A
Tensile modulus (GPa)	1000	500–1000	100–500	1000[a] 36.5[b]	14
Specific gravity (g/cm^3)	0.8–1.8	3.5	1.8–2.1	1.9–2.3	1.7
Thermal conductivity (W/m.K)	2000–6000	900–2320	21–180	298[a] 2.2[b]	0.4
Thermal expansion coefficient (K^{-1})	Negligible	$(1–3) \times 10^{-6}$	$\sim 1 \times 10^{-6}$	-1×10^{-6}[a] 2.9×10^{-5}[b]	6.2×10^{-5}
Cost (US$/g)	0.3–10	>100	0.02–0.1	0.002–0.01	40–70

[a]In-plane
[b]*c*-Axis

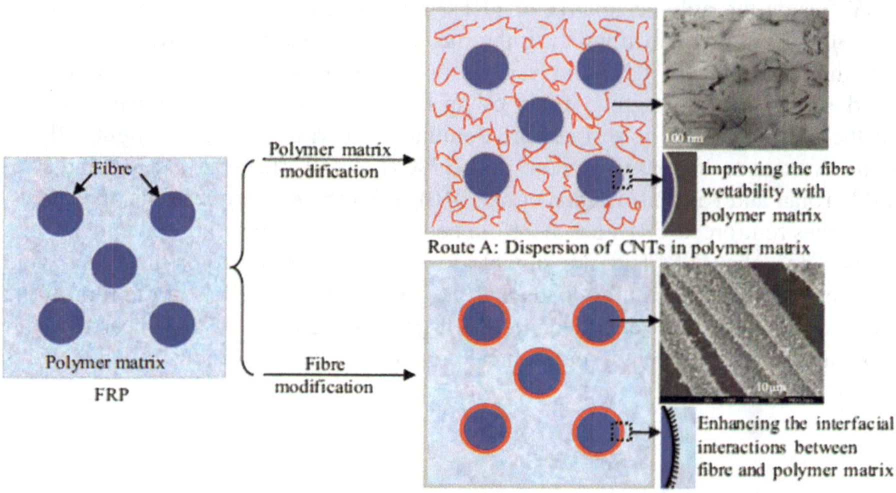

Fig. 3.12 Enhancing the performance of FRP by utilising CNTs based on the physical properties of the FRP (Ma and Zhang 2014)

FRP is a type of composite material consisting of continuous fibres and the polymer matrix. The long continuous fibres give rise to strength and stiffness, whereas the polymer matrix ensures the stiffness, out-of-plane and delamination strengths, and fracture toughness of the material. Based on these two physical characteristics of the FRPs, the CNTs can either be added into the polymer resins by dispersion to increase wettability or by deposition of CNTs on the surface of the fibre (Fig. 3.12) (An et al. 2012; Davis et al. 2011; Gao and Pan 2012). The latter method employs the large surface area of the CNT to improve the interfacial interactions between the matrix and the fibres (Ma and Zhang 2014).

The use of CNTs has shown some improvements in the mechanical properties of materials which result in higher fatigue resistance, shear strength and fracture toughness compared to the conventional composites reinforced by microscale materials (Geng et al. 2008). For CNT to be effective in reinforcing composites, requirements such as strong interfacial contact between the polymer matrix and CNTs, suitable dispersion technique, and good alignment of high aspect ratio CNTs must be satisfied. Significant reductions in laminate strength and stiffness are caused by the propagation and formation of interlaminar cracks in FRPs. Delamination occurs when there is a reduced interfacial contact between the matrix and fibre. Moreover, out-of-plane tensile loads to in-plane compressive loads and low-velocity impact in the local transverse position can also cause delamination. Challenges in synthesising the CNTs in FRPs are proven to be the most difficult part. Contradicting results by using various techniques to enhance the properties of materials have also been reported (Ng et al. 2013). Particularly the issue of non-uniform dispersion of CNTs inside the polymer matrix of FRPs will result in poor reinforcement effect. Nevertheless, studies in improving the mechanical properties of FRP have shown that the dispersion of CNTs in various polymers increase the matrix-dominated interlaminar shear strengths (ILSS) by up to 45%. Whereas, by depositing CNTs on the surface of the fibre, the fibre-dominated interfacial shear strengths (IFSS) were reported to increase by more than 30% (Ma and Zhang 2014; Ma and Kim 2011; Khan and Kim 2011; Qian et al. 2010). With these improvements, the carbon nanotubes reinforced FRPs are therefore better at handling bending force and cyclic loads. The enhanced FRPs are suitable to be applied in the rotor blades of wind turbines that experience high loads and fatigue effects, as well as at the root of the blade which constantly experiencing the high bending moment and shear force. A summary of improvement in mechanical properties with the introduction of CNTs is shown in Table 3.4.

3.3.2 Nanocomposite Coatings for Rotor Blades

One of the challenges for wind turbines is the fouling of rotor blades with insects or formation of ice, as well as material erosion due to environmental effects (abrasive dust particles or diffusion of moisture/water vapour). Fouling on the surface of the blade can cause deterioration of blade health and may lead to blade failure. Moreover, the aerodynamic efficiency of the wind turbine can be affected due to the alterations on the surfaces of the airfoil.

As the installation of wind turbines in the cold-climate market continues to rise, the formation of ice on the blade's surface will be more frequent. The effects of icing include increased cyclic loads and rotational vibration, reduced energy output, higher noise levels, increased safety and health risks, and reduced lifespan of turbine components. Figure 3.13 shows the power reduction curve due to ice on the leading

Table 3.4 Mechanical properties of various FRPs with CNTs (Ma and Zhang 2014; Ma and Kim 2011; Khan and Kim 2011; Qian et al. 2010)

Testing method	Matrix	Fibre	CNT (wt%)	Improvement (%)
Single fibre fragmentation	Epoxy	Carbon fibre	–	>475
	Epoxy	Carbon fibre w/ CNT bucky paper	–	31
	Epoxy	Alumina fibre	–	69
	Epoxy	Woven carbon fibre	–	30
	Polymethylmethacrylate	Silica fibre	–	80–150
Short beam shear	Epoxy	Glass fibre	1.00	~7
	Epoxy	Carbon fibre	0.25	27
	Epoxy	Woven glass fibre	2.00	33
	Epoxy	Glass fibre fabrics	0.30	~5
	Epoxy	Glass fibre	0.30	20
	Epoxy	Glass fibre	0.015	45
	Epoxy	Glass fibre	0.30	16
	Vinyl ester-epoxy resin	Glass fibre	0.10	11
	Vinyl ester	Woven glass fibre	0.10	20–45

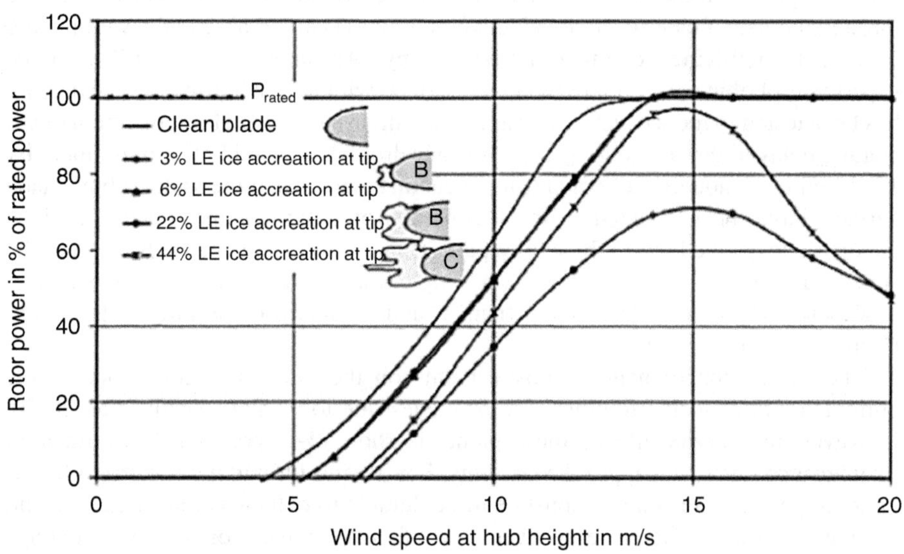

Fig. 3.13 Various ice shapes are affecting the power curves of a 300 kW wind turbine (Fakorede et al. 2016)

edge of turbine blades. Therefore, solutions to the icing events must be investigated to minimise failures and enhance turbine performance (Dalili et al. 2009). There exists passive ice protection technique to prevent ice accumulation through the use of anti-icing coatings. Nanocomposite coatings are thought to be well suited for

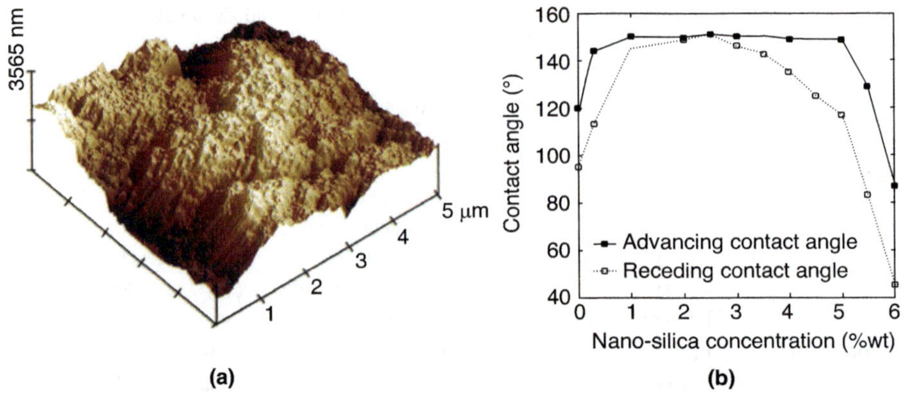

Fig. 3.14 (**a**) Atomic force microscopy image of wind turbine blade surface coated with silica nanoparticles; (**b**) effect of different nano-silica concentrations on the contact angle of a water drop on a blade surface (Karmouch and Ross 2010)

anti-icing coatings due to their very high contact angles with water and the ability to absorb and dissipate higher impact energy caused by the repeated impact of particulate matter (higher resistant to erosion). Also, there is a type of nanocomposite that has been proposed that would benefit from the elevated temperature of the surface through the use of electrically conductive particles (Dalili et al. 2009). A promising solution to anti-icing coating is introduced by (Arianpour et al. 2013) with the superhydrophobic nanocomposite coating doped with titanium powder. The coating has been tested to reduce water adhesion strength on a surface. The application of the coating delayed the freezing process of water droplets in a −15 °C environment by 12–13 min. In another study of anti-icing coating, samples of the wind turbine blade surface have been painted with a superhydrophobic coating made of silica nanoparticles embedded in the epoxy paint (Fig. 3.14a). Tests have shown that the superhydrophobic surfaces have a very high water contact angle (approximately 152°, Fig. 3.14b), a hysteresis less than 2° and a water drop sliding angle of 0.5° (Karmouch and Ross 2010).

The application of nanocomposite paints on the blade surfaces is one of the current solutions to the fouling of rotor blades due to erosion (Dalili et al. 2009). However, the service life of these paints is short (3–5 years) and require high maintenance costs to reapply the coatings. For a wind turbine with a design life of up to 20 years, these coatings must be more durable to reduce maintenance cost and turbine downtime. Therefore, to overcome this matter, the fatigue life of coatings based on polymeric materials has been explored. The fatigue S-N curves of polypropylene nanocomposite, talc-filled polypropylene, and polypropylene are shown in Fig. 3.15. The horizontal axis corresponds to the number of cycles to failure (N), whereas the vertical axis corresponds to the maximum cyclic stress (S). It is shown that the nanocomposite filler can influence the fatigue life of the coating by delaying the initiation of fatigue at higher stress levels (Zhou et al. 2005). Moreover, based on literature, the use of nanofillers with large aspect ratios into polymers can increase

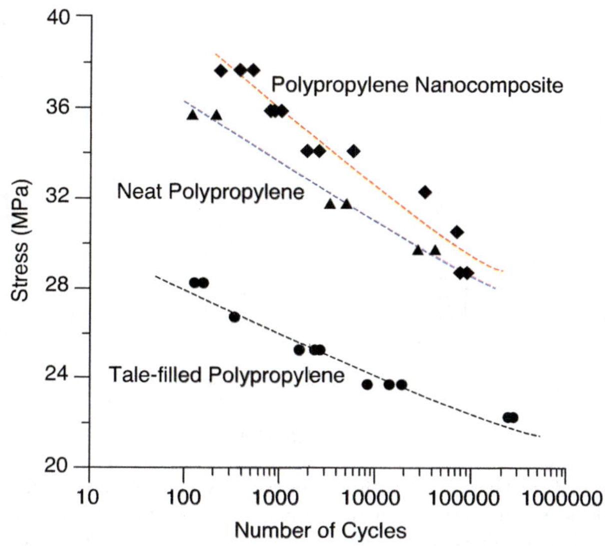

Fig. 3.15 Fatigue curves of composites talc-filled polypropylene and polypropylene nanocomposite, and neat polypropylene (PP) (Zhou et al. 2005)

the material's resistance to erosion. A study based on the water vapour permeability of 0.2 wt% CNT reinforced polycaprolactone (PCL) nanocomposites was shown to decrease from 1.51 to 1.08 g mm/m^2.day (Khan et al. 2013). It is also known that glass fibres used as reinforcements in FRP structures of wind turbine blades are vulnerable to water vapour due to its internal chemical structures. Therefore, epoxy-based nanocomposites coating that contains CNTs and graphene were applied on glass fibres and exposed to the ageing process in alkali solution. Comparisons between different coatings have shown that the fibres with CNT/epoxy and graphene/epoxy coatings maintained about 77% and 87% of their original strength after 5 days of ageing, which were higher than the fibres with an epoxy coating (Fig. 3.16). Based on these results, one can conclude that nanocomposite coatings are crucial in determining the barrier properties of fibres in which the modification of coatings with nanoparticles would delay the erosion effects on the surface of wind turbine blades.

Multifunctional nanocomposite coating material has also been investigated for its viability on turbine blades. Carbon nanofiber (CNF) paper-based nanocomposite coating material was tested for its impact-friction resistance, vibrational damping, and superhydrophobicity properties. The superhydrophobicity of the coating material was measured based on water contact angle. A value higher than 160° was recorded which was attributed to the Si-O bond and the polyhedral oligomeric silsesquioxane (POSS) molecules that were attached covalently to the nanofiber surface (Liang et al. 2011). Meanwhile, the wear and coefficient of friction (COF) result from tribometer tests of various samples of the multifunctional nanocomposite material are shown in Fig. 3.17. The wear rate of the sample G2 (CNFs/gGraphite) shows the lowest value due to the self-lubricating effect of the graphite particles

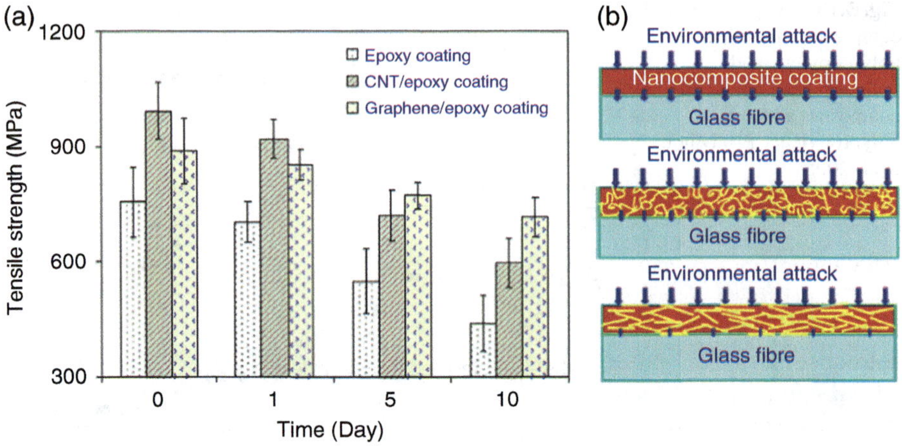

Fig. 3.16 Glass fibres with various coatings. (**a**) Tensile strength vs. day graphs of glass fibres after the ageing process in alkaline solution; (**b**) illustration of moisture and alkali ions diffusion through various coatings: bottom-graphene/epoxy coating; middle-CNT/epoxy coating; top-neat epoxy coating

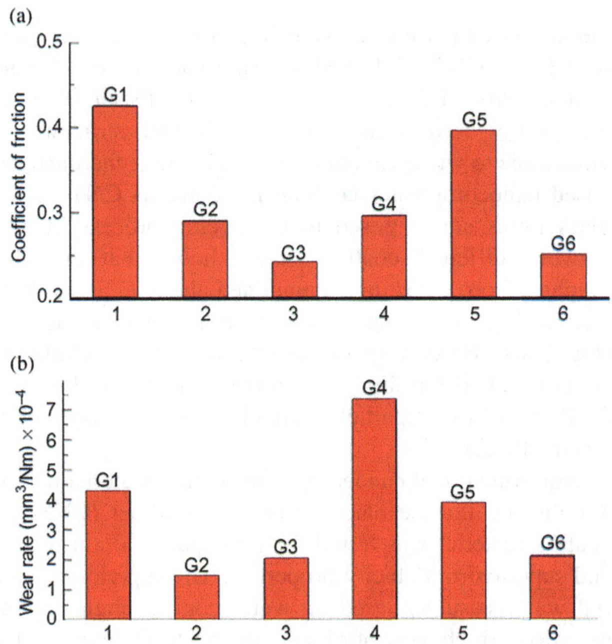

Fig. 3.17 Experimental data of (**a**) coefficient of friction, and (**b**) wear rate measurements. *G1* Pure CNFs, *G2* CNFs/gGraphite, *G3* CNFs/gGraphite/SCF, *G4* CNFs/gGraphite/SCF, *G5* CNFs/Nano-TiO_2/nNano-Al_2O_3, *G6* CNFs/gGraphite/SCF/nNano-TiO_2/nNano-Al_2O_3 (Liang et al. 2011)

(1.78×10^{-4} mm^3/Nm). In general, the nanocomposite materials and the addition of short carbon fibres and graphite result in low wear rate and COF. This would provide a better sand/dust erosion mitigation for turbine blades with effective surface protection (Liang et al. 2011). Moreover, the nanocomposite material showed a

300% increase in vibrational damping at the second and third mode frequencies compared to baseline composite. This would ensure better durability and stability of the wind turbine blades during operation.

3.4 Conclusion

The applications of nanotechnology on wind turbine technology, in particular, the turbine blades have been reviewed. It has been shown that nanomaterials can be used to increase the performance of blade materials through structural reinforcement, introduces anti-icing properties on surface blades, enhance the barrier performance of wind turbine blades through the use of nanoparticles, and improve the damping ratios of wind turbine laminates for increased structural stability, durability and dynamic response.

References

An Q, Rider AN, Thostenson ET (2012) Electrophoretic deposition of carbon nanotubes onto carbon-fiber fabric for production of carbon/epoxy composite with improved mechanical properties. Carbon 11(50):4130–4133

Ancona D, McVeigh J (2001) Wind turbine-materials and manufacturing fact sheet. US Department of Energy: Princeton Energy Resources International for the Office of Industrial Technologies

Arianpour F, Farzaneh M, Kulinich SA (2013) Hydrophobic and ice-retarding properties of doped silicone rubber coatings. Appl Surf Sci 265:546–552

Barakati, SM (2007) Wind turbine system: history, structure and dynamic model. Handbook of renewable energy technology, edited by Ahmed F Zobaa and Ramesh C. Bansa, 21–51. Singapore: World Scientific Publishing. https://doi.org/10.1142/9789814289078_0002

Bethune DS et al (1993) Cobalt-catalysed growth of carbon nanotubes with single-atomic-layer walls. Nature 363:605–607

Brown KA, Brooks R (2010) Design and analysis of vertical axis thermoplastic composite wind turbine blade. Plastics Rubber Compos 39(3–5):111–121

Campbell S (2015) Annual blade failures estimated at around 3,800. https://www.windpowermonthly.com/article/1347145/annual-blade-failures-estimated-around-3800

Castorrini A et al (2016) Computational analysis of wind-turbine blade rain erosion. Comput Fluids 141:175–183

Chen J et al (2013) Structural optimization study of composite wind turbine blade. Mater Des 46:247–255

Chong WT et al (2017) Cross axis wind turbine: pushing the limit of wind turbine technology with complementary design. Appl Energy 207:78–95

Chou JS et al (2013) Failure analysis of wind turbine blade under critical wind loads. Eng Fail Anal 27:99–118

Cox K, Echtermeyer A (2012) Structural design and analysis of a 10MW wind turbine blade. Energy Procedia 24:194–201

Dal Monte A et al (2017) Proposal for a coupled aerodynamic–structural wind turbine blade optimization. Compos Struct 159:144–156

Dalili N, Edrisy A, Carriveau R (2009) A review of surface engineering issues critical to wind turbine performance. Renew Sustain Energy Rev 13:428–438

Davis DC et al (2011) A strategy for improving mechanical properties of a fiber reinforced epoxy composite using functionalized carbon nanotubes. Compos Sci Technol 8(71):1089–1097

Edelstein WA et al (2003) Wind energy. A report prepared for the Panel on Public Affairs (POPA). American Physical Society

Elliot DL et al (1986) Wind energy resource atlas of the United States. Department of Energy, Pacific Northwest Laboratory, Richland, WA

Fagan EM et al (2017) Physical experimental static testing and structural design optimisation for a composite wind turbine blade. Compos Struct 164:90–103

Fakorede O et al (2016) Ice protection systems for wind turbines in cold climate: characteristics, comparisons and analysis. Renew Sust Energ Rev 65:662–675

Gao Y, Pan L (2012) Fatigue of nanotube-reinforced carbon fiber epoxy composites. Adv Mater Res 510:753–756

Geng Y et al (2008) Effects of surfactant treatment on mechanical and electrical properties of CNT/epoxy nanocomposites. Compos A: Appl Sci Manuf 39(12):1876–1883

Golfman Y (2012) Hybrid anisotropic materials for wind power turbine blades. CRC Press, Boca Raton

Griffith DT, Resor BR, Ashwill TD (2012) Challenges and opportunities in large offshore rotor development: Sandia 100-meter blade research. In: AWEA WINDPOWER Conference and Exhibition

Hamdan A et al (2014) A review on the micro energy harvester in Structural Health Monitoring (SHM) of biocomposite material for Vertical Axis Wind Turbine (VAWT) system: a Malaysia perspective. Renew Sust Energ Rev 35:23–30

Hameed MS, Afaq SK, Shahid F (2015) Finite element analysis of a composite VAWT blade. Ocean Eng 109:669–676

Iijima S (1991) Helical microtubes of graphitic carbon. Nature 354:56–58

Iijima S, Ichihashi T (1993) Single-shell carbon nanotubes of 1-nm diameter. Nature 363:603–605

Karmouch R, Ross GG (2010) Superhydrophobic wind turbine blade surfaces obtained by a simple deposition of silica nanoparticles embedded in epoxy. Appl Surf Sci 257:665–669

Khan SU, Kim JK (2011) Impact and delamination failure of multiscale carbon nanotube-fiber reinforced polymer composites: a review. Int J Aeronaut Space Sci 12(2):115–133

Khan RA et al (2013) Mechanical and barrier properties of carbon nanotube reinforced PCL-based composite films: effect of gamma radiation. J Appl Polym Sci 127(5):3962–3969

Kim KC et al (2014) Experimental and numerical study of the aerodynamic characteristics of an archimedes spiral wind turbine blade. Energies 7(12):7893–7914

Kong CD, Lee HS, Kim MW (2011) Aerodynamic and structural design of a high efficiency small scale composite vertical axis wind turbine blade. In: International Conference on Composite Materials. Jeju, Korea

Liang F et al (2011) Multifunctional nanocomposite coating for wind turbine blades. Int J Smart Nano Mater 2(3):120–133

Ma PC, Kim JK (2011) Carbon nanotubes for polymer reinforcement. CRC Press, Boca Raton

Ma PC, Zhang Y (2014) Perspectives of carbon nanotubes/polymer nanocomposites for wind blade materials. Renew Sust Energ Rev 30:651–660

Martinez E et al (2009) Life cycle assessment of a multi-megawatt wind turbine. Renew Energy 34 (3):667–673

Mishnaevsky L Jr (2012) Composite materials for wind energy applications: micro-mechanical modeling and future directions. Comput Mech 50(2):195–207

Mishnaevsky L et al (2017) Materials for wind turbine blades: an overview. Materials 10(11):1285

Monroy Aceves C et al (2012) Design methodology for composite structures: a small low air-speed wind turbine blade case study. Mater Des 36:296–305

Ng KW, Lam WH, Pichiah S (2013) A review on potential applications of carbon nanotubes in marine current turbines. Renew Sust Energ Rev 28:331–339

Nijssen RPL, Brøndsted P (2013) Chapter 6: Fatigue as a design driver for composite wind turbine blades. In: Nijssen RPL and Brøndsted P (eds) Advances in Wind Turbine Blade Design and Materials. Cambridge: Woodhead Publishing, pp. 175–209

Nijssen R, de Winkel GD (2016) Chapter 5: Developments in materials for offshore wind turbine blades. In: Ng C and Ran L (eds) Offshore Wind Farms: Technologies, Design and Operation. Elsevier Science & Technology, pp. 85–104

Pourrajabian A et al (2016) Aero-structural design and optimization of a small wind turbine blade. Renew Energy 87:837–848

Qian H et al (2010) Carbon nanotube-based hierarchical composites: a review. J Mater Chem 20:4751–4762

Ragheb M (2012) Aerodynamics of rotor blades. http://mragheb.com/NPRE%20475%20Wind%20Power%20Systems/Aerodynamics%20of%20Rotor%20Blades.pdf

Schubel PJ, Crossley RJ (2012) Wind turbine blade design. Energies 5(9):3425–3449

Song Q (2012) Design, fabrication, and testing of a new small wind turbine blade. University of Guelph, Guelph, ON

Tan CS, Maragatham K, Leong YP (2013) Electricity energy outlook in Malaysia. IOP Conf Ser Earth Environ Sci 16:012126

Thostenson ET, Ren ZF, Chou TW (2001) Advances in the science and technology of carbon nanotubes and their composites: a review. Compos Sci Technol 61(13):1899–1912

Xu B, Lu F, Song G (2018) Experimental study on anti-icing and deicing model wind turbine blades with continuous carbon fiber sheets. J Cold Reg Eng 31(1):04017024

Yang ZJ et al (2014) Design and verification of a small-scale lift-type vertical axis wind turbine composite blade. International Conference on Materials for Renewable Energy and Environment (ICMREE). Chengdu, China

Zeng J, Song B (2017) Research on experiment and numerical simulation of ultrasonic de-icing for wind turbine blades. Renew Energy 113:706–712

Zhou Y et al (2005) Experimental study on thermal and mechanical behavior of polypropylene, talc/polypropylene and polypropylene/clay nanocomposites. Mater Sci Eng A 402(1–2):109–117

Zobaa AF, Bansal RC (2011) Handbook of renewable energy technology. World Scientific, Singapore

Chapter 4
Nanomaterials: Electromagnetic Wave Energy Loss

Gan Jet Hong Melvin, Yaofeng Zhu, and Qing-Qing Ni

4.1 Introduction

The development and advancement of technology related to the utilization of electromagnetic (EM) wave energy is rapidly growing and progressing fast. For example, communication appliances and electronic tools in commercial, industrial, scientific, and military fields, such as navigation systems, computers, mobile phones, radar technology and wireless network systems (Huo et al. 2009; Qin and Brosseau 2012; Melvin et al. 2014a, b, 2017a).

Even though these technologies are convenient, excessive or unnecessary EM radiation has restricted their further progress and is becoming a serious crisis. EM radiation can cause severe interruption of electronic systems, device malfunction, generate false images, pollute the environment, or harm the health of human beings especially expectant mothers and children, and so on (Huo et al. 2009; Qin and Brosseau 2012; Melvin et al. 2014a, b, 2017a; Yusoff et al. 2002). These are some of the reasons why the usage of self-generated EM radiation devices such as mobile phones, wireless computer, and the devices in the same category are strictly prohibited in certain areas, which include hospitals, banks, petrol stations, and inside airplanes (Melvin et al. 2017a; Yusoff et al. 2002).

G. J. H. Melvin (✉)
Material and Mineral Research Unit (MMRU), Faculty of Engineering, Universiti Malaysia Sabah, Jalan UMS, Kota Kinabalu, Sabah, Malaysia
e-mail: melvin.gan@ums.edu.my

Y. Zhu
Key Laboratory of Advanced Textile Materials and Manufacturing Technology, Ministry of Education, Zhejiang Sci-Tech University, Hangzhou, People's Republic of China
e-mail: yfzhu@zstu.edu.cn

Q.-Q. Ni
Department of Functional Machinery and Mechanics, Shinshu University, Ueda, Tokida, Japan
e-mail: niqq@shinshu-u.ac.jp

S. Siddiquee et al. (eds.), *Nanotechnology: Applications in Energy, Drug and Food*,
https://doi.org/10.1007/978-3-319-99602-8_4

Therefore, it is essential to cut off and to defend human beings and electronic devices from excessive or unnecessary EM radiation exposure. Accordingly, the demand for EM wave absorber or microwave absorber materials have increased and received much attention to effectively solve the problem of exposure to EM radiation. Furthermore, EM wave absorbers are also necessary in various applications between GHz frequency ranges, in accordance with the development of technologies, such as electronic fields, telecommunication fields, and so on (Pan et al. 2016; Yin et al. 2014). The process of EM wave absorption can be referred to the EM wave energy loss, where the EM wave energy is depleted and then transformed into other energy, such as heat, and in result the EM wave cannot be reflected or permeated through the absorber materials (Huo et al. 2009; Qin and Brosseau 2012).

An ideal EM wave absorber material should be lightweight, strongly absorb EM waves, possess tunable absorption frequency, and multifunctionality. Nanomaterials have received significant attention as among essential candidates for EM wave absorber materials. In contrast to the corresponding bulk materials, nanomaterials have relatively lower density and higher specific surface area, which allow them to show high performance as an effective EM wave absorber. For instance, due to their lower relative density and larger specific surface area, will leads to numerous active atoms at the nanomaterials surface, which results in large interfacial dielectric loss induced by interface polarization, that will contribute to the high EM wave absorption properties (Huo et al. 2009).

Generally, EM wave absorber consist a filler material inside a matrix material. The filler which can be one or more constituents play an important role to do most the absorbing. Nanomaterials such as nanoparticles and hybrid nanomaterials are among excellent candidates as the filler. The matrix material is chosen for its physical properties such as temperature resistance, weatherability, etc. Commonly, polymer matrix is chosen because they are light weight, cost efficient, and flexible.

Research related to EM wave absorbers started in the 1930s (Seville 2005), and first patent was in 1936 in Netherlands (Machinerieen 1936). The absorber was a quarter-wave resonant type using carbon black as a lossy resistive material and titanium dioxide for its high permittivity in order to reduce the thickness. During World War II, Germany navy has developed several EM wave absorbers (Jaumann absorber, etc.) for the technology of radar camouflage for submarines (Seville 2005). The material had a layered structure and was based on graphite particles and other semiconductive materials embedded in a rubber matrix. In the United States, Salisbury screen that was invented by W. W. Salisbury in 1952 (Salisbury 1952), was one of the first concepts in radar absorbent material, later known as "stealth technology", used to prevent enemy radar detection of military vehicles. Later, led by Halpern at MIT Radiation Laboratory, materials known as Halpern Anti Radiation Paint (HARP) was developed (Halpern 1960; Halpern et al. 1960). In the following years and after few decades (on to today), the development of EM wave absorbers is still developing and some of them are already used for practical applications and commercial products. In recent trends, without doubt, nanomaterials play a great role as the prominent materials that can lead to high performance EM wave absorber.

4.2 Electromagnetic Wave Energy Loss

EM wave absorption ability is highly associated with EM wave energy loss. In the energy loss process, EM wave energy will be dissipated into other form of energy, such as heat, and in result the EM wave cannot be reflected or permeated through the absorber materials. The incident wave through the absorber materials undergoes the following phenomenon: penetration, reflection, and absorption, as shown in Fig. 4.1.

In order to produce an excellent EM wave absorber material, following two basic requirements, but not limited to, must be fulfilled (Huo et al. 2009; Zhu et al. 2013; Qing et al. 2011):

1. Concept of matched impedance: the characteristic impedance of an absorber should be almost equal to that of the free space to achieve zero reflection on the front surface of the material, which requires the complex permittivity close to complex permeability.
2. The absorber materials can absorb incident waves as much as possible inside the absorbers, which requires the materials to exhibit strong magnetic or/and dielectric loss.

Two main material parameters that highly associated with the EM wave absorption are complex permittivity ε (real part ε' and imaginary part ε'', $\varepsilon = \varepsilon' - j\varepsilon''$) and complex permeability μ (real part μ' and imaginary part μ'', $\mu = \mu' - j\mu''$). The real part of permittivity and permeability represent the energy storage, while the imaginary part of permittivity and permeability are related to dielectric loss or energy dissipation within a material resulting from conduction, resonance, and relaxation mechanisms. The EM absorption properties of a material backed by a perfect electric conductor (metal plate), as shown in Fig. 4.2, typically investigated using the reflection loss (*R.L.*, dB) that is calculated according to the transmission line theory as follows (Huo et al. 2009; Qin and Brosseau 2012; Melvin et al. 2014a, b, 2017a; Yusoff et al. 2002):

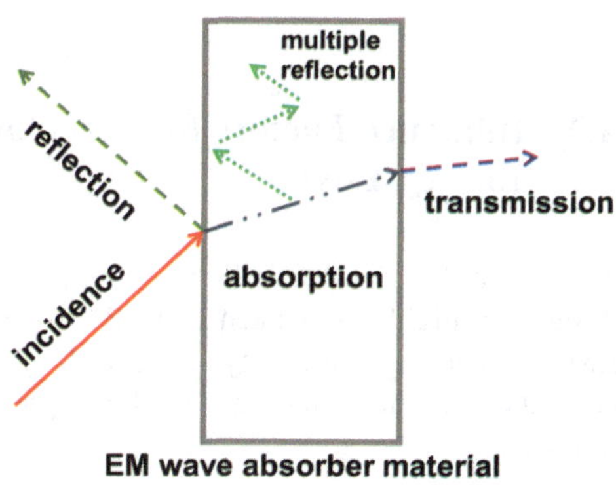

Fig. 4.1 General processes of an incident EM wave through an EM wave absorber material

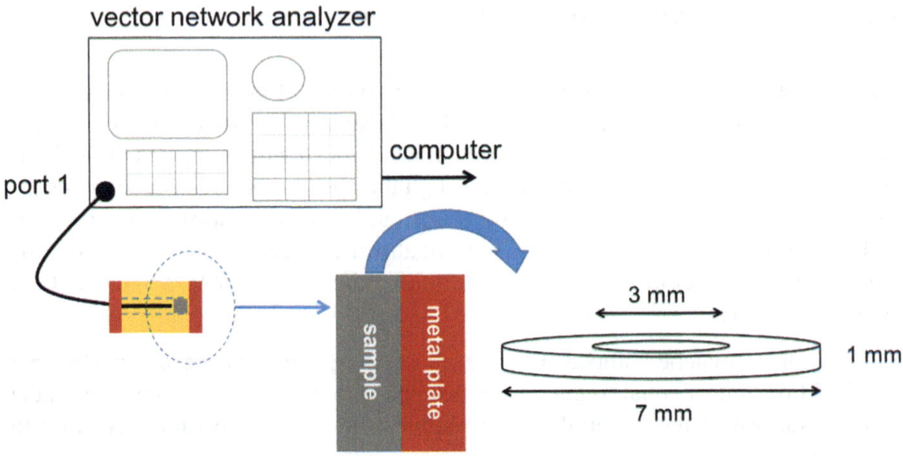

Fig. 4.2 Experimental setup to evaluate EM wave absorption properties. The sample is designed into toroidal shape with an outer diameter of 7.0 mm, inner diameter of 3.0 mm, and thickness of 1.0 mm

$$R.L. = 20 \log\left|\frac{Z_{\mathrm{in}} - 1}{Z_{\mathrm{in}} + 1}\right| \tag{4.1}$$

where the normalized input impedance (Z_{in}) is given by the formula,

$$Z_{\mathrm{in}} = \sqrt{\frac{\mu_r}{\varepsilon_r}} \tan\mathrm{h}\left[j\left(\frac{2\pi f d}{c}\right)\sqrt{\mu_r \varepsilon_r}\right] \tag{4.2}$$

where $\varepsilon_r = \varepsilon' - j\varepsilon''$, $\mu_r = \mu' - j\mu''$, f is the EM wave frequency (Hz), d is the thickness of the absorber (m), and c is the velocity of light in free space (m/s). Typically, the value of -10 dB and -20 dB for *R.L.* represent that 90% and 99%, respectively, EM waves are absorbed by the materials. The energy depletion mechanism associated majorly with the dielectric loss and magnetic loss.

4.3 Influence Factors for Electromagnetic Wave Energy Loss

The material's structure, morphology, and size; electrical and magnetic properties; design, are highly associated with EM wave energy loss, thus influence greatly their EM wave absorption ability. In order to gain excellent EM wave absorption property, the materials must be well designed to achieve the concept of matched impedance.

4.3.1 Nanoeffect

Nanoscale materials were endowed with outstanding electric and magnetic characteristics as a result of their particular size, surface, and quantum tunnel effect. Nanomaterials such as nanoparticles and hybrid nanocomposites are among essential candidates as EM wave absorbing materials. For instance, the relative density of nanomaterials is lower and their specific surface area is larger than those of the corresponding bulk materials. Consequently, there are a great number of active atoms at the nanomaterial surface, which has a large interfacial dielectric loss induced by interface polarization and will allows multi reflection to progress (Huo et al. 2009; Melvin et al. 2014a, b, 2017a).

Furthermore, hybrid nanocomposites such as carbon nanotubes (CNTs) decorated with nanoparticles on their surface, will absorb energy through electron hopping and shows enhanced EM absorption bandwidth (Qin and Brosseau 2012; Melvin et al. 2014a, b, 2017a). The combination of nanoparticles with CNTs can integrate the properties of these two components to form hybrid nanocomposites for use as EM wave absorber materials. Moreover, in the case where the particle size is below the skin depth, the eddy current loss can be induced which can further develop the stability of wave absorption property. Generally, the skin depth of a particular material is about 1 μm at frequency of 10 GHz, which prove that nanomaterials will possess great EM wave absorption properties at broad frequency (Huo et al. 2009).

4.3.2 Heterogeneity

Heterogeneous system, such as nanoparticles attached onto another nanomaterial's surface, or commonly referred as hybrid nanomaterials or nanocomposites, play a significant role in order to obtain high EM wave absorption performance. The combination of two or more appropriate nanomaterials will lead to a far more complex dielectric relaxation by additional dielectric interfaces and higher polarization charges at the interface between the nanomaterials (Zhu et al. 2015; Ni et al. 2015b; Bhattacharya and Das 2013). Furthermore, orientational and space charge polarization also contributed to the significant EM wave absorption performance. The space charge polarization can be reflected to be highly associated with the heterogeneity presents at the interface between the nanomaterials, meanwhile the bound charges (dipoles) present in the hybrid nanomaterials or nanocomposites are related to the orientational polarization (Bhattacharya and Das 2013).

Furthermore, another factor that also comes along with the heterogeneity of the nanomaterials is the weight fraction or content of absorbing materials. Generally, low weight fraction of absorbing nanomaterials will only produce low or insufficient absorption performance. In contrast, at higher weight fraction, enhanced complex permittivity and permeability will be able to contribute to the enhanced *R.L.* when

EM wave transmits through the absorbing nanomaterials. Simultaneously, the connectivity between the nanomaterials will increase and the EM wave propagation paths inside the nanocomposites become more complex, as the content increase (Zhu et al. 2015). These made them to be able to polarize repeatedly at high frequency electric field, which causes the electric energy dissipated into other form of energy such as heat energy.

4.3.3 Fundamental EM Parameters

4.3.3.1 Magnetic Loss

Three major energy losses will be produced when magnetic materials are penetrated with EM waves: eddy current loss, magnetic hysteresis loss, and residual loss. Eddy current loss occurs in the case if a conductor material was put in an alternating magnetic field, where a close induced current will be produced inside the material, and further will dissipate the energy (Huo et al. 2009). Furthermore, eddy current loss is also affected by other factors, such as orientation, grain size, surface roughness, morphology of material, and so on. Meanwhile, magnetic hysteresis loss is induced by the irreversible domain movement and magnetic moment rotation of magnetic material (Huo et al. 2009). Moreover, magnetic loss except the two losses mentioned before can be referred as residual loss. At low frequency, the residual loss was caused mainly by magnetic after-effect loss, which includes thermal fluctuation or the hysteresis of some electrons and ions moving to equilibrium position relative to the diffusion of applied magnetic field. This kind of loss is determined by the amplitude of alternating magnetic field and relaxation time of materials. Conversely, at high frequency, the residual loss is affected by size resonance, ferromagnetic resonance, natural resonance and/or domain wall resonance (Huo et al. 2009; Yamada and Otsuki 1997).

Examples of magnetic nanoparticles are such as materials that commonly consist of magnetic elements such as iron, nickel, cobalt, and their chemical compounds. Cui et al. synthesized nanocomposites consist of magnetite (Fe_3O_4)/polyaniline and the maximum *R.L.* is −31.3 dB at 9 GHz, when the thickness is 3 mm (Cui et al. 2012). Ali et al. fabricated $MgO/CuFe_2O_4$ nanocomposites which exhibited maximum *R.L.* of −25.35 dB at 8.38 GHz (Ali et al. 2017). There are many more papers and reports presented, where magnetic materials were applied for EM wave absorber.

4.3.3.2 Dielectric Loss

Interacting with EM waves, dielectric material dissipate electric energy which then transformed into other energy such as thermal energy, and the energy is refereed as

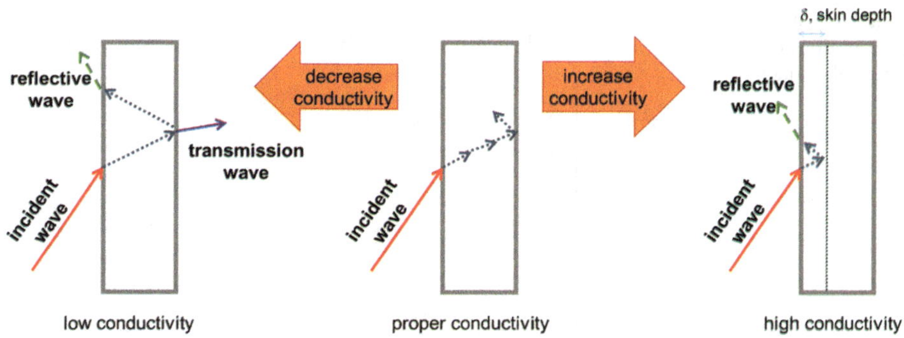

Fig. 4.3 EM wave transmission model for materials with different conductivity

dielectric loss. Basically, the mechanisms of dielectric loss can be referred to conductance loss, dielectric relaxation loss, and resonance loss (Huo et al. 2009).

EM wave absorber material with certain electric conductivity will produce conductance current. In the case an alternating electric field operated on it, the current would dissipate the energy into heat. The conductance loss is majorly determined by the electric conductivity of materials. However, a suitable electric conductivity of material must be considered to obtain great EM wave absorption effect. Figure 4.3 demonstrates the relationship between conductivity and EM wave transmission through absorber material. The impedance of material with high conductivity is relatively small in contrast to air, thus the skin depth is small and nearly most of EM wave will be reflected by the absorbers. Generally, carbon materials can be considered as a good electrical conductor materials. Carbon-based material such as carbon black (conductor) was incorporated into matrix and used as EM wave absorber since it was first patented in Netherlands in 1936 (Machinerieen 1936). Carbon black remained as the favorite filler for absorber materials for a few decades. Afterwards, with the discovery of CNTs, which are excellent electric conductor in the 1990s, besides their application for various fields, they are also have been noticed as an exceptional candidate for EM wave absorber materials. High electrical conductivity of CNTs makes them capable of dissipating electrostatic charges or shielding EM radiation. Ting et al. (2012) fabricated nanocomposites using CNTs and conducting polymer, polyaniline. By utilizing conducting materials, their materials showed enhanced EM wave absorption. Furthermore, metallic particle such as Ag nanoparticles is also a good conductor, abundant, easy to prepare, and inexpensive (Zhang et al. 2010; Yang et al. 2008; Dong et al. 2012). Ag nanoparticles decoration would have a beneficial effect on the electrical conductivity of CNTs because the inherent electrical conductivity of Ag is superior to the CNTs without Ag (Zhang et al. 2010). Ramesh et al. (2009) investigated the EM wave absorption properties of Ag nanoparticles embedded in thin polymer films. The films exhibited appreciable EM wave absorption in the 8–12 GHz range. Meanwhile, Zhao et al. (2008a, b) examined the EM absorption properties of CNTs filled with Ag nanowires. The maximum *R.L.* of −19.2 dB at 7.8 GHz was obtained.

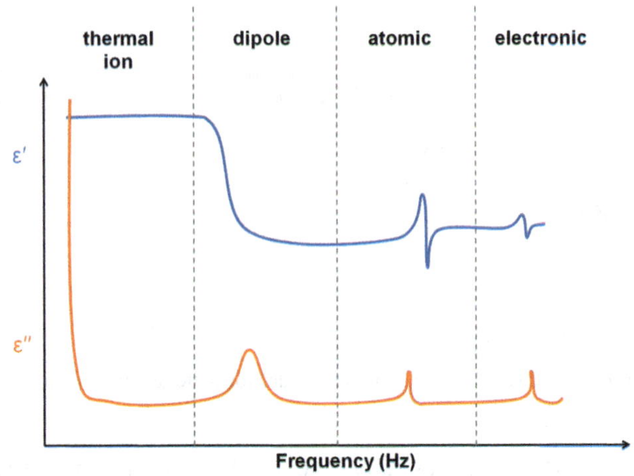

Fig. 4.4 Various polarizations over a broad range of frequency

Under the application of electric field, materials would be polarized, and if the change of polarization is slower than that of electric field, the dielectric relaxation loss will be produced. The most common polarizations from low frequency are: thermal ion polarization, dipole rotation polarization, ion (atomic) polarization, and electronic displacement polarization (Huo et al. 2009; Hashimoto et al. 2012). Figure 4.4 exhibits various polarizations over a broad frequency. The time for thermal ion and dipole rotation polarization is about 10^{-8}–10^{-2} s; meanwhile, for electronic displacement and ion polarization, the time is very short (energy loss at high frequency, ~10^{15} Hz), about 10^{-15}–10^{-14} s. As a result, at high frequency, the electronic displacement and ion polarization play as an essential role in relaxation loss.

The resonance loss originated from resonance effect induced by vibration of atoms, ions, or electron inside of the particular materials. Moreover, it is produced at the scope of infrared to ultraviolet frequency.

4.3.3.3 Complex Permittivity and Complex Permeability

Complex permittivity and complex permeability are basic parameters associated with the EM wave absorption properties of a material. Furthermore, dielectric loss factor (tan $\delta_\varepsilon = \varepsilon''/\varepsilon'$) and magnetic loss factor (tan $\delta_\mu = \mu''/\mu'$) also play a great role to reveal the intrinsic reasons for the EM wave absorption.

From tan δ_ε, the bigger the ε'', better absorption effect can be expected. Thus, materials with high permittivity are often chosen as the candidate for absorber materials. However, the reflection part of wave is relatively large for a material with too high permittivity (Zhang et al. 2007), which indicates that material with proper permittivity must be selected according to practical need. In the case of polymer-based nanocomposites, where the filler is nanomaterials and the matrix is polymer, low dielectric loss polymer is preferable for the absorption performance, which can allow more waves to transmit into the absorber material (Yusoff et al. 2002).

From tan δ_μ and magnetic loss mechanisms, we also can interpret that the bigger the μ'' and the smaller the μ', the larger the magnetic loss of materials can be obtained. However, according to the concept of matched impedance, when the permeability is equal to the permittivity, there is no reflection and the EM wave absorption effect is the finest.

4.3.4 Design of Materials

Generally, the EM wave absorption ability can be manipulated physically by controlling the weight fraction composition, thickness, and layer structure of the absorber materials. For instance, higher weight fraction of filler in the matrix of a particular absorber material will produce higher absorption ability. Although higher absorption ability can be achieved by this method, some problems are inevitable: cost performance, dispersion and homogeneity problems, electric conductivity percolation threshold (reflected wave), and so on. By adjusting the thickness of absorber materials, higher absorption ability can be obtained and most of the cases, the absorption peak will shift to lower frequency. The *R.L.* will increase as the thickness increase, until a certain frequency, and then it will decrease gradually. However, it is difficult to prepare light weight and flexible absorber materials by this way, and to fulfill the demands for thin absorber material.

In order to gain higher absorption ability and broad absorption frequency, multilayer structure of absorber material is an effective method. Figure 4.5 shows the double-layer structure of EM wave absorber.

Basically double-layer structure includes layers of: impedance matching layer (matching layer), EM wave loss layer (absorption layer), and reflective layer (metal/conductive plate), in which the matching layer can allow the EM wave transmission without (ideal) or with very minimal reflection by adjusting the complex permittivity and permeability of nanomaterials; the EM wave is depleted in the absorption layer,

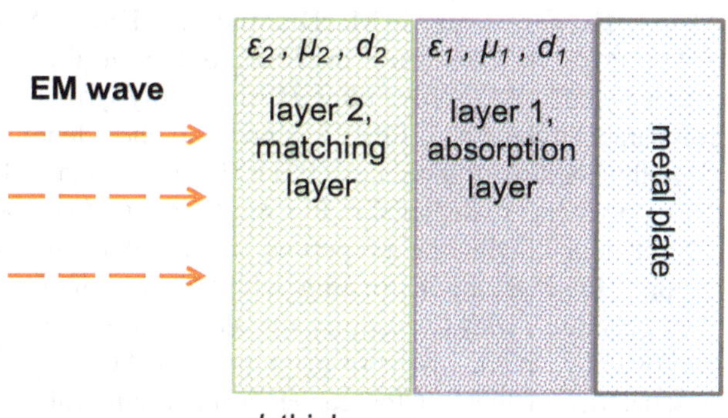

Fig. 4.5 Multilayer EM wave absorber

where most of EM wave energy loss occurs in here, composing of high dielectric or magnetic loss nanomaterials; the reflective layer acts as a reflector to make a small quantity of EM transmission wave back to the absorption layer (Huo et al. 2009; Qin and Brosseau 2012; Melvin et al. 2015; Ni et al. 2015a; Wu et al. 2012). By implementing this specific structure design, the absorbers are able to exhibit enhanced EM wave absorption ability and broad absorption frequency. Furthermore, by manipulating the thickness of matching and absorption layer, absorption at various frequencies can be achieved.

4.4 Electromagnetic Wave Absorber Materials and Their Fabrication Process

Nowadays, there are various kinds of EM wave absorbers that has been received great attention. For instance, but not limited to, various conducting (e.g. conducting polymer, metals, graphite, carbon black, CNTs, graphene), dielectric (e.g. $BaTiO_3$ or TiO_2), and magnetic (e.g. γ-Fe_2O_3, Fe_3O_4, $BaFe_{12}O_{19}$) materials (Gomes 2012).

4.4.1 Electromagnetic Wave Absorber Based on Intrinsically Conducting Polymers (ICPs)

Intrinsically conducting polymers (ICPs) are naturally conducting due to the existance of a conjugated π electron system in their structure (Bhadra et al. 2009; Shirakawa 2001; MacDiarmid 2001; Heeger 2001b). Some of ICPs characteristics are low energy optical transition, low ionization potential, and a high electron affinity (Unsworth et al. 1992). The facts that their electrical conductivity can be exploited by adjusting certain parameters such as oxidation state, doping level, morphology, and chemical structure, makes them dominant candidate for numerous techno-commercial applications. Related to the discovery of conducting polymers, Alan J. Heeger, Alan G. MacDiarmid and Hideki Shirakawa were awarded the Nobel Prize in Chemistry in 2000. Some of the representatives of ICPs are polyacetylene (PA), polyaniline (PANI), polypyrrole (PPy), polythiophene (PTh), polyparaphenylene (PPP), polyparaphenylenevinylene (PPV), poly (3,4-ethylenedioxythiophene) (PEDOT), and so on. The chemical structures of these polymers are depicted in Fig. 4.6 (Park and Lee 2005). The undoped polymers exhibit poor conducting properties and listed in the insulator or semiconductors region. However, upon treating with dopants and/or subjecting to chemical or electrochemical redox reactions, the electrical conductivity of the polymers can be improved by several orders of magnitude, as shown in Fig. 4.7. Generally speaking, conductivity of these ICPs reaches a maximum value ~10^3 S/m (Baeriswyl et al.

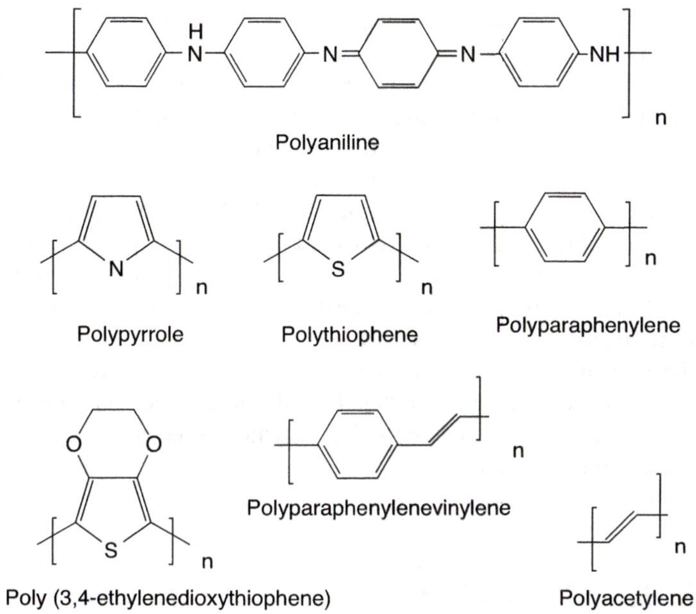

Fig. 4.6 Chemical structures of several kinds of conducting polymers (Park and Lee 2005)

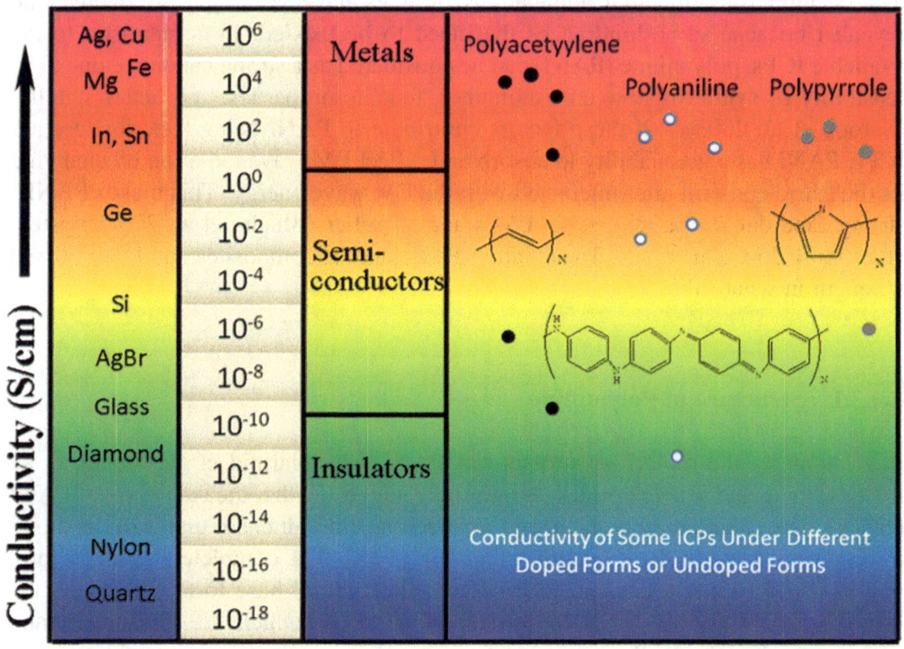

Fig. 4.7 Conductivity of some conductive polymers in comparison to typical metals, semiconductors or insulators (Gomes 2012)

1992; Cao et al. 1992; Chiang et al. 1978a, b; Chandrasekhar and Naishadham 1999; Ellis 1986; Nalwa 1997).

The intrinsic conductivity of conjugated polymers in the EM wave field makes them a valuable candidate to be utilized for absorbing/shielding material. Specifically, dependence of their conductivity on frequency has stimulated many scientific works to implement these phenomenon to EM wave applications (Coleman and Petcavich 1978; Karasz et al. 1985; Saini et al. 2009). Moreover, the unique properties like tunable conductivity, adjustable permittivity/permeability via synthetic means, low density, non-corrosiveness, nominal cost, facile processing, and controllable EM attributes, further reinforce their candidature as EM wave absorbing material in the stealth technology (Nalwa 1997). Furthermore, conducting polymers appear to be one of the few materials capable of displaying dynamic EM wave absorption behavior, which are called "intelligent stealth materials", attributed to the reversible electrical properties of conducting polymers affected by redox doping/de-doping processes (Gomes 2012).

4.4.2 Intrinsically Conducting Polymer of Polyaniline

To the best of our knowledge, ICPs have numerous potential applications. Nevertheless, numerous potential applications of ICPs have yet to be discovered and revealed because some hindrances that need to be tackled. In comparison to the available ICPs, polyaniline (PANI) can be considered as a strong candidate due to its feasibility of synthesis, low cost monomer, tunable properties, and better stability (Saroop et al. 2003). All these factors contribute to PANI being superior to other ICPs. PANI has a great ability to absorb and reflect EM wave radiation by changing its dielectric constant after interacted with the EM wave energy. This makes PANI a strong candidate to be utilized as EM wave absorber (Bhadra et al. 2008; Epstein et al. 1994; Joo et al. 1995). The detail of PANI and its composites will be discussed at length in what follows.

4.4.2.1 Structure of Polyaniline

In relations to the molecular structure of PANI, MacDiarmid et al. (Huang et al. 1986; Chiang and MacDiarmid 1986) initially suggested that the base form of PANI consists of an alternating reduced and oxidized repeat unit chain that is divided into three states based on the oxidation state ($0 < y < 1$). The completely reduced form and oxidized form are designated the 'leucoemeraldine' base form (LEB) and the 'pernigraniline' form (PEN), the oxidation states of which are equal to 1 and 0, respectively. The 'half-oxidized' form ($y = 0.5$) is called the 'emeraldine base' form (EB). The molecular formula of PANI with various oxidation states and protonic acid doping process is shown in Fig. 4.8 (Wan 2009).

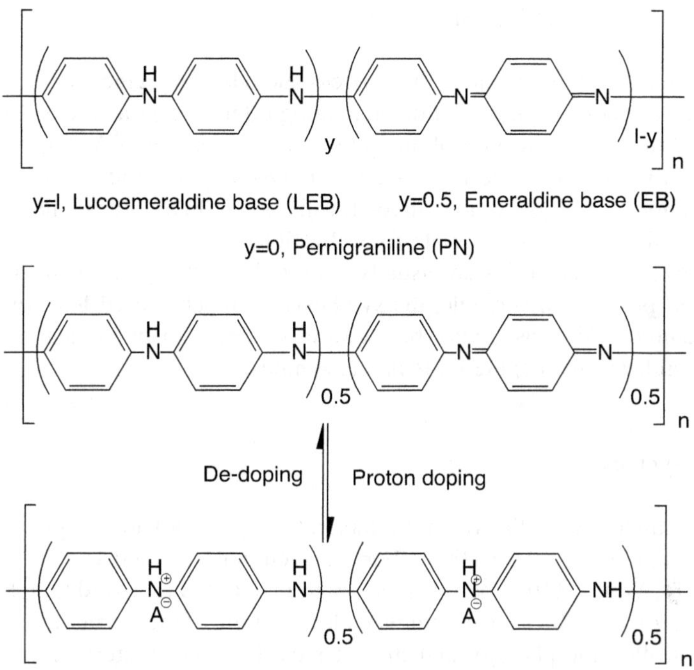

Fig. 4.8 Scheme of the molecular formula of PANI with different oxidation states and reversible switching from the insulating EB form to the conducting emeraldine salt form by a doping/de-doping process (Wan 2009)

4.4.2.2 Synthesis of Polyaniline

Various methods including physical routes like electrospinning (Pinto et al. 2003; Zhou et al. 2003; Kahol and Pinto 2004), mechanical stretching (He et al. 2001), doping induced solution route (He et al. 2003), and chemical approaches are utilized in order to synthesis PANI, and can be generally categorized into the chemical polymerized way and the electrochemical polymerized way, which is briefly described as follows (Harun et al. 2007).

Chemical Polymerization

Through the chemical polymerization, aniline monomers react with an excess amount of an oxidant in an appropriate solvent, for example acid. Instantly, the polymerization will start and requires continuous stirring. One of the main benefits of chemical polymerization is the possibility of mass-production at a practical cost (Toshima and Hara 1995; Chao and March 1988).

Electrochemical Polymerization

Electrochemical polymerization, the process includes engaging both counter and reference electrodes, into the solution containing diluted monomer and electrolyte in a solvent. After the application of an appropriate voltage, the PANI film spontaneously starts to form on the working electrolyte. One important feature of the electropolymerization technique is the direct formation of PANI films that are highly conductive, simple, and appropriate for utilization.

The above two categories are usually occurred for the synthesis of the conventional PANI powder. Meanwhile, the synthesis approaches used for preparation of nano-structured PANI, as briefly below, are divided into two categories: (1) the template synthesis and (2) the no-template synthesis.

Template Synthesis

Mainly, nano-structured PANI can be fabricated by using hard-templates and soft-templates approaches (Wan 2008). For hard-template method (Parthasarathy and Martin 1994; Wan 2009), a porous membrane is required as a hard template, which guides the growth of the nanostructures within the pore of the membrane. As a result, a well-controlled morphology and diameter of the nanostructures can be gained, which is the advantages of hard template method in comparison with different methods. Nevertheless, the removal of template is necessary after polymerization in order to gain pure nanostructures. The post-process of template removal results in unavoidable complex synthesis process, and in the same time destroys the nanostructures.

Conversely, the soft-template method is another influential approach to synthesize nano-structured PANI (Wan 2009). Comparatively, for this approach a porous membrane is not needed, and is related with a self-assembly process. Hydrogen bonding, π-π stacking, van der Waals forces, and electrostatic interactions are used as the driving forces (Boal et al. 2000) in the self-assembly process. In comparison to the other molecular interactions, hydrogen bonding is the major driving force in the self-assembly process (Kosonen et al. 2000). Furthermore, a variety of surface micelles, liquid-crystalline phases, surfactants, colloidal particles, structure directing molecules, and aniline oligomers have also been utilized as the soft-template to synthesize PANI nanostructure (Carswell et al. 2003; Huang et al. 2002; Wei et al. 2002; Zhang and Wan 2002; Qiu et al. 2001; Wang and Caruso 2001; Bartlett et al. 2001; Wang et al. 2004).

No-Template Synthesis

The method for synthesizing one-dimensional nano-structured PANI without any templates, hard or soft, was named "no-template synthesis". This synthesis method can be subdivided interfacial polymerization, radiolytic synthesis, rapid mixing reaction method and sonochemical synthesis, et al. (Virji et al. 2004; Pillalamarri et al. 2005; Huang and Kaner 2004; Jing et al. 2006).

4.4.2.3 Synthesis of Other Intrinsically Conducting Polymers

Since the first ICP, the development keeps progressing and then polyacetylene (PA) was synthesized by Shirakawa et al. (1977). They showed that the conductivity of PA could be improved through chemical doping and in reality it can be changed from an insulator to a metal like conductor. Continuation from the research related to PA, various polymers such as polypyrrole (PPy), polythiophene, polyparaphenylene, and polypara-phenylenevinylene, and their derivatives, have been synthesized and studied. Simultaneously, it is found that their potential applications in wide broad area such as electrical, electronics, thermoelectric, electrochemical, electromagnetic, electromechanical, electroluminescence, electrorheological, chemical, membrane, and sensors (Unsworth et al. 1992; Schoch 1994; Angelopoulos 2001; Gospodinova and Terlemezyan 1998). PPy is another inexpensive, easy to synthesis, readily available conducting polymer that has interesting properties and abundant applications. For example, Xu et al. (2008) prepared micro-structured Ni/PPy core/shell composites by in situ chemical oxidative polymerization of pyrrole monomer in the presence of Ni powder. It is important to notice that the ternary Debye relaxations for enhanced dielectric loss induced by PPy coatings. Duchet et al. (1998) prepared nanostructures composed of PPy by chemically synthesizing PPy within the pores of micro-porous and nano-porous Particle Track-etched Membranes (PTM).

4.4.3 Electromagnetic Wave Absorber Based on Inorganic Nanomaterials

Inorganic nano-structured EM wave absorbers have received steadily growing interest because of their fascinating properties such as unique absorbing EM wave energy, high magnetic and electric loss. According to inorganic nano-structured EM wave absorbers loss mechanism, it can be divided into three types: magnetic, dielectric, and conducting nanomaterials. Furthermore, it is worth to notice that the combination of two or more nanomaterials can integrate the properties of the two components or more to form hybrid nanocomposites for use as promising EM wave absorbers.

4.4.3.1 Magnetic Nanomaterials

Magnetic nanoparticles, due to their unique magnetic and electronic properties, have received great attention. The magnetic nanoparticles include metals such as iron, cobalt, nickel and their alloys, oxides (γ-Fe_2O_3, Fe_3O_4, cobalt oxide), iron cobalt oxide ferrite, barium ferrite, and so on. Magnetite (Fe_3O_4), an important member of spinel-ferrite, possesses both complex permittivity and permeability, which are two important factors in EM wave absorption. Zou et al. (2010) prepared magnetic Fe_3O_4 particles from copper/iron ore cinder by precipitation oxidization method. It was demonstrated that the Fe_3O_4 particles could be used as an effective EM wave absorbing material. M-type barium ferrite with hexagonal molecular structure ($BaFe_{12}O_{19}$) is another promising material for EM wave absorbing because of its fairly large magneto-crystalline anisotropy, relatively large magnetization, and excellent chemical stability. Xu et al. (2007) successfully synthesized PANI/ $BaFe_{12}O_{19}$ nanocomposites with novel coralloid structures through one-step in situ polymerization method. The reflection loss of PANI in 2–18 GHz was essentially enhanced by $BaFe_{12}O_{19}$.

4.4.3.2 Dielectric Nanomaterials

Barium titanate ($BaTiO_3$) ferroelectric ceramic material is one of typical dielectric material. The perovskite-type structure of $BaTiO_3$ is widely used in multiple areas, due to its high dielectric constant, good ferroelectric properties and nonlinear optical properties. Especially, the excellent permittivity and polarization properties of $BaTiO_3$ are the two key factors for them to become a good candidate for EM wave absorption materials and have received considerable attention from many researchers over recent years. Jing et al. (2009) reported that the flake-shaped $BaTiO_3$ was prepared by sol-gel method and found it has excellent EM absorption properties. Chen et al. (2008) reported that the EM wave absorption property of $BaTiO_3$ prepared by the sol-gel method. It was found that the higher content of $BaTiO_3$ in composite yielded the good reflection loss in the X and Ku bands. $BaTiO_3$ nanotubes also showed their potential as high potential EM wave absorber, where their hollow structure allows condition for EM wave energy loss that leads to higher EM wave absorption performance (Zhu et al. 2012).

4.4.3.3 Conducting Nanomaterials

Among the conducting nanomaterials, carbonaceous materials due to lightweight and unique properties make them a promising candidate for EM wave absorption application. Wang et al. (2008) reported that nanocrystalline carbon black/barium titanate compound particle was synthesized by sol-gel method. It was found that the compound particle is mainly a kind of electric and dielectric lossy materials and exhibits excellent EM wave absorption performance. Li et al. (2012) reported that porous carbon fibers and carbon nanofibers were prepared from polyacrylonitrile/ polymethylmethacrylate blend fibers as precursors. It was found that the EM wave absorbing performance of the porous carbon fibers is superior to that of the carbon nanofibers. Alternatively, carbon materials derived from waste materials including agricultural wastes such as rice husks (RHs), which is low cost, also have shown their potential to be developed into outstanding EM wave absorber (Melvin et al. 2017b). Mixture of flattened graphene and corrugated graphene layer, including few and multilayer graphene were obtained from RHs heat treated at high temperature, and they showed considerable good EM wave absorption capability.

4.4.4 Intrinsically Conducting Polymer Based Nanocomposites

An extensive comparison of properties between different available shielding/absorbing materials showed that no single phase material can consist all the characteristics of EM wave absorption in order to obtain preferred ability under various environments and numerous applications (Gomes 2012). Consequently, few trials have also been conducted to explore the valuable characteristics of above materials by producing strategic combinations, such as admixtures, blends and composites (Ajayan et al. 2000; Cao et al. 1995; Chung 2001; Colaneri and Schacklette 1992; Dhawan et al. 2003; Gangopadhyay et al. 2001; Gupta and Choudhary 2011; Moon et al. 2010; Saini et al. 2011; Zhang et al. 2011). Among these options, inorganic/ICP nanocomposites have been highlighted due to their remarkable characteristics and wide broad applications (Chandrasekhar and Naishadham 1999; Freund and Deore 2007; Heeger 2001a). Recently, based on the availability of numerous nanomaterials and the availability to exploit their electrical and EM properties through design, has leads to further research to explore for the suitable materials for EM wave absorber and other applications (Ajayan et al. 1994; Baughman et al. 2002; Geim 2009; Moniruzzaman and Winey 2006; Rozenberg and Tenne 2008). Especially, inorganic/ICP nanocomposites have attracted massive scientific consideration due to eminent set of properties and various favorable applications.

4.4.4.1 Composite Types

ICP as Filler

ICPs can be used as a filler to design good flexibility and compatibility with various polymer matrices in order to form composites. Such composites are formed either by solution processing or by melt phase mixing/blending. Works done by Zhang et al. (2012) utilizing polypyrrole nanotubes as filler, the novel polypyrrole nanotubes/

polyaniline composite films are prepared via a facile solvent evaporation method. The electrical conductivity and mechanical ductility is significantly increased by the addition of polypyrrole nanotubes filler.

ICP as Matrix Polymer

The exploitation of diverse ICPs as nanocomposite matrix offers several benefits such as design flexibility, good filler incorporation-ability, specific interactions with fillers and EM wave non-transparency. The incorporation of various conducting, dielectric or magnetic nanoparticles in conducting polymer matrices can be achieved either by ex-situ physical mixing processes or by in-situ polymerization (Gomes 2012). Li et al. (2009) reported the synthesis of flake-like polypyrrole/$SrFe_{12}O_{19}$ composites via in situ polymerization. Yang et al. (2009) reported the synthesis of polyaniline/Fe_3O_4 nanocomposites by polymerizing aniline in the presence of Fe_3O_4 nanoparticles upon the use of H_2O_2 as oxidation.

4.4.4.2 Composite Formation Techniques

Physical Mixing Processes

Commonly, physical mixing processes are where nanoparticles, synthesized in an external synthesis step, are added or mixed to resin (organic solution), usually followed by a solvent evaporation method and drying or mechanical mixing. This approach was used by Musikhin et al. (2002) to generate polymer-dielectric nanocrystal composites. He et al. (2011) reported the synthesis of PANI/carbonyl iron powder (CIP)/Fe_3O_4 composite by mechanical mixing PANI/CIP composite with PANI/Fe_3O_4 composite.

In Situ Polymerization Method

This approach carried out the polymerization under the controlled conditions and in the presence of specific dopants and fillers. Typically, during the pre-polymerization process, monomers are generally adsorbed over dispersed nanofiller particles. The polymerization was started by addition of specific initiator/oxidant and allowed to proceed till reaction gets completed leading to formation of ICP based nanocomposite. Chen et al. (2010) reported the synthesis of multi-walled carbon nanotubes-polyaniline composites through in situ chemical polymerization. Lu et al. (2009) reported the facile synthesis of highly regulated core-shell Fe_3O_4/polypyrrole microspheres is achieved using a surfactant directed chemical oxidation polymerization in aqueous solution.

4.5 Conclusions

In this chapter, EM wave energy loss, where higher EM wave energy loss represents higher EM wave absorption capability, was discussed focusing on various nanomaterials. The affecting factors were elaborated, which is highly associated with nanoeffects, heterogeneity, fundamental EM parameters, and so on. Furthermore, conducting, dielectric, magnetic, and hybrid nanomaterials which are strong candidate to be utilized as effective and promising EM wave absorber materials were reviewed. The capability to modulate the absorption and bandwidth to suit various applications in different frequency bands of the nanomaterials/ nanocomposites indicates that they could be an excellent EM wave absorber.

References

Ajayan PM, Stephan O, Colliex C, Trauth D (1994) Aligned carbon nanotube arrays formed by cutting a polymer resin-nanotube composite. Science 265(5176):1212–1214

Ajayan PM, Schadler LS, Giannaris C, Rubio A (2000) Single-walled carbon nanotube–polymer composites: strength and weakness. Adv Mater 12(10):750–753.https://doi.org/10.1002/(SICI)1521-4095(200005)12:10<750::AID-ADMA750>3.0.CO;2-6

Ali K, Iqbal J, Jana T, Ahmad N, Ahmad I, Wan D (2017) Enhancement of microwaves absorption properties of $CuFe_2O_4$ magnetic nanoparticles embedded in MgO matrix. J Alloy Compd 696:711–717. https://doi.org/10.1016/j.jallcom.2016.10.220

Angelopoulos M (2001) Conducting polymers in microelectronics. IBM J Res Dev 45(1):57–75. https://doi.org/10.1147/rd.451.0057

Baeriswyl D, Campbell DK, Mazumdar S (1992) An overview of the theory of π-conjugated polymers. In: Kiess HG (ed) Conjugated conducting polymers. Springer, Berlin

Bartlett PN, Birkin PR, Ghanem MA, Toh CS (2001) Electrochemical syntheses of highly ordered macroporous conducting polymers grown around self-assembled colloidal templates. J Mater Chem 11(3):849–853. https://doi.org/10.1039/B006992M

Baughman RH, Zakhidov AA, De Heer WA (2002) Carbon nanotubes—the route toward applications. Science 297(5582):787–792. https://doi.org/10.1126/science.1060928

Bhadra S, Singha NK, Khastgir D (2008) Semiconductive composites from ethylene 1-octene copolymer and polyaniline coated nylon 6: studies on mechanical, thermal, processability, electrical, and EMI shielding properties. Polym Eng Sci 48(5):995–1006. https://doi.org/10.1002/pen.21025

Bhadra S, Khastgir D, Singha NK, Lee JH (2009) Progress in preparation, processing and applications of polyaniline. Prog Polym Sci 34(8):783–810. https://doi.org/10.1016/j.progpolymsci.2009.04.003

Bhattacharya P, Das CK (2013) In situ synthesis and characterization of $CuFe_{10}Al_2O_{19}$/MWCNT nanocomposites for supercapacitor and microwave-absorbing applications. Ind Eng Chem Res 52(28):9594–9606. https://doi.org/10.1021/ie4005783

Boal AK, Ilhan F, DeRouchey JE, Thurn-Albrecht T, Russell TP, Rotello VM (2000) Self-assembly of nanoparticles into structured spherical and network aggregates. Nature 404(6779):746–748. https://doi.org/10.1038/35008037

Cao Y, Smith P, Heeger AJ (1992) Counter-ion induced processibility of conducting polyaniline and of conducting polyblends of polyaniline in bulk polymers. Synth Met 48(1):91–97. https://doi.org/10.1016/0379-6779(92)90053-L

Cao Y, Qiu J, Smith P (1995) Effect of solvents and co-solvents on the processibility of polyaniline: I. Solubility and conductivity studies. Synth Met 69(1):187–190. https://doi.org/10.1016/0379-6779(94)02412-R

Carswell AD, O'Rea EA, Grady BP (2003) Adsorbed surfactants as templates for the synthesis of morphologically controlled polyaniline and polypyrrole nanostructures on flat surfaces: from spheres to wires to flat films. J Am Chem Soc 125(48):14793–14800. https://doi.org/10.1021/ja0365983

Chandrasekhar P, Naishadham K (1999) Broadband microwave absorption and shielding properties of a poly(aniline). Synth Met 105(2):115–120. https://doi.org/10.1016/S0379-6779(99)00085-5

Chao TH, March J (1988) A study of polypyrrole synthesized with oxidative transition metal ions. J Polym Sci A: Polym Chem 26(3):743–753. https://doi.org/10.1002/pola.1988.080260306

Chen X, Wang G, Duan Y, Liu S (2008) Electromagnetic characteristics of barium titanate/epoxide resin composites in X and Ku bands. J Alloys Compd 453(1):433–436. https://doi.org/10.1016/j.jallcom.2006.11.092

Chen WY, Ghule AV, Chang JY, Ling YC (2010) Morphology and dopant influence electrical properties and stability of multiwalled carbon nanotube-polyaniline composites. Curr Nanosci 6 (1):59–68. https://doi.org/10.2174/157341310790226324

Chiang JC, MacDiarmid AG (1986) 'Polyaniline': Protonic acid doping of the emeraldine form to the metallic regime. Synth Met 13(1–3):193–205. https://doi.org/10.1016/0379-6779(86)90070-6

Chiang CK, Gau SC, Fincher CR Jr, Park YW, MacDiarmid AG, Heeger AJ (1978a) Polyacetylene, $(CH)_x$: n-type and p-type doping and compensation. Appl Phys Lett 33(1):18–20. https://doi.org/10.1063/1.90166

Chiang CK, Druy MA, Gau SC, Heeger AJ, Louis EJ, MacDiarmid AG, Park YW, Shirakawa H (1978b) Synthesis of highly conducting films of derivatives of polyacetylene,(CH)x. J Am Chem Soc 100(3):1013–1015. https://doi.org/10.1021/ja00471a081

Chung DDL (2001) Electromagnetic interference shielding effectiveness of carbon materials. Carbon 39(2):279–285. https://doi.org/10.1016/S0008-6223(00)00184-6

Colaneri NF, Schacklette LW (1992) EMI shielding measurements of conductive polymer blends. IEEE Trans Instrum Meas 41(2):291–297. https://doi.org/10.1109/19.137363

Coleman MM, Petcavich RJ (1978) Fourier transform infrared studies on the thermal degradation of polyacrylonitrile. J Polym Sci B Polym Phys 16(5):821–832. https://doi.org/10.1002/pol.1978.180160507

Cui C, Du Y, Li T, Zheng X, Wang X, Han X, Xu P (2012) Synthesis of electromagnetic functionalized Fe_3O_4 microspheres/polyaniline composites by two-step oxidative polymerization. J Phys Chem B 116(31):9523–9531. https://doi.org/10.1021/jp3024099

Dhawan SK, Singh N, Rodrigues D (2003) Electromagnetic shielding behaviour of conducting polyaniline composites. Sci Technol Adv Mater 4(2):105–113. https://doi.org/10.1016/S1468-6996(02)00053-0

Dong RX, Liu CT, Huang KC, Chiu WY, Ho KC, Lin JJ (2012) Controlling formation of silver/carbon nanotube networks for highly conductive film surface. ACS Appl Mater Interfaces 4 (3):1449–1455. https://doi.org/10.1021/am2016969

Duchet J, Legras R, Demoustier-Champagne S (1998) Chemical synthesis of polypyrrole: structure–properties relationship. Synth Met 98(2):113–122. https://doi.org/10.1016/S0379-6779(98)00180-5

Ellis JR (1986) Handbook of conducting polymers. In: Skotheim TA (ed) Commercial applications of intrinsically conducting polymers. Marcel Dekker, New York

Epstein AJ, Joo J, Kohlman RS, Du G, MacDiarmid AG, Oh EJ, Min Y, Tsukamoto J, Kaneko K, Pouget JP (1994) Inhomogeneous disorder and the modified Drude metallic state of conducting polymers. Synth Met 65(2–3):149–157. https://doi.org/10.1016/0379-6779(94)90176-7

Freund MS, Deore BA (2007) Self-doped conducting polymers. Wiley, Hoboken

Gangopadhyay R, De A, Ghosh G (2001) Polyaniline-poly (vinyl alcohol) conducting composite: material with easy processability and novel application potential. Synth Met 123(1):21–31. https://doi.org/10.1016/S0379-6779(00)00573-7

Geim AK (2009) Graphene: status and prospects. Science 324(5934):1530–1534. https://doi.org/10.1126/science.1158877

Gomes ADS (2012) New polymers for special applications. InTech Prepress, Novi Sad

Gospodinova N, Terlemezyan L (1998) Conducting polymers prepared by oxidative polymerization: polyaniline. Prog Polym Sci 23(8):1443–1484. https://doi.org/10.1016/S0079-6700(98)00008-2

Gupta A, Choudhary V (2011) Electromagnetic interference shielding behavior of poly (trimethylene terephthalate)/multi-walled carbon nanotube composites. Compos Sci Technol 71(13):1563–1568. https://doi.org/10.1016/j.compscitech.2011.06.014

Halpern O (1960) Method and means for minimizing reflection of high frequency radio waves. US Patent 2923934

Halpern O, Johnson MHJ, Wright RW (1960) Isotropic absorbing layers. US Patent 2951247

Harun MH, Saion E, Kassim A, Yahya N, Mahmud E (2007) Conjugated conducting polymers: a brief overview. Acad J 2:63–68

Hashimoto T, Spiess HW, Takenaka M (2012) Polymer characterization. In: Matyjaszewski K, Moller M (eds) Polymer science: a comprehensive reference, vol 2. Elsevier, Amsterdam

He HX, Li CZ, Tao NJ (2001) Conductance of polymer nanowires fabricated by a combined electrodeposition and mechanical break junction method. Appl Phys Lett 78(6):811–813. https://doi.org/10.1063/1.1335551

He C, Tan Y, Li Y (2003) Conducting polyaniline nanofiber networks prepared by the doping induction of camphor sulfonic acid. J Appl Polym Sci 87(9):1537–1540. https://doi.org/10.1002/app.11599

He Z, Fang Y, Wang X, Pang H (2011) Microwave absorption properties of PANI/CIP/Fe_3O_4 composites. Synth Met 161(5):420–425. https://doi.org/10.1016/j.synthmet.2010.12.020

Heeger AJ (2001a) Nobel lecture: semiconducting and metallic polymers: the fourth generation of polymeric materials. Rev Mod Phys 73(3):681. https://doi.org/10.1103/RevModPhys.73.681

Heeger AJ (2001b) Semiconducting and metallic polymers: the fourth generation of polymeric materials (Nobel lecture). Angew Chem Int Ed 40(14):2591–2611. https://doi.org/10.1002/1521-3773(20010716)40:14<2591::AID-ANIE2591>3.0.CO;2-0

Huang J, Kaner RB (2004) Nanofiber formation in the chemical polymerization of aniline: a mechanistic study. Angew Chem 116(43):5941–5945. https://doi.org/10.1002/ange.200460616

Huang WS, Humphrey BD, MacDiarmid AG (1986) Polyaniline, a novel conducting polymer. Morphology and chemistry of its oxidation and reduction in aqueous electrolytes. J Chem Soc Faraday Trans 1 82(8):2385–2400. https://doi.org/10.1039/F19868202385

Huang L, Wang Z, Wang H, Cheng X, Mitra A, Yan Y (2002) Polyaniline nanowires by electropolymerization from liquid crystalline phases. J Mater Chem 12(2):388–391. https://doi.org/10.1039/B107499G

Huo J, Wang L, Yu H (2009) Polymeric nanocomposites for electromagnetic wave absorption. J Mater Sci 44:3917–3927. https://doi.org/10.1007/s10853-009-3561-1

Jing X, Wang Y, Wu D, She L, Guo Y (2006) Polyaniline nanofibers prepared with ultrasonic irradiation. J Polym Sci A: Polym Chem 44(2):1014–1019. https://doi.org/10.1002/pola.21217

Jing L, Wang G, Duan Y, Jiang Y (2009) Synthesis and electromagnetic characteristics of the flake-shaped barium titanate powder. J Alloys Compd 475(1):862–868. https://doi.org/10.1016/j.jallcom.2008.08.038

Joo J, Oh EJ, Min G, MacDiarmid AG, Epstein AJ (1995) Evolution of the conducting state of polyaniline from localized to mesoscopic metallic to intrinsic metallic regimes. Synth Met 69(1):251–254. https://doi.org/10.1016/0379-6779(94)02438-5

Kahol PK, Pinto NJ (2004) An EPR investigation of electrospun polyaniline-polyethylene oxide blends. Synth Met 140(2):269–272. https://doi.org/10.1016/S0379-6779(03)00370-9

Karasz FE, Capistran JD, Gagnon DR, Lenz RW (1985) High molecular weight polyphenylene vinylene. Mol Cryst Liq Cryst 118(1):327–332. https://doi.org/10.1080/00268948508076234

Kosonen H, Ruokolainen J, Knaapila M, Torkkeli M, Jokela K, Serimaa R, Brinke GT, Bras W, Monkman AP, Ikkala O (2000) Nanoscale conducting cylinders based on self-organization of hydrogen-bonded polyaniline supramolecules. Macromolecules 33(23):8671–8675. https://doi.org/10.1021/ma0010783

Li Q, Zhang C, Wang Y, Li B (2009) Preparation and characterization of flake-like polypyrrole/$SrFe_{12}O_{19}$ composites with different surface active agents. Synth Met 159(19):2029–2033. https://doi.org/10.1016/j.synthmet.2009.07.025

Li G, Xie T, Yang S, Jin J, Jiang J (2012) Microwave absorption enhancement of porous carbon fibers compared with carbon nanofibers. J Phys Chem C 116(16):9196–9201. https://doi.org/10.1021/jp300050u

Lu X, Mao H, Zhang W (2009) Fabrication of core-shell Fe_3O_4/polypyrrole and hollow polypyrrole microspheres. Polym Compos 30(6):847–854. https://doi.org/10.1002/pc.20666

MacDiarmid AG (2001) "Synthetic metals": a novel role for organic polymers (Nobel lecture). Angew Chem Int Ed 40(14):2581–2590. https://doi.org/10.1002/1521-3773(20010716)40:14<2581::AID-ANIE2581>3.0.CO;2-2

Machinerieen NV (1936) Dispositif et procédépour l"amélioration de dispositifs de production et de réception d' ondesélectriques ultra-courtes. FR Patent 802728

Melvin GJH, Ni Q-Q, Suzuki Y, Natsuki T (2014a) Microwave-absorbing properties of silver nanoparticle/carbon nanotube hybrid nanocomposites. J Mater Sci 49(14):5199–5207. https://doi.org/10.1007/s10853-014-8229-9

Melvin GJH, Ni Q-Q, Natsuki T (2014b) Electromagnetic wave absorption properties of barium titanate/carbon nanotube hybrid nanocomposites. J Alloy Compd 615:84–90. https://doi.org/10.1016/j.jallcom.2014.06.191

Melvin GJH, Ni Q-Q, Natsuki T, Wang Z, Morimoto S, Fujishige M, Takeuchi K, Hashimoto Y, Endo M (2015) Ag/CNT nanocomposites and their single- and double-layer electromagnetic wave absorption properties. Synth Met 209:383–388. https://doi.org/10.1016/j.synthmet.2015.08.017

Melvin GJH, Ni Q-Q, Wang Z (2017a) Performance of barium titanate@carbon nanotube nanocomposite as an electromagnetic wave absorber. Phys Status Solidi A 214:1600541. https://doi.org/10.1002/pssa.201600541

Melvin GJH, Wang Z, Ni Q-Q, Siambun NJ, Rahman MM (2017b) Electromagnetic wave absorption properties of rice husks carbonized at 2500°C. AIP Conf Proceed 1901(1):020002. https://doi.org/10.1063/1.5010439

Moniruzzaman M, Winey KI (2006) Polymer nanocomposites containing carbon nanotubes. Macromolecules 39(16):5194–5205. https://doi.org/10.1021/ma060733p

Moon IK, Lee J, Ruoff RS, Lee H (2010) Reduced graphene oxide by chemical graphitization. Nat Commun 1:73. https://doi.org/10.1038/ncomms1067

Musikhin S, Bakueva L, Sargent EH, Shik A (2002) Luminescent properties and electronic structure of conjugated polymer-dielectric nanocrystal composites. J Appl Phys 91(10):6679–6683. https://doi.org/10.1063/1.1470239

Nalwa HS (1997) Handbook of organic conductive molecules and polymers. Wiley, New York

Ni Q-Q, Melvin GJH, Natsuki T (2015a) Double-layer electromagnetic wave absorber based on barium titanate/carbon nanotube nanocomposites. Ceram Int 41(8):9885–9892. https://doi.org/10.1016/j.ceramint.2015.04.065

Ni Q-Q, Zhu Y-F, Yu L-J, Fu Y-Q (2015b) One-dimensional carbon nanotube@barium titanate@polyaniline multiheterostructures for microwave absorbing application. Nanoscale Res Lett 10(1):174. https://doi.org/10.1186/s11671-015-0875-6

Pan H, Yin X, Xue J, Cheng L, Zhang L (2016) In-situ synthesis of hierarchically porous and polycrystalline carbon nanowires with excellent microwave absorption performance. Carbon 107:36–45. https://doi.org/10.1016/j.carbon.2016.05.045

Park SM, Lee HJ (2005) Recent advances in electrochemical studies of π-conjugated polymers. Bull Kor Chem Soc 26(5):697–706. https://doi.org/10.5012/bkcs.2005.26.5.697

Parthasarathy RV, Martin CR (1994) Synthesis of polymeric microcapsule arrays and their use for enzyme immobilization. Nature 369(6478):298–301. https://doi.org/10.1038/369298a0

Pillalamarri SK, Blum FD, Tokuhiro AT, Story JG, Bertino MF (2005) Radiolytic synthesis of polyaniline nanofibers: a new templateless pathway. Chem Mater 17(2):227–229. https://doi.org/10.1021/cm0488478

Pinto NJ, Johnson AT Jr, MacDiarmid AG, Mueller CH, Theofylaktos N, Robinson DC, Miranda FA (2003) Electrospun polyaniline/polyethylene oxide nanofiber field-effect transistor. Appl Phys Lett 83(20):4244–4246. https://doi.org/10.1063/1.1627484

Qin F, Brosseau C (2012) A review and analysis of microwave absorption in polymer composites filled with carbonaceous particles. J Appl Phys 111:061301

Qing Y, Zhou W, Luo F, Zhu D (2011) Optimization of electromagnetic matching of carbonyl iron/$BaTiO_3$ composites for microwave absorption. J Magn Magn Mater 323(5):600–606. https://doi.org/10.1016/j.jmmm.2010.10.021

Qiu H, Wan M, Matthews B, Dai L (2001) Conducting polyaniline nanotubes by template-free polymerization. Macromolecules 34(4):675–677. https://doi.org/10.1021/ma001525e

Ramesh GV, Sudheendran K, Raju KC, Sreedhar B, Radhakrishnan TP (2009) Microwave absorber based on silver nanoparticle-embedded polymer thin film. J Nanosci Nanotechnol 9 (1):261–266. https://doi.org/10.1166/jnn.2009.J041

Rozenberg BA, Tenne R (2008) Polymer-assisted fabrication of nanoparticles and nanocomposites. Prog Polym Sci 33(1):40–112. https://doi.org/10.1016/j.progpolymsci.2007.07.004

Saini P, Choudhary V, Singh BP, Mathur RB, Dhawan SK (2009) Polyaniline-MWCNT nanocomposites for microwave absorption and EMI shielding. Mater Chem Phys 113 (2):919–926. https://doi.org/10.1016/j.matchemphys.2008.08.065

Saini P, Choudhary V, Singh BP, Mathur RB, Dhawan SK (2011) Enhanced microwave absorption behavior of polyaniline-CNT/polystyrene blend in 12.4-18.0 GHz range. Synth Met 161 (15):1522–1526. https://doi.org/10.1016/j.synthmet.2011.04.033

Salisbury WW (1952) Absorbent body for electromagnetic waves. US Patent 2599944

Saroop M, Ghosh AK, Mathur GN (2003) Polyaniline based conductive polymers–an overview. Int J Plast Technol 7:41–61

Schoch KF (1994) Update on electrically conductive polymers and their applications. IEEE Electric Insulat Mag 10(3):29–32. https://doi.org/10.1109/57.285420

Seville P (2005) Review of radar absorbing materials (No. DRDC-TM-2005-003), Defense Research and Development Atlantic Dartmouth, Canada

Shirakawa H (2001) The discovery of polyacetylene film: the dawning of an era of conducting polymers (Nobel lecture). Angew Chem Int Ed 40(14):2574–2580. https://doi.org/10.1002/1521-3773(20010716)40:14<2574::AID-ANIE2574>3.0.CO;2-N

Shirakawa H, Louis EJ, MacDiarmid AG, Chiang CK, Heeger AJ (1977) Synthesis of electrically conducting organic polymers: halogen derivatives of polyacetylene, (CH)x. J Chem Soc Chem Commun 16:578–580. https://doi.org/10.1039/C39770000578

Ting TH, Jau YN, Yu RP (2012) Microwave absorbing properties of polyaniline/multi-walled carbon nanotube composites with various polyaniline contents. Appl Surf Sci 258 (7):3184–3190. https://doi.org/10.1016/j.apsusc.2011.11.061

Toshima N, Hara S (1995) Direct synthesis of conducting polymers from simple monomers. Prog Polym Sci 20(1):155–183. https://doi.org/10.1016/0079-6700(94)00029-2

Unsworth J, Lunn BA, Innis PC, Jin Z, Kaynak A, Booth NG (1992) Technical review: conducting polymer electronics. J Intel Mat Syst Str 3(3):380–395. https://doi.org/10.1177/1045389X9200300301

Virji S, Huang J, Kaner RB, Weiller BH (2004) Polyaniline nanofiber gas sensors: examination of response mechanisms. Nano Lett 4(3):491–496. https://doi.org/10.1021/nl035122e

Wan M (2008) Conducting polymers with micro or nanometer structure. Springer, Berlin

Wan M (2009) Some issues related to Polyaniline micro-/nanostructures. Macromol Rapid Commun 30(12):963–975. https://doi.org/10.1002/marc.200800817

Wang D, Caruso F (2001) Fabrication of polyaniline inverse opals via templating ordered colloidal assemblies. Adv Mater 13(5):350–354. https://doi.org/10.1002/1521-4095(200103)13:5<350::AID-ADMA350>3.0.CO;2-X

Wang X, Liu N, Yan X, Zhang W, Wei Y (2004) Alkali-guided synthesis of polyaniline hollow microspheres. Chem Lett 34(1):42–43. https://doi.org/10.1246/cl.2005.42

Wang G, Chen X, Duan Y, Liu S (2008) Electromagnetic properties of carbon black and barium titanate composite materials. J Alloys Compd 454(1):340–346. https://doi.org/10.1016/j.jallcom.2006.12.077

Wei Z, Zhang Z, Wan M (2002) Formation mechanism of self-assembled polyaniline micro/nanotubes. Langmuir 18(3):917–921. https://doi.org/10.1021/la0155799

Wu H, Wang L, Guo S, Shen Z (2012) Double-layer structural design of dielectric ordered mesoporous carbon/paraffin composites for microwave absorption. Appl Phys A Mater Sci Process 108(2):439–446. https://doi.org/10.1007/s00339-012-6906-6

Xu P, Han X, Jiang J, Wang X, Li X, Wen A (2007) Synthesis and characterization of novel coralloid polyaniline/$BaFe_{12}O_{19}$ nanocomposites. J Phys Chem C 111(34):12603–12608. https://doi.org/10.1021/jp073872x

Xu P, Han X, Wang C, Zhou D, Lv Z, Wen A, Wang X, Zhang B (2008) Synthesis of electromagnetic functionalized nickel/polypyrrole core/shell composites. J Phys Chem B 112 (34):10443–10448. https://doi.org/10.1021/jp804327k

Yamada S, Otsuki E (1997) Analysis of eddy current loss in Mn–Zn ferrites for power supplies. J Appl Phys 81(8):4791–4793. https://doi.org/10.1063/1.365465

Yang GW, Gao GY, Wang C, Xu CL, Li HL (2008) Controllable deposition of Ag nanoparticles on carbon nanotubes as a catalyst for hydrazine oxidation. Carbon 46(5):747–752. https://doi.org/10.1016/j.carbon.2008.01.026

Yang C, Du J, Peng Q, Qiao R, Chen W, Xu C, Shuai Z, Gao M (2009) Polyaniline/Fe_3O_4 nanoparticle composite: synthesis and reaction mechanism. J Phys Chem B 113 (15):5052–5058. https://doi.org/10.1021/jp811125k

Yin X, Kong L, Zhang L, Cheng L, Travitzky N, Greil P (2014) Electromagnetic properties of Si–C–N based ceramics and composites. Int Mater Rev 59(6):326–355

Yusoff AN, Abdullah MH, Ahmad SH, Jusoh SF, Mansor AA, Hamid SAA (2002) Electromagnetic and absorption properties of some microwave absorbers. J Appl Phys 92(2):876–882. https://doi.org/10.1063/1.1489092

Zhang L, Wan M (2002) Synthesis and characterization of self-assembled polyaniline nanotubes doped with D-10-camphorsulfonic acid. Nanotechnology 13(6):750. https://doi.org/10.1088/0957-4484/13/6/311

Zhang XF, Dong XL, Huang H, Lv B, Lei JP, Choi CJ (2007) Microstructure and microwave absorption properties of carbon-coated iron nanocapsules. J Phys D Appl Phys 40 (17):5383–5387. https://doi.org/10.1088/0022-3727/40/17/056

Zhang W, Li W, Wang J, Qin C, Dai L (2010) Composites of polyvinyl alcohol and carbon nanotubes decorated with silver nanoparticles. Fibers Polym 11(8):1132–1136. https://doi.org/10.1007/s12221-010-1132-3

Zhang HB, Yan Q, Zheng WG, He Z, Yu ZZ (2011) Tough graphene-polymer microcellular foams for electromagnetic interference shielding. ACS Appl Mater Interfaces 3(3):918–924. https://doi.org/10.1021/am200021v

Zhang WD, Xiao HM, Fu SY (2012) Preparation and characterization of novel polypyrrole-nanotube/polyaniline free-standing composite films via facile solvent-evaporation method. Compos Sci Technol 72(15):1812–1817. https://doi.org/10.1016/j.compscitech.2012.08.002

Zhao DL, Li X, Shen ZM (2008a) Electromagnetic and microwave absorbing properties of multi-walled carbon nanotubes filled with Ag nanowires. Mater Sci Eng B 150(2):105–110. https://doi.org/10.1016/j.mseb.2008.04.004

Zhao DL, Li X, Shen ZM (2008b) Microwave absorbing property and complex permittivity and permeability of epoxy composites containing Ni-coated and Ag filled carbon nanotubes. Compos Sci Technol 68(14):2902–2908. https://doi.org/10.1016/j.compscitech.2007.10.006

Zhou Y, Freitag M, Hone J, Staii C, Johnson AT Jr, Pinto NJ, MacDiarmid AG (2003) Fabrication and electrical characterization of polyaniline-based nanofibers with diameter below 30 nm. Appl Phys Lett 83(18):3800–3802. https://doi.org/10.1063/1.1622108

Zhu YF, Zhang L, Natsuki T, Fu YQ, Ni QQ (2012) Facile synthesis of $BaTiO_3$ nanotubes and their microwave absorption properties. ACS Appl Mater Interfaces 4(4):2101–2106. https://doi.org/10.1021/am300069x

Zhu Y-F, Fu Y-Q, Natsuki T, Ni Q-Q (2013) Fabrication and microwave absorption properties of $BaTiO_3$ nanotube/polyaniline hybrid nanomaterials. Polym Compos 34(2):265–273. https://doi.org/10.1002/pc.22409

Zhu Y-F, Ni Q-Q, Fu Y-Q (2015) One-dimensional barium titanate coated multi-walled carbon nanotube heterostructures: synthesis and electromagnetic absorption properties. RSC Adv 5(5):3748–3756. https://doi.org/10.1039/C4RA11784K

Zou Z, Xuan AG, Yan ZG, Wu YX, Li N (2010) Preparation of Fe_3O_4 particles from copper/iron ore cinder and their microwave absorption properties. Chem Eng Sci 65(1):160–164. https://doi.org/10.1016/j.ces.2009.06.003

Chapter 5
Emerging Nanotechnology for Third Generation Photovoltaic Cells

Biju Mani Rajbongshi and Anil Verma

5.1 Introduction

The Sun is ultimate source of energy, which emits radiation in the entire electromagnetic spectrum from gamma rays to radio waves. Solar energy in its original form provides heat and light for the life to be sustained on the earth. The radiant energy that is incident on the top of the atmosphere, is not fall on the ground, this energy is modified and attenuated by the atmospheric gases due to scattering and absorption processes. Part of this irradiance, about 30% is reflected to the space by the atmosphere and 20% of solar radiation is absorbed by clouds and molecules in the air on earth's surface. The remaining portion reaches the surface directly and/or after undergoing multiple scattering by air molecules and suspended particles. The absorption of radiation in the atmosphere increases as the length of the air mass increases. The path length through the atmosphere is described as Air Mass (AM), which is defined as Air Mass = 1/Cosθ, θ is the angle of Sun with overhead. When the Sun is directly overhead, air mass is one and represented as AM1 and the spectrum outside the atmosphere is AM0. Due to the earth atmospheric variables, the energy reaching the earth surfaces gets attenuated and hence the air mass changes. The terrestrial solar spectrum after atmospheric disturbance is AM1.5 (Fig. 5.1a) corresponding to the angle of incidence solar radiation 48° (Green 1981). The irradiance that reaches the earth's surface is restricted to wavelength range; 290–2500 nm and 0.1% of the solar energy reaches on the earth surface can power the need of whole world. However, there is an enormous gap between our current use of solar energy and its immense potential. It is attributed mainly to low efficiency of the harvesting process/ technology, which leads to high cost. There are

B. M. Rajbongshi (✉) · A. Verma (✉)
Sustainable Environergy Research Lab (SERL), Department of Chemical Engineering, Indian Institute of Technology Delhi, Delhi, India
e-mail: anilverma@iitd.ac.in

S. Siddiquee et al. (eds.), *Nanotechnology: Applications in Energy, Drug and Food*,
https://doi.org/10.1007/978-3-319-99602-8_5

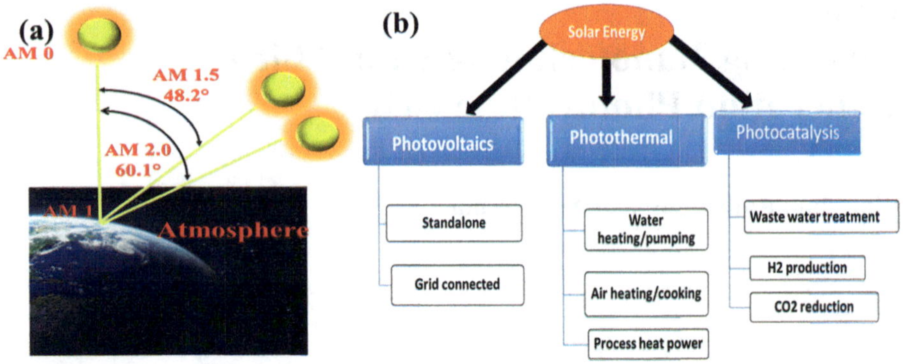

Fig. 5.1 (**a**) Solar air masses and (**b**) different process of harnessing and application of solar energy (Source: Reference and taken a copyright permission)

three different routes for harnessing solar energy efficiently: (i) solar thermal, (ii) solar photocatalysis, and (iii) solar photovoltaic (Fig. 5.1b). All the three routes are the area of current vigorous research due to a number of technological challenges in the conversion efficiency of each of the processes. To overcome these challenges, it is required to develop (i) new cost-effective and environment-friendly technologies to convert sunlight into storable and dispatchable forms, (ii) new materials to assist in efficient utilization of sunlight, and (iii) new techniques to harness the full spectrum of solar radiation.

Renewable energy is strongly demanded due to shortage of conventional fuel supplies and unstable oil prices. The concern over the environment would be markedly narrowed with a cleaner and renewable energy source. Photovoltaic technology can be the great contributor for future energy production. Photovoltaic effect was first developed by Becquerel in 1839 for selenium but modern era of photovoltaic technology started on 1954 by the development of silicon solar cell with 6% efficiency, which was further improved to 14% under terrestrial sunlight (El Chaar et al. 2011). Significant technological improvements can be seen in material, architecture and overall cost in recent years. As per material point of view and technological time scale, development of solar cells can be categorized into three generations (first, second and third) as depicted by Fig. 5.2. The majority of solar cells available in the market are single junction silicon solar cell, known as first generation solar cell characterized by high areal production costs and moderate efficiency. These technologies limit their energy conversion efficiency to 31% according to Shockley–Queisser efficiency limit. Fixed band gap of silicon wafer (1.11 eV) is the origin of photon loss and hence reduction of device efficiency. The second generation devices based on amorphous silicon (a-Si), cadmium telluride (CdTe) and Copper Indium (Gallium) Diselenide (CIS, CIGS) thin film share the same performance restrictions as conventional Si devices but promise to lower the cost compared to first generation devices. But these films suffer from higher non-radiative recombination losses due to the lower film quality and shows modest efficiency. Third generation PVs are designed to get advantages of both the first and second generation devices.

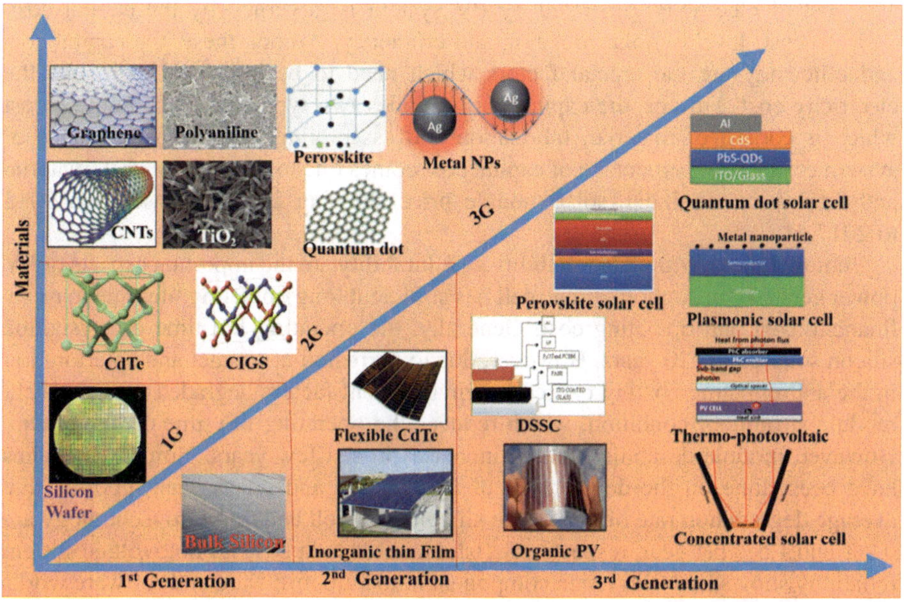

Fig. 5.2 Development of solar cell structures and materials (Source: Reference and taken a copyright permission)

Application of wide band gap materials can overcome the photon losses and enhances the efficiency above the Shockley–Queisser efficiency limit through the following four methods: multi-junction cells, intermediate-band cells, hot carrier cells and solar spectrum conversion. One of the greatest advantages of third-generation thin-film devices is that they can be fabricate using solution-based, low-temperature roll-to-roll manufacturing methods, incorporating conventional printing techniques on flexible substrates, which are economical alternative for solar cells fabrication for next few years. The lower material and fabrication cost of these third generation devices can give the promise of reducing the price per watt of PVs.

Photovoltaic system can be used independently (off grid and grid condition) as well as it can be used in integrated with other power generating system. This power is applicable to communication in earth and space, remote power, lighting, remote monitoring, water pumping etc. To make applicability of this technology feasible and wider, the main factors needs to be address are efficiency, cost, and operating life time and system reliability. The trend of photovoltaic cell and module efficiency is increased from 1993 to 2019, which can be seen in the latest report by Green (2018). So far, the highest efficiency of the photovoltaic solar cell has achieved 26.7% for single junction terrestrial silicon and 38.8% for five junctions solar cell under global AM1.5 spectrum (IEC60904–3: 2008, ASTM G-173-03 global). Similarly, the terrestrial module efficiency achieved 24.4% for crystalline silicon solar cell. Moreover, emerging technology concentrating photovoltaic solar cell has achieved 29.3% for GaAs (Green 2018).

A few years ago cost of electricity generated by PV technology was four times higher than the fossil fuel but now cost of PV system shows decreasing trend.

The cost of electricity generated by PV system is governed by the capital cost, variable cost, level of solar radiation and efficiency. Among these, the capital cost and efficiency are the crucial factors which need to be improved to reduce the electricity cost. Further, the capital cost includes two type of cost (i) module cost which is consist of material, fabrication and assembling cost and (ii) balance of system cost. Large percentage of capital cost comes due to the module cost, material and fabrication. In India, solar module price showing 29% decrease from 2015 to 2017.

Photovoltaic cell/module reliability and durability are the important aspects of PV power generation. A photovoltaic cell has sufficient long operating life time to repay financial cost and operating cost. Generally, the operating life time of crystalline silicon solar cell is 20 years. However, due to corrosion at contact and deterioration in the antireflecting coating caused by environment factors degraded the solar cell/module at outdoor condition, which reduces the operating life time of the system. However, module durability getting increases in last few years. Numerous studies have been done on the degradation of PV module and it has been reported that average degradation rate of crystalline silicon solar cell and thin film technology are 0.7% and 1.5% per year, respectively (Ishii and Masuda 2017). Photovoltaic system reliability shows how all system component hardware, firmware and software works under specified environment and condition which are designed for.

5.2 Photovoltaic Energy Conversion

Solar energy conversion consists of two main process, light absorption; where photons of solar radiation excited the electrons of absorbing material to the higher energy state and conversion; where the excited electronic energy converted to the useable form of energy like thermal, electrical and chemical energy. Photovoltaic energy conversion from solar energy is a direct conversion of electromagnetic energy in the form of current and voltage. Solar cell is the building block of solar PV system.

5.2.1 Basic of Photovoltaic Cell

Photovoltaic cell or Solar cell is a PN junction diode illuminated by solar radiation with a material of single band gap, E_g. The photons energy from the Sun with energy lower than the band gap of the material cannot give any contribution to electron generation. The photons with energy higher than the band gap energy E_g can generate the electric charge as a cell output and the rest of the energy wasted as heat. Thus, absorption and conversion of the solar energy depends on the semiconductor material properties in p-n junction. Hence, the physics of solar cell is dependent on the physics of semiconductor materials.

5.2.1.1 Energy Level

Electrons associated with isolated atoms have well defined set of discrete energy levels. When several atoms are brought together, the energy level of atoms forms into bands according to the band theory of solids. In the equilibrium state of crystal all the electrons are in the lowest energy state. At low temperature, certain energy level will be occupied by two electrons, this energy level is known as the Fermi level (E_f). At high temperature, there is finite probability of electron occupation in the state of higher energy than E_f (conduction band) and the state of energy lower than E_f (valance band) having probability of being empty. From this it can be define metal, semiconductor and insulator of solids (Fig. 5.3). In metal, E_f lies within the allowed band because of overlapping bands. In insulator, there is one band fully occupied by electron with high energy gap to the next highest energy band. At low temperature semiconductor behave like insulator and electron lies in the lower energy state. At higher temperature, electron will occupy in the next higher band.

In intrinsic semiconductor, thermal excitation can creates free electrons in the conduction band and free holes in the valance band. However, the conductivity of the semiconductor can be increase adding extra electrons and holes in the conduction band and valance band, respectively through imposing impurity to create donor and acceptor imperfection. Depending on the impurities in the host material p-type and n-type semiconductor can be formed for p–n junction. Impurities which are differed by valance of one atom from the host atom can create shallow acceptor or donor energy levels and if the impurity differ valance atom more than one atom form the host atom can create deep acceptor or donor energy level. In case of silicon the pentavalent phosphorous is commonly used as donor impurity (n-type Si) and trivalent boron is used as acceptor impurity (p-type Si) (Singh 2015).

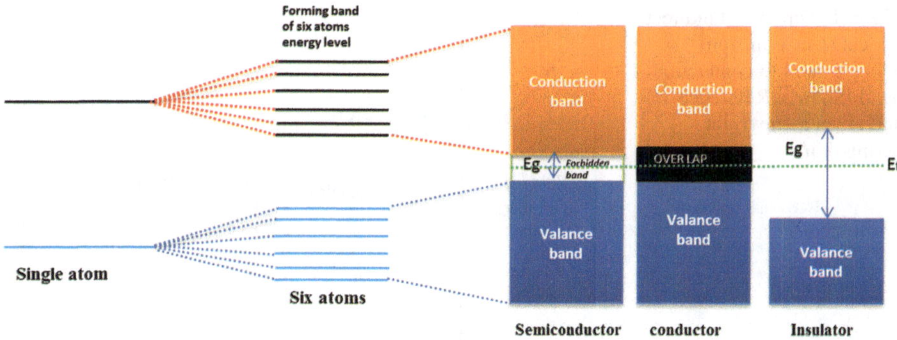

Fig. 5.3 Energy level for conductor, insulator and semiconductor (Source: Reference and taken a copyright permission)

5.2.1.2 Interaction of Light with Semiconductor and Optical Band Gap

Light is absorbed by the semiconductor material under illumination and the photon energy excites the electrons from lower energy state to higher energy state. Since, the generation of free electrons and holes is necessary for photovoltaic effect optical absorption by the material is an important factor. The magnitude of absorption is describing the absorption constant (α_λ) which is a function of wavelength and its gives the idea of how far below the surface of the cell a particular wavelength of light get absorbed. The intrinsic absorption in the semiconductor materials happens through two different process; direct and indirect absorption (Fig. 5.4). Direct and indirect absorption depends on the energy transition between the conduction and valance band. In direct transition, the minimum of conduction and maximum of valance band occurs at same wave vector with minimum energy for transition to occur but in the indirect band gap minimum of conduction band and maximum of valance band will have different wave vector with maximum energy for transition to occur. In direct band gap material absorption constant is high and it's directly related to the photon energy. Absorption constant (α_λ) is increases rapidly with the increase of photon energy. So, for the photovoltaic application to absorb the whole solar spectrum direct band gap materials need to be 1–3 μm thickness. Indirect band gap shows relatively low value of absorption constant, so large thickness (20–50 μm) of the material must be used. Many of the solar cell materials are direct band gap material. Most commonly used solar cell material Silicon and Germanium is indirect absorber materials (Fahrenbruch and Richard 2012).

In order to utilized the power from photons interaction in a semiconductor, three process required to complete, (i) the photons has to be absorbed and electrons holes must be generated, (ii) the electrons holes must be separated, (iii) the charge carrier must be transport to load before they recombine and loss their potential energy.

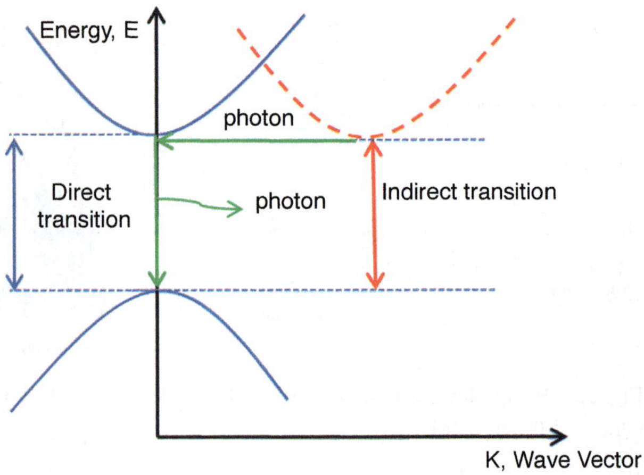

Fig. 5.4 Direct and indirect optical transition from valance band to conduction band (Source: Reference and taken a copyright permission)

5.2.1.3 Transport Characteristics

After appropriate energy of photons gets absorbed, electrons and holes are generated and these charges must be separated. Sometimes, the photoexcited electrons in the conduction band may get recombine through different recombination process. Under radiative recombination process, electrons from the excited energy state make a transition to lowest energy state emitting the energy as light. Radiative recombination occurs rapidly in direct band gap materials. In Augur recombination process, electrons recombine with the valance band holes by giving the excess energy to other electron. The second electron again relaxes back to the original state by emitting phonon. Sometimes, recombination occurs due to the traps and defect states. The impurity and defects can create some energy levels, where electrons of conduction band relax and then recombine with the holes in the valance band. Recombination can occur at surface very efficiently because surface represents severe defects in the crystal structure. Other important terms related to the charge transport are carrier life time and diffusion length (Sze and Ng 2006). Any recombination mechanism can help to define the carrier life time which is again defined as the average life time of electrons and holes before recombination.

$$\tau_e = \frac{\Delta n}{U} \quad \tau_h = \frac{\Delta p}{U} \tag{5.1}$$

where U is the net recombination rate, Δn and Δp are the disturbance of the respective carriers from their equilibrium value n_o and p_o. Other terms that are used in charge transport is the diffusion length which is defined as the average distance a carrier diffuses before recombination. Crystal imperfection due to dopant, vacancies, dislocation and surface act as recombination canters, which reduces the carrier life time and diffusion length. In typical solar cell the charge carriers are separated and transport to the load by use of P–N junction. P–N junction provides an electric field that sweeps the electrons and holes in the opposite direction (Green 1981).

5.2.1.4 P–N Junction

The basic device requirement for the photovoltaic energy conversion is to separate the charge carriers to the useful load. For this electronic asymmetry in the semiconductor structure is required and the asymmetry should such that electrons and holes should get separated from each other. Combination of p-type material and n-type material and hence formation of P-N junction can give this kind of asymmetry where p-type material have large hole density and n-type material have large electron density. This asymmetry can create built in electric field in the junction. Inherent asymmetry in P–N junction encourage electrons to flow from P-side to N-side and generated holes flow from N-side to P-side, which separated the electrons and holes, finally deliver this charges to the external circuit for the load. When a p-type semiconductor and n-type semiconductor thermally diffuses to form P–N junction, naturally, electrons and holes will flow in opposite direction to each other. However, electron leaving from the n-type semiconductor leaves in a charge imbalance and creates ionized donor. Similarly, holes will create ionized acceptor in the junction. In this way a layer of positive and negative ionized impurities is form in the P–N junction. This charge region in P–N junction is called the space charge region or depletion region. These charges will create an electric field which will prevent the natural diffusion. In this case, P–N junction will be in the equilibrium state. Without any external potential or energy this system will be in thermal equilibrium with only one Fermi level. However, the electric field in the space charge region can affect the Fermi level. Due to this the band diagram of P–N junction is not same as that of isolated material, though the individual materials are not affected. In n-type material Fermi level is closed to the conduction band and in p-type material it is closed to the valance band. This energy difference of Fermi level can create the band bending at the deletion region (Neamen 1997).

Under non equilibrium condition, P–N junction behaves differently. This non equilibrium can create through applying some external potential or solar radiation. Biasing in the P–N junction can show the scenario under non equilibrium condition. P–N junction at forward bias condition, the Fermi level will be disturbed at the junction and it will be same as equilibrium away from the junction. At this condition, Fermi levels are known as quasi Fermi level at P and N-side. Due to forward bias, electron diffusion current increases exponentially while minority electrons drift current remains unaffected. Similar effects can be seen for the holes. At forward bias condition, diffusion current is larger than drift current, so the net current will flow from p-side to n-side (at the direction of diffusion of electrons and holes). At revers bias condition, the potential energy barrier height between p and n-side increases and it prohibits the diffusion of charges, but drift current will be remain unchanged. Since, the diffusion current is negligible; net current will be the drift current (Singh 2015).

5.2.1.5 P–N Junction Under Illumination of Light: Solar Cell

Under illumination electron hole pairs will be generated at the space charge region and quasi neutral region. The electron hole pairs immediately move due to the electric field, electron to the N-side and hole to the P-side. In quasi neutral region also electron holes are generated and in their random motion some of the minority carrier will come near to the space charge region. In this way minority electrons will go from N-side to the P-side leaving behind positive charge carrier and similarly for the minority holes will come from N-side to P-side leaving negative charge (Fig. 5.5a). Such, there will be net increase in the positive and negative charge in P-side and N-side, respectively. This builds up a potential difference in the P–N junction due to interaction of light which called photovoltaic effect. This generated photovoltage biases the junction in forward bias mode and the light generated drift current (J_L) will flow from N-side to P-side. However, diffusion current will flow in

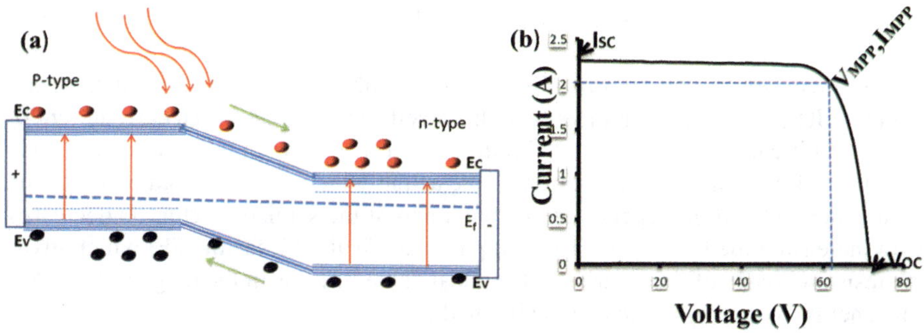

Fig. 5.5 (**a**) P–N junction under illumination and (**b**) I–V characteristics of solar cell (Source: Reference and taken a copyright permission)

opposite direction to the drift current. Since, light generated current is higher than the diffusion current so, net current will flow from N-side to P-side. This current is opposite in the generated photo voltage and hence the diode I–V curve (Fig. 5.5b) shifted to the fourth quadrant with negative power which implies that the power is extracted from the device (Singh 2015).

The relationship between the current and voltage in an ideal p–n junction is given by

$$\mathrm{J_d} = \mathrm{J_o}\left[\exp\frac{eV}{Kt} - 1\right] \tag{5.2}$$

where, V is the applied voltage across the junction; e, the electric charge; k, Boltzmann constant; T, absolute temperature; J_o, is the saturation current. Under illumination, the net current is the difference between the light generated current and diode diffusion current.

$$\begin{aligned} \mathrm{J} &= \mathrm{J_L} - \mathrm{J_d} \\ &= \mathrm{J_L} - \mathrm{J_o}\left[\exp\frac{eV}{Kt} - 1\right] \\ &= \mathrm{J_L} - \mathrm{J_o}\{\exp[\mathrm{e(V + JRs)/kt}] - 1\} \end{aligned} \tag{5.3}$$

R_S, is the internal series resistance (Green 1981).

5.2.2 *Solar Cell Performance Parameters*

Solar cell performances are compared with four performance parameters; open circuit voltage, short circuit current, fill factor and efficiency. It can be see that the open circuit voltage V_{oc} for ideal cell is

$$V_{oc} = (KT/e)\ln\left[(J_L/J_d) + 1\right] \quad (5.4)$$

So, open circuit voltage linearly decreases with increasing temperature. Open circuit voltage is the maximum voltage that a cell can generate, when current is zero. Short circuit current (I_{sc}) is the maximum current that flows in the solar cell at open circuit condition. Short circuit current is the actually the light generated current. The light generated current depends on the band gap of the semiconductor, its decreases with increasing the band gap. The third parameter is the fill factor (FF) which gives the resistive losses of the solar cell. The ideal value of FF is in the range 0.7 to 0.85. The energy conversion efficiency is defined as

$$\eta = \frac{V_{OC} I_{SC} FF}{P_{in}} \quad (5.5)$$

short circuit current decreases with increasing the band gap and open circuit voltage increases with the increase of band gap. So, there is an optimum band gap to get the highest absorption (Singh 2015, Fonash 2012).

5.3 Nanotechnology in Photovoltaic Cell

Nanotechnology or nanoscaled technology is the combination of two words nano and technology where, nano refers to the size of 0.1 μm to100 nm. Nanotechnology studies the properties of material at atomic and molecular scale. Nanotechnology applied in the different device by modulating the size, shape, properties and functionality. Application of nanotechnology in the device depends on different material properties at nanoscale.

1. ***Surface properties***: At nanoscale, their physical and chemical property of the materials depends on its surface properties. In nanomaterial a larger fraction of the atoms is at the surface which influences physical properties. Atoms and molecules in nanomaterial are unstable, they have high surface energy. Not only the size of a nanomaterial is important, but also its shape. Extension of the surface area depends on the shape of the material. So, different structured materials are used in the devices to increase the surface area specially those device where the reaction takes place on the surface (Hahn 2003).
2. ***Electric properties***: Some nanomaterials exhibit electrical properties that are absolutely exceptional. Their electrical properties are related to their unique structure. Two of these are fullerenes and carbon nanotubes. Carbon nanotubes can behave as conductors or semiconductors depending on their nanostructure (Cao 2004).
3. ***Optical properties***: Optical absorption and emission properties of a nano-sized semiconductor can be tuned by controlling its crystallite size. Nanomaterial generally shows peculiar optical properties with their different nanostructure.

Some nanomaterials display very different optical properties, such as colour and transparency, compared to bulk materials. One of the effects is localised surface plasmon resonance (LSPR) metallic nanostructure. One of the consequences of the LSPR effect, metal nanoparticles shows very strong visible absorption due to the resonant coherent oscillation of the plasmons. LSPR energy is sensitive to the shape and size of the nanoparticle. As a result, colloids of metal nanoparticles such as gold or silver can display colours such as red, purple or orange depending on the shape, size and surrounding media of the nanoparticles. Tuning the size of the semiconductor nanocrystal, band gap can be tuned and, therefore, the wavelength absorbed/emitted by the crystal. As a result, the same material emits different colours depending on its size (Flory et al. 2011). Reflectance can be reduced using nanostructured like moth eye which reduces the reflectance from the silicon surface (Tsakalakos 2010).

4. ***Magnetic properties***: Nanostructuring of bulk magnetic materials can be used to design the soft or hard magnets with improved properties. In general, the magnetic behaviour of a material depends on the structure of the material and on its temperature. When the size of these domains reaches the nanoscale, these materials show new properties due to quantum confinement, for example the giant magnetoresistance effect (GMR). This is a fundamental nano-effect which is now being used in modern data storage devices (Pelecky and Diandra 1996).

Nanostructured materials are generally, nanoporous, nanocrystalline, nanocomposite and hybrid materials. Nanoporous materials have nano-sized pores; a nanocrystalline material consists of many nano-sized crystalline domains; a nanocomposite material contains two or more phase-separated components, and hybrid materials are made of a combination of organic and inorganic components interconnected at a molecular level. Nanotechnology can help in solving different problems of human in the domain of energy, climate change or fatal diseases. In renewable energy technology, nanotechnology is getting tremendous attention in different application (Luisa and Sutherland 2013). In photovoltaics technology, nanotechnologies offer the possibility to apply alternative materials and fabrication methods to produce cells with tailored absorption properties in order to absorb a larger portion of the solar energy spectrum. Numerous types of nanomaterials are being investigated for the application in dye sensitized, Quantum dot sensitized, plasmonic and organic solar cell to increase the efficiency and trying to overcome the energy challenge. In solar thermal energy conversion, nanotechnologies can be used to fabricate complex nano-structured mirrors and lenses to optimise solar thermal collection. Furthermore, nanoporous aerogels are used as transparent and thermally isolating materials for the cover material of solar collectors. Nanotechnology is also getting attention in hydrogen production through photocatalysis. Photocatalysis involves harvesting solar photons in a semiconductor surface and subsequent conversion of these photons to electronic energy which induces the desired water-splitting reaction. Photoexcited electron and hole has a high tendency to recombine. Nanostructures help to minimize the recombination by increasing the electron life time over which charges have to survive and be transported to the surface. For hydrogen storage, material should also have outstanding insulating and pressurisation properties in order to avoid hydrogen leakage. This problem can potentially be solved using nanotechnology to develop new materials with exceptional properties in terms of strength and density. Nanotechnology can offer some indirect energy saving solutions by developing materials with better properties.

Although existence of solar cell crosses over 50 years but still it cannot compete the conventional energy sources due to the hindering factor, efficiency and cost. Low efficiency of the solar cell is because of different losses due to the material properties and technological losses. Different types of losses in the performance parameters which reduces the efficiency are

1. *Short circuit current loss (I_{sc})*: Silicon is quite reflective which causes most of the incident light reflected and reduces the short circuit current. Cell thickness is another factor to reduce the short circuit current. If the thickness of the cell is not optimized light will be pass through the material and energy will be lost. Another source of I_{sc} loss is due to the recombination of the electrons and holes in the bulk semiconductor and interface. Short circuit current is dependent on the spectral mismatch losses. High energy photon than the band gap of the photovoltaic materials will be loss giving heat to the cell and low energy photon gives transmission losses.
2. *Open circuit voltage loss*: *(V_{oc})* Main losses in open circuit voltage are due to the recombination (augur recombination), lower the recombination higher will be V_{oc}. Additionally recombination at depletion region also decreases V_{oc}.
3. *Fill factor loss*: Series and shunt resistance reduces the fill factor. Depletion region recombination also can reduce the fill factor.

These types of losses in the performance parameters are the fundamental losses in the solar cell and nanotechnology can gives the promise to enhance the efficiency and reduces the cost (Singh 2015; Wolf 1971).

5.3.1 Nanotechnology in Improving the Efficiency

Nanotechnology can helps to overcome current performance barriers and substantially improves the conversion of solar energy into electricity. Incorporation of nanotechnology can reduces all the fundamental losses (Fig. 5.6).

1. ***Spectral mismatch***: Spectral mismatches in solar cell can be improved using up and down conversion nanomaterials of solar spectrum. Different nanomaterials have been investigated for up and down converter of solar spectrum. In an up conversion process, photons with energy lower than the band gap of the solar cell material are converted to higher energy photons and in down conversion, photons with higher energy converted to lower energy photons. Some nanoparticles can emit ultraviolet/visible/near-infrared light under near-infrared excitation and this unique property is used for the up conversion. There are three different process of

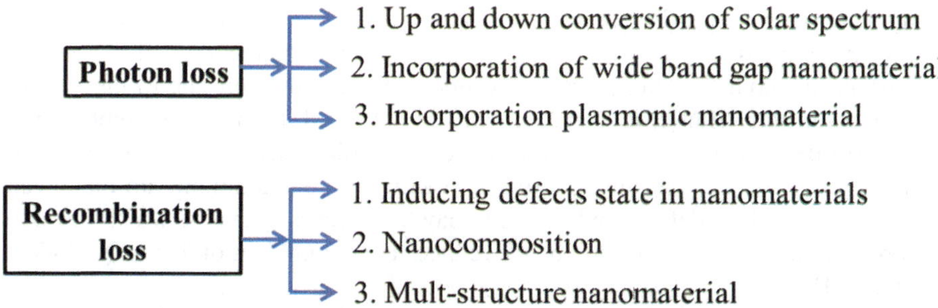

Fig. 5.6 Different Nanotechnology to enhance photon harvesting and reduce recombination losses (Source: Reference and taken a copyright permission)

up conversion; use of rare-earth-doped micro- and nano-crystals, which can work above 800 nm, triplet–triplet annihilation up-conversion, where the triplet states of two organic molecules interact with each other, exciting one molecule to its emitting state to produce fluorescence, and up-conversion in quantum nanostructures, where two quantum dots with different band gaps separated by a barrier (Shang et al. 2015). In down conversion, a single high energy photon absorbed by the down converting material results in the emission of two visible photons and hence increases the quantum efficiency. The lanthanide or rare-earth elements are generally used for up and down converter of solar spectrum because of its unique property and rich energy level structures of these ions, which allows for efficient spectral conversion, mediated by resonant energy transfer between neighbouring lanthanide ions. Most of the up and down converter nanomaterial have been investigated in different types of solar cell in recent years (Van Sark et al. 2012; Richards et al. 2012; De Wild et al. 2011).

In conventional silicon solar cell, the useful wavelength of light is above 700 nm due to its band gap. Photon harvesting can be improves using semiconductor with wide band gap so that its band gap be tuned according to the solar spectrum. In third generation solar cell, different types of semiconductor are used with tuned band gap. There different processes of tunning the band gap are doping, co-doping, and compositization. These types of material are used in dye sensitized solar cell. Plasmonic nanomaterial can also reduce the photon loss in solar cell. Due to the surface Plasmon resonance, spectral absorption spectra can be increase (Piyadasa et al. 2017).

2. ***Recombination losses***: Nanotechnology plays an important role in the reduction of recombination losses. Different processes reported for this are shown in Fig. 5.6. Incorporation of defects in the material can create a mid-band gap in between the host material. This defects level can act as electron collecting centre which reduces the immediate recombination of electrons with holes. Defects are generally created though doping and vacancies inside the material. Nanocompostion of material can help in electron transport after excitation in third generation solar cell. Two nanomaterials with different energy levels can create modification of energy levels which helps in the reduction of recombination (Wang et al. 2017). Use of 1D nanomaterials show advantages for the

application in solar cells because they can trap photons more effectively with the proper geometrical configuration for the exciton generation and serves as electron path in the axial direction, which favours electron collection due to the shorter collection time (Han et al. 2012). In silicon solar cell, the major recombination occurs due to the rear surface and wafer edges. Surface passivation by thin film of variety of material like silicon dioxide, silicon nitride and aluminium oxide can reduce the surface defects and this film may be electrically charge which helps in repealing the electron hole and hence reduces the recombination losses (Reinders et al. 2017).

3. ***Reflection losses***: To increase the efficiency and reduce the reflection losses various anti-reflecting coating are investigated in solar cells. Nano and micro-structures with broadband light-trapping capability can significantly reduce the front surface reflection. Surface texturing on the silicon surface can reduce the reflection losses. Recently plasma texturing (nanotexturing) on the silicon surface getting attention. Nanostructured metallic back reflectors are used to reduce the reflection losses (Reinders et al. 2017). In CdTe based solar cell, silica and zirconia are used as antireflecting coating (Womack et al. 2017). Metal oxides TiO_2, ZnO are used as antireflecting coating in silicon solar cell (Shi et al. 2012; Lee et al. 2008). Different nanotechnology have been reported like MgF_2 nano-particle (Reddy and Sekhar 2018) and its composite with MgO (Horst et al. 2018), biomimetic nanostructures ZnS, ZnSe (Chan et al. 2017) graphene oxide (Yang 2017), carbon nanotube (Assadi and Hanaei 2017) showing better performance in different solar cell structure.

5.3.2 Nanotechnology in Reducing the Cost

The most promising application of nanotechnology is the reduction of cost. Facile synthesis of nanomaterial can reduce the processing cost compared to the first and second generation solar cell. Silicon solar cell needs 1000 times more light absorbing material than third generation solar cell. Another approach to reduce the cost is to minimize the use of active material compared to entire wafer required for the bulk silicon-based first generation PVs. Developments in nano-tech solar cells via nanotubes, quantum dots, and hot carriers can reduce the cost of PV cells. Plastic solar cell technology can reduce the cost because it uses tiny nanorods dispersed within in a polymer. This type of cell is cheaper to manufacture than conventional ones for two main reasons: first, these plastic cells are not made from silicon, which is expensive and second, manufacturing of these cells does not require expensive equipment instead it can be synthesized in a beaker not like conventional silicon based solar cells. Thin film is a cost-effective solution because it uses a cheap support onto which the active component is applied as a thin coating. As a result less material is required and hence reduces the cost (Jacoby 2016).

5.3.3 Nanotechnology in Improving Stability

Device stability is obviously an important consideration for photovoltaic technologies as the longer the photovoltaic system operates, the lower is the total cost. In conventional silicon soar cell, encapsulation is used to protect the solar cells form moisture which provides optical and electrical transmissivity and hence increases reliability. Different encapsulating materials that have been used in PV module are ethylene vinyl acetate, silicones, thermoplastic silicone elastomer, polyvinyl butyral and ionomers, more specially ethylene ionomer (Peike et al. 2013).

In dye sensitized solar cell, instability issue comes from the liquid electrolyte. To avoid these many solid and quasi solid state materials like p-type semiconductor (CuI, CuBr, or CuSCN) or organic hole transporting materials (polyaniline, polypyrol, polythyophene), polymer electrolyte (PEDOT, Spiro-OMeTAD, P_3HT) are used. However, their conversion efficiencies are not comparable with those of the liquid solar cells. In quasi-solid-state DSSC, ionic liquid electrolytes (MPII, NMBI, GuSCN, NMBI, NMBI, BMIPF6) showing better choice to increase stability due to their unique physicochemical properties such as high thermal stability, negligible vapour pressure, relatively high ionic conductivity, and good stability. Different hole transporting materials have been developed to fabricate solid state dye sensitized solar cell with enhanced stability (Zhang et al. 2018a). For long-term stabilities of CoS, FeS_2, TiS_2/PEDOT:PPS, multiwall CNTs (MWCNT), and reduced graphene oxide (RGO) has been reported for counter electrode stability (Sining et al. 2015; Krebs and Spanggaard 2005; Bin et al. 2006). It has been reported that improving nanoscale morphology and increasing crystallinity of the semiconducting polymer can improve the stability of polymer soar cell (Dong et al. 2008).

The most investigated lead halide perovskite is $CH_3NH_3PbI(MAPbI_3)$ with tolerance factor 0.91. Replacing the MA organic cation with a smaller size (MA: 2.17 Å) by HC (FA: 2.53 Å) can push the tolerance factor of 0.99, which could enhance the thermal stability. In $FAPbI_3$, doping could leads to the enhancement of the thermal stability. Reducing the size of the X site could also enlarge the tolerance factor and enhance the device stability. The smaller size of Br (1.96 Å) and Cl (1.81 Å) compared with I (2.2 Å) could be a good candidate for replacing or doping I site. In n–i–p structure, Spiro-OMeTAD has been used as the hole transport layer, which is very sensitive to water and reduces device stability. Several materials that been used to replace Li-salt doped Spiro-OMeTAD are CuSCN and Cu. The hole transport layer, such as PEDOT: PSSS is generally used and this layer can easily absorb water. It also has acidic properties, which leads to the instability of the devices. The most successful results are based on using ZnO and NiOx as the hole transport layer to improve the device stability. Recently, incorporation of buffer layer of Al_2O_3 has been reported for long term stability (Guarnera et al. 2015; Qin et al. 2017; You et al. 2016). Engineering the band alignment of the quantum dot layers in quantum dot solar cell through the use of different ligand treatments can improve the stability.

5.4 Recent Progress in Nanotechnologies for Photovoltaic Cells

Nanotechnology helps us to make solar energy more economical. Nanotechnology or sometimes referred as "third generation PV" is used in order to help increase conversion efficiency of solar cell since energy band-gap can be controlled by nanoscale components. Most of the solar cell technologies presently on the market are conventional silicon solar cell i.e., the first generation solar cell. However, material cost still effecting on this technology. Second generation solar cell, which is thin film based somehow reduces the material cost. Evolution of new materials and technology prospects for thin film cells based on new concepts will be appear to be quite good. Engineered nanomaterials like introduction of one or more energy levels within the bandgap (intermediate band solar cell) and multiple exciton generation (MEG), antenna to rectify the submicron wavelength (antenna–rectifier solar cells) are the focusing area of new and improved nanotechnology of current generation solar cells. These three advanced nanotechnology can boost the solar cell devices efficiency and at the same time minimizes the cost.

1. ***Intermediate band (IB) solar cell***
   The main physics behind intermediate band solar cell is to create one or more energy levels within the band gap of the PV materials, so that they can absorb photons as like a single band gap through the transition from the valance band to intermediate band (Fig. 5.7a). In IB process, transitions takes place from valance bands to the intermediate band and finally from the intermediate band to the conduction band. By this process the electron density on the conduction band increases. The theoretical efficiency limit of Intermediate band solar cell is 63% whereas, for the single band solar cell is 41% (Mlinar 2013).

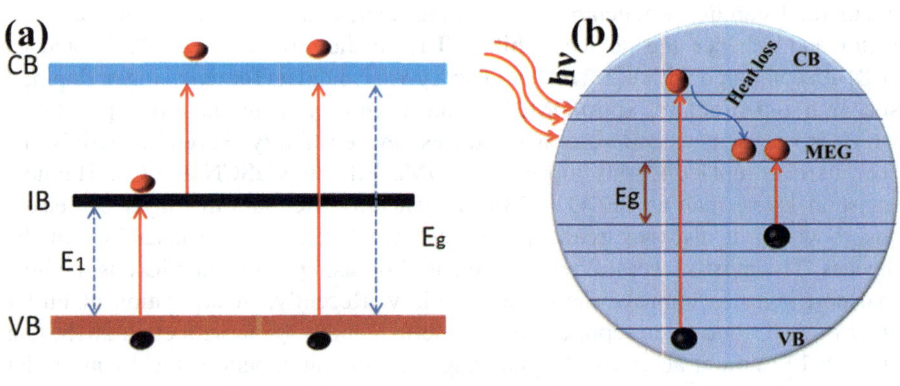

Fig. 5.7 Band diagram of (**a**) Intermediate band solar cell and (**b**) Multiple Exciton Generation (Source: Reference and taken a copyright permission)

2. ***Multiple exciton generation***
   According to the Shockley and Queisser limit for maximum thermodynamic efficiency of solar cell, one photon can generate one electron-hole pair (exciton). Multiple exciton generation exceeds this limit and it's basic idea to generate more than one exciton from a single photon. Through the impact of ionization, high energy photon (around twice the band gap) can generate more than one exciton close to the band gap energy. The photon with higher energy absorbed by the materials creates a hot exciton with excess energy. This excess energy prior to cooling converted to kinetic energy which further generates an exciton (Fig. 5.7b). MEG not only improves the solar cell efficiency but also reduces thermal loss in the device. MEG is possible in bulk material like Si, bS, PbSe, Ge, InSb, but it is very effective in semiconducting nano crystal (Beard and Randy 2008).

3. ***Antenna–rectifier solar cell***
   Another new technology is the antenna–rectifier system where an antenna and rectifier used to boost the solar cell efficiency. Here, the wave nature of light is used by the optical nano-antenna and rectify it which is applicable to the longer wavelength to the radio and microwave. Metal and semiconductor are used as antenna, however, recently semiconductor antenna has getting more attention because of light absorption. In antenna rectifier solar cell, the theoretical efficiency is 100%, because semiconductor band gap is not playing any role. Besides this, this system can operate whole day and the fabrication cost is minimum compared to conventional solar cell (Garret and Grover 2013; Sharma et al. 2015).

   In recent years, research and development has included many improvements in nanotechnology in the different types of solar cell which can be discussed below.

5.4.1 Organic and Polymer Solar Cells

Organic solar cell (OPV) is a new technology and still it is in developing stage. Due to the form flexibility and low processing temperature, researcher focusing on the improvement of this technology (Fig. 5.8b) and the highest efficiency of the organic solar cell has received 11.2% (Green 2018). Organic solar cells consist of two electrodes, one of the electrodes is optically transparent and the active organic layer(s) is sandwiched in between. Indium tin oxide (ITO) is frequently used as anode material due to its optical transparency and electrical conductivity. Low work function metals, for example, Al, Mg, Ag, Ca, and their alloys are used as cathode. The work function difference between the two electrodes generates an internal potential known as built in potential in the cell which gives driving force and helps for the collection of the charge carriers at the respective electrodes. These active layers materials may be molecules, oligomer and polymer. Active layer

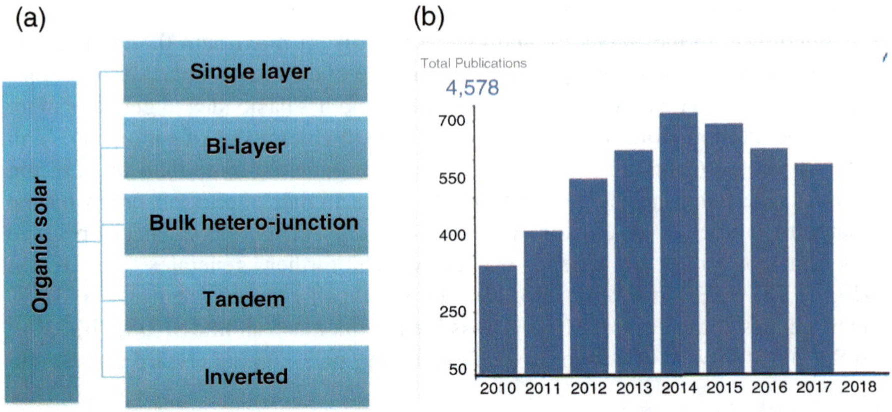

Fig. 5.8 (**a**) Different structure of organic solar cells and (**b**) published paper from 2010 to 2018 on organic bulk heterojunction solar cell (Source: Reference and taken a copyright permission or include weblink and accessed specific date and year)

consist of a donor to generate electron and an acceptor layer which accept the electron from donor layer. Depending on the architecture organic solar cell can be divided in five different structure of organic solar cell (Fig. 5.8a).

5.4.1.1 Materials for Donor

The commercialized organic solar cell was first discovered by Tang with Copper phthalocyanine (CuPc) as the donor layer and 3,4,9,10-perylenetetracarboxylic-bis-benzimidazole (PTCBI) as the acceptor layer with efficiency 1%. Due to their outstanding light harvesting properties, Phthalocyanines and their homologues are most studies in organic solar cell as donor material. Subphthalocyanines are found to be superior to other phthalocyanines based due to better energy level alignment and large open circuit voltage. Fabrication of polymer based organic solar cell makes the system flexible and cost effective. Many polymer based material are used as active material among them P_3HT widely used. It is conducting polymer and it is the excitation of the π-orbital electron in P_3HT that gives the photovoltaic effect in the blend. Few of the other efficient polymer based electron donor materials are 2,1,3-benzothiadiazole (BT), pyrrolo[3,4-c]pyrrole-1,4-dione (DPP) derivatives and benzo [1,2-b;4,5-b0]dithiophene-based polymers (BDT) (Benanti and Venkataraman 2006; Heremans et al. 2009).

5.4.1.2 Materials for Acceptor

Fullerenes and their derivatives are well known and most successful electron acceptor materials. Fullerene (C_{60}) exhibits interesting chemical and physical properties. Another most promising material is $PC_{60}BM$ and its corresponding C_{70} derivative

($PC_{70}BM$). These materials are extensively used as acceptors in OPV (Hoppe and Sariciftci 2004). Besides this Indene–fullerene are used as electron acceptor materials in OPVs. Recently, use of non-fullerene small molecules have emerged as promising electron acceptors with power conversion efficiency 12% (Hou et al. 2018).

5.4.1.3 Electrode Materials

A low sheet resistance materials (<10 Ω/sq) is used to transport the photo-generated carriers with little losses. Many metals are generally used for the non-transparent back contact. The materials for the transparent front electrodes are much more limited. Indium tin oxide (ITO) is the most prominent transparent electrode for OPV devices. Recently, ZnO doped with group III elements, i.e. B, Ga and Al are used as alternative transparent electrode materials. Other material to replace ITO are carbon nanotubes, graphene, highly conductive polymers, metal grids, metal nano-meshes, optically thin metal layers alone or thin metals in combination with other metal oxides (Kaur et al. 2014; Abdulrazzaq et al. 2013; Li et al. 2012).

Beside this active layer and electrode, many new technologies and materials have been developed to enhance the efficiency. Optical spacer is new addition to the organic solar cell. The optical spacer effect is a concept to optimize the absorption within the active layer. A transparent layer is inserted between the reflecting electrode and the active layer. Depending on the thickness and refractive index of the transparent optical spacer layer, the maximum absorption can be shifted. TiO_x and ZnO_x has been used as optical spacer P_3HT:PCBM solar cell (Kim et al. 2006; Kyaw et al. 2013). An interfacial layer has been included in organic solar cell to protect the organic material from physical and chemical interaction with metal electrode. Some metals like Al, Mg/Ag, Ca/Al, Ca/Ag, Ba/Al, Au are use as interfacial layer. PEDOT:PSS is most commonly used interfacial layer in organic solar cell. p-type interfacial layers such as sulfonated poly(diphenylamine), polyaniline (PANI) and polyaniline–poly(styrene sulfonate) (PANI–PSS) has also been used to improve the device efficiency. The main challenges in the organic solar cell are poor light absorption, charge transport and instability. Therefore, to enhance the efficiency, the possible solution is to design, synthesis, and application of new materials with suitable band gaps, appropriate HOMO and LUMO energy levels, and high charge carrier motilities (Steim et al. 2010). Some of efficient organic solar cell recently reported and its device components are given in the following Table. 5.1.

Table 5.1 Some reported results on organic solar cell

Devices	Efficiency	Ref.
ITO/PEDOT:PSS/BTTR:$PC_{71}BM$/(PFN)/Al	8%	Ye et al. (2018)
ITO/(PFN)/H-TiO_2/PTB7-Th:$PC_{71}BM$/MoO_3/Ag	9.12%	Nam et al. (2018)
ITO/PEDOT:PSS/PBDB-T:F–F/F–Cl/F–B/PDINO/Al	9.59%	Wang et al. (2018c)
ITO/ZnO/PFN/PTZ1:ITIC/MoO_3/Al	9.20%	Xiao et al. (2018)
ITO/ZnO/PDB-T-SF:IT-4F/MoO_3/Al	13.1%	Zhao et al. (2017)

5.4.2 *Perovskite Solar Cells*

Perovskite solar cells (PSC) have attracted attention and been considered to open up a new and promising avenue for next generation photovoltaic development because of their simple structure, low production cost and superb photovoltaic performance since 2009. In following years, researcher are focusing on the fabrication of Perovskite solar cells which can be seen in the number of publications (Fig. 5.9c) and this cell shows maximum efficiency 22.1% in 2016 (Fig. 5.9b) (Lim et al. 2018). A PSC typically consist of a transparent electrode, usually conductive oxide, charge transport layers, such as electron transport layer (ETL) and hole transport layer (HTL), an absorber layer which is organometallic halide perovskite sandwiched between the charge transport layers, and a counter electrode which is typically metal. Organolead halogen 3D Perovskites refer to a kind of semiconductors with (ABX_3) structures where A is a monovalent cation (organic or Cs^+), B is a divalent metal (Cu^{2+}, Ni^{2+}, Co^{2+}, Fe^{2+}, Mn^{2+}, Cr^{2+}, Pd^{2+}, Cd^{2+}, Ge^{2+}, Sn^{2+}, Pb^{2+}, Eu^{2+}). The electronic properties of the perovskite, such as band gap and absorbing range, could be easily tailored. In addition to 3D perovskites, other perovskite structures such as layered 2D perovskites are also used. So far, $CH_3NH_3PBX_3$ (x = cl, Br, I) has coming as widely used perovskite materials, for PV applications. They exhibit: (i) higher diffusion lengths of electron and hole, (ii) lower surface recombination rate and favourable grain boundary effects, (iii) high optical absorptivity, (iv) lower structural defect tolerance, (v) electrically clear defect, comprising point defects and grain boundaries. So far, the studied materials for perovskite solar cells are $CH_3NH_3PbI_3$, $CH_3NH_3PbI_{3-x}Cl_x$,$CH_3NH_3PbBr_3$, $CH_3NH_3Pb(I_{1-x}Brx)_3$, $HC(NH_2)_2PbI_3$, HC $(NH_2)_2Pb(I_{1-x}Brx)_3$ and $CH_3NH_3SnI_3$. Two typical structures can be constructed; mesoscopic nanostructure and planar structure based upon its properties (Zhang et al. 2016; Djurišić et al. 2017; Ansari Haque et al. 2018). In the mesoscopic nanostructure, charge transport improves by employing nanomaterials, such as ZnO nanorods, 3D-TiO_2 nanoparticles/ITO nanowire composites, and TiO_2 nanoparticle/graphene composites (Mali et al. 2015; Son et al. 2014). The interface engineering of mesoscopic materials can control charge transport and recombination. Besides, these structures, hole transporting material (HTM) free structure are

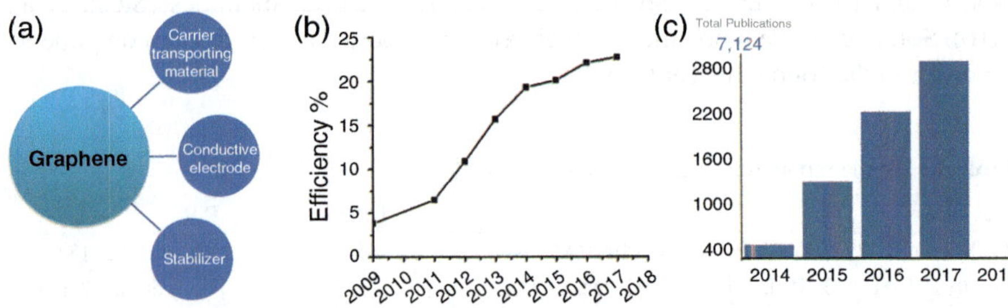

Fig. 5.9 (**a**) Application of graphene in PSC (**b**) Efficiency trend from 2009 to 2017 and (**c**) Published paper on perovskite solar cell (Source: Reference and taken a copyright permission or include weblink and accessed specific date and year)

Table 5.2 Some of the recently reported results on perovskite solar cell

Devices	Efficiency	Ref.
FTO/bl-TiO_2/$MAPbI_3$/spiro-OMeTAD/Au	17.2%	Huang et al. (2017)
FTO/TiO_2/ZrO_2/AB-$MAPbI_3$/graphite/carbon	15.6%	Hu et al. (2018)
(FTO)/ZnO–MgO–EA+/$MAPbI_3$/SpiroOMeTAD/Au	18.3%	Cao et al. (2018)
FTO/cp-TiO_2/CsBr-doped mp-TiO_2/Rb/Cs/$FA_{0.85}MA_{0.15}$ /spiro-OMeTAD/Au	21%	Seo et al. (2018)
FTO/TiO_2-$CuInS_2$-NAs/$CH_3NH_3PbI_3$/Spiro-MeOTAD/Au	11.7%	Gao et al. (2018)

also reported because perovskite possesses hole transporting properties. The first demonstration of the HTM-free structure with an FTO/2D sheet TiO_2/$CH_3NH_3PbI_3$/Au configuration showed a PCE of 5.5% (Zhou et al. 2015) and further it was improves to 8% using thinner TiO_2 nanoparticle film instead of TiO_2 nano-sheet.

Recently, Carbon based materials have been showing promising material for the application in perovskite solar cell due to its unique muliti-functional behaviour like hole conducting material (Habisreutinger et al. 2014) and counter electrode (Zhang et al. 2018b). Graphene has been successfully used as electrode, charge collecting material as well as stabilizing material in perovskite solar cell (Fig. 5.9a). Zhu et al. has reported the enhancement of power conversion efficiency from 8.81% to 10.15% due to incorporation of graphene quantum dot as electron extracting layer of PSC (Zhu et al. 2014). Some of the recent improvement in the device structure and the efficiency is given in the Table 5.2.

5.4.3 Quantum Dot Solar Cell

Quantum dots (QDs) are three-dimensionally confined semiconductor nanocrystals which physico-chemical interaction of the ligands interacts with different surface species. Quantum rods (QRs) have two-dimensional confinement and similarly one-dimensional confined nanocrystals are often called quantum plates, sheets, or wells.

After first proposal by Nozik in Nozik 2002, on the application of QDs in solar cell (Nozik 2002), this technology is getting more attention (Fig. 5.10b). Because of its unique properties like, (i) the energy gaps of the QDs and its light absorption spectra can be tuned by tuning the size of QDs, (ii) QDs have large absorption coefficients resulting from the quantum confinement effect, (iii) QDs have the potential to multiple exciton generation (MEG) with one single photon absorption, QD's are used in different solar cell (Fig. 5.10a). Different QDs materials that have been used in solar cell such as PbS, PbSe, PbTe, CdSe, InAs, Si, carbon nanotube to get advantages like (1) to sensitize wide-band gap semiconductors; (2) intimate contact with electron and/or hole conducting polymers; and (3) increases

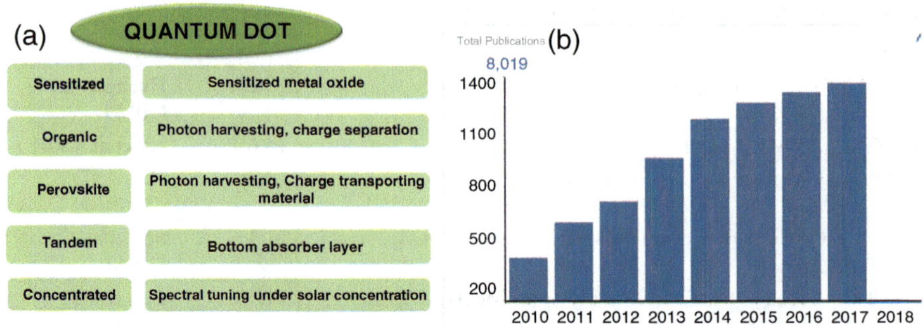

Fig. 5.10 (**a**) Application of quantum dot in different types of solar cell (**b**) Published paper on Quantum dot solar cell (Source: Reference and taken a copyright permission or include weblink and accessed specific date and year)

electron/hole conductivity (Kamat 2008). The unique properties of QDs coupled with attractive concept of interband solar cell (IBSCs) can leads to intensive research efforts on QD-IBSCs. Self-assembled quantum dots (SAQDs) and colloidal quantum dots (CQDs) have the potential to enhance the photogeneration of carriers through the QD energy level or band. Both are different in each other by synthesis process. Self-assembled quantum dots follows epitaxy growth and colloidal quantum dots follows solution based processing. However, they follow the same inter band device architecture. In recent years, two types of CQDSCs-QD-sensitized solar cells (QDSSCs) and quantum dot heterojunction solar cells (QDHSCs) have been developed rapidly. Typically, QDSSCs consist of a QD sensitized photoelectrode and a counter electrode separated by a liquid electrolyte. Wide band gap metal oxide nanostructures (such as nanoparticulate, nanorods, nanowires (NWs), nanotubes, and inverse opal) are used as the photoelectrodes in QDSSCs. A number of QDs, such as CdS, CdSe, CdTe, PbS, CdSeTe, and core–shell structure or double layered QDs, such as CdSe/CdTe, ZnTe/CdSe, CdS/CdSe, PbS/CdS, have been applied as sensitizers in QDSSCs (Rühle et al. 2010). In QDHSCs two different cell structures have been reported planar depleted hetero-junction quantum dot bulk heterojunction (HQDSCs) solar cells. Planar depleted hetero-junction is composed of a transparent conduction electrode (ITO or FTO), a metal oxide semiconductor, absorbing QD thin film with a thickness from a few 10–100 nm, and a metal electrode. So, the thickness of the QD film, and hence optical absorption, is limited by the depletion length and the diffusion length. In bulk HQDSCs nanostructured, metal oxides are applied and CQDs are interpenetrated into these nanostructured metal oxide. Such a bulk junction structure shows extension of the depletion region, which results in a great enhancement in optical absorption, charge separation, and collection efficiency. So, the morphology and thickness of the nanostructures of the metal oxides also plays key factors for photovoltaic performances of the bulk HQDSCs (Zheng et al. 2016; Semonin et al. 2012; Sogabe et al. 2016). The main reason for the poor photovoltaic performance of the QDSCs is due to the interfacial recombination occurring at (1) the QD surface and/or QD–QD interface; (2) the QD/metal electrode interface; and (3) the QD/metal oxide interface (Mora-Sero et al. 2009). Research is

Table 5.3 Some recently reported results on quantum dot solar cell

Devices	Efficiency	Ref.
Al/ZnO/PbS-I-/PbS-EDT/NiO/ITO	9.7%	Wang et al. (2018a)
ITO/ZnMgO/PbS-PbI_2/(Ag-doped PbS EDT)/Au	10.6%	Jeong et al. (2018)
Al/Graphene/P-SQDs/n-Si	16.2%	Kim et al. 2018
TiO_2/ZnS/CdS/ZnCdS/CdSe/ZnS/polysulfide/Cu_2S	6.05%	Wu et al. (2018)
Al/MoO_3/P_3HT-PCBM/C-QDS/ZnO/ITO	9.64%	Zhang et al. (2018c)

focusing on the improvement of efficiency and some strategies of interface engineering have been explored to reduce the recombination. For reduction of recombination (1), QDs surface passivation by organic and inorganic ligand or, during the synthesis process. For recombination (2), band allingment at interface is showing a good result of about efficiency 8.55% (Chuang et al. 2014). Lastly to reduce the recombination at QD/metal oxide interface (3), surface passivation of metal oxide nanostructure is very important (Wei et al. 2018). To boost overall device performance multiple exciton generation, plasmonic nanomaterial nanowire- quantum dot composite, semiconductor-carbon nanomaterial implemented on quantum dot solar cells (Kamat 2013). Some of the QDs solar cells with good device efficiency are given below in Table 5.3.

5.4.4 *Dye Sensitized Solar Cell*

Dye-sensitized solar cell (DSSC) is composed of nano-crystalline semiconductor oxide film electrode, dye sensitizers, electrolytes, counter electrode and transparent conducting substrate. Due to the simplicity of the synthesis procedure, tunability of light absorption, sensitivity to diffused light and ability to design flexible solar panels make semiconductor nanostructure an important candidate in solar cell. Different approaches have been used to enhance the performance of DSSC in different components as shown in Fig. 5.11. Widely used semiconductor oxides in dye-sensitized solar cell include TiO_2, ZnO, SnO_2, Nb_2O_5. To prepare the semiconductor film electrode with uniform-size, high-special surface area, and porous structure are the focuses of research. To obtain effective photoanodes, a variety of nanostructured has been reported including nanorod, nanotube, nanosheet, mesoporous structures, and 3D hierarchical architectures which improve the efficiency due to its unique properties (Ye et al. 2015). For example, 1D semiconductor nanostructures exhibit excellent charge transport properties and 3D mesoporous nano/microspheres, gives larger surface area (>100 m^2/g). Doping ion (e.g. F, I, Mg, Nb and Cu) in semiconductor metal oxides have been reported for DSSCs, to reduce recombination and to extend electron lifetime in photoanodes. Decoration of plasmonic nanomaterial in metal oxide film is also extensively used method to enhance the device efficiency. Improvement of light absorption in DSSC, up and

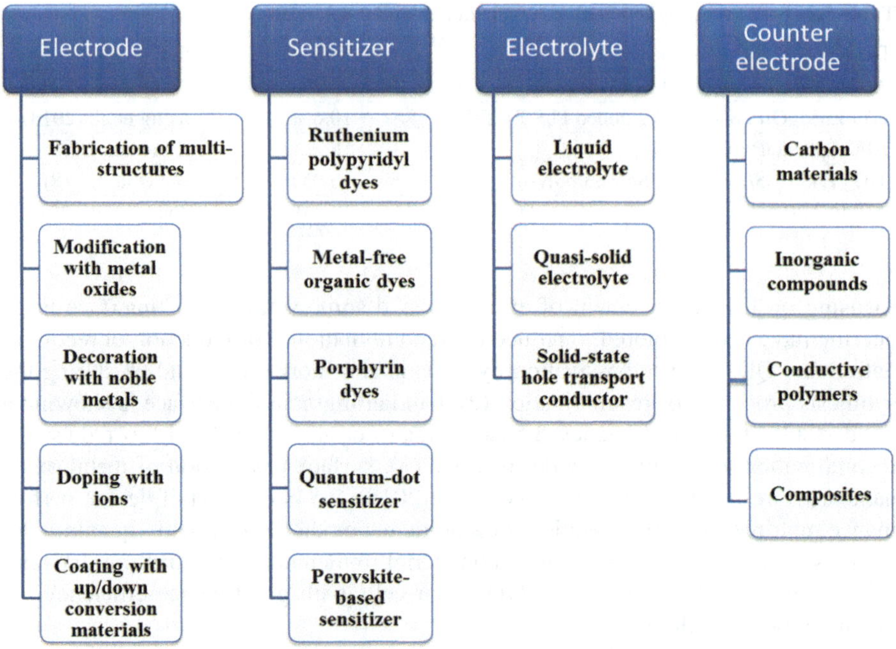

Fig. 5.11 Different approaches for the improvement of DSSC performance (Ye et al. 2015)

down spectral converter is showing good results. Use of Yb^{3+}–Tm^{3+}/Yb^{3+}–Ho^{3+}-doped $NaYF_4$ (Zou et al. 2012) and Er^{3+}/Yb^{3+}-doped LaF (Shan and Demopoulos 2010) up converter and $LaVO_4$:Dy^{3+} and YF_3:Eu^{3+} down converter has been reported in DSSC (Fig. 5.11).

Dye sensitizers serve as the solar energy absorber in DSC. The sensitizers used in DSSC were divided into two types: organic dye and inorganic dye according to the structure. Inorganic dye includes metal complex, such as polypyridyl complexes of ruthenium and osmium, metal porphyrin, phthalocyanine and inorganic quantum dots, while organic dye includes natural organic dye and synthetically organic dye. Recently, different sensitizer has been developed and used in dye sensitized solar cell like Ruthenium polypyridyl dyes (Gao et al. 2008), Metal-free organic dyes (Mishra et al. 2009), Porphyrin dyes (Mathew et al. 2014), and Quantum-dot sensitizer (Teymourinia et al. 2018). Sensitization with one single dye was restricted to some wavelength according to their properties. Co-sensitization of several dyes with different spectral is another good approach to enhance the light absorption for wide-band-gap semiconductor. Electrolyte plays an important role in conversion efficiency and stability of the solar cells. The electrolyte used in DSSC is divided into three types; liquid electrolyte, quasi-solid state electrolyte, and solid electrolyte. Organic solvent electrolytes were widely used in dye-sensitized solar cells for their low viscosity, fast ion diffusion, high efficiency, easy to design. However, the solar cells based on organic electrolyte have the disadvantages such as less long-term

Table 5.4 Some of recently reported results on dye sensitized solar cells

Devices	Efficiency	Ref.
FTO/TiO_2/electrolyte/B-doped carbon nanotubes-Pt/ITO	7.17%	Yeh et al. (2018)
ITO/1D-ZnO/CuO/Dye/I^-–I_3^-/Pt/ITO	6.18%	Kilic et al. (2018)
FTO/TiO_2/dye/I_3^-–I^-/Pt/FTO/QD	7.8%	Teymourinia et al. (2018)
FTO/TiO_2/blue dye (R6)/ Co(II/III) tris(bipyridyl)-based redox electrolyte/Pt/FTO/	12.6%	Ren et al. (2018)
FTO/TiO_2/Ruthenium dye/sulfonated block ionomer/Pt/FTO	7%	Al-Mohsin et al. (2018)

Table 5.5 Some of the reported results on plasmonic solar cell

Devices	Efficiency	Ref.
FTO/Au@SiO_2-TiO_2/N749/I_3^-–I^-/Pt/ITO	7.7%	Zheng et al. (2018a)
ITO/PEDOT:PSS/PTB_7:$PC_{70}BM$-Au nanorods/LiF/A	8.43%	Zheng et al. (2018a)
FTO/TiO_2-Ag/CdS/CdSe/polysulfide/Cu_2S	6%	Wang et al. (2018b)
Al/LiF/PCBM/$CH_3NH_3PbI_{3-x}Cl_x$/AuAgANCs PEDOT: PSS/ITO	16.76%	Sun et al. 2018
FTO/TiO_2/TiO_2NPs/NTs-Ag@TiO_2NPs/N719/I_3^-–I^-/Pt/ FTO	10.6%	Rho et al. 2018b

stability, difficulty in robust sealing and leakage of electrolyte due to the volatility of organic solvent. Ionic liquid electrolytes were developed in recent years due to its good chemical and thermal stability, negligible vapour pressure, non-flammability, high ionic conductivity and high solubility for organic or inorganic materials. Recently, some attempts were made to improve the long-term stability by using a hole transporting materials. Several inorganic P-type materials (e.g. CuI/CuSCN and $CsSnI_3$) and organic polymers (e.g. PEDOT, P_3HT) have been successfully used in solid-state DSSCs (SS-DSSCs) (Gorlov and Kloo 2008; Fujishima and Zhang 2006). However, their conversion efficiencies are not comparable with those of the liquid solar cells. Materials with high conductivity for charge transport, good electrocatalytic activity and excellent stability are used a counter electrode. Generally, noble metals Pt, Au, Ag are used. However, recently other materials also have been developed to replace the noble metal and make the system cost effective. Carbon materials (porous carbon, carbon nanotubes (CNTs) and grapheme), inorganic compounds (CoS_2, $CuInS_2$, Cu_2ZnSnS_4, Co_9S_8, Sb_2S_3, Cu_2S and $CoMoS_4$) and Conductive polymers (polyaniline, PEDOT, polypyrrole) have been reported in DSSC (Grätzel 2003). The efficiency of DSSC has received 12% (Green 2018) and to further improve the efficiency, improve material properties in different cell structure also been investigated. DSSC is integrated with pervoksite solar cell and different interband solar cell (Quantum dot, Tendam solar). Recently, compound parabolic concentrator in coupled with DSSC to generate more power from small size devices (Selvaraj et al. 2018). Some of the recent developments in DSSC are listed in Table 5.4.

5.4.5 Plasmonic Solar Cell

Plasmonics is an emerging field of nanotechnology that makes use of the nanoscale properties of metals. Surface plasmon is the oscillation of charge density at the metal surface and exists at the metal-dielectric interface. Surface plasmon resonance can be created by exciting the metal surface with coming photon flux. Large local electric field due to surface plasmon resonance at metal surface may cause light scattering and generate huge charge carriers. Use of plasmonics in solar cells getting tremendous interests among the researchers and its can be seen in the increasing number of publications over the last few years (Fig. 5.12b).

Use of plasmonic nanoparticles can help solar cells in three different ways (Fig. 5.12a). First, metallic nanoparticles on the surface of solar cell behaves as sub-wavelength scattering element and helps in trap freely propagating the light waves from Sun to the absorbing semiconductor thin film, by folding the light into a thin absorber layer, which increases the optical path length. Second, increasing light concentration by the plasmon nanoparticle. Here, metallic nanoparticles embedded in the semiconductor can be behave as ‘antenna’ for the incident sunlight that stores the incident energy in a localized surface plasmon mode which increases the effective absorption cross section. Third, an uneven metallic film on the back surface of a thin photovoltaic absorber layer can couple sunlight into surface plasmon polaritons (SPP) at the metal-semiconductor interface (Mandal and Sharma 2016; Atwater Harry and Polman 2010). These SPPs excited at the metal-semiconductor interface helps efficiently trapping and managing light in the semiconductor layer. A different metal nanostructure, such as nanoparticles, nanorods or nanowires has been used for the excitation of localized surface plasmons.

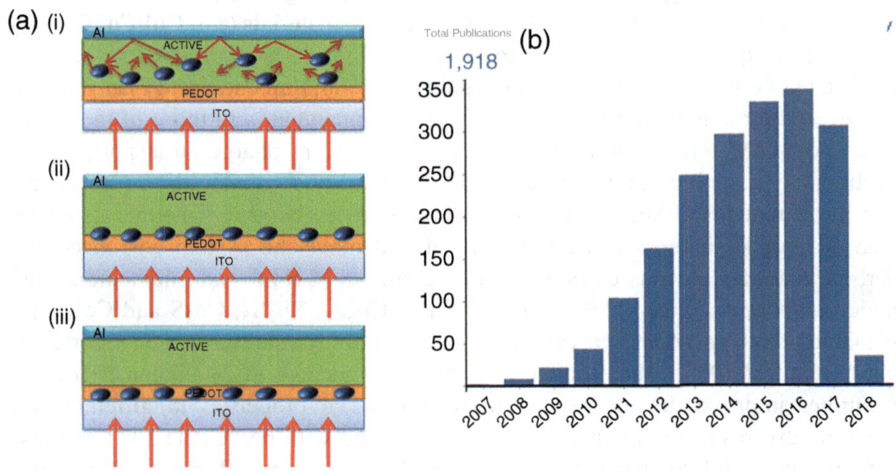

Fig. 5.12 (**a**) Plasmonic nanoparticle in three different location in the PV material. (**b**) Paper published on plasmonic solar cell from 2007 to 2018 (Source: Reference and taken a copyright permission or include weblink and accessed specific date and year)

The unique advantages of plasmonic nanoparticles have been employed in different types of solar cell to enhance the efficiency. In dye sensitized solar cell, interests on plasmon assisted spectral absorption are growing up. It is shown that the surface plasmon resonance in noble metal nanoparticles such as Ag, Cu and Au nanoparticles enhances light absorption thereby increasing the photocurrent and hence the device efficiency (Ding 2011). Recently, it has been shown that in comparison to using individual metal nanoparticles, the light harvesting efficiency is enhanced by incorporating two metal nanoparticles into DSSC which can give broader absorption region (Lim et al. 2017). Sometimes the incorporation of various metal nanoparticles can show lower power conversion efficiency because of corrosion of metal nanoparticles, increased charge recombination and back reaction of photo-generated carriers. To overcome these, research is giving the attention to use core-shell structures of different shape and size (Qi et al. 2011). Incorporating plasmonic nanoparticles such as gold (Au) and silver (Ag), has been used in organic photovoltaic (OPV) devices. These can improve the optical absorption by increasing the optical thickness of the organic absorber layer while keeping its physical thickness small, increases charge carrier transport, and hence the device performances in OPV. With the incorporation of different metallic nanostructures in different components of device architecture as like active layers, buffer layers, electrodes, or between adjacent layers of OPVs, different plasmonic mechanisms follows on their spatial arrangement across the device cross-section (Bajpai et al. 2017). Coupling light into solar plasmon polaritons could also solve the problem of light absorption in quantum-dot solar cells. In temdam solar cell, metal contact with plasmonic nanostructure layer couples with different semiconductor shows different spectral bands in the solar spectrum (Pillai and Green 2010; Vangelidis et al. 2018; Rho et al. 2018a).

5.4.6 Solar Thermophotovoltaic

Solar thermophotovoltaic (STPV) systems is getting attention to absorb and convert the broadband solar radiation spectrum into a narrowband thermal emission spectrum and tuned the spectrum to the spectral response of a photovoltaic cell (Fig. 5.13b). In STPV, sunlight is converted into electricity by absorbing solar photons as heat, which is then emitted as thermal radiation; this radiation creates electron-hole pairs via a low-band gap photovoltaic (PV) medium. STPV shows different advantages like high-efficiency by harnessing the entire solar spectrum, compactness because of their solid-state nature and storing energy using thermal or chemical means which make its wide scale application. The basic device structure of STPV is shown in Fig. 5.13a.

Efficiency of STPV depends on the selective absorber and selective emitter. Solar absorber converts solar radiation into thermal energy. Selective solar absorber should absorb broadband solar radiation and supress re-radiation at high temperature. Several types of selective solar absorbers that are suitable for STPV applications (a) metal-

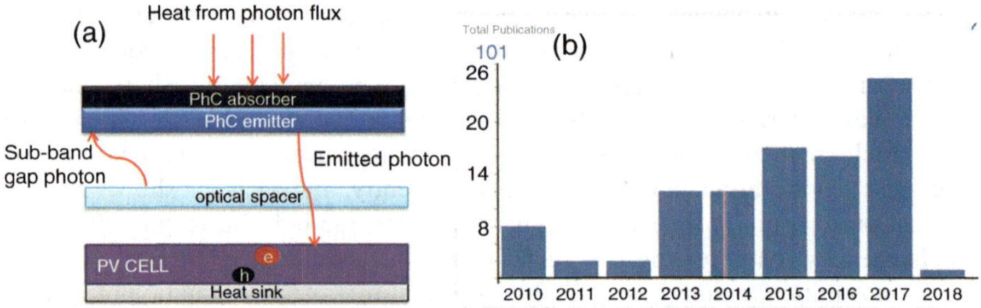

Fig. 5.13 (**a**) Device structure of thermophotovoltaics (**b**) Published paper on thermophotovoltaics from 2010 to 2018 (Source: Reference and taken a copyright permission or include weblink and accessed specific date and year)

dielectric composites (ceramic binders Al_2O_3, SiO_2, AlON and ZrO_2), (b) semiconductor–metal tandems (c-Si, a-Si, Ge, PbS) plasmonic absorbers (c) and one-dimensional (1D)/two- dimensional (2D)/three-dimensional (3D) photonic crystals (PhCs). Thermal emitters should efficiently harvest heat energy generated from concentrated solar radiation and at the same time it should be mechanically and thermally stable under temperatures as high as 1500 °C for long periods of time. Rare-earth elements such as erbium, thulium, ytterbium, samarium, and holmium are promising material for selective emitter. Semiconductor materials with a band significantly lower than 1 eV gap are used for TPV cells. Apart from silicon several material like copper indium diseenide, Germanium silicon/germanium, GaSb has been used for photocell material for solar thermophotovoltaics. The main technological challenge to achieve high TPV system efficiency is matching the spectral sensitivity of the photocells with the radiation spectrum of the emitter. Theoretical efficiency of solar thermophotovoltaics is 80% but practically efficiency is not so high. Presently, research is focusing on the theoretical modeming for different absorber and emitter in STPV (Bitnar et al. 2013; Zhou et al. 2016; Coutts 2001).

5.5 Future Research Needs and Recommendations

Although third generation solar cell is exhibiting better efficiency compared to the second generation solar cell but the performance is well below the silicon solar cell. Further, a lot of limitations in efficiency and stability at outdoor conditions are there in large scale applications with competitive efficiency levels conventional silicon solar cells. Future research needs to focus on the following area to overcome the barrier for large scale application of third generation photovoltaic to replace silicon cell/modules without compromising the cost.

1. Perovskite and multi-junction solar cell, showing a good performance with device efficiency comparable to conventional silicon solar cell but at the same time stability of these technologies is the barrier for the commercialization. Perovskite

solar cells at the present status lasts only for some months at the outdoor testing condition however silicon cell lasts for 25 years. Finding new materials for encapsulation, charge conductor in perovskite solar cell requires interdisciplinary research.
2. Mini module of DSSC and organic solar cell has been developed with module efficiency 10.7% and 9.7%, respectively, but the stability of this modules only about 1000 h. Future research needs to find new materials for better performance as well as to study actual life time and degradation mechanism.
3. Current research trend shows the use of nanotechnology to shape up the solar spectrum using the up and down converter material. However, thermophotovoltaic (TPV) and concentrated photovoltaics (CPV) solar cells can be another promising technology. Thermophotovoltaic can use the IR wavelength for energy harvesting and concentrated photovoltaic can reduce the cell area. Both the systems can reduce the system cost. Though TPV and CPV have been developed, but improvement of materials can make it more efficient.
4. Solar cell generally operates under light illumination at AM1.5, but the climatic conditions are not same through over the year. So, research needs to find out new generation of solar cell applicable to all weather condition. Recently, it has been reported that graphene can attract the positively charged ions, such as sodium, calcium, and ammonium, resulting in separated layers of positive and negative ions. This property of graphene is used in dye sensitized solar cell and it shows conversion efficiency 9.8%, from slightly salty water (Tang et al. 2016). This opens up a new scope to develop other materials to fabricate all-weather solar cell operated by Sun and rain.
5. Considering the clean energy production, bio-based solar cell will be the next future solar cell. Most of the materials that have been used in solar cell are toxic and harsh to the environment. Replacing those materials with bio-based material will be good approach for the future research on solar cell.

5.6 Conclusions

Photovoltaic is a growing technology and the excellent area for the application for nanotechnology. The performance and cost of the device can be controlled using nanostructured and nanotechnology. New materials of different size and structure have been developed to improve the efficiency of solar cells without any compromise the cost. Nanotechnology has tremendous potential to tune the optical properties of materials used in solar cells. Using advanced nanostructures in solar cell, it is possible to manipulate the overall absorption of solar cell and even fundamental properties like recombination rates. Semiconductor nanoparticles applied in a low temperature printing process that results in low cost solar cells. Now the advancement of nanotechnology in third generation solar cell, it can be installed as a coating on windows or other building materials.

References

Abdulrazzaq OA, Saini V, Bourdo S, Dervishi E, Biris AS (2013) Organic solar cells: a review of materials, limitations, and possibilities for improvement. Part Sci Technol 31(5):427–442

Al-Mohsin HA, Mineart KP, Armstrong DP, El-Shafei AA, Spontak RJ (2018) Quasi-Solid-State Dye- Sensitized Solar Cells Containing a Charged Thermoplastic Elastomeric Gel Electrolyte and Hydrophilic/phobic Photosensitizers. Solar RRL, 2:1700145

Ansari Haque MI, Qurashi A, Nazeeruddin MK (2018) Frontiers, opportunities, and challenges in perovskite solar cells: a critical review. J Photochem Photobiol C Photchem Rev 35:1–24

Assadi MK, Hanaei H, (2017) Transparent Carbon Nanotubes (CNTs) as Antireflection and Self-cleaning Solar Cell Coating. In: Engineering Applications of Nanotechnology. Springer, 101–114

Atwater Harry A, Polman A (2010) Plasmonics for improved photovoltaic devices. Nat Mater 9 (3):205

Bajpai M, Srivastava R, Dhar R (2017) Effect of plasmonic enhancement of light absorption on the efficiency of polymer solar cell. In: Recent trends in materials and devices. Springer, Cham, New York, pp 315–317

Beard MC, Randy JE (2008) Multiple exciton generation in semiconductor nanocrystals: Toward efficient solar energy conversion. Laser Photonics Rev 2(5):377–399

Benanti TL, Venkataraman D (2006) Organic solar cells: An overview focusing on active layer morphology. Photosynth Res 87(1):73–81

Bin L, Wang L, Kang B, Wang P, Qiu Y (2006) Review of recent progress in solid-state dye-sensitized solar cells. Sol Energy Mater Sol Cells 90(5):549–573

Bitnar B, Durisch W, Holzner R (2013) Thermophotovoltaics on the move to applications. Appl Energy 105:430–438

Cao G (2004) Nanostructures & nanomaterials: synthesis, properties & applications. Imperial College Press, London

Cao J, Wu B, Chen R, Wu Y, Hui Y, Mao BW, Zheng N (2018) Efficient, Hysteresis Free, and Stable Perovskite Solar Cells with ZnO as Electron Transport Layer: Effect of Surface Passivation. Advanced Materials 30:1705596

Chan L, DeCuir EA Jr, Fu R, Morse DE, Gordon MJ (2017) Biomimetic nanostructures in ZnS and ZnSe provide broadband anti-reflectivity. J Opt 19(11):114007

Chuang CHM, Brown PR, Bulović V, Bawendi MG (2014) Improved performance and stability in quantum dot solar cells through band alignment engineering. Nat Mater 13(8):796

Coutts TJ (2001) An overview of thermophotovoltaic generation of electricity. Sol Energy Mater Sol Cells 66(1–4):443–452

De Wild J, Meijerink A, Rath JK, Van Sark WGJHM, Schropp REI (2011) Upconverter solar cells: materials and applications. Energy Environ Sci 12:4835–4848

Ding I (2011) Plasmonic dye-sensitized solar cells. Adv Energy Mater 1(1):52–57

Djurišić AB, Liu FZ, Tam HW, Wong MK, Ng A, Surya C, Chen W, He ZB (2017) Perovskite solar cells–an overview of critical issues. Prog Quantum Electron 53:1–37

Dong S, Pootrakulchote N, Li R, Guo J, Wang Y, Zakeeruddin SM, Grätzel M, Wang P (2008) New efficiency records for stable dye-sensitized solar cells with low-volatility and ionic liquid electrolytes. J Phys Chem C 112(44):17046–17050

El Chaar L, Lamont LA, El Zein N (2011) Review of photovoltaic technologies. Renew Sust Energ Rev 15:2165–2175

Fahrenbruch AL, Richard HB (2012) Fundamental of solar cell: photovoltaic solar energy conversion. Academic Press, Cambridge, MA ISBN-0-12-247680-8

Flory F, Escoubas L, Berginc G (2011) Optical properties of nanostructured materials: a review. J Nanophotonics 5:052502

Fonash S (2012) Solar cell device physics. Elsevier, New York

Fujishima A, Zhang XT (2006) Solid-state dye-sensitized solar cells. In: Nanostructured materials for solar energy conversion. Elsevier, New York, pp 255–273

Gao F, Wang Y, Shi D, Zhang J, Wang M, Jing X, Humphry-Baker R, Wang P, Zakeeruddin SM, Grätzel M (2008) Enhance the optical absorptivity of nanocrystalline TiO_2 film with high molar extinction coefficient ruthenium sensitizers for high performance dye-sensitized solar cells. J Am Chem Soc 130(32):10720–10728

Gao F, Dai H, Pan H, Chen Y, Wang J, Chen Z (2018) Performance enhancement of perovskite solar cells by employing TiO_2 nanorod arrays decorated with $CuInS_2$ quantum dots. J Colloid Interface Sci 513:693–699

Garret M, Grover S (2013) Rectenna solar cells, 4th edn. Springer, New York

Gorlov M, Kloo L (2008) Ionic liquid electrolytes for dye-sensitized solar cells. Dalton Trans 20:2655–2666

Grätzel M (2003) Dye-sensitized solar cells. J Photochem Photobiol C Photchem Rev 4(2):145–153

Green MA (1981) Solar cells: operating principles, technology, and system applications. Prentice Hall, Englewood Cliffs, NJ ISBN: 0138222703

Green MA (2018) Solar cell efficiency tables (version 51). Prog Photovolt Res Appl 26:3–12

Guarnera S, Abate A, Zhang W, Foster JM, Richardson G, Petrozza A, Snaith HJ (2015) Improving the long-term stability of perovskite solar cells with a porous Al_2O_3 buffer layer. J Phys Chem Lett 6(3):432–437

Habisreutinger SN, Leijtens T, Eperon GE, Stranks SD, Nicholas RJ, Snaith HJ (2014) Carbon nanotube/polymer composites as a highly stable hole collection layer in perovskite solar cells. Nano Lett 14(10):5561–5568

Hahn H (2003) Unique features and properties of nanostructured materials. Adv Eng Mater 5 (5):277–284

Han N, Wang F, Johnny C (2012) One-dimensional nanostructured materials for solar energy harvesting. Nanomater Energy 1(1):4–17

Heremans P, Cheyns D, Rand BP (2009) Strategies for increasing the efficiency of heterojunction organic solar cells: material selection and device architecture. Acc Chem Res 42(11):1740–1747

Hoppe H, Sariciftci NS (2004) Organic solar cells: an overview. J Mater Res 19(7):1924–1945

Horst S, Wang J, Wilkinson SJ (2018) Durable MGO-MGF2 composite film for infrared anti-reflection coatings. US Patent Application No. 15/450,647

Hou J, Inganäs O, Friend RH, Gao F (2018) Organic solar cells based on non-fullerene acceptors. Nat Mater 17(2):119

Hu Y, Zhang Z, Mei A, Jiang Y, Hou X, Wang Q, Du K, Rong Y, Zhou Y, Xu G, Han H (2018) Improved performance of printable perovskite solar cells with bifunctional conjugated organic molecule. Adv Mater 30(11). https://doi.org/10.1002/adma.201705786

Huang A, Lei L, Zhu J, Yu Y, Liu Y, Yang S, Bao S, Cao X, Jin P (2017) Achieving high current density of perovskite solar cells by modulating the dominated facets of room-temperature DC magnetron sputtered TiO_2 electron extraction layer. ACS Appl Mater Interfaces 9 (3):2016–2022

Ishii T, Masuda A (2017) Annual degradation rates of recent crystalline silicon photovoltaic modules. Prog Photovoltaics 25:953–967

Jacoby M (2016) The future of low-cost solar cells. Chem Eng News 94(18):30–35

Jeong YJ, Song JH, Jeong S, Baik SJ (2018) PbS Colloidal Quantum Dot Solar Cells With Organic Hole Transport Layers for Enhanced Carrier Separation and Ambient Stability. IEEE J Photovoltaics 99:1–6

Kamat PV (2008) Quantum dot solar cells. Semiconductor nanocrystals as light harvesters. J Phys Chem C 112(48):18737–18753

Kamat PV (2013) Quantum dot solar cells. The next big thing in photovoltaics. J Phys Chem Lett 4 (6):908–918

Kaur N, Singh M, Pathak D, Wagner T, Nunzi JM (2014) Organic materials for photovoltaic applications: Review and mechanism. Synth Met 190:20–26

Kilic B, Turkdogan S, Astam A, Baran SS, Asgin M, Gur E, Kocak Y (2018) Interfacial engineering of CuO nanorod/ZnO nanowire hybrid nanostructure photoanode in dye-sensitized solar cell. J Nanopart Res 20(1):11

Kim JY, Kim SH, Lee HH, Lee K, Ma W, Gong X, Heeger AJ (2006) New Architecture for high efficiency polymer photovoltaic cells using solution based titanium oxide as an optical spacer. Adv Mater 18(5):572–576

Kim JM, Kim S, Shin DH, Seo SW, Lee HS, Kim JH, Jang CW, Kang SS, Choi SH, Kwak GY, Kim KJ (2018) Si-quantum-dot heterojunction solar cells with 16.2% efficiency achieved by employing doped-graphene transparent conductive electrodes. Nano Energy 43:124–129

Krebs FC, Spanggaard H (2005) Significant improvement of polymer solar cell stability. Chem Mater 17(21):5235–5237

Kyaw AKK, Wang DH, Wynands D, Zhang J, Nguyen TQ, Bazan GC, Heeger AJ (2013) Improved light harvesting and improved efficiency by insertion of an optical spacer (ZnO) in solution-processed small-molecule solar cells. Nano Lett 13(8):3796–3801

Lee YJ, Ruby DS, Peters DW, McKenzie BB, Hsu JW (2008) ZnO nanostructures as efficient antireflection layers in solar cells. Nano Lett 8(5):1501–1505

Li G, Zhu R, Yang Y (2012) Polymer solar cells. Nat Photonics 6(3):153

Lim SP, Lim YS, Pandikumar A, Lim HN, Ng YH, Ramaraj R, Bien DCS, Abou-Zied OK, Huang NM (2017) Gold–silver@ TiO_2 nanocomposite-modified plasmonic photoanodes for higher efficiency dye-sensitized solar cells. Phys Chem Chem Phys 19(2):1395–1407

Lim EL, Yap CC, Jumali MHH, Teridi MAM, Teh CH (2018) A mini review: can graphene be a novel material for perovskite solar cell applications? Nano-Micro Lett 10(2):27

Luisa HF, Sutherland D (2013) Nanotechnologies: principles, applications, implications and hands-on activities. Publications Office of the European Union, Luxembourg

Mali SS, Shim CS, Park HK, Heo J, Patil PS, Hong CK (2015) Ultrathin atomic layer deposited TiO_2 for surface passivation of hydrothermally grown 1D TiO_2 nanorod arrays for efficient solid-state perovskite solar cells. Chem Mater 27(5):1541–1551

Mandal P, Sharma S (2016) Progress in plasmonic solar cell efficiency improvement: a status review. Renew Sust Energ Rev 65:537–552

Mathew S, Yella A, Gao P, Humphry-Baker R, Curchod BF, Ashari-Astani N, Tavernelli I, Rothlisberger U, Nazeeruddin MK, Grätzel M (2014) Dye-sensitized solar cells with 13% efficiency achieved through the molecular engineering of porphyrin sensitizers. Nat Chem 6 (3):242

Mishra A, Fischer MK, Bäuerle P (2009) Metal free organic dyes for dye sensitized solar cells: from structure: property relationships to design rules. Angew Chem Int Ed 48(14):2474–2499

Mlinar V (2013) Engineered nanomaterials for solar energy conversion. Nanotechnology 24 (4):042001

Mora-Sero I, Gimenez S, Fabregat-Santiago F, Gomez R, Shen Q, Toyoda T, Bisquert J (2009) Recombination in quantum dot sensitized solar cells. Acc Chem Res 42(11):1848–1857

Nam M, Huh JY, Park Y, Hong YC , Ko DH (2018) Interfacial Modification Using Hydrogenated TiO_2 Electron Selective Layers for High Efficiency and Light Soaking Free Organic Solar Cells. Advanced Energy Materials 8:1703064

Neamen DA (1997) Semiconductor physics and devices, 3rd edn. McGraw-Hill, New York

Nozik AJ (2002) Quantum dot solar cells. Phys E 14:115–120

Peike C, Hädrich I, Wei KA, Dürr I (2013) Overview of PV module encapsulation materials. Photovol Int 19:85–92

Pelecky L, Diandra L, Rieke RD (1996) Magnetic properties of nanostructured materials. Chem Mater 8(8):1770–1783

Pillai SA, Green MA (2010) Plasmonics for photovoltaic applications. Sol Energy Mater Sol Cells 94(9):1481–1486

Piyadasa A, Wang S, Gao PX (2017) Band structure engineering strategies of metal oxide semiconductor nanowires and related nanostructures: a review. Semicond Sci Technol 32 (7):073001

Qi J, Dang X, Hammond PT, Belcher AM (2011) Highly efficient plasmon-enhanced dye-sensitized solar cells through metal@ oxide core–shell nanostructure. ACS Nano 5 (9):7108–7116

Qin X, Zhao Z, Wang Y, Wu J, Jiang Q, You J (2017) Recent progress in stability of perovskite solar cells. J Semicond 38(1):011002

Reddy K, Sekhar C (2018) Broad band antireflective coatings using novel in-situ synthesis of hollow MgF_2 nanoparticles. Sol Energy Mater Sol Cells 176:259–265

Reinders L, Ang Ã, Verlinden P, Freundlich A (2017) Photovoltaic solar energy: from fundamentals to applications. John Wiley & Sons, Hoboken, NJ

Ren Y, Sun D, Tsao HN, Yuan Y, Zakeeruddin SM, Wang P, Grätzel M (2018) A stable blue photosensitizer for color palette of dye-sensitized solar cells reaching 12.6% efficiency. J Am Chem Soc 140(7):2405–2408

Rho WY, Yang HY, Kim HS, Son BS, Suh JS, Jun BH (2018a) Recent advances in plasmonic dye-sensitized solar cells. J Solid State Chem 258:271–282

Rho WY, Kim HS, Chung WJ, Suh JS, Jun BH, Hahn YB (2018b) Enhancement of power conversion efficiency with TiO_2 nanoparticles/nanotubes-silver nanoparticles composites in dye-sensitized solar cells. Appl Surf Sci 429:23–28

Richards BS, Aruna I, MacDougall SKW Hueso JM (2012) Up-and down-conversion materials for photovoltaic devices. In: Photonics for solar energy systems IV, 8438. International Society for Optics and Photonics, Bellingham, WA, p 843802

Rühle S, Shalom M, Zaban A (2010) Quantum dot sensitized solar cells. ChemPhysChem 11 (11):2290–2304

Selvaraj P, Baig H, Mallick TK, Siviter J, Montecucco A, Li W, Paul M, Sweet T, Gao M, Knox AR, Sundaram S (2018) Enhancing the efficiency of transparent dye-sensitized solar cells using concentrated light. Sol Energy Mater Sol Cells 175:29–34

Semonin OE, Luther JM, Beard MC (2012) Quantum dots for next-generation photovoltaics. Mater Today 15(11):508–515

Seo JY, Uchida R, Kim HS, Saygili Y, Luo J, Moore C, Kerrod J, Wagstaff A, Eklund M, McIntyre R, Pellet N (2018) Boosting the efficiency of perovskite solar cells with csbr modified mesoporous TiO_2 beads as electron selective contact. Adv Funct Mater 28:1705763

Shan GB, Demopoulos GP (2010) Near infrared sunlight harvesting in dye sensitized solar cells via the insertion of an up-converter TiO_2 nanocomposite layer. Adv Mater 22(39):4373–4377

Shang Y, Shuwei H, Chunhui Y, Guanying C (2015) Enhancing solar cell efficiency using photon upconversion materials. Nano 5:1782–1809

Sharma A, Singh V, Bougher TL, Cola BA (2015) A carbon nanotube optical rectenna. Nat Nanotechnol 10(12):1027

Shi E, Zhang L, Li Z, Li P, Shang Y, Jia Y, Jinquan W (2012) TiO_2-coated carbon nanotube-silicon solar cells with efficiency of 15%. Sci Rep 2:884

Singh SC (2015) Solar photovoltaics: fundamentals, technologies and applications. PHI Learning Pvt. Ltd., New Delhi

Sining Y, Lund PD, Hinsch A (2015) Stability assessment of alternative platinum free counter electrodes for dye-sensitized solar cells. Energy Environ Sci 8(12):3495–3514

Sogabe T, Shen Q, Yamaguchi K (2016) Recent progress on quantum dot solar cells: a review. J Photonics Energy 6(4):040901

Son DY, Im JH, Kim HS, Park NG (2014) 11% efficient perovskite solar cell based on ZnO nanorods: an effective charge collection system. J Phys Chem C 118(30):16567–16573

Steim R, Kogler FR, Brabec CJ (2010) Interface materials for organic solar cells. J Mater Chem 20 (13):2499–2512

Sun Z, Xiahou Y, Cao T, Zhang K, Wang Z, Huang P, Zhu K, Yuan L, Zhou Y, Song B, Xia H (2018) Enhanced p-i-n type perovskite solar cells by doping AuAg@ AuAg core-shell alloy nanocrystals into PEDOT: PSS layer. Org Electron 52:309–316

Sze SM, Ng KK (2006) Physics of semiconductor devices. John Wiley & Sons, Hoboken, NJ

Tang Q, Zhang H, He B, Yang P (2016) An all-weather solar cell that can harvest energy from sunlight and rain. Nano Energy 30:818–824

Teymourinia H, Salavati-Niasari M, Amiri O, Farangi M (2018) Facile synthesis of graphene quantum dots from corn powder and their application as down conversion effect in quantum dot-dye-sensitized solar cell. J Mol Liq 251:267–272

Tsakalakos L (2010) Nanotechnology for photovoltaics. CRC Press, Boca Raton, FL

Van Sark WGJHM, Meijerink A, Schropp REI (2012) Solar spectrum conversion for photovoltaics using nanoparticles. In: Third generation photovoltaics. InTech, London

Vangelidis I, Theodosi A, Beliatis MJ, Gandhi K, Laskarakis A, Patsalas P, Logothetidis S, Silva SRPP, Lidorikis E (2018) plasmonic organic photovoltaics: unraveling plasmonic enhancement for realistic cell geometries. ACS Photonics 5(4):1440–1452

Wang N, Liu M, Tan H, Liang J, Zhang Q, Wei C, Zhao Y, Sargent EH, Zhang X (2017) Compound homojunction: heterojunction reduces bulk and interface recombination in ZnO photoanodes for water splitting. Small 13(10):1603527

Wang R, Wu X, Xu K, Zhou W, Shang Y, Tang H, Chen H, Ning Z (2018a) Highly efficient inverted structural quantum dot solar cells. Adv Mater 30(7):1704882

Wang Y, Zhang Q, Huang F, Li Z, Zheng YZ, Tao X, Cao G (2018b) In situ assembly of well-defined Au nanoparticles in TiO_2 films for plasmon-enhanced quantum dot sensitized solar cells. Nano Energy 44:135–143

Wang Y, Zhang Y, Qiu N, Feng H, Gao H, Kan B, Ma Y, Li C, Wan X, Chen Y (2018c) A halogenation strategy for over 12% efficiency nonfullerene organic solar cells. Adv Energy Mater 8:1702870

Wei H, Li D, Zheng X, Meng Q (2018) Recent progress of colloidal quantum dot based solar cells. Chin Phys B 27(1):018808

Wolf M (1971) A new look at silicon solar cell performance. Energy Convers 11(2):63–73

Womack G, Kaminski PM, Ali A, Isbilir K, Gottschalg R, Walls JM (2017) Performance and durability of broadband antireflection coatings for thin film CdTe solar cells. J Vac Sci Technol A 35:021201

Wu Q, Hou J, Zhao H, Liu Z, Yue X, Peng S, Cao H (2018) Charge recombination control for high efficiency CdS/CdSe quantum dot co-sensitized solar cells with multi-ZnS-layer. Dalton Trans 47(7):2214–2221

Xiao B, Song J, Guo B, Zhang M, Li W, Zhou R, Liu J, Wang HB, Zhang M, Luo G, Liu F (2018) Improved photocurrent and efficiency of non-fullerene organic solar cells despite higher charge recombination. J Mater Chem A 6:957–962

Yang Y (2017) Broadband graphene oxide anti-reflection coating on silicon nanostructures. In: Frontiers in optics. Optical Society of America, Washington, DC

Ye M, Wen X, Wang M, Iocozzia J, Zhang N, Lin C, Lin Z (2015) Recent advances in dye-sensitized solar cells: from photoanodes, sensitizers and electrolytes to counter electrodes. Mater Today 18(3):155–162

Ye C, Wang Y, Bi Z, Guo X, Fan Q, Chen J, Ou X, Ma W, Zhang M (2018) High-performance organic solar cells based on a small molecule with thieno [3, 2-b] thiophene as π-bridge. Org Electron 53:273–279

Yeh MH, Leu YA, Chiang WH, Li YS, Chen GL, Li TJ, Chang LY, Lin LY, Lin JJ, Ho KC (2018) Boron-doped carbon nanotubes as metal-free electrocatalyst for dye-sensitized solar cells: heteroatom doping level effect on tri-iodide reduction reaction. J Power Sources 375:29–36

You J, Meng L, Song TB, Guo TF, Yang MY, Chang WH, Hong Z (2016) Improved air stability of perovskite solar cells via solution-processed metal oxide transport layers. Nat Nanotechnol 11:75–81

Zhang F, Wang S, Li X, Xiao Y (2016) Recent progress of perovskite solar cells. Curr Nanosci 12 (2):137–156

Zhang J, Marina F, Anders H, Gerrit B (2018a) Solid-state dye-sensitized solar cells. In: Molecular devices for solar energy conversion and storage. Springer, Singapore, pp 151–185

Zhang J, Meng Z, Guo D, Zou H, Yu J, Fan K (2018b) Hole-conductor-free perovskite solar cells prepared with carbon counter electrode. Appl Surf Sci 430:531–538

Zhang R, Zhao M, Wang Z, Wang Z, Zhao B, Miao Y, Zhou Y, Wang H, Hao Y, Chen G, Zhu F (2018c) Solution-processable ZnO/carbon quantum dots electron extraction layer for highly efficient polymer solar cells. ACS Appl Mater Interfaces 10(5):4895–4903

Zhao W, Li S, Yao H, Zhang S, Zhang Y, Yang B, Hou J (2017) Molecular optimization enables over 13% efficiency in organic solar cells. J Am Chem Soc 139(21):7148–7151

Zheng Z, Ji H, Yu P, Wang Z (2016) Recent progress towards quantum dot solar cells with enhanced optical absorption. Nanoscale Res Lett 11(1):266

Zheng YZ, Tao X, Zhang JW, Lai XS, Li N (2018a) Plasmonic enhancement of light-harvesting efficiency in tandem dye-sensitized solar cells using multiplexed gold core/silica shell nanorods. J Power Sources 376:26–32

Zhou X, Bao C, Li F, Gao H, Yu T, Yang J, Zhu W, Zou Z (2015) Hole-transport-material-free perovskite solar cells based on nanoporous gold back electrode. RSC Adv 5(72):58543–58548

Zhou Z, Sakr E, Sun Y, Bermel P (2016) Solar thermophotovoltaics: reshaping the solar spectrum. Nanophotonics 5(1):1–21

Zhu Z, Ma J, Wang Z, Mu C, Fan Z, Du L, Bai Y, Fan L, Yan H, Phillips DL, Yang S (2014) Efficiency enhancement of perovskite solar cells through fast electron extraction: the role of graphene quantum dots. J Am Chem Soc 136(10):3760–3763

Zou W, Visser C, Maduro JA, Pshenichnikov MS, Hummelen JC (2012) Broadband dye-sensitized upconversion of near-infrared light. Nat Photonics 6(8):560

Chapter 6
Nanotechnology: Emerging Opportunities for Fuel Cell Applications

Wai Yin Wong and Nabila A. Karim

6.1 Fuel Cell

Fuel cells, which are promising energy devices generated from alternative energy, convert chemical energy from fuel to electricity through an electrochemical reaction. The main component in which the electrochemical reaction occurs is called membrane electrolyte assembly (MEA), which consists of an anode, a cathode and a membrane. Oxidation reaction occurs at the anode side, whereas reduction reaction occurs at the cathode side. The membrane separates the anode and cathode and assists in proton transfer from anode to cathode to complete the electrochemical circuit. The electrons from oxidation reaction in anode flow through the external circuit to generate electricity.

Fuel cells, which operate on the basis of various types of fuels at the anode with different operating temperatures, consist of many types. Polymer electrolyte membrane fuel cell (PEMFC) is a type of fuel cell that uses hydrogen gas as fuel to form reaction products of protons and electrons (Eq. 6.1). This reaction, called hydrogen oxidation reaction, occurs at the anode side. The proton from the hydrogen oxidation permeates through the polymer membrane to the other side, called the cathode side. Meanwhile, the electron from hydrogen oxidation is attracted to and withdraws to the cathode side through the electrical circuit. The other reaction that occurs at the cathode side is oxygen reduction reaction (ORR) (Eq. 6.2). This electrochemical reaction needs oxygen from an external source and reacts with it to form water. Figure 6.1 shows the schematic of PEMFC (Li et al. 2014). The complete reaction of both hydrogen oxidation reaction (HOR) and oxygen reduction reaction (ORR) in Eq. (6.3) generates electricity.

W. Y. Wong (✉) · N. A. Karim
Fuel Cell Institute, Universiti Kebangsaan Malaysia, Bangi, Selangor, Malaysia
e-mail: waiyin.wong@ukm.edu.my; nabila.akarim@ukm.edu.my

S. Siddiquee et al. (eds.), *Nanotechnology: Applications in Energy, Drug and Food*,
https://doi.org/10.1007/978-3-319-99602-8_6

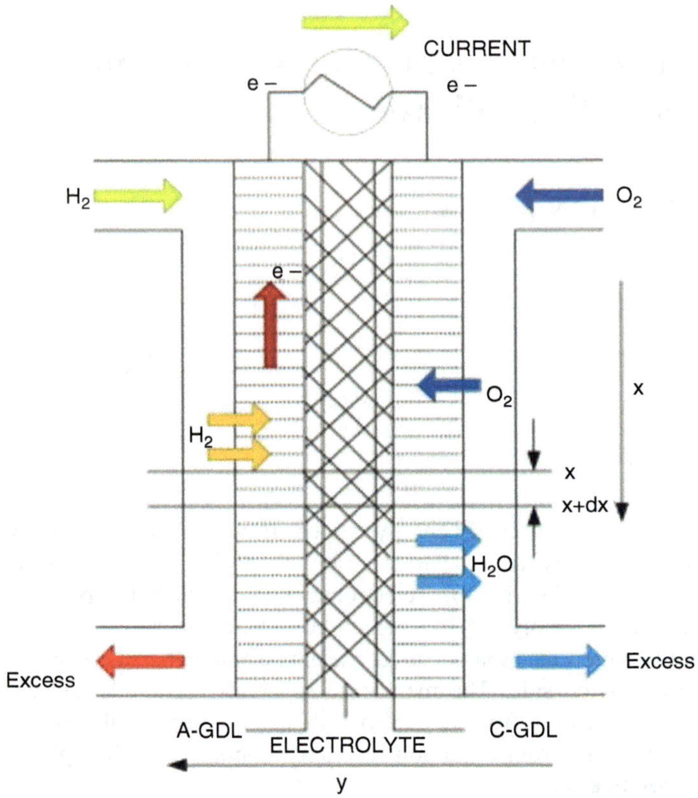

Fig. 6.1 Schematic diagram of PEMFC (Reprinted from Li et al. 2014. Copyright with permission from Elsevier)

$$H_2 \rightarrow 2H^+ + 2e^- \tag{6.1}$$

$$O_2 + 4H^+ + 4e^- \rightarrow 2H_2O \tag{6.2}$$

$$2H_2 + O_2 \rightarrow H_2O \tag{6.3}$$

Direct liquid fuel cells (DLFC), which have fuel variants such as methanol, ethanol, formic acid and glycerol, are called direct methanol fuel cell (DFMC), direct ethanol fuel cell (DEFC), direct formic acid fuel cell (DFAFC) and direct glycerol fuel cell (DGFC), respectively. The advantages of DLFC include high power density, ease of charging and low environmental effect. These fuels react as oxidation reactions to generate carbon dioxide and other by-products as well as protons and electrons. The chemical equations for the oxidation reaction of methanol, ethanol, formic acid and glycerol are shown in Eqs. (6.4), (6.5), (6.6) and (6.7), respectively. The reduction of oxygen in the cathode in DLFC also occurs as shown in Eq. (6.2), similar to that in PEMFC. Different electrical applications require the use of different DLFC. The complete reaction of fuel oxidation and ORR produces electricity.

$$CH_2OH + H_2O \rightarrow 6H^+ + 6e^- + CO_2 \tag{6.4}$$

$$C_2H_5OH + 3H_2O \rightarrow 12H^+ + 12e^- + 2CO_2 \tag{6.5}$$

$$HCOOH \rightarrow CO_2 + 2H^+ + 2e^- \tag{6.6}$$

$$CH_2OHCHOHCH_2OH + 20OH^- \rightarrow 3CO_3^{2-} + 14H_2O + 14e^- \tag{6.7}$$

Catalysts are generally used in both anode and cathode to facilitate the electrochemical reactions. A catalyst is a material or substance that increases a chemical reaction without being destroyed or decomposed, and an electrocatalyst is a catalyst involved in an electrochemical reaction. A catalyst should have high-density active sites, stability, electrical conductivity and moderate surface adsorption. However, catalysts in fuel cells face technical challenges such as low catalytic activity, high cost, and poor durability and reliability in fuel cell electrode. Activity and stability of the electrochemical reaction depends significantly on particle size, catalyst composition, shape, morphology and surface structure. Therefore, nanoparticles have been widely used to address such issues.

6.2 Nanostructured Electrocatalysts for Various Types of Fuel Cells

6.2.1 Polymer Electrolyte Membrane Fuel Cell (PEMFC)

Platinum (Pt) is a catalyst characterised by high selectivity in both reactions in PEMFC. High platinum costs have resulted in studies to reduce platinum catalyst intensive use. The strategies to reduce platinum loading by synthesising the platinum nanoparticle and high dispersal at the surface of catalyst support such as carbon support or other materials to maximise the active surface area. Conventional black Pt is used at the beginning of Pt use as catalyst in PEMFC. Black commercial Pt has a large size structure, and commonly used loading is 4 mg/cm^2 (Acres et al. 1997). Using platinum black has produced power as much as 30 W/cm^2. Platinum particles have been synthesised and supported on high-surface carbon support to reduce the use of platinum by loading as low as 0.1 mg/cm^2. The diameter of synthesised platinum catalyst for fuel cell applications usually range from 1.9 nm to 5 nm. Using a combination method of galvanostatic pulsed electrodeposition and pulsed ultrasonication with high power has produced Pt nanoparticles supported on carbon black with particle size between 5 and 10 nm (Karousos et al. 2017). The synthesis method is used to increase the integration between catalyst and support enhanced catalytic activity.

Pt catalyst is used for both anode and cathode sides in PEMFC. Alloy with ruthenium (Ru) and molybdenum (Mo) has also been introduced as an electrocatalyst in PEMFC to reduce Pt loading. High possibilities of trace gases such as carbon monoxide (CO), sulphur dioxide (SO_2), nitrogen oxide (NO_x) and ammonia are noted where these gases are prone to poisoning the surface of catalysts. The use of alloy metal, bimetallic or ternary Pt-based metal reduces the poisoning effect on the catalyst surface. Ru and Mo metals can dissociate water molecule and reduce the adsorption of CO, respectively. PtRu and PtMo are prepared by bubbling CO to form carbonyl compounds as metal precursors and are supported in a single-walled carbon nanotube (SWCNT) (Maya-Cornejo et al. 2018). The semi-spherical morphology of nanoparticle catalysts with good distribution has particle sizes of 10 and 13 nm for PtRu and PtMo, respectively. PtRu/SWCNT has an open-circuit voltage (OCV) which is 1.037 V lower than that of PtMo/SWCNT (1.044 V) and Pt/SWCNT (1.010 V). However, PtRu/SWCNT has high power-density single-cell performance which is 132 mW/cm^2, whereas PtMo/SWCNT and Pt/SWCNT only have 107 and 66 mW/cm^2, respectively. The high performance of PtRu is due to its resistance to harmful gases.

Pt catalyst for PEMFC not only has disadvantages towards traces gases that poisoning surface Pt, but a weak interaction between Pt and carbon support catalyst could lead to non-uniform distribution of Pt and low durability of the catalyst. Tantalum-doped titania is used as catalyst support for Pt to overcome these problems (Anwar et al. 2017). The Pt nanoparticle catalyst support of tantalum-doped titania (Pt/Ta–TiO_2) is synthesised using modified sol-gel method with calcination temperature at 800 °C. The synthesised Ta–TiO_2 has an average particle size of 24.2 nm, whereas Pt nanoparticles have an average size of 3.5 nm and are evenly distributed on the catalyst support, as shown in Fig. 6.2. The Ta–TiO_2 support for the Pt nanoparticle increases electrical conductivity as shown by a comparison with Pt/C using electrochemical active surface area (ECSA) analysis. After analysis of 10,000 cycles, Pt/Ta–TiO_2 ECSA decreases from 36.5 to 26.9 m^2/g, whereas Pt/C ECSA decreases from 62.3 to 30.5 m^2/g. The loss of ECSA is at a low 25.3% in Pt/Ta–TiO_2, whereas Pt/C loses more than half of the original value. Other catalysts such as Ni–Nb–Pt–Fe (Sánchez et al. 2016), Pt–SnO_2–Nb (Shahgaldi and Hamelin 2015), Pt–Ag (Esfandiari et al. 2016), Pt–niobium carbide (Nabil et al. 2017) and PtRuNi (Moreno et al. 2007) show increased activity of reaction in PEMFC.

6.2.2 Direct Methanol Fuel Cell (DMFC)

A promising opportunity in green technology, DMFC is characterised by high-energy conversion efficiency, low pollutant emission, low operating temperature and ease in handling liquid fuel. Bimetallic PtRu alloy is currently the best catalyst for methanol oxidation for DMFC. In a previous study, PtRu prepared with different atomic ratios by chemical impregnation was followed by ethylene glycol to investigate the effect of catalytic activity of methanol oxidation reaction (Hsieh et al.

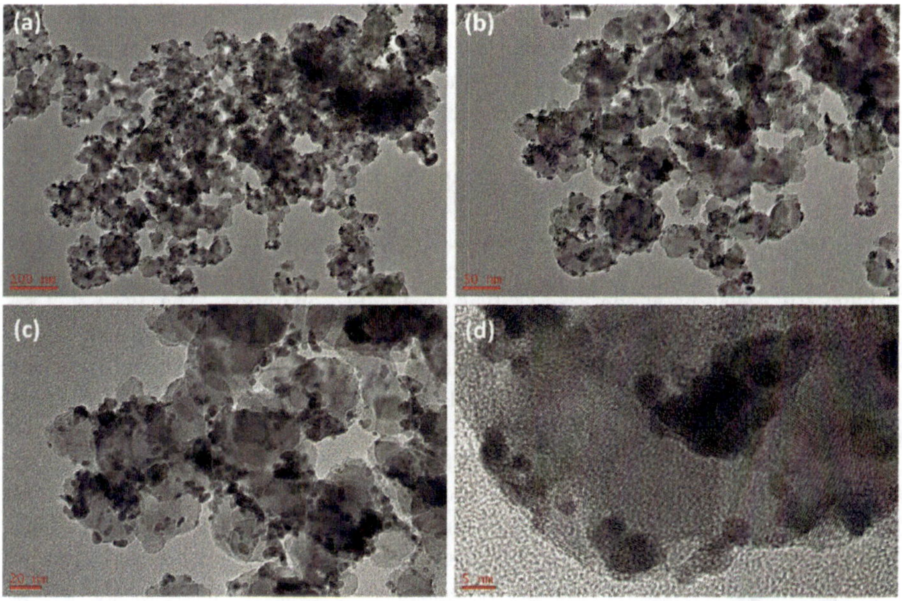

Fig. 6.2 TEM image of Pt supported on Ta–TiO_2 at different magnification; (**a**) 100 nm; (**b**) 50 nm; (**c**) 20 nm and (**d**) 5 nm. (Reprinted from Anwar et al. 2017. Copyright with permission from Elsevier)

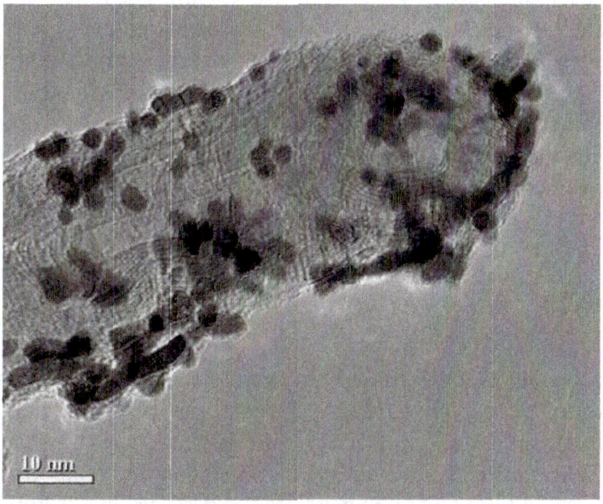

Fig. 6.3 High magnification of Bright-field HR-TEM micrographs of PtRu/CNT electrocatalyst (Reprinted from Hsieh et al. 2009. Copyright with permission from Elsevier)

2009). The synthesised PtRu is supported on carbon nanotube (CNT), as shown in Fig. 6.3. Four types of PtRu, namely, $Pt_{100}Ru_0$, $Pt_{75}Ru_{25}$, $Pt_{50}Ru_{50}$ and $Pt_{25}Ru_{75}$ show that decreasing the atomic ratio of Ru decreases the particle size from 4.34 to 2.77 nm. $Pt_{75}Ru_{25}$ possesses the highest ECSA value at 582 cm^2/mg, compared with $Pt_{100}Ru_0$, $Pt_{50}Ru_{50}$ and $Pt_{25}Ru_{75}$. However, catalyst $Pt_{50}Ru_{50}$ has the highest ratio of forward anodic peak current to reverse anodic current, which refers to the tolerance towards surface poisoning. The ratio for catalyst $Pt_{50}Ru_{50}$ is 4.62 compared with $Pt_{100}Ru_{0}$, $Pt_{75}Ru_{25}$ and $Pt_{25}Ru_{75}$ with ratios of 1.12, 2.12 and undetected (peak current density is extremely small), respectively. Other studies use various supports for PtRu such as silica template (Jiang and Kucernak 2009) and nitrogen-doped CNT (Chetty et al. 2009) which can contribute to increasing the catalytic activity of methanol oxidation reaction.

Ternary metal such as Pt–Co–Cr has also been investigated for methanol oxidation reaction. The idea of searching for ternary metal comes from research works that believe Ru metal in anode is dissolved in DMFC operation. The $Pt_{30}Co_{30}Cr_{40}$ ternary metal has mass activity of 10.3 $A/g_{noble\ metal}$ compared with PtRu/C which has only 4.03 $A/g_{noble\ metal}$ after a 20-h chronoamperometry test (Jeon et al. 2009a). Other ternary alloys for methanol oxidation include PtNiCr (Jeon et al. 2009b), PtRuFe (Lee et al. 2009) and $PtRuWO_x$ (Huang et al. 2009).

Core-shell Ru–Pd structure is prepared for methanol oxidation reaction in DMFC. Core-shell electronic effect and bifunctional mechanism for methanol oxidation reaction has been investigated (Kübler et al. 2018). The core-shell terrace illustrated in Fig. 6.4 shows the best performance compared with the core shell only. As the XRD pattern shows, the Pd particles have a face-centred cubic (fcc) crystal system, whereas Ru/C has a typical hexagonal close-packed system. The core-shell XRD patterns of these two crystal systems are exhibited. The Ru/C has a particle size of 3.7 nm, and the first layer coating of Pd, namely, Ru@1Pd/C, has a particle size which increases to 4.0 nm. Increasing the layer coating of Pd, namely, Ru@2Pd/C, causes the particle size to grow to 4.6 nm. The terrace coating process, namely, core-shell terrace has caused the particle size to average approximately 4.7 nm. The core-shell terrace achieves the highest catalytic activity by having mass densities of 0.5, 0.6 and 0.7 V compared with those of RHE which are 15.5, 32.5 and 38.9 A/g_{Pd}, respectively. The satisfactory performance results from efficient Pd utilisation and having more active sites of Pd and stronger electronic effect of the core-shell terrace compared with alloy metal.

6.2.3 *Direct Ethanol Fuel Cell (DEFC)*

Ethanol has lower toxicity than methanol and is easily produced from biomass such as sugarcane, corn and agricultural waste. However, the C–C bond cleavage problem in the oxidation of ethanol in DEFC is its main drawback. As stated previously, the Pt metal catalyst is prone to poisoning by other trace gases such as CO. The possibility of CO to be generated as a by-product is high in the oxidation of ethanol. Thus, Pt metal must be alloyed with other metals. Core-shell nanoparticles of Ni–Pt/C with several atomic ratios are synthesised using ethylene glycol as a reducing agent (Fetohi et al. 2017). The XRD pattern shows that no nickel peak was detected for all core-shell Ni–Pt samples as the Pt-rich shell coated on Ni nanoparticles. However, if the loading of nickel increases, the peak of Pt broadens. The diffraction

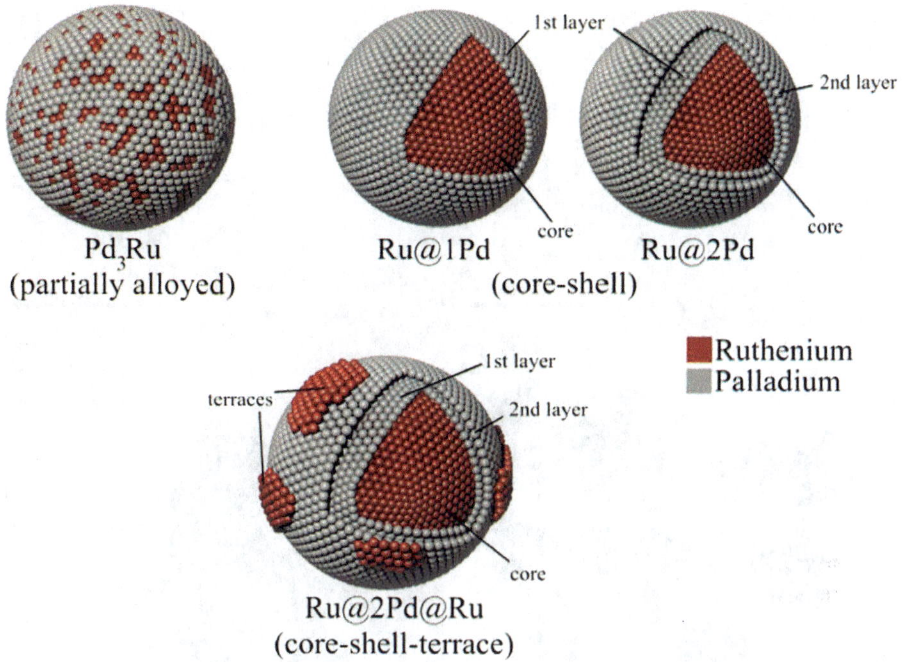

Fig. 6.4 Perspective view on the 3-d model of idealized spherical nanoparticles showing the expected structural composition of alloyed Pd_3Ru, core-shell structured Ru@1Pd with 1 monolayer of Pd, core-shell structured Ru@2Pd with two monolayers of Pd and the core-shell terrace structured Ru@2Pd@Ru (Reprinted from Kübler et al. 2018. Copyright with permission from Elsevier)

peaks of Pt values at 40.20°, 45.64° and 67.57° refer to Pt (111), Pt (200) and Pt (220), respectively. The size particles of core-shell Ni-Pt range from 1.72 nm to 2.88 nm depending on the ratio of Pt:Ni. This value is small compared with the range of 10–13 nm reported in another study which synthesised alloy Pt–Ni. The ECSA for core-shell Ni–Pt is 24.05 m^2/g for the ratio of Pt:Ni and is thrice larger than Pt/C with ECSA of only approximately 5.65 m^2/g.

Alloy PtAu nanoflowers synthesised by using facile one-pot method have high catalytic activity of ethanol oxidation reaction for DEFC compared with that of commercial Pt/C (Li et al. 2017). Catalytic activity is increased by using a flowerlike structure because its unique structure results in lower coordination atoms, steps, corners and interior boundaries. As shown in Fig. 6.5, the PtAu nanoflowers have a size distribution of 40–50 nm. Elemental mapping analysis shows that the Pt and Au metals distributed evenly on the nanoflower structures. The XRD analysis also shows that Pt has a face-centred cubic at degrees of 41.4°, 48.2°, 70.5°, 81.11° and 85.38° which refer to (111), (200), (220), (211) and (222), respectively. The catalyst of PtAu nanoflowers which have different atomic ratio shows increasing mass activity by the following flow order: Pt_3Au_1 (2891 mA/mg) < Pt_1Au_3

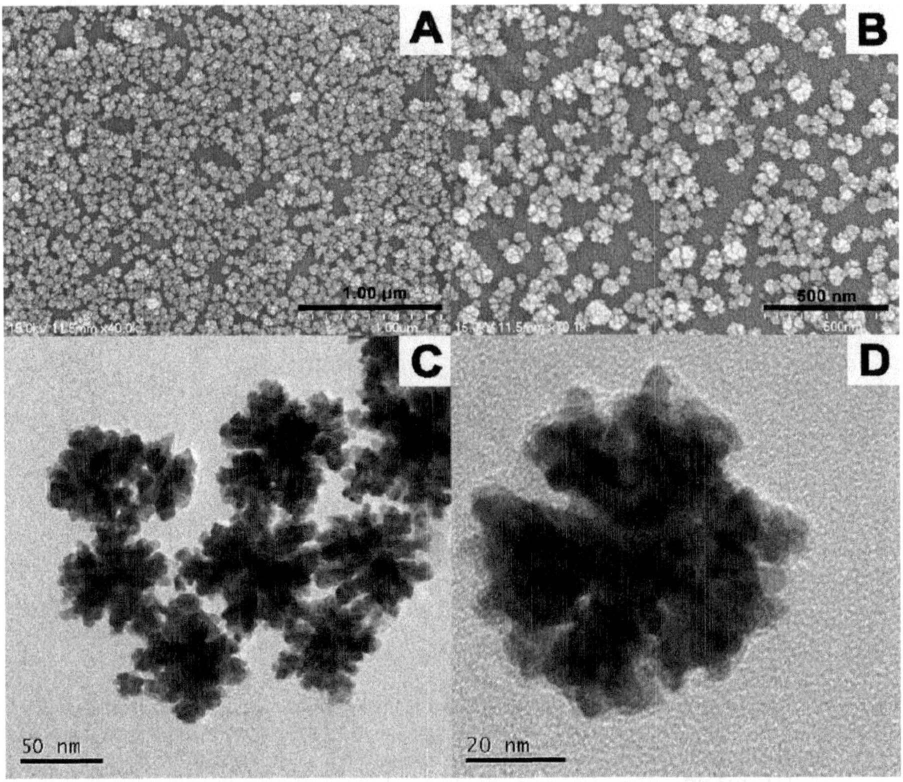

Fig. 6.5 SEM image (**a** and **b**) and TEM image (**c** and **d**) with different magnification of flower-like Pt_1Au_1 nanocrystals (Reprinted from Li et al. 2017. Copyright with permission from Elsevier)

(3838 mA/mg) < Pt_1Au_1 (6482 mA/mg). The mass activity of Pt_1Au_1 is 28.3 times higher than that of commercial Pt/C, which has mass activity of only 227 mA/mg. However, after 500 cycles, the mass activity of Pt_1Au_1 becomes 1941.29 mA/mg.

Cubic nanoparticle of Pt with (100) orientation combined with SnO_2 supported on carbon has been used as an electrocatalyst for ethanol oxidation reaction in DEFC (Antoniassi et al. 2017). Metal oxide has unique properties which could increase catalyst activity in ethanol oxidation reaction. The Pt–SnO_2/C has a particle size of approximately 5 nm. The XRD confirms that the Pt has diffraction peaks at 39°, 46°, 67°, 81° and 85° for (111), (200), (220), (311) and (222), respectively, whereas SnO_2 has a cassiterite phase at diffraction rates of 33° and 52°. A single-cell performance test is evaluated by polarisation and power density curves, as shown in Fig. 6.6. The power density using the Pt–SnO_2/C catalyst for DEFC is 58 mW/cm^2 which is high compared with the 13 mW/cm^2 of Pt/C. The product from ethanol oxidation reaction analysed by gas chromatography (GC) shows high selectivity product to form CO_2.

Pd metal has also been used for electrocatalyst in ethanol oxidation reaction because of advantages compared with Pt metal such as greater abundance and

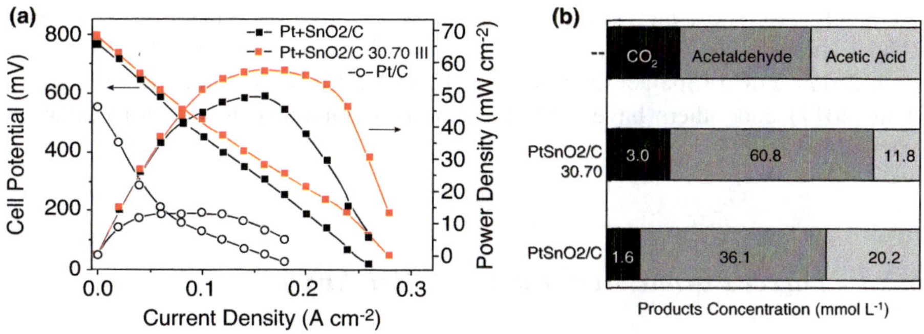

Fig. 6.6 (**a**) Polarization and power density curves and (**b**) products distribution analyzed by GC (Reprinted from Antoniassi et al. 2017. Copyright with permission from Elsevier)

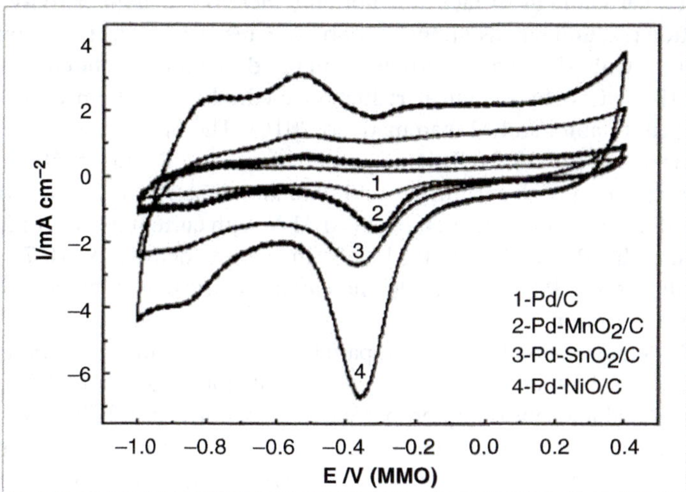

Fig. 6.7 Cyclic voltammograms of Pd/C and different Pd-metal oxide/C electrocatalyst in 0.5 M NaOH solution at 50 mV/s (Reprinted from Abdel Hameed 2017b. Copyright with permission from Elsevier)

high activity. Synthesised Pd–metal oxide/C as ethanol electro-oxidation reaction catalyst for DEFC uses reducing agent of ethylene glycol with modified microwave-assisted polyol process (Abdel Hameed 2017a). A combination of Pd–MnO_2/C, Pd–SnO_2/C and Pd–NiO/C produces particle sizes of 3.5, 3.2 and 3.0 nm, respectively. The particle size of Pd/C is originally 5.5 nm. The addition of metal oxide reduces the particle size of Pd–metal oxide/C. The ECSA of the catalyst is analysed by using cyclic voltammetry in 0.5 M NaOH solution, as shown in Fig. 6.7. The ECSA values of Pd–MnO_2/C, Pd–SnO_2/C and Pd–NiO/C are 28.12, 64.20 and 107.65 m^2/g, respectively. The uniform distribution of Pd–NiO/C shows increasing ECSA and more active ethanol oxidation reaction in DEFC. Other metal oxides, such as V_2O_5

and RuO_2, also show increasing catalytic activity with a combination of Pd metal (Abdel Hameed 2017b). Moreover, catalysts such as core-shell Cu–Pt (El-Khatib et al. 2017), PdCu (Shafaei Douk et al. 2017), Au–MnO_2 nanorod (Karuppasamy et al. 2017) and others have catalytic activity comparable to ethanol oxidation reaction for DEFC.

6.2.4 Direct Formic Acid Fuel Cell (DFAFC)

DFAFC has clean power sources with non-toxicity similar to other DLFC also and non-flammable fuel compared with PEMFC. Until today, Pd-based catalyst is the best catalyst for formic acid oxidation reaction in DFAFC. Pd nanoparticles supported on carbon high surface area not only increase catalytic activity of formic acid oxidation reaction but also improve corrosion resistance, electrical conductivity and porosity. Synthesis Pd nanoparticles supported on nanostructured carbon black composite (Pd–NCB) by chemical reduction method use sodium borohydride as reducing agent (Habibi and Mohammadyari 2016). The uniform distribution of Pd on NCB show that Pd and NCB have particle sizes of 4.8 and 25–60 nm, respectively. The oxidation peak for formic acid oxidation reaction in CV analysis shows that using Pd–NCB nanoparticle catalyst is 0.31 V with current density of 24.90 mA/cm^2, whereas the Pd catalyst is 0.51 V with current density of 10.78 mA/cm^2. Support catalyst for Pd nanoparticle increases electrical conductivity of formic acid oxidation reaction.

The PhH_x nanocatalyst has been prepared by treating commercial Pd black with n-butylamine in solvothermal condition at different atomic ratios of Pd:H to increase the efficiency and anti-poisoning property and ensure low cost (Zhang et al. 2018). The changes in electronic structure by functionalisation of surface structure of Pd, which becomes PdH_x, increase the mass and specific activities of the catalyst, as shown in Table 6.1. The table shows the changes in the Pd:H ratio and temperature, thereby leading to an increase in mass and specific activities of the catalyst for formic acid oxidation reaction in DFAFC.

Table 6.1 Catalyst activities and peak potentials of Pd and PdH_x nanocatalyst with different H content for formic acid electrooxidation (Reprinted from Zhang et al. 2018. Copyright with permission from Elsevier)

Samples	Mass activity (mA/cm^2)	Specific activity (mA/cm^2)	Peak potential (V)
Pd Black	0.34 ± 0.08	1.02 ± 0.07	0.22 ± 0.02
$PdH_{0.10}$	0.80 ± 0.09	2.54 ± 0.06	0.20 ± 0.03
$PdH_{0.29}$	0.87 ± 0.07	2.76 ± 0.08	0.17 ± 0.02
$PdH_{0.33}$	0.97 ± 0.08	4.87 ± 0.04	0.15 ± 0.03
$PdH_{0.43}$-150 °C	1.06 ± 0.06	5.12 ± 0.05	0.04 ± 0.02
$PdH_{0.43}$-250 °C	1.06 ± 0.07	5.11 ± 0.05	0.04 ± 0.02

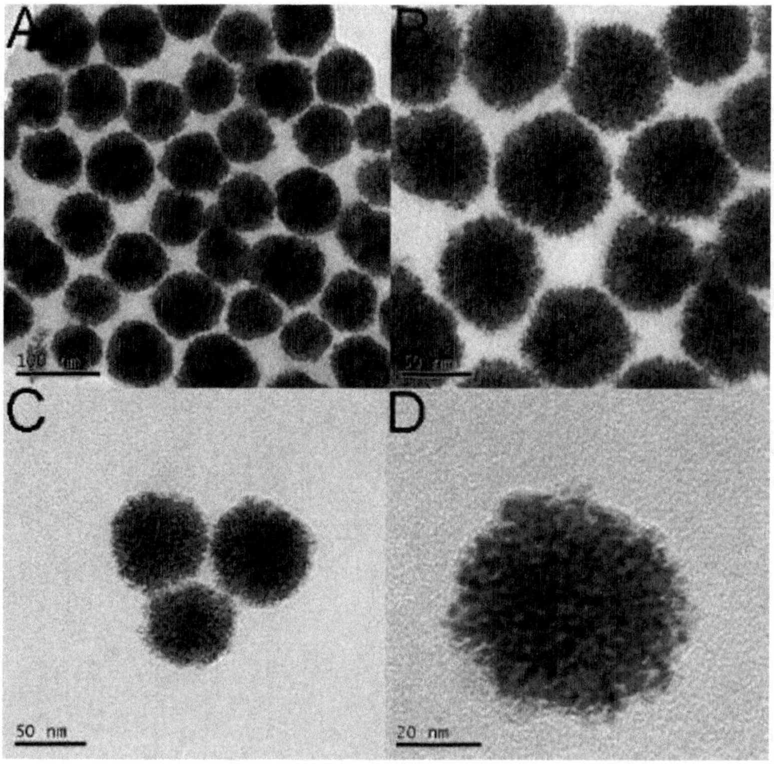

Fig. 6.8 TEM images of Pt_1Ru_1 NC with different magnifications (**a**) 100 nm, (**b**) 50 nm, (**c**) 50 nm and (**d**) 20 nm (Reprinted from Xu et al. 2018. Copyright with permission from Elsevier)

Nanodendrite structure, which has a unique capability due to its characteristics such as high porosity, percentage of surface atoms and interconnected sub-branches suitable in oxidation reaction for DFAFC and alloying Pd metal with other metals such as cobalt (Co), iron (Fe) and nickel (Ni), has attracted growing attention (Ma et al. 2017). The alloy PdCo has a particle size of 5–7 nm in diameter, whereas the PdCo nanodendrite has a particle size of approximately 32.3 nm. PdFe nanodendrite and nanodendrite have particle sizes of 4–6 and 33.3 nm, respectively. The small size and porous structure can increase catalytic activity and mass diffusion and transport for reactant and products. Increasing ECSA mainly results from the nanostructure of dendrite; PdCo nanodendrite and Pd/C have ECSA of 29.3 and 18.0 m^2/g, respectively. The peak current density from PdCo nanodendrites remarkably increases to 2467.7 A/g, whereas Pd/C has a peak current density of only 698.3 A/g. The nanodendelion-like structure of alloy metal PtRu has a tiny nanoseed with diameter of 2–5 nm, as shown in Fig. 6.8 (Xu et al. 2018). PtRu nanodendelion-like structure has high mass and specific activities of 1857.4 mA/mg and 18.3 mA/cm^2, respectively, which are higher than that of commercial Pt/C.

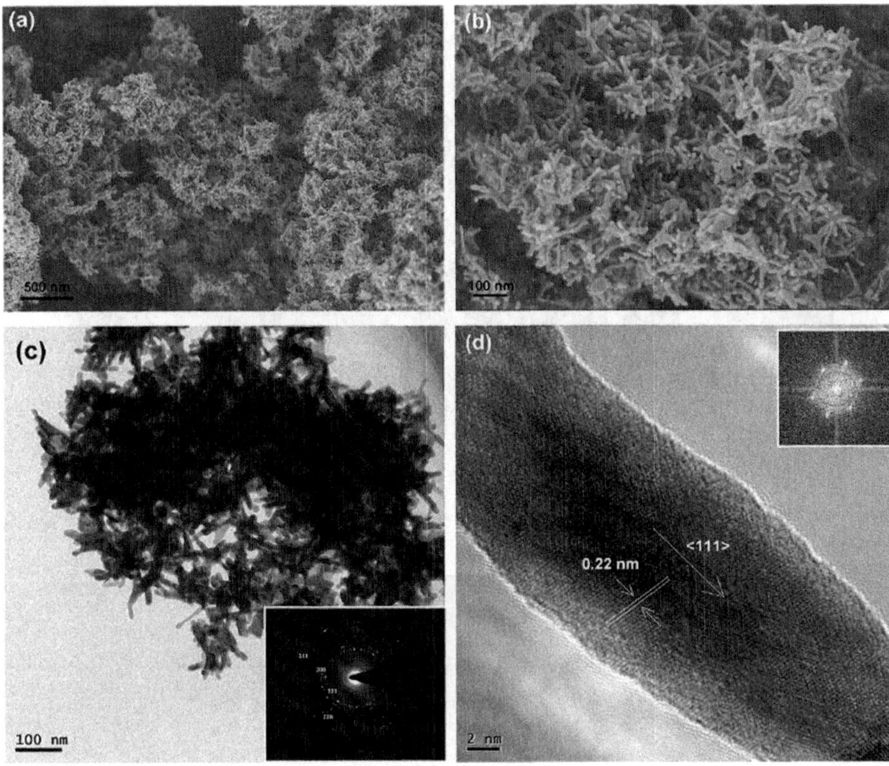

Fig 6.9 (**a** and **b**) SEM, (**c**) TEM and (**d**) HRTEM images of $Pd_{51}Cu_{49}$. Insert of (**c**): SAED pattern of $Pd_{51}Cu_{49}$. Insert of (**d**): corresponding FFT pattern (Reprinted from Yang et al. 2016. Copyright with permission from Elsevier)

A 1D structure has demonstrated efficient catalytic activity in the long term by inhibiting particle migration, Ostwald ripening and aggregation. Pd–Cu alloy 3D network, which assembles nanorod synthesis by facile one-pot hydrothermal reaction, has high potential to enhance the catalytic activity oxidation reaction of formic acid for DFAFC (Yang et al. 2016). Two atomic ratios, $Pd_{51}Cu_{49}$ and $Pd_{76}Cu_{24}$, have been tested in the oxidation reaction of formic acid. The $Pd_{51}Cu_{49}$ nanorod is 15 nm in diameter and grows along the (111) direction because of the low surface energy of (111) crystal planes, as shown in Fig. 6.9. The diffraction peak for both $Pd_{51}Cu_{49}$ and $Pd_{76}Cu_{24}$ located between pure fcc Pd and Cu indicate the formation of a single-phase uniform alloy structure. The ECSA for both $Pd_{51}Cu_{49}$ and $Pd_{76}Cu_{24}$ according to Pd basis are 33.2 and 12.7 m^2/g, respectively. The increase in ECSA for $Pd_{51}Cu_{49}$ is much higher than that of Pd black, which only has ECSA of 11.5 m^2/g. Other catalysts for formic acid oxidation reaction, such as core-shell Ru-Pd (Zhang et al. 2017b), CuPdN (Jia et al. 2016), PtPdCu (Ye et al. 2017), Pd–carbon nanobowls (Jia et al. 2017) and Pd–WC (Guo et al. 2016) show significant capability for DFAFC.

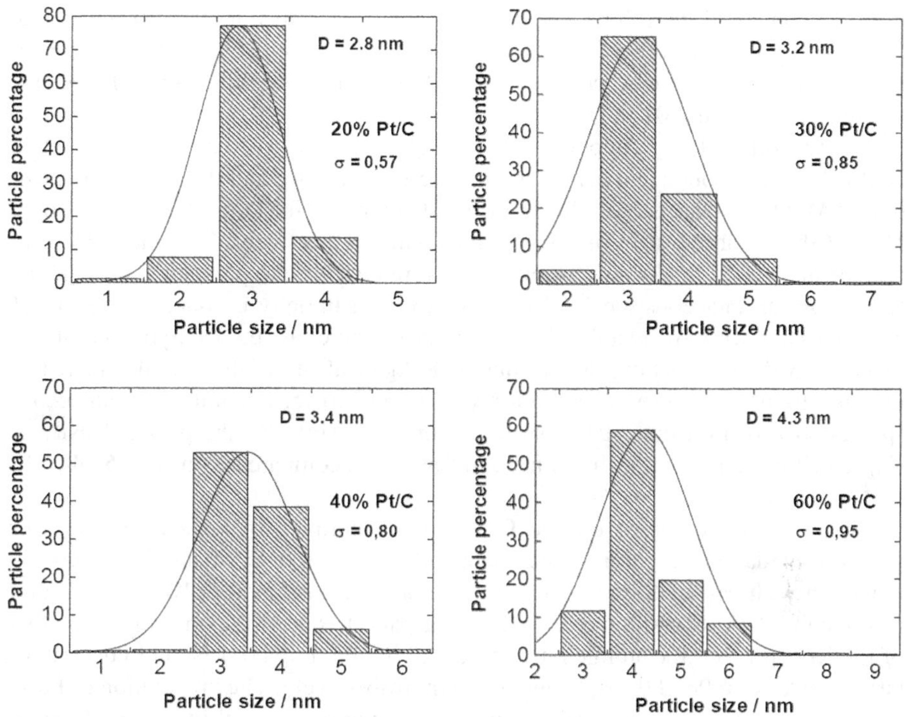

Fig. 6.10 Histogram of the Pt particle size distribution of the different catalyst (Reprinted from Ferreira Frota et al. 2017. Copyright with permission from Elsevier)

6.2.5 Direct Glycerol Fuel Cell (DGFC)

The Pt catalyst has also been used for oxidation of glycerol in DGFC. Using the catalyst along with catalyst support is determined by weight percent (wt.%), which indicates the amount of metal (usually 10–60 wt.%) as confirmed by thermogravimetric analysis. The mass transport of reactants and release of the products affected by the change of metal weight percent modifies the thickness of the catalytic area. Production of Pt supported on carbon black (Pt/C) with various Pt loading has resulted in particles with different sizes. All the Pt/C produced have typical peaks associated with Pt fcc 39.8° (111), 46.2° (200), 67.5° (220), 81.3° (311) and 85.8° (222). The loadings of platinum on carbon black as shown in Fig. 6.10 with 20, 30, 40 and 60 wt.% have average particle sizes of 2.8, 3.2, 3.4 and 4.3 nm, respectively. The increase of particle size results from heterogeneous growth of the Pt nanoparticle as the carbon surface available is less than the number of Pt nanoparticles. Ratio current density of forward scan to backward scan in cyclic voltammetry analysis refers to the effect of catalyst poisoning in the reaction of glycerol oxidation. High ratios show resistance to catalyst poisoning. Small Pt nanoparticles show high-activity glycerol oxidation with value onset potential of

0.472 V compared with that of RHE but has low resistance to catalyst poisoning. Single-cell power performance using catalyst loading of 60 wt.% of Pt/C has power with a value of 35 mW/cm^2 compared with 20 wt.% of Pt/C, which has 20 mW/cm^2 (Ferreira Frota et al. 2017).

Preparation of PtAg supported on MnO_x/C for glycerol oxidation reaction in DGFC has been conducted using colloid reaction for PtAg and thermal decomposition of MnO_x/C (Garcia et al. 2017). Catalysts prepared for Pt/C, PtAg/C and PtAg/MnO_x/C have particle sizes of 2.7, 4.6 and 4.8 nm, respectively. The ratio of forward current density to backward current density represents the poisoning of catalytic surface by carbonaceous species. The ratio of Pt/C is initially 1.2 but increases to 7.1 after alloying with Ag metal. Alloying of PtAg increases the catalytic activity of glycerol oxidation reaction. Supporting the PtAg catalyst on MnO_x/C also increases the ratio to 10.3 because of the high probability of lower amount of carbonaceous species adsorbed on the catalyst surface. For PtAg/MnO_x/C, the power density of single-cell measurement is 102.8 mW/cm^2 at 90 °C compared with only 59.4 mW/cm^2 for Pt/C.

Similarity properties of Pd and Pt lead to exploration of Pd and Pd alloy for glycerol oxidation reaction. PdNi metal alloy prepared by metal ion chemical reduction with borohydride has shown potential as a catalyst for glycerol oxidation reaction in DGFC (Carvalho et al. 2017). The particle size of PdNi is larger than that of Pd depending on the atomic ratio of Pd and Ni. Pd, PdNi, Pd_2Ni and Pd_3Ni have particle sizes of 6.96, 8.93, 8.77 and 9.41 nm, respectively. The preparation of PdSn supported on functionalised carbon immobilises the alloy metal for glycerol oxidation reaction (Wang et al. 2016a). The synthesised PdSn has crystallite size ranging from 4.4 to 5.6 nm compared with Pd/C with only 8.2 nm. Figure 6.11 shows the CV, chronoamperometries (CAs) and mass density of PdSn catalyst. The catalyst of Pd_3Sn has highest mass density in oxidation reaction of glycerol compared with atomic ratios of PdSn and Pd catalysts. The CA analysis also shows that Pd_3Sn has the highest Pd mass-based current density after 3000 s of approximately 0.01324 A/mg_{Pd} compared with PdSn, Pd_2Sn and Pd, which have 0.00732, 0.00541 and 0.00893 A/mg_{Pd}, respectively. Other studies focus on alloying Pd with other metals, such as PdAu (Yahya et al. 2017), Au-CeO_2 (Yuan et al. 2016) and PdCu (Maya-Cornejo et al. 2014), which also show high catalytic activity for glycerol oxidation reaction.

6.3 Oxygen Reduction Reaction Electrocatalysts

Cube-shaped Pt nanoparticles with surface structure of (100) supported on carbon Vulcan has also been synthesised using various platinum loadings, as shown in Fig. 6.12. The average nanoparticle size for all loadings is 9.2–9.7 nm along with increasing platinum loading from 10 to 50 wt.%, but higher loading results in the agglomeration of the catalyst. In cyclic voltammetry analysis of oxygen reduction reaction where the catalysts have high activity, the ORR onset potential shifts to a

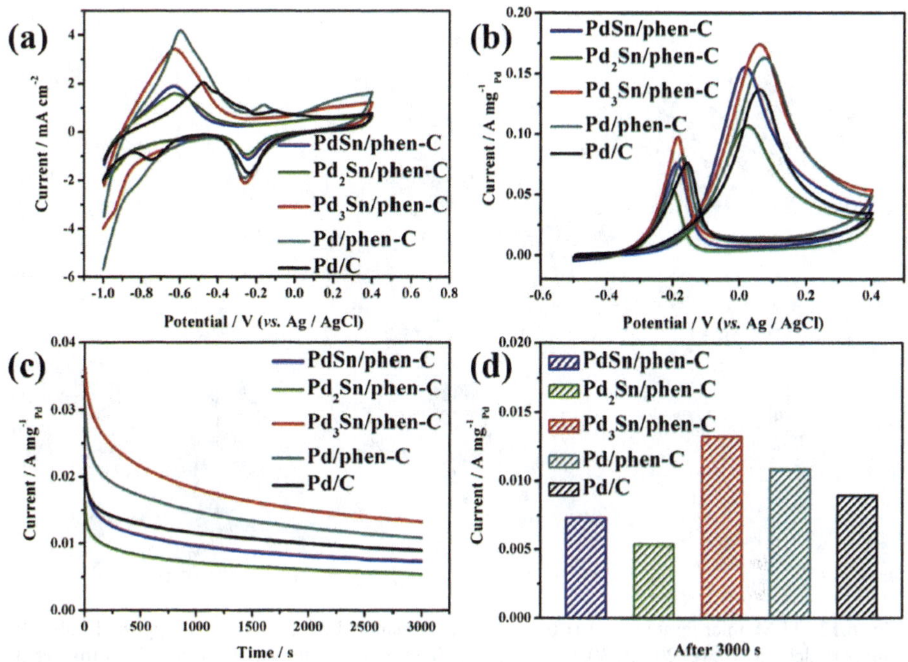

Fig. 6.11 (**a**) 1 CVs of Pd_xSn_y/phen-C, Pd/phen-C and Pd/C electrocatalysts in N_2 saturated 0.1 M KOH solution; (**b**) GOR in N_2-saturated 0.1 M KOH + 0.5 M glycerol solution; (**c**) CAs of the GOR at −0.1 V (vs. Ag/AgCl) on the corresponding catalysts for 3000 s; (**d**) current of catalysts after 3000 s (Reprinted from Wang et al. 2016a. Copyright with permission from Elsevier)

positive value from 0.74 V to 0.85 V for 10 wt.% and 50 wt.% of P/C, respectively. However, the specific and mass activities for all catalysts at different loadings are independent of loading. The catalysts of cubic nanocube Pt/C have mass activity with average value of 59.2 mA/mg (Jukk et al. 2017).

Core-shell structure can reduce the loading of Pt without reducing its catalytic activity in ORR (Davies et al. 2017). The core-shell structure of oxide–Pt has optimised the minimum Pt shell coverage that can maintain its high activity and stability for ORR. The tin-doped titania is used for the support. As the Pt coverage reduces, the onset of ORR also reduces in CV analysis with saturated O_2 gases of 0.5 M $HClO_4$ at 5 mV/s. After 200 cycles, the core shell with oxide support has improved stability compared with the Pt catalyst only.

Pt nanowire has also been produced for juga ORR in PEMFC. Pt nanowires overcome the problem in Pt nanoparticles, whereas nanoparticles tend to agglomerate and dissolve during the operation of PEMFC owing to their high surface energies. The single-crystal Pt nanowires grown on porous carbon nanofibers (PCNFs) along the <111> direction have a diameter of 4 nm, and the distance between the <111> planes is 0.25 nm. The Pt nanowires have low ECSA with value of 42.8 m^2/g but have high onset potential (0.884 V). The mass and specific

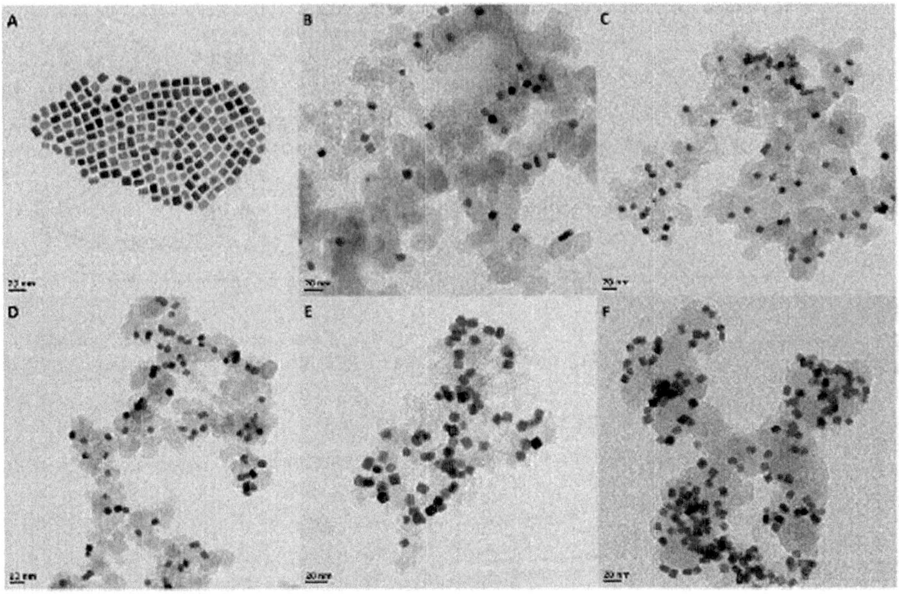

Fig. 6.12 TEM micrographs of (**a**) unsupported Pt nanocubes, and carbon-supported cubic Pt nanoparticles: (**b**) 10, (**c**) 20, (**d**) 30, (**e**) 40 and (**f**) 50 wt. % Pt/C catalyst (Reprinted from Jukk et al. 2017. Copyright with permission from Elsevier)

activities of Pt nanowires at 0.80 V are high, namely, 57.3 mA/mg and 0.134 mA/cm ^{2}, respectively. In comparison, commercial Pt/C has 24.7 mA/mg and 0.054 mA/cm2 at 0.80 V for mass and specific activities, respectively (Wang et al. 2016b). Adding a specific sulphur dope on graphene-supported Pt nanowires has resulted in mass and specific activities at 0.9 V versus RHE, 182 mA/mgPt and 662 μA/cm^2, respectively (Hoque et al. 2016). The Pt nanowires with diameters ranging from 4 to 16 nm could be enhanced in doping other elements.

Alloy metal nanoparticles have also been used not only to increase the activity of ORR but also to reduce the usage of Pt metal. The combination of Pt and palladium (Pt–Pd) supported in carbon denoted as Pt–Pd/C has been used for ORR in fuel cells. The synthesised Pt-Pd/C has a nanoparticle size of 5.3 nm compared with Pd/C which has a size of 5.1 nm (Liao et al. 2017). Owing to the similarity in size and structure between Pd and Pt, the Pt–Pd/C is effectively dispersed on carbon support carried out by high-angle annular dark field-scanning transmission electron microscopy, as shown in Fig. 6.13. The ECSA and specific activity of Pt–Pd/C, namely, 28.56 m2/g and 3.36 mA/cm^2, respectively, are higher than those of Pt only. In comparison, other studies synthesised decorating Pt on the Pd nanoflowers (Pt–PdNF) using underpotential deposition method (Zhang et al. 2017a). The Pd nanosheet from PdNF is 4–8 nm thick and several hundred nanometers long. The Pt mass activity on ORR is 0.73 mg/μgPt at 0.9 V versus RHE, whereas specific activity is 0.57 mA/cm^2, which is high, compared with that of commercial Pt/C for

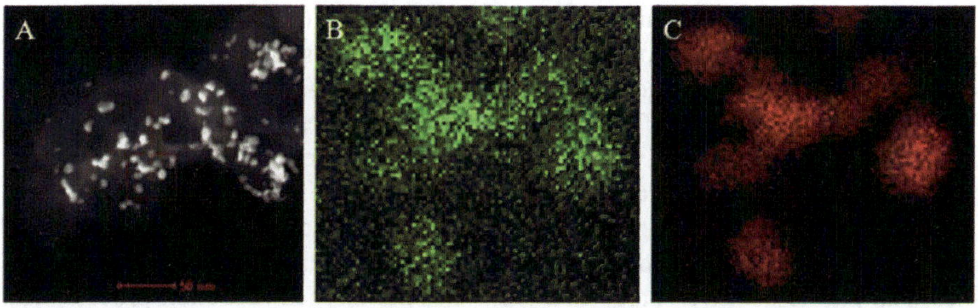

Fig. 6.13 HAADF-STEM image and elemental mapping for Pt–Pd nanoparticles in the Pt–Pd/C (Pt:Pd = 1:100). (**a**) HAADF-STEM image of Pt–Pd nanoparticles. EDX mappings of (**b**) Pt (Geatches et al. 2012), (**c**) Pd (red) (Reprinted from Liao et al. 2017. Copyright with permission from Elsevier)

both values. Other nanostructures of Pt–Pd, such as nanodendrite (Xiong et al. 2017) for ORR, have also been synthesised and discussed.

6.4 Metal Free Carbon Nanostructured Electrocatalysts

In the past, efforts to reduce the cost of electrodes include reducing Pt loading and developing Pt- and metal-free catalysts. The ultimate response to this challenge is to replace the scarce and expensive platinum, but this objective is far from being achieved. Nonetheless, the major challenges in the Pt-free metal catalysts arise from their lower intrinsic activity and metal dissolution in the long run. Thus, utilisation of metal-free carbon materials is attempted to solve the metal dissolution issue. Carbon nanostructures have been actively investigated because of their advantage in having a large surface area per volume, which can either allow the deposition of nanosize metal particles or provide larger surface-active sites for reaction. Modification at the atomic level through doping is performed to alter the properties of the pristine carbon nanostructures that provide active sites for catalytic reaction to be conducted.

In the pristine form, carbon nanostructures are inert to chemical reaction because of the uniform electron distribution in the carbon matrix. The high electrical conductivity of carbon nanostructures is due to the presence of pi-cloud in the graphene layers. 2D graphene, single-walled or multiwalled CNTs are advantageous for application in the fuel cell electrodes used to facilitate the electron transfer process upon chemical reaction. Such carbon nanostructures can be further tailored based on their electronic properties to produce catalysts that can function in the oxygen reduction process in fuel cells. Introduction of foreign atoms into the carbon matrix would create surface defects that alter the electronic properties of the carbon nanostructures which would possess catalytic activity. Modification at the atomic level of the nanostructures could be achieved via chemical vapour deposition, pyrolysis or arc discharge with the introduction of a n- or p-type dopant. The concept of doping carbon nanostructures with non-metal elements (such as boron, nitrogen or phosphorus) from group 13 or 15 in the periodic table is similar to that in silicon as semiconducting material. The unique properties in the n- or p-doped carbon is that the creation of the surface defects on carbon nanostructures as a result of the electron disturbance on the pi-cloud in the carbon ring provides reaction sites that allow the adsorption of reactants such as oxygen. This condition serves as the first step in the process of oxygen reduction reaction on the catalyst surface. In most studies, greater attention was paid to N-doped carbon materials owing to ease in creating more active sites with partial positive charge that facilitates the oxygen reduction process with promising performance.

6.4.1 Nitrogen-Doped Carbon Nanotubes

Typically, nitrogen and boron have been used as n- and p-type dopants, respectively, to perturb the homogenous pi-cloud in the CNTs to provide catalytic effect for oxygen reduction reaction in fuel cell cathode. For the n-type dopant, the substitution of nitrogen atom with one excess valence electron into the aromatic ring results in the increase in the localised density of states at the Fermi level and creates active sites for oxygen reduction reaction. From the first-principle molecular dynamics simulation, the adsorption of oxygen molecule on the nitrogen atom doped into the CNTs is unfavourable due to the repulsive force between lone pair electrons of oxygen and nitrogen atoms. The higher electronegativity in the nitrogen atom results in perturbation in the electron delocalisation within the carbon ring, where the nitrogen draws the electrons towards itself and creates a partial positive charge on the carbon atom adjacent to it. The shift in the electron distribution induces a dipole moment on the oxygen molecules near the surface of the carbon atom and initiates the oxygen reduction process on the carbon active sites through adsorption.

The effect of doping alters the physical structure of the pristine CNT and is closely related to the catalytic effect of the NCNT. The doping of nitrogen induces a surface defect of the CNTs through the formation of pentagon–heptagon pairs in the hexagonal graphitic network of the carbon ring. Inspection from Raman spectra suggests that the degree of surface defect of CNTs increases with the increase in nitrogen doping level. Different nitrogen precursors in the synthesis of nitrogen-doped CNTs result in various degrees of surface defects (Wong et al. 2013). Figure 6.14 shows that the nitrogen-doped CNTs differ in tube thickness and surface defect level as a result of using different nitrogen precursors. The tube thickness and structure are correlated to the chemical states of nitrogen dopant in the carbon matrix. Nitrogen can be doped into the carbon matrix via four bonding states, namely, pyridinic-N, quaternary-N/graphitic-N, pyrrole-N and pyridinic-N oxides.

Experimentally, the formation of different bonding states of nitrogen in the carbon matrix results from the change in synthesis temperature and nitrogen precursor. X-ray photoelectron spectroscopy analysis, as shown in Fig. 6.15, indicates that with high pyridinic-N and the presence of oxides on the NCNT surface, the tube

Fig. 6.14 Different tube morphology of NCNT synthesis from nitrogen precursors of (**a**) aniline, (**b**) diethylamine and (**c**) ethylenediamine (Reprinted from Wong et al. 2013. Copyright with permission from Elsevier)

Fig. 6.15 XPS spectra of NCNT synthesis from nitrogen precursors of (*a*) aniline, (*b*) diethylamine and (*c*) ethylenediamine (Reprinted from Wong et al. 2013. Copyright with permission from Elsevier)

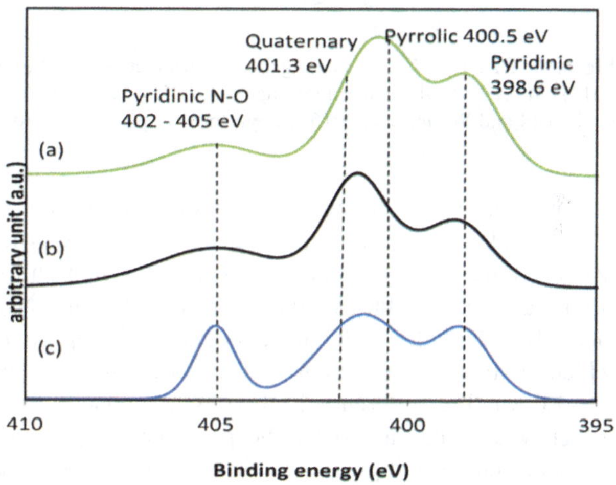

surface appears more corrugated and smaller in tube diameter. This condition is caused by the selective adsorption of nitrogen on the zigzag-shaped nanotube edge to create a curvature defect through the formation of pentagon-type defects. More incoming nitrogen dopants on the CNT lead to saturation at the tube edge or tube end and inhibit the growth of the nanotube. The pyridinic-like structure on the nanotube wall eventually decreases the wall stability. Thus, shorter tubes, smaller diameters and highly corrugated tubes are obtained. The protruding edges on the tube surface are believed to serve as the active sites for oxygen reduction.

The catalytic effect of NCNT has been investigated for oxygen reduction reaction in both alkaline and proton exchange membrane fuel cells. Collaboratively, NCNT shows comparable catalytic activity to Pt catalyst in alkaline electrolyte. Low intrinsic activity was observed in the acidic electrolyte which mimics PEMFC conditions. Figure 6.16 shows the rotating ring disk voltammetry on NCNT under oxygen saturated condition in (a) acidic (Wong et al. 2015) and (b) alkaline electrolyte (Wong et al. 2014). The oxygen reduction activities in both electrolytes are differentiated by the onset potential and presence of current limiting plateau. In alkaline medium, a well-defined current limiting plateau was observed when the

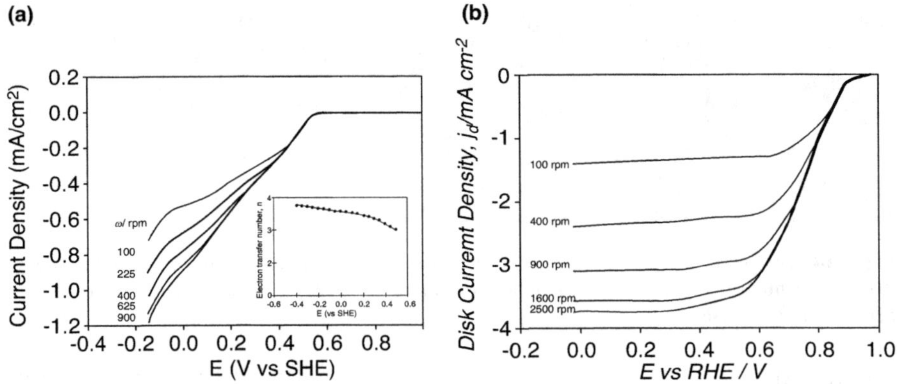

Fig. 6.16 RRDE on NCNT in oxygen-saturated electrolyte of (**a**) 0.5 H_2SO_4 solution (Wong et al. 2014) (**b**) 0.1 M KOH solution (Wong et al. 2015), at scan rate of 5 mV/s (Reprinted from Wong et al. 2014 and Wong et al. 2015. Copyright with permission from Elsevier)

potential was swept below 0.3 V with the onset potential recorded at 0.9 V versus RHE. No clear current limiting plateau was observed when the test was performed in acidic medium. This result indicates that a non-uniform rate of charge transfer occurred on the surface of the electrode when the NCNT was exposed to acidic electrolyte. In addition, the onset potential was approximately 0.55 V versus SHE, which is much lower than that reported using platinum as catalyst (approx. 0.9–1.0 V). The difference in the oxygen reduction kinetics in various pH values is believed to be caused by the presence of quinone species on the surface of aromatic ring. The higher catalytic effect of NCNT in alkaline electrolyte is likely attributed to the surface-confined quinone species that can serve as active catalysts (Jürmann et al. 2007). This condition can be observed when some catalytic activity was exhibited on a pristine CNT sample.

The low catalytic activity in acid electrolyte is possibly attributed to the protonation of quinone free radical species during the reduction process. However, the in-depth mechanism remains unclear. For fuel cell application, the cathodic reaction is highly desirable to proceed via a four-electron transfer pathway in which it would generate the cell potential closest to the ideal value of 1.23 V for a single stack of fuel cell. The standard thermodynamic electrode potentials for ORR processes in various electrolytes are shown in Eqs. (6.8)–(6.13). Nonetheless, the actual potential obtained experimentally is always lower than the ideal value which is influenced by the kinetic rate of the reaction that is dependent on the type of electrode, catalyst and electrolyte pH.

In acidic electrolyte:

$$O_2 + 4H^+ + 4e^- \rightarrow H_2O \quad E^0 = 1.23\,V \tag{6.8}$$

$$O_2 + 2H^+ + 2e^- \rightarrow H_2O_2 \quad E^0 = 0.70\,V \tag{6.9}$$

$$H_2O_2 + 2H^+ + 2e^- \rightarrow 2H_2O \quad E^0 = 1.76\,V \tag{6.10}$$

In alkaline electrolyte:

$$O_2 + H_2O + 4e^- \rightarrow 4OH^- \quad E^0 = 0.40\,V \tag{6.11}$$

$$O_2 + H_2O + 2e^- \rightarrow HO_2^- + OH^- \quad E^0 = -0.065\,V \tag{6.12}$$

$$HO_2^- + H_2O + 2e^- \rightarrow 3OH^- \quad E^0 = 0.867\,V \tag{6.13}$$

The reaction order, selectivity and kinetic data of the reaction are experimentally obtained via rotating ring electrode (RDE) or rotating ring disk electrode (RRDE) techniques. Koutecky–Levich plot is used to obtain the information on the mass diffusion constant and kinetic limiting current based on the Koutecky–Levich equation as follows (Eq. 6.14):

$$\frac{1}{i} = \frac{1}{i_k} + \frac{1}{B_L\omega^{0.5}} \tag{6.14}$$

where i = disc current, i_k = limiting kinetic current, BL = mass diffusion constant and ω = rotation rate (rpm). For NCNT catalyst, the K–L plots in both alkaline and acidic electrolyte in Fig. 6.17 have shown that the reaction is governed by mixed-diffusion process, which is first order with respect to the oxygen concentration.

From the RRDE data, the electron transfer number, n, and hydrogen peroxide selectivity (% H_2O_2) can be calculated based on Eqs. (6.15) and (6.16), respectively, as follows:

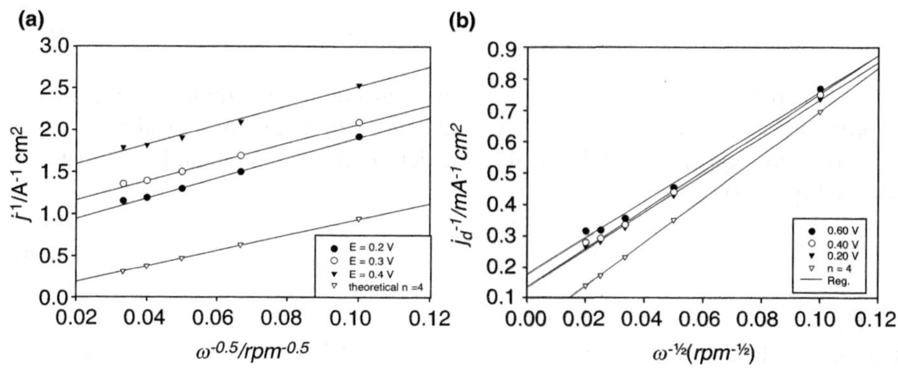

Fig. 6.17 Koutecky–Levich plot of NCNT in (**a**) 0.5 M H_2SO_4 solution and (**b**) 0.1 M KOH solution, under oxygen-saturated condition (Reprinted from Wong et al. 2014 and Wong et al. 2015. Copyright with permission from Elsevier)

$$n = \frac{4i_d}{i_d + {}^{i_r}/_{N}} \tag{6.15}$$

$$\%H_2O_2 = \frac{4-n}{2} \times 100\% \tag{6.16}$$

where $i_d = i =$ disc current, $i_r =$ ring current, N = ring collection efficiency.

The n value of 3.9 obtained in alkaline electrolyte using NCNT was found to be close to that obtained using conventional Pt catalysts (n = 3.95). This result also indicates a four-electron transfer pathway that was dominant in this process to produce OH^-. On the other hand, a much lower value was obtained when NCNT was investigated in acidic electrolyte. At a low overpotential, the maximum n value obtained was 3.6 given that the reaction occurs at a temperature of 35 °C. At room temperature, the n value obtained was only 2.5, indicating that a two-electron transfer occurred in this electrolyte to produce H_2O_2 as side products, which were found unfavourable for the fuel cell system. Based on Tafel plot at low over potential region, the reaction was under pure kinetic control. The Tafel slope used to determine the number of electron transfer in the rate determining step fell in the region of 70–85 mV/dec, which is close to 60 mV/dec (obtained on the Pt catalysts). This result shows that a two-electron transfer dominates the rate-determining step. Nonetheless, such a higher value on NCNT was due to the intermediate adsorption that occurred during the process. This fact can be related to a postulated phenomenon in which at higher pH value, oxygen adsorption was initiated with the formation of semiquinone radical anion on the NCNT surface, followed by a disproportionation reaction of super peroxide ions into peroxide and hydroxide ions.

The oxygen reduction on the NCNT in high pH values follows the steps from Eqs. (6.17) to (6.19) (Maldonado et al. 2006):

$$O_2 + H_2O + 2e^- \rightarrow HO_2^- + OH^- \tag{6.17}$$

$$HO_2^- + H_2O + 2e^- \rightarrow 3OH^- \tag{6.18}$$

$$HO_2^- \rightarrow 0.5O_2 + OH^- \tag{6.19}$$

At low pH value, the Tafel slope obtained in the low overpotential region was approximately 120 mV/dec. This value indicated a first electron transfer process to the adsorbed oxygen molecules as the rate-determining step which is postulated to follow Eq. (6.20):

$$O_{2,ads} + H^+ + e^- \rightarrow HO_{2,ads} \tag{6.20}$$

This result indicates a sluggish reaction that occurs on the NCNT surface comparable to that of the Pt/C catalyst in which the latter shows a second electron transfer process in this overpotential region as the rate-determining step as shown in Eq. (6.21):

$$O_{2,ads} + 2H^{+} + 2e^{-} \rightarrow H_2O_{2,ads} \tag{6.21}$$

The different reaction pathway of NCNT in the low pH value from Pt/C catalyst is attributed to the weaker adsorption energy of oxygen on the NCNT active sites without the presence of valence d-band found in Pt.

The major constraint in developing N-doped materials as ORR catalyst is related to the unclear origin of the active sites and the nitrogen doping level. As mentioned, various precursors would lead to different morphologies of carbon nanostructures such as NCNT and N-graphene and would result in different performance. Generally, the nitrogen precursors used are pyridine, melamine, ammonia, aliphatic amines and acetonitrile. The molecules are typically small. Upon exposure to high temperature in the range of 600–1000 °C with the presence of metal catalysts (such as ferrocene, metal phthalocyanine and others), the particles are deposited on the metal surface and grow in axial direction to form nanotubes. The presences of carbon and nitrogen elements in the precursors cause the formation of N–C bonds in various bonding states and produce NCNT. Despite the difficulty in predicting the role of different bonding states as active sites, the effect of nitrogen doping level shows a correlation with the ORR performance. Geng et al. (2011) showed that with increased nitrogen content within the CNT matrix using melamine and ferrocene as precursors, the ORR activity increased regardless of the electrolyte medium examined. A study by Chen et al. (2011) using pyridine as precursor showed that the electron transfer number increased with nitrogen doping content, thereby leading to higher water selectivity, similar to the findings reported by Higgins et al. (2011) who used aliphatic amines as precursors.

With the extensive works and positive findings, NCNT was found to be a promising metal-free catalyst for alkaline fuel cell cathode. This material has the advantage of affordability because the metal dissolution issues can be discarded owing to the non-existence of the metals. Optimisation on the catalyst loading is crucial in obtaining power density on a par with Pt/C to enable operations in the long run.

6.4.2 *Nitrogen-Doped Graphene*

The emergence of research on heteroatom doping on graphene began after the successful synthesis of NCNT in 2009 for various applications as alternative to rare and expensive noble metals such as Pt. N-doped graphene is the unrolled sheet of NCNT. It can be produced through in-situ synthesis or post-treatment of pristine graphene materials. For in-situ synthesis, several methods have been adopted, similar to that for NCNT, which include chemical vapour deposition (CVD), ion implantations and others. Nitrogen containing precursors such as ammonia, urea and melamine were often used for nitrogen doping into the graphene matrix for both methods. Similar to NCNT, most studies on nitrogen-doped graphene showed

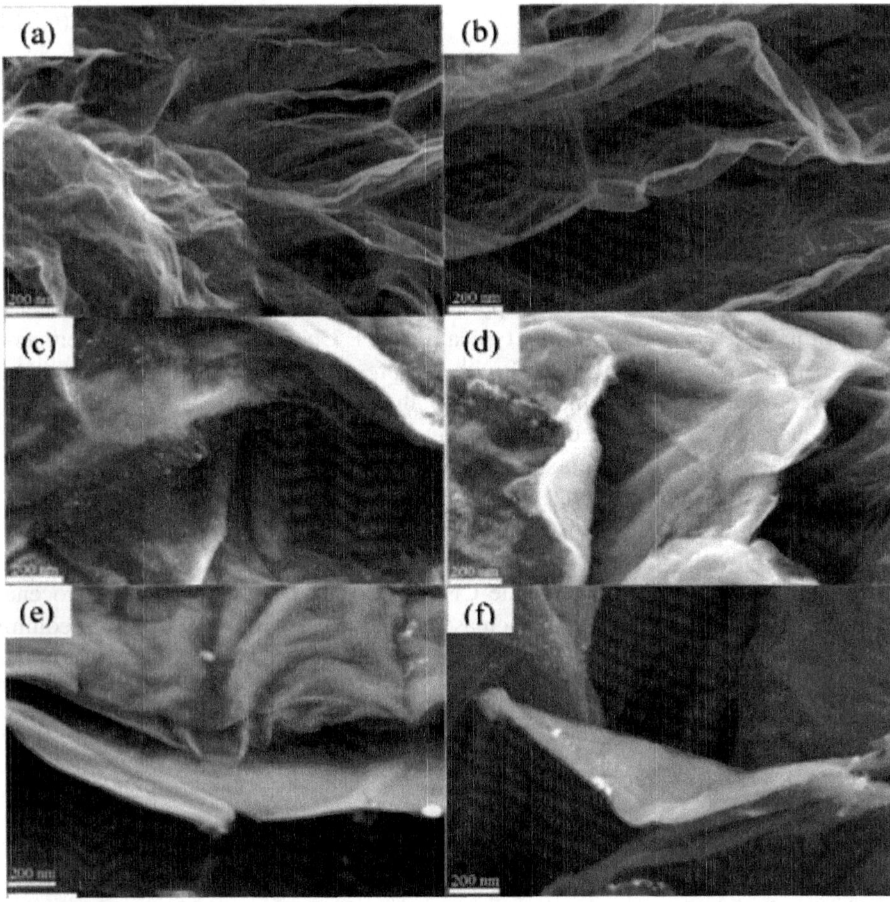

Fig. 6.18 FESEM images of nitrogen-doped graphene prepared from graphene oxide: melamine with weight ratio of (**a**) 3:1, (**b**) 1:1, (**c**) 1:3, and graphene: urea with weight ratio of (**d**) 3:1, (**e**) 1:1, (**f**) 1:3 (Reprinted from Soo et al. 2016. Copyright with permission from Elsevier)

superior performance for oxygen reduction reaction in alkaline condition. Only a few reports have shown its activity in acidic medium.

Soo et al. (2016) conducted a study to compare the ORR activity of N-doped graphene in alkaline medium using various nitrogen precursors of melamine and urea and GO as graphene source. The results showed that melamine yielded N-doped multilayer graphene, which is more transparent, in comparison to a bulk, aggregated layered structure obtained with urea, as shown in Fig. 6.18. Despite different morphologies obtained, the ORR activities were found comparable. By increasing the graphitic-N content using the same precursor, an improvement in ORR activity was obtained in this study. Lu et al. (2017) presented an in-situ chemical oxidation polymerisation followed by pyrolysis to produce 3D NG from o-phenylenediamine (OPD) with silica colloids (SiO_2) as template. Their study showed that the

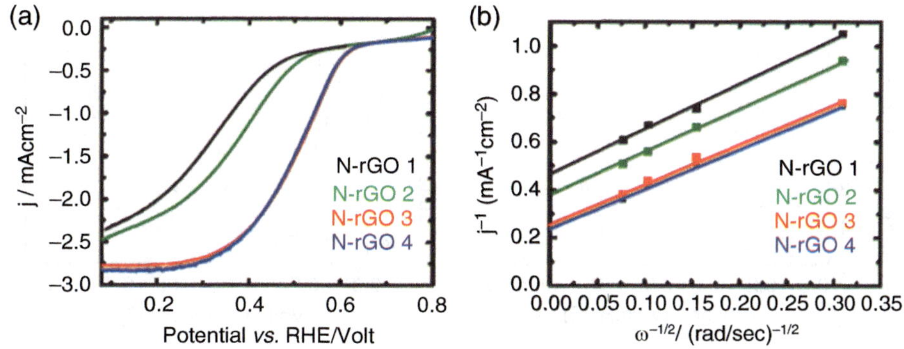

Fig. 6.19 (**a**) Polarisation curves for N-rGO prepared by different concentration of hydrazine hydrate, respectively in oxygen saturated 0.5 M H_2SO_4 electrolyte. (**b**) Corresponding Koutecky-Levich plots at 0.27 V (Reprinted/adapted by permission from Springer Nature: On the electrocatalytic activity of nitrogen-doped reduced graphene Oxide: Does the nature of nitrogen really control the activity towards oxygen reduction? by Sourav Bag and Retna Raj, Copyright (2016))

3D-NG@SiO2-2-900 catalysts exhibit the half wave potential of only 0.06 V lower than 6 wt% Pt/C. This sample with mass ratio (OPD:SiO2) of 1:2 pyrolysed at 900 °C possessed highest porosity, surface defect and mass transmission channel. The nitrogen bonding states in this sample are mainly pyridinic-N and pyrrolic-N. Both synthesis methods can produce catalytic active N-doped graphene, but the active sites are debatable.

Depending on the synthesis route and precursors, some limited studies showed that N-doped graphene possesses some catalytic activity in acidic medium. Sourav and Retna (Bag and Raj 2016) conducted a study using ammonia and hydrazine hydrate as precursor to produce N-doped rGO through a simple post-treatment method at low temperature. They revealed that the samples could catalyse ORR in an acidic medium, with onset potential of 0.55 V versus RHE obtained, similar to that obtained in NCNT, as shown in Fig. 6.19. The group claimed that the presence of pyridinic-N is responsible for improving the onset potential and limiting the current density. The doping of pyridinic-N onto the carbon matrix helps to increase the 2*p* state of C-N whilst reducing the C–C electron density. A study by Lai et al. (2012) showed that graphitic nitrogen determines the limiting current density, and pyridinic nitrogen is important in improving the onset potential.

Shui et al. (2015) demonstrated the ORR performance of N-doped graphene and N-doped vertically aligned CNT in acidic medium. N-doped graphene was shown to be inferior to N–CNT. Interestingly, the researchers tailored the two materials to be composites of N–G–CNT, and the ORR activity in acidic medium was found to exceed the performance of Pt/C. Figure 6.20 shows the electrocatalytic activities of these materials in a half cell test.

When tested in a single cell, the materials achieved a peak current density of 300 W g^{-1}, as displayed in Fig. 6.21. This result shows that hybriding CNT and graphene with proper nitrogen doping could overcome the limitation of the

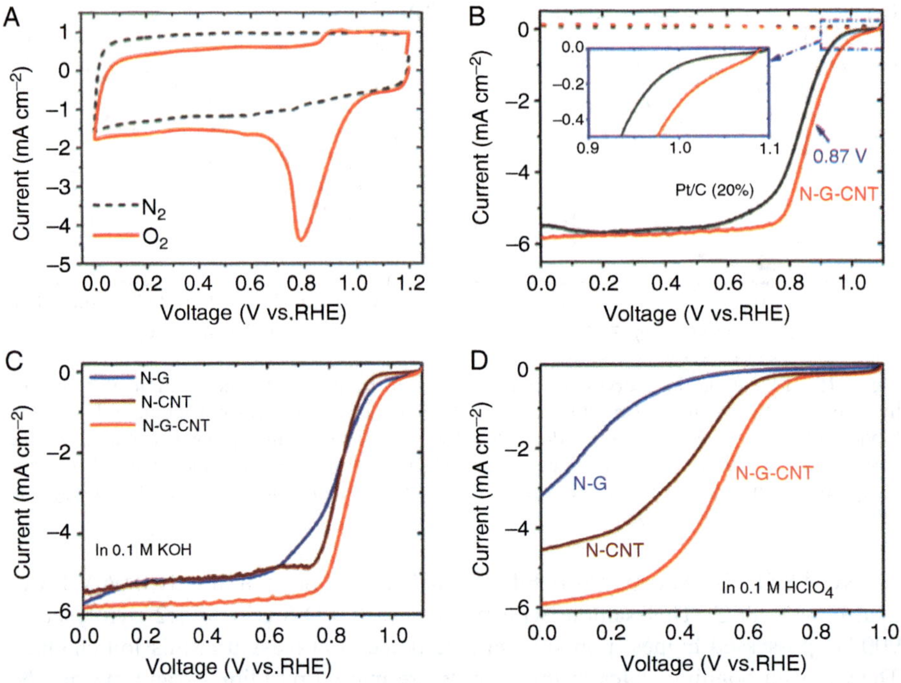

Fig. 6.20 Electrocatalytic activities of metal-free carbon nanocatalysts in half-cell tests (**a**) Cyclic voltammetry of N–G–CNT in O_2- and N_2-saturated 0.1 M KOH. (**b**) Linear sweep voltammetry (LSV) curves of N–G–CNT and Pt/C through RRDE in O2-saturated 0.1 M KOH solution, scan rate of 10 mV/s and 1600 rpm. (**c** and **d**) LSV of N–G, N–CNT and N–G–CNT in (**c**) 0.1 M KOH and (**d**) 0.1 M $HClO_4$ (Shui et al. 2015)

Fig. 6.21 Single cell performance test in PEMFC using N–G–CNT as cathodic catalyst (Shui et al. 2015)

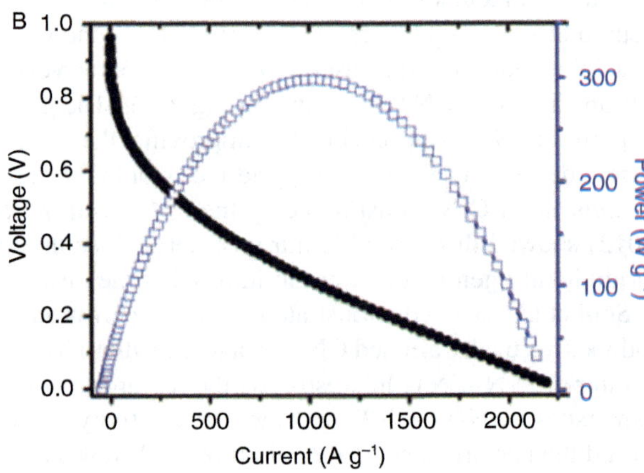

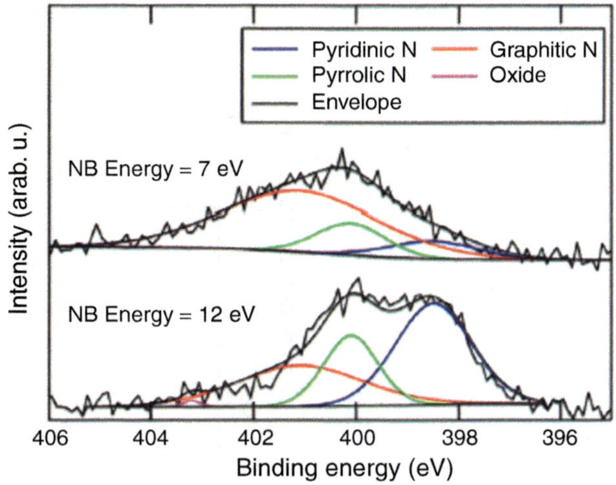

Fig. 6.22 N 1s XPS spectra of nitrogen-doped graphene produced using different neutral beam energy (Reprinted from Okada et al. 2016. Copyright with permission from Elsevier)

respective N-doped compounds possibly because of synergistic effects between CNT and graphene.

As mentioned, no consensus has been reached on the origin of the active sites for ORR on the N-doped nanocarbons owing to the various possible factors that result in complexity in analysing this aspect such as use of precursors, temperature and multilayers structure. The common consensus is that graphitic-N and pyridinic-N are the active sites. Various techniques were employed to tailor the nitrogen bonding states experimentally on the graphene surface and was tested for ORR activity to further understand this condition.

Okada et al. (2016) provided insights into the role of pyridinic-N and graphitic-N in catalysing the ORR in alkaline medium. The researchers adopted a selective doping method using different neutral beam energy in a plasma chamber to create defects on a single-layer graphene sheet and introduce graphitic-N and pyridinic-N bonding states, as shown in Fig. 6.22.

A linear sweep voltammogram in Fig. 6.23 indicates that both types of nitrogen bonding states catalysed the ORR process via different pathways in which graphitic-N could facilitate the reduction process in two steps, whereas pyridinic-N involved only one step. They concluded with the electron transfer number of $n = 3.3$ obtained on graphitic-N, compared with $n = 2.6$ for pyridinic-N. Thus, graphitic-N is more suitable for ORR when using a monolayer of graphene. Nonetheless, various findings were obtained when multilayer graphene was doped with nitrogen.

Another interesting technique used to investigate the active sites was conducted by Xing et al. (2014). The group studied the chemical composition change before and after ORR using synchrotron-based X-ray photoelectron spectroscopy analyses to determine the active sites on N-doped multilayer graphene. They revealed that pyridinic-N should be responsible as the active sites given the evidence that the oxygen reduction intermediate $OH_{(ads)}$ are chemically attached to the carbon atoms adjacent to the pyridinic-N. Figure 6.24 shows the evident attachment of $OH_{(ads)}$ on

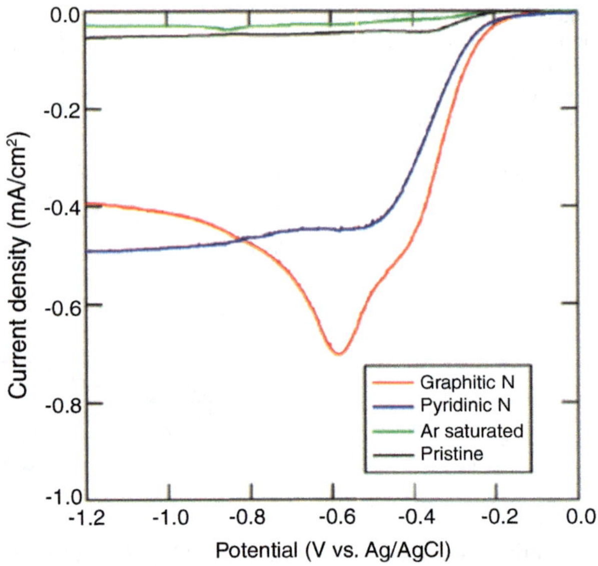

Fig. 6.23 LSV of nitrogen-doped graphene with different nitrogen bonding states (Reprinted from Okada et al. 2016. Copyright with permission from Elsevier)

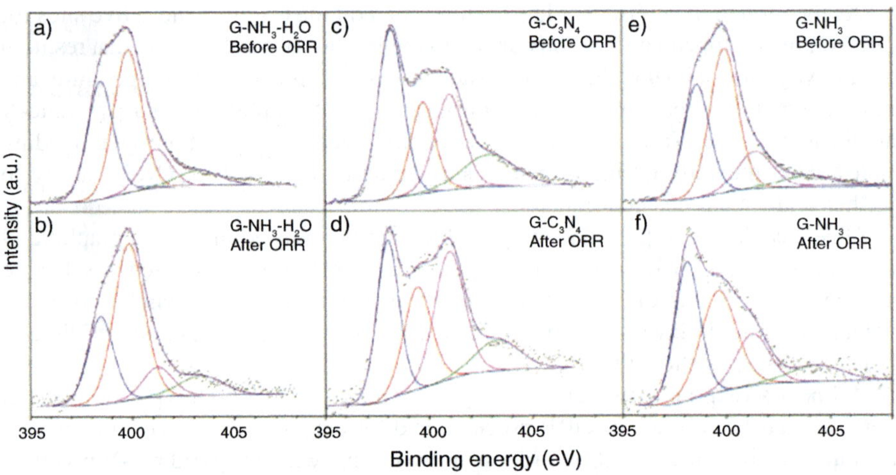

Fig. 6.24 Nitrogen 1s XPS spectra of the three multilayer graphene samples of (**a**) G-NH_3-H_2O before ORR, (**b**) G-NH_3-H_2O after ORR (**c**) G-C_3N_4 before ORR (**d**) G-C_3N_4 after ORR (**e**) G-NH_3 before ORR and (**f**) G-NH_3 after ORR. The least-square fitted peaks are pyridinic nitrogen at 398.5 eV (blue), "pyrrolic" nitrogen at 399.8 eV (red), graphitic nitrogen at 401.2 eV (purple) and nitrogen oxide at 403 eV (green) (Xing et al. 2014)

pyridinic-N with the increase in pyrrolic-N and decrease in pyridinic-N after ORR. Aside from showing bonding states that influence the catalytic activity, the microstructure of the N-doped graphene has significant effects on ORR activity. This result leaves room for studying how the creation of surface defects of graphene would alter the microstructure of the surface, and more innovative techniques may be required to provide further insights.

Density functional theory (DFT) calculations have been widely used to simplify the process on identifying the active sites and elucidating ORR mechanisms. Attempts have been made to examine both alkaline and acidic mediums using this computational molecular modelling approach. In most studies, DFT calculations focus on graphitic-N as nitrogen bonding states on graphene owing to suggestions that it is likely to exhibit higher ORR activity than that on pyridinic-N. As mentioned, O_2 can be reduced via two different reduction pathways. Four-electron transfer was proposed as the predominant pathway in alkaline medium in which O_2 is completely reduced to two OH^-, whereas two-electron transfer pathway serves as the competing reaction in which O_2 is partially reduced to OOH^-. The latter pathway would lead to a lower potential because of inadequate use of O_2. At present, metal-based ORR catalysts are ranked in terms of ORR activity based on the volcano curve plot and are described as a function of $O_{(ads)}$ adsorption energy. Although Pt sits on top of the curve, it remains as the most active ORR catalyst to date. Thus, obtaining quantitative data on the N-doped graphene was simpler to use in evaluating the ORR activity following the aforementioned curve.

In general, two O_2 reduction mechanisms were proposed regardless of electron transfer pathway: (i) associative and (ii) dissociative mechanism.

The associative mechanism can be described from Eqs. (6.22) to (6.26):

$$O_2 + * \rightarrow O_{2(ads)} \tag{6.22}$$

$$O_{2(ads)} + H_2O + e^- \rightarrow OOH_{(ads)} + OH^- \tag{6.23}$$

$$OOH_{(ads)} + e^- \rightarrow O_{(ads)} + OH^- \tag{6.24}$$

$$O_{(ads)} + H_2O + e^- \rightarrow OH_{(ads)} + OH^- \tag{6.25}$$

$$OH_{(ads)} + e^- \rightarrow OH^- + * \tag{6.26}$$

where * denotes free adsorption sites on the N-graphene surface.

The dissociative mechanism can be described from Eqs. (6.27–6.29):

$$\frac{1}{2}O_2 + * \rightarrow O_{(ads)} \tag{6.27}$$

$$O_{(ads)} + H_2O + e^- \rightarrow OH_{(ads)} + OH^- \tag{6.28}$$

$$OH_{(ads)} + e^- \rightarrow OH^- + * \tag{6.29}$$

Yu et al. (2011) conducted the DFT calculations on graphitic N-doped graphene with N concentration of 8.33% as obtained from their previous experiment, which showed satisfactory ORR activity. In their study, they simulated two structures differentiated by the location where N was doped, as shown in Fig. 6.25. In the first structure (S1), one N atom is separated by two C atoms. In the second structure

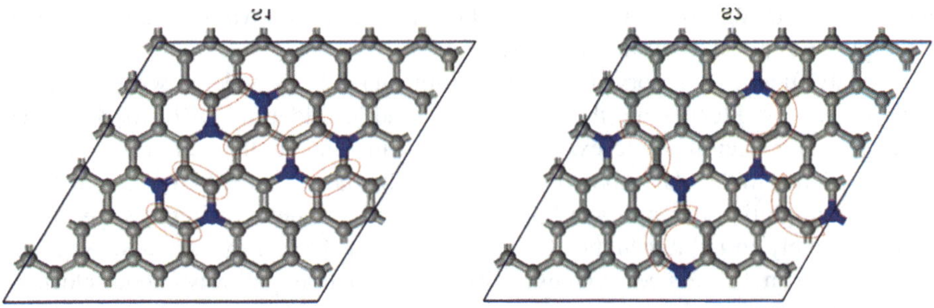

Fig. 6.25 Simulated N-graphene structure with different nitrogen doping location (Reprinted from Yu et al. 2011. Copyright with permission from Elsevier)

(S2), one N atom is separated by three C atoms. The group modelled the N-doped graphene surface with the presence of water layers on its surface because the reaction involves heterogeneous liquid–gas phase reaction on solid catalysts. Water plays an important role which would affect the stability of the intermediates and reaction barriers.

The simulation illustrates that both structures would provide similar reaction mechanisms of ORR, as shown in Fig. 6.26. The group found that the dissociative mechanisms were not favourable because of the very high activation energy (1.56 eV) required to dissociate $O_{2(ads)}$ into $O_{(ads)}$, whereas the effective barrier for associative mechanism was appreciably lower (0.56 eV). In the real alkaline fuel cell condition, 1.56 eV is not reasonable for the dissociation process to be realised. The calculation also showed that the O–O bond breaking of $OOH_{(ads)}$ was exothermic with an effective barrier of 0.56 eV, whereas $OOH_{(ads)}$ desorption was unfavourable with an energy loss of 0.22 eV (endothermic). This result correlates with the experimental results that only minimal peroxide is produced and suggested that the ORR mainly proceed with a four-electron transfer pathway in alkaline media. Through the kinetic analysis, the rate-determining step for the ORR reaction was the removal of $O_{(ads)}$, similar to the postulated results on Pt <1 1 1>. Thus, the overall ORR activity was ruled by the rate of $OH_{(ads)}$ removal.

On the other hand, an earlier study by Okamoto (2009) provided insights on the ORR mechanism of N-doped graphene in acidic media. N-doped graphene has low intrinsic activity in acidic media. This work showed that reduction of oxygen via four- and two-electron pathway to produce H_2O and $H{=}O_2$, respectively, was feasible. Further study by Zhang et al. (2013) determined the favourable pathway in acidic media. The group modelled a similar 6x6 graphene supercell with one nitrogen doped into the matrix with graphitic-N bonding state. They calculated the reaction energy and activation barrier for each possible reaction step involved in both four- and two-electron pathways. Figure 6.27 shows a schematic of the ORR pathways of N-doped graphene in acidic medium. The reaction energy and activation barrier are presented in Table 6.2.

The calculations agree with findings from Okamoto et al. (2009) in which both pathways are feasible. However, the two-electron pathway was found more

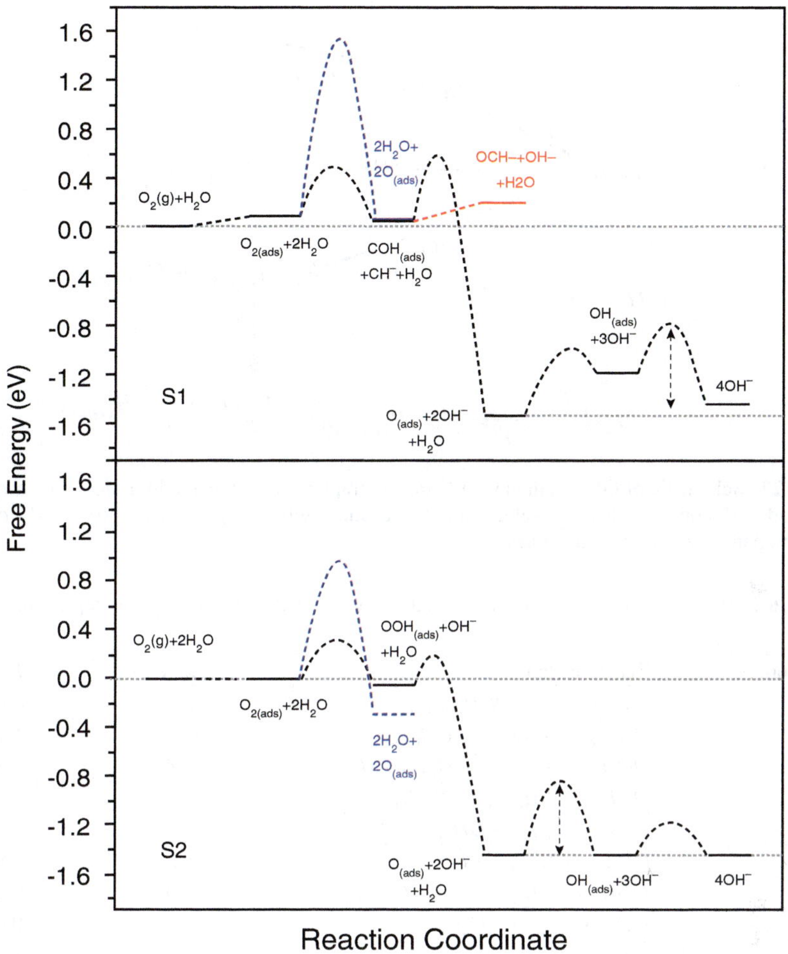

Fig. 6.26 Free energy diagram for ORR on S1 and S2 under the condition of 0.04 V and pH = 14. The black line indicates the intermediates and energy barriers of associative mechanism while the blue line indicates that of dissociative mechanism. The red line indicates the formation of OOH– (Reprinted from Yu et al. 2011. Copyright with permission from Elsevier)

favourable as observed from the low activation barrier of 0.09 eV from the reduction of OOH into H_2O_2, with −2.21 eV energy released. This result shows that H_2O_2 molecules can be formed easily. The group also deduced the rate-determining step for both pathways. For the two-electron pathway, step (i) was shown to be the rate-determining step with highest activation barrier (0.63 eV) calculated among steps (i), (vii) and (viii). For the four-electron pathway, the rate-determining step was step (vi) with an activation barrier of 0.82 eV. Comparison of the two activation barriers shows that the two-electron pathway was concluded as more energetically favourable.

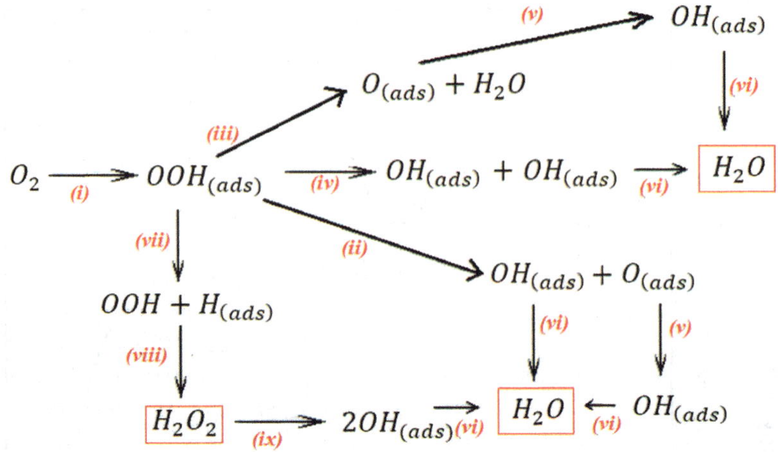

Fig. 6.27 Schematic of ORR pathway on N-doped graphene in acidic medium. Reaction steps (*i*), (*vii*) and (*viii*) corresponds to two-electron pathway and reaction steps (*i*)–(*vi*) corresponds to four-electron pathway via different routes

Table 6.2 Reaction energy (E_{rea}) and activation barrier (E_a) of each reaction step (Zhang et al. 2013)

Step no.	Reaction step	E_{rea} (eV)	E_a (eV)
(i)	$O_2 + H_{(ads)} \rightarrow OOH_{(ads)}$	−0.96	0.63
(ii)	$OOH_{(ads)} \rightarrow O_{(ads)} + OH_{(ads)}$	0.25	1.18
(iii)	$OOH_{(ads)} + H_{(ads)} \rightarrow O_{(ads)} + H_2O$	−2.99	0.55
(iv)	$OOH_{(ads)} + H_{(ads)} \rightarrow 2OH_{(ads)}$	−2.91	0.72
(v)	$O_{(ads)} + H_{(ads)} \rightarrow OH_{(ads)}$	−2.21	0.54
(vi)	$OH_{(ads)} + H_{(ads)} \rightarrow H_2O$	−2.23	0.82
(vii)	$OOH_{(ads)} + H_{(ads)} \rightarrow OOH + H_{(ads)}$	0.16	0.23
(viii)	$OOH + H_{(ads)} \rightarrow H_2O_2$	−2.21	0.09
(ix)	$H_2O_2 \rightarrow 2OH_{(ads)}$	−0.48	1.39

Thus, experimental and theoretical studies summarise that the two-electron pathway was more favourable in acidic medium, whereas the four-electron pathway was preferred in alkaline medium. This result is consistent with the experimental work which showed that N-doped CNT and N-doped graphene are more active catalysts in alkaline medium and are suitable for alkaline fuel cell applications.

6.4.3 Boron-Doped Carbon Nanostructures

In contrast to nitrogen, doping of boron would result in electron deficiency within the aromatic ring and create a partial positive charge on boron and a partial negative charge on adjacent carbon, thereby creating active sites with different surface charge

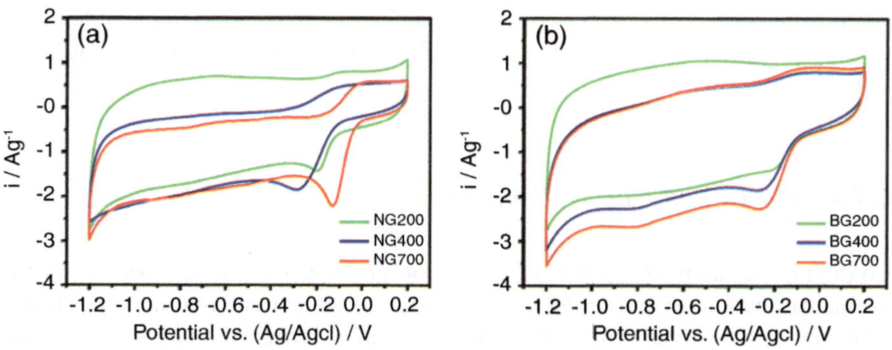

Fig. 6.28 CV of (**a**) N-doped and (**b**) B-doped graphene in O_2-saturated 0.1 KOH electrolyte (Reprinted from Xu et al. 2014. Copyright with permission from Elsevier)

densities. An earlier work by Yang et al. (2011), who were motivated by the success of N-doped carbon in catalysing ORR activity, attempted to dope electron-deficient boron into CNT to examine its performance. With similar difference in electronegativity between boron (2.04) and carbon (2.55) to nitrogen (3.04) and carbon, B-doped carbon was predicted to possess activity towards ORR. The performance of BCNT was reported to be relatively low compared with that of Pt/C, recording highest electron transfer number of 2.5 with highest boron content of 2.24 at percentage assigned as B3CNTs. The group performed theoretical DFT calculations to further reveal the origin of the electrocatalytic activity. The B–C sigma bonds were being polarised and enabled the adsorption of oxygen molecule on the positively charged boron atom.

Despite the claim that boron-doped carbon compounds can act as metal-free catalysts, fewer studies were found compared with that of N-doped carbon, mainly due to the lower ORR activity of boron-doped carbon. A comparative study was performed by Xu et al. (2014) on the performance of B- and N-doped graphene on ORR activity in alkaline electrolyte, as shown in Fig. 6.28. N-doped graphene exhibited an onset potential of 0.0 V versus Ag/AgCl, more positive than that on B-doped graphene at −0.1 V, thereby indicating that the ORR occurs more easily on the N-doped graphene. In most studies, the boron precursors such as boric acid and boron oxides contain oxide elements. Through high-temperature pyrolysis or thermal annealing, the doped boron would form the bonding states of BCO_2 and BC_2O aside from B_{subs}–C bond. Thus, the CV curve shows that B-doped graphene appeared to be more capacitive towards ORR, which is attributed to the presence of oxygen atoms in the B-doped carbon compound.

A later work by Vineesh et al. (2015) used rhombohedral boron carbide (B4C) as boron source to dope into graphene. This work successfully showed an average electron transfer number of 3.9 towards ORR, with high current density of 16.32 mA cm^{-2} at 0.6 V and high methanol tolerance in alkaline medium. The ORR activity was comparable to that of Pt/C. From the XPS spectra, no oxygen was found bonded with boron and carbon. B_{subs}-C, B_4C and BC_3 were detected on the graphene surface. Thus, searching for suitable boron precursors is crucial in strategically tuning the electronic properties of B-doped compound to achieve desirable activity.

6.4.4 Co-Doped Carbon Nanostructures

Co-doping carbon nanostructures with two or more heteroatoms has become one of the main directions in modifying their properties to achieve catalytic activities higher than the single-doped carbon nanostructures. In principle, co-doping of various heteroatoms would disturb the electron distribution to a greater extent, thereby inducing a greater dipole moment between the carbon and heteroatoms. As a result, the oxygen adsorption could occur more readily because of the stronger interaction between the oxygen atom and positively charged atom on the carbon nanostructures. Several attempts (Choi et al. 2013; Zehtab Yazdi et al. 2015; Zheng et al. 2013) using different approaches such as two-step CVD methods to dope nitrogen and boron sequentially into the carbon matrix and microwave-hydrothermal method to dope both atoms simultaneously have been shown to successfully improve the ORR activity compared with that of the simply doped one. However, few of these findings were found less active than Pt/C. Wang et al. (2012) constructed a BCN graphene via thermal annealing of GO with boric acid and ammonia with different doping levels and bonding states. Based on DFT calculation, doping at intermediate level would result in lower energy gap and higher charge density, which is shown to possess higher ORR activity. The small HOMO–LUMO gap means that the structure has low kinetic stability and high chemical reactivity, in which it is energetically favourable to allow the electrons to occupy and leave the orbital with ease, thereby allowing the formation of activated complexes for ORR to precede. Interestingly, $B_{12}C_{77}N_{11}H_{26}$ possesses higher onset potential than Pt/C, indicating superior ORR performance. We conclude that carbon nanostructures are beneficial materials that can be used to tailor the structure at the atomic level through simple experimental techniques to achieve the desired physicochemical properties for applications such as fuel cell catalysts.

6.5 Future Perspective, Challenges and Opportunities

Significant improvements particularly on the performance of various types of fuel cells were achieved at the experimental stage with the implementation of nanomaterials in the components of fuel cells, particularly at the electrode. Unprecedented opportunities are blooming with the accelerating improvement of nanomaterials synthesis methods. The introduction of nanomaterials opens new insight on the properties changes due to the quantum effect in nanosized materials that enable the utilisation of many types of noble-free elements in catalysing the anodic and cathodic reaction in fuel cells. For instance, cluster of atoms such as macromolecules, fibre-like nanosized carbon such as nanorods and nanotubes, and nanocrystalline Pt- and non-Pt metals have been found beneficial in the electrocatalysis process. In addition, carbon allotropes in nanostructured forms such as carbon nanotubes, carbon nanofibers and graphene, are found as excellent as catalyst support which can also be modified into metal-free catalysts. This shows that nanotechnology would be one of the promising solution towards the technical barrier exists in fuel cell technology mainly hindered by the high component cost due to high utilisation of expensive Pt.

Despite the many advantages, there are two major challenges needs to be resolved for practical applications. Firstly, the dissolution issues of nanosized metals in real fuel cell operation which eventually deteriorate the performance. Secondly, the mass production of nanosized materials with desired ordered structures that will promise consistent results in all batches. These issues are foreseen to be solved with the future advancement of nanotechnology. Looking at the ongoing progress for the discovery of the new nanomaterials, there exists opportunities to engineer new forms of nanostructure such as nanocage and nanoframe that may be possible to solve for the catalyst dissolution and deactivation issues. In addition, it is possible to further tune the carbon support such as carbon nanotubes and graphene on its electronic structure through doping to provide stronger anchor with the nanocrystalline that may reduce the possibility of nanoparticle dissolutions. For the latter challenge, this requires further development on nanotechnology that allows the materials synthesis through automation at the nanoscale.

6.6 Conclusion

Nanostructured materials are crucial in improving the electrocatalytic activity in fuel cell electrodes and have the following main effects: (1) lower catalyst loading required because of larger specific surface area of metal nanoparticles for catalytic reactions, (2) carbon nanostructures support homogenous deposition of metal nanoparticles and (3) the materials act as catalytic sites upon heterogeneous doping onto the carbon nanostructures. Further efforts are expected in tailoring the nano-structure compounds to achieve the benchmarked activity for Pt/C.

References

Abdel Hameed RM (2017a) Enhanced ethanol electro-oxidation reaction on carbon supported Pd-metal oxide electrocatalysts. J Colloid Interface Sci 505:230–240. https://doi.org/10.1016/j.jcis.2017.05.095

Abdel Hameed RM (2017b) Facile preparation of Pd-metal oxide/C electrocatalysts and their application in the electrocatalytic oxidation of ethanol. Appl Surf Sci 411:91–104. https://doi.org/10.1016/j.apsusc.2017.03.108

Acres GJK et al (1997) Electrocatalysts for fuel cells. Catal Today 38:393–400. https://doi.org/10.1016/s0920-5861(97)00050-3

Antoniassi RM, Silva JCM, Oliveira Neto A, Spinacé EV (2017) Synthesis of Pt+SnO2/C electrocatalysts containing Pt nanoparticles with preferential (100) orientation for direct ethanol fuel cell. Appl Catal B Environ 218:91–100. https://doi.org/10.1016/j.apcatb.2017.06.031

Anwar MT, Yan X, Shen S, Husnain N, Zhu F, Luo L, Zhang J (2017) Enhanced durability of Pt electrocatalyst with tantalum doped titania as catalyst support. Int J Hydrog Energy 42:30750–30759. https://doi.org/10.1016/j.ijhydene.2017.10.152

Bag S, Raj CR (2016) On the electrocatalytic activity of nitrogen-doped reduced graphene oxide: does the nature of nitrogen really control the activity towards oxygen reduction? J Chem Sci 128 (3):339–347. https://doi.org/10.1007/s12039-016-1034-z

Carvalho LL, Colmati F, Tanaka AA (2017) Nickel–palladium electrocatalysts for methanol, ethanol, and glycerol oxidation reactions. Int J Hydrog Energy 42:16118–16126. https://doi.org/10.1016/j.ijhydene.2017.05.124

Chen Y, Wang J, Liu H, Banis MN, Li R, Sun X, Sham TK, Ye S, Knights S (2011) Nitrogen doping effects on carbon nanotubes and the origin of the enhanced electrocatalytic activity of supported pt for proton-exchange membrane fuel cells. J Phys Chem C 115(9):3769–3776. https://doi.org/10.1021/jp108864y

Chetty R et al (2009) PtRu nanoparticles supported on nitrogen-doped multiwalled carbon nanotubes as catalyst for methanol electrooxidation. Electrochim Acta 54:4208–4215. https://doi.org/10.1016/j.electacta.2009.02.073

Choi CH, Chung MW, Kwon HC, Park SH, Woo SI (2013) B, N- and P, N-doped graphene as highly active catalysts for oxygen reduction reactions in acidic media. J Mater Chem A 1 (11):3694–3699. https://doi.org/10.1039/C3TA01648J

Davies JC, Hayden BE, Offin L (2017) Stabilising oxide core—platinum shell catalysts for the oxygen reduction reaction. Electrochim Acta 248:470–477. https://doi.org/10.1016/j.electacta.2017.07.132

El-Khatib KM, Hameed RMA, Amin RS, Fetohi AE (2017) Core–shell structured Cu@Pt nanoparticles as effective electrocatalyst for ethanol oxidation in alkaline medium. Int J Hydrog Energy 42:14680–14696. https://doi.org/10.1016/j.ijhydene.2017.04.118

Esfandiari A, Kazemeini M, Bastani D (2016) Synthesis, characterization and performance determination of an Ag@Pt/C electrocatalyst for the ORR in a PEM fuel cell. Int J Hydrog Energy 41:20720–20730. https://doi.org/10.1016/j.ijhydene.2016.09.097

Ferreira Frota E, Silva de Barros VV, de Araújo BRS, Gonzaga Purgatto Â, Linares JJ (2017) Pt/C containing different platinum loadings for use as electrocatalysts in alkaline PBI-based direct glycerol fuel cells. Int J Hydrog Energy 42:23095–23106. https://doi.org/10.1016/j.ijhydene.2017.07.125

Fetohi AE, Amin RS, Hameed RMA, El-Khatib KM (2017) Effect of nickel loading in Ni@Pt/C electrocatalysts on their activity for ethanol oxidation in alkaline medium. Electrochim Acta 242:187–201. https://doi.org/10.1016/j.electacta.2017.05.022

Garcia AC, Ferreira EB, Silva de Barros VV, Linares JJ, Tremiliosi-Filho G (2017) PtAg/MnOx/C as a promising electrocatalyst for glycerol electro-oxidation in alkaline medium. J Electroanal Chem 793:188–196. https://doi.org/10.1016/j.jelechem.2016.11.053

Geatches DL, Clark SJ, Greenwell HC (2012) Iron reduction in nontronite-type clay minerals: modelling a complex system. Geochim Cosmochim Acta 81:13–27. https://doi.org/10.1016/j.gca.2011.12.013

Geng D, Liu H, Chen Y, Li R, Sun X, Ye S, Knights S (2011) Non-noble metal oxygen reduction electrocatalysts based on carbon nanotubes with controlled nitrogen contents. J Power Sources 196(4):1795–1801. https://doi.org/10.1016/j.jpowsour.2010.09.084

Guo S-D, Hu X-C, Yang J-G, Chen H, Zhou Y (2016) Palladium nanoparticles supported on hollow mesoporous Tungsten carbide microsphere as electrocatalyst for formic acid oxidation. J Fuel Chem Technol 44:698–702. https://doi.org/10.1016/S1872-5813(16)30034-2

Habibi B, Mohammadyari S (2016) Palladium nanoparticles/nanostructured carbon black composite on carbon–ceramic electrode as an electrocatalyst for formic acid fuel cells. J Taiwan Inst Chem Eng 58:245–251. https://doi.org/10.1016/j.jtice.2015.05.033

Higgins D, Chen Z, Chen Z (2011) Nitrogen doped carbon nanotubes synthesized from aliphatic diamines for oxygen reduction reaction. Electrochim Acta 56(3):1570–1575. https://doi.org/10.1016/j.electacta.2010.11.003

Hoque MA et al (2016) Optimization of sulfur-doped graphene as an emerging platinum nanowires support for oxygen reduction reaction. Nano Energy 19:27–38. https://doi.org/10.1016/j.nanoen.2015.11.004

Hsieh C-T, Lin J-Y, Yang S-Y (2009) Carbon nanotubes embedded with PtRu nanoparticles as methanol fuel cell electrocatalysts. Physica E 41:373–378. https://doi.org/10.1016/j.physe.2008.08.060

Huang T, Zhang D, Xue L, Cai W-B, Yu A (2009) A facile method to synthesize well-dispersed PtRuMoOx and PtRuWOx nanoparticles and their electrocatalytic activities for methanol oxidation. J Power Sources 192:285–290. https://doi.org/10.1016/j.jpowsour.2009.03.037

Jeon MK, Cooper JS, McGinn PJ (2009a) Investigation of PtCoCr/C catalysts for methanol electro-oxidation identified by a thin film combinatorial method. J Power Sources 192:391–395. https://doi.org/10.1016/j.jpowsour.2009.02.087

Jeon MK, Zhang Y, McGinn PJ (2009b) Effect of reduction conditions on electrocatalytic activity of a ternary PtNiCr/C catalyst for methanol electro-oxidation. Electrochim Acta 54:2837–2842. https://doi.org/10.1016/j.electacta.2008.11.027

Jia J, Shao M, Wang G, Deng W, Wen Z (2016) Cu3PdN nanocrystals electrocatalyst for formic acid oxidation. Electrochem Commun 71:61–64. https://doi.org/10.1016/j.elecom.2016.08.009

Jia N, Shi Y, Zhang S, Chen X, Chen P, An Z (2017) Carbon nanobowls supported ultrafine palladium nanocrystals: a highly active electrocatalyst for the formic acid oxidation. Int J Hydrog Energy 42:8255–8263. https://doi.org/10.1016/j.ijhydene.2016.12.136

Jiang J, Kucernak A (2009) Synthesis of highly active nanostructured PtRu electrocatalyst with three-dimensional mesoporous silica template. Electrochem Commun 11:623–626. https://doi.org/10.1016/j.elecom.2008.12.040

Jukk K, Kongi N, Tammeveski K, Arán-Ais RM, Solla-Gullón J, Feliu JM (2017) Loading effect of carbon-supported platinum nanocubes on oxygen electroreduction. Electrochim Acta 251:155–166. https://doi.org/10.1016/j.electacta.2017.08.099

Jürmann G, Schiffrin DJ, Tammeveski K (2007) The pH-dependence of oxygen reduction on quinone-modified glassy carbon electrodes. Electrochim Acta 53(2):390–399. https://doi.org/10.1016/j.electacta.2007.03.053

Karousos DS, Desdenakis KI, Sakkas PM, Sourkouni G, Pollet BG, Argirusis C (2017) Sonoelectrochemical one-pot synthesis of Pt—carbon black nanocomposite PEMFC electrocatalyst. Ultrason Sonochem 35:591–597. https://doi.org/10.1016/j.ultsonch.2016.05.023

Karuppasamy L, Chen CY, Anandan S, Wu JJ (2017) High index surfaces of Au-nanocrystals supported on one-dimensional MoO_3-nanorod as a bi-functional electrocatalyst for ethanol oxidation and oxygen reduction. Electrochim Acta 246:75–88. https://doi.org/10.1016/j.electacta.2017.06.040

Kübler M, Jurzinsky T, Ziegenbalg D, Cremers C (2018) Methanol oxidation reaction on core-shell structured Ruthenium-Palladium nanoparticles: relationship between structure and electrochemical behavior. J Power Sources 375:320–334. https://doi.org/10.1016/j.jpowsour.2017.07.114

Lai L et al (2012) Exploration of the active center structure of nitrogen-doped graphene-based catalysts for oxygen reduction reaction. Energy Environ Sci 5:7936–7942. https://doi.org/10.1039/C2EE21802J

Lee KR, Jeon MK, Woo SI (2009) Composition optimization of PtRuM/C (M=Fe and Mo) catalysts for methanol electro-oxidation via combinatorial method. Appl Catal B Environ 91:428–433. https://doi.org/10.1016/j.apcatb.2009.06.011

Li D, Yu Y, Jin Q, Gao Z (2014) Maximum power efficiency operation and generalized predictive control of PEM (proton exchange membrane) fuel cell. Energy 68:210–217. https://doi.org/10.1016/j.energy.2014.02.104

Li S et al (2017) Monodispersed porous flowerlike PtAu nanocrystals as effective electrocatalysts for ethanol oxidation. Appl Surf Sci 422:172–178. https://doi.org/10.1016/j.apsusc.2017.05.246

Liao M et al (2017) Highly active Pt decorated Pd/C nanocatalysts for oxygen reduction reaction. Int J Hydrog Energy 42:24090–24098. https://doi.org/10.1016/j.ijhydene.2017.08.003

Lu X, Li Z, Yin X, Wang S, Liu Y, Wang Y (2017) Controllable synthesis of three-dimensional nitrogen-doped graphene as a high performance electrocatalyst for oxygen reduction reaction. Int J Hydrog Energy 42(27):17504–17513. https://doi.org/10.1016/j.ijhydene.2017.02.090

Ma Y et al (2017) A general strategy to the synthesis of carbon-supported PdM (M = Co, Fe and Ni) nanodendrites as high-performance electrocatalysts for formic acid oxidation. J Energy Chem 26:1238–1244. https://doi.org/10.1016/j.jechem.2017.10.024

Maldonado S, Morin S, Stevenson KJ (2006) Structure, composition, and chemical reactivity of carbon nanotubes by selective nitrogen doping. Carbon 44(8):1429–1437. https://doi.org/10.1016/j.carbon.2005.11.027

Maya-Cornejo J, Arjona N, Guerra-Balcázar M, Álvarez-Contreras L, Ledesma-García J, Arriaga LG (2014) Synthesis of Pd-Cu bimetallic electrocatalyst for ethylene glycol and glycerol oxidations in alkaline media. Procedia Chem 12:19–26. https://doi.org/10.1016/j.proche.2014.12.036

Maya-Cornejo J, Garcia-Bernabé A, Compañ V (2018) Bimetallic Pt-M electrocatalysts supported on single-wall carbon nanotubes for hydrogen and methanol electrooxidation in fuel cells applications. Int J Hydrog Energy 43:872–884. https://doi.org/10.1016/j.ijhydene.2017.10.097

Moreno B, Chinarro E, Pérez JC, Jurado JR (2007) Combustion synthesis and electrochemical characterisation of Pt–Ru–Ni anode electrocatalyst for PEMFC. Appl Catal B Environ 76:368–374. https://doi.org/10.1016/j.apcatb.2007.06.012

Nabil Y, Cavaliere S, Harkness IA, Sharman JDB, Jones DJ, Rozière J (2017) Novel niobium carbide/carbon porous nanotube electrocatalyst supports for proton exchange membrane fuel cell cathodes. J Power Sources 363:20–26. https://doi.org/10.1016/j.jpowsour.2017.07.058

Okada T, Inoue KY, Kalita G, Tanemura M, Matsue T, Meyyappan M, Samukawa S (2016) Bonding state and defects of nitrogen-doped graphene in oxygen reduction reaction. Chem Phys Lett 665(Supplement C):117–120. https://doi.org/10.1016/j.cplett.2016.10.061

Okamoto Y (2009) First-principles molecular dynamics simulation of O2 reduction on nitrogen-doped carbon. Appl Surf Sci 256(1):335–341. https://doi.org/10.1016/j.apsusc.2009.08.027

Sánchez M, Pierna AR, Lorenzo A, Del Val JJ (2016) Effect of cocatalyst and composition on catalytic performance of amorphous alloys for ethanol electrooxidation and PEMFCs. Int J Hydrog Energy 41:19749–19755. https://doi.org/10.1016/j.ijhydene.2016.06.064

Shafaei Douk A, Saravani H, Noroozifar M (2017) Novel fabrication of PdCu nanostructures decorated on graphene as excellent electrocatalyst toward ethanol oxidation. Int J Hydrog Energy 42:15149–15159. https://doi.org/10.1016/j.ijhydene.2017.04.280

Shahgaldi S, Hamelin J (2015) The effect of low platinum loading on the efficiency of PEMFC's electrocatalysts supported on TiO_2–Nb, and SnO_2–Nb: an experimental comparison between active and stable conditions. Energy Convers Manag 103:681–690. https://doi.org/10.1016/j.enconman.2015.06.050

Shui J, Wang M, Du F, Dai L (2015) N-doped carbon nanomaterials are durable catalysts for oxygen reduction reaction in acidic fuel cells. Sci Adv 1(1):e1400129. https://doi.org/10.1126/sciadv.1400129

Soo LT, Loh KS, Mohamad AB, Daud WRW, Wong WY (2016) Effect of nitrogen precursors on the electrochemical performance of nitrogen-doped reduced graphene oxide towards oxygen reduction reaction. J Alloy Comp 677(Supplement C):112–120. https://doi.org/10.1016/j.jallcom.2016.03.214

Vineesh TV, Kumar MP, Takahashi C, Kalita G, Alwarappan S, Pattanayak DK, Narayanan TN (2015) Bifunctional electrocatalytic activity of boron-doped graphene derived from boron carbide. Advanced Energy Mater 5(17):1500658. https://doi.org/10.1002/aenm.201500658

Wang S, Zhang L, Xia Z, Roy A, Chang DW, Baek J-B, Dai L (2012) BCN graphene as efficient metal-free electrocatalyst for the oxygen reduction reaction. Angew Chem Int Ed 51 (17):4209–4212. https://doi.org/10.1002/anie.201109257

Wang W, Kang Y, Yang Y, Liu Y, Chai D, Lei Z (2016a) PdSn alloy supported on phenanthroline-functionalized carbon as highly active electrocatalysts for glycerol oxidation. Int J Hydrog Energy 41:1272–1280. https://doi.org/10.1016/j.ijhydene.2015.11.017

Wang Y, Luo H, Li G, Jiang J (2016b) Highly active platinum electrocatalyst towards oxygen reduction reaction in renewable energy generations of proton exchange membrane fuel cells. Appl Energy 173:59–66. https://doi.org/10.1016/j.apenergy.2016.04.019

Wong WY, Daud WRW, Mohamad AB, Kadhum AAH, Loh KS, Majlan EH (2013) Influence of nitrogen doping on carbon nanotubes towards the structure, composition and oxygen reduction reaction. Int J Hydrog Energy 38(22):9421–9430. https://doi.org/10.1016/j.ijhydene.2013.01.189

Wong WY, Daud WRW, Mohamad AB, Kadhum AAH, Loh KS, Majlan EH, Lim KL (2014) The impact of loading and temperature on the oxygen reduction reaction at nitrogen-doped carbon nanotubes in alkaline medium. Electrochim Acta 129(Supplement C):47–54. https://doi.org/10.1016/j.electacta.2014.02.084

Wong WY, Daud WRW, Mohamad AB, Loh KS (2015) Effect of temperature on the oxygen reduction reaction kinetic at nitrogen-doped carbon nanotubes for fuel cell cathode. Int J Hydrog Energy 40(35):11444–11450. https://doi.org/10.1016/j.ijhydene.2015.06.006

Xing T, Zheng Y, Li LH, Cowie BCC, Gunzelmann D, Qiao SZ, Huang S, Chen Y (2014) Observation of active sites for oxygen reduction reaction on nitrogen-doped multilayer graphene. ACS Nano 8(7):6856–6862. https://doi.org/10.1021/nn501506p

Xiong X, Chen W, Wang W, Li J, Chen S (2017) Pt-Pd nanodendrites as oxygen reduction catalyst in polymer-electrolyte-membrane fuel cell. Int J Hydrog Energy 42:25234–25243. https://doi.org/10.1016/j.ijhydene.2017.08.162

Xu X, Yuan T, Zhou Y, Li Y, Lu J, Tian X, Wang D, Wang J (2014) Facile synthesis of boron and nitrogen-doped graphene as efficient electrocatalyst for the oxygen reduction reaction in alkaline media. Int J Hydrog Energy 39(28):16043–16052. https://doi.org/10.1016/j.ijhydene.2013.12.079

Xu H, Yan B, Li S, Wang J, Wang C, Guo J, Du Y (2018) One-pot fabrication of N-doped graphene supported dandelion-like PtRu nanocrystals as efficient and robust electrocatalysts towards formic acid oxidation. J Colloid Interface Sci 512:96–104. https://doi.org/10.1016/j.jcis.2017.10.049

Yahya N, Kamarudin SK, Karim NA, Masdar MS, Loh KS (2017) Enhanced performance of a novel anodic PdAu/VGCNF catalyst for electro-oxidation in a glycerol fuel cell. Nanoscale Res Lett 12:605. https://doi.org/10.1186/s11671-017-2360-x

Yang L, Jiang S, Zhao Y, Zhu L, Chen S, Wang X, Wu Q, Ma J, Ma Y, Hu Z (2011) Boron-doped carbon nanotubes as metal-free electrocatalysts for the oxygen reduction reaction. Angew Chem Int Ed 50(31):7132–7135. https://doi.org/10.1002/anie.201101287

Yang F, Zhang Y, Liu P-F, Cui Y, Ge X-R, Jing Q-S (2016) Pd–Cu alloy with hierarchical network structure as enhanced electrocatalysts for formic acid oxidation. Int J Hydrog Energy 41:6773–6780. https://doi.org/10.1016/j.ijhydene.2016.02.145

Ye W et al (2017) Pt4PdCu0.4 alloy nanoframes as highly efficient and robust bifunctional electrocatalysts for oxygen reduction reaction and formic acid oxidation. Nano Energy 39:532–538. https://doi.org/10.1016/j.nanoen.2017.07.025

Yu L, Pan X, Cao X, Hu P, Bao X (2011) Oxygen reduction reaction mechanism on nitrogen-doped graphene: a density functional theory study. J Catal 282(1):183–190. https://doi.org/10.1016/j.jcat.2011.06.015

Yuan W, Zhang J, Shen PK, Li CM, Jiang SP (2016) Self-assembled CeO2 on carbon nanotubes supported Au nanoclusters as superior electrocatalysts for glycerol oxidation reaction of fuel cells. Electrochim Acta 190:817–828. https://doi.org/10.1016/j.electacta.2015.12.152

Zehtab Yazdi A, Fei H, Ye R, Wang G, Tour J, Sundararaj U (2015) Boron/nitrogen Co-doped helically unzipped multiwalled carbon nanotubes as efficient electrocatalyst for oxygen reduction. ACS Appl Mater Interfaces 7(14):7786–7794. https://doi.org/10.1021/acsami.5b01067

Zhang J, Wang Z, Zhu Z, (2013) A density functional theory study on oxygen reduction reaction on nitrogen-doped graphene. J Mol Model 19(12):5515–5521.

Zhang B, Yu J, Tang H, Du L, Li C, Liao S (2017a) Platinum-decorated palladium-nanoflowers as high efficient low platinum catalyst towards oxygen reduction. Int J Hydrog Energy 42:22909–22914. https://doi.org/10.1016/j.ijhydene.2017.07.135

Zhang X-J, Zhang J-M, Zhang P-Y, Li Y, Xiang S, Tang H-G, Fan Y-J (2017b) Highly active carbon nanotube-supported Ru@Pd core-shell nanostructure as an efficient electrocatalyst toward ethanol and formic acid oxidation. Molec Catal 436:138–144. https://doi.org/10.1016/j.mcat.2017.04.015

Zhang J et al (2018) Stable palladium hydride as a superior anode electrocatalyst for direct formic acid fuel cells. Nano Energy. https://doi.org/10.1016/j.nanoen.2017.11.075

Zheng Y, Jiao Y, Ge L, Jaroniec M, Qiao SZ (2013) Two-step boron and nitrogen doping in Graphene for enhanced synergistic catalysis. Angew Chem Int Ed 52(11):3110–3116. https://doi.org/10.1002/anie.201209548

Chapter 7
Nanotechnology-Based Drug Delivery Systems: Past, Present and Future

Riana Awang Saman and Mohammad Iqbal

7.1 Introduction

Conventional method of drugs delivery is problematic due to the low efficiency rate, poor biodistribution and selectivity. Those limits perhaps can be overcome by using controlled drug delivery system. Controlled drug delivery system works by transporting the drug to the target action site, thus lowering side effects to other tissues. The delivered drugs can be protected against rapid degradation, providing more concentrated drugs available in the target tissue by means requiring low dosage of drug (Wilczewska et al. 2012). The technology provides means of bypassing liver as to avoid the first pass metabolism. Cell-specific targeting is a way of attaching drugs to their specifically designed carriers.

Development in nanotechnology has pointed out the potential of nanoparticles with the size less than 100 nm, as drug carriers. Nanoparticles are ranging 10–1000 nm in size. Dissolve, entrapping, encapsulation or attaching drugs to the matrixes of nanoparticles. Nanoparticles are gaining interest due to their large surface area, and their ability to adsorb and carry other compounds.

7.2 Nanocarriers Used in Drug Delivery System

Optimized nanocarriers are easily taken up by cells compare to macromolecules. Such nanocarriers used in drug delivery systems are polymers, liposomes, dendrimers, solid lipid nanoparticles, etc. For a successful targeted therapy, the mean of the drug conjugation to its nanocarriers and its targeting strategy are

R. Awang Saman · M. Iqbal (✉)
Biotechnology Research Institute, Universiti Malaysia Sabah, Kota Kinabalu, Sabah, Malaysia
e-mail: miqbal@ums.edu.my

S. Siddiquee et al. (eds.), *Nanotechnology: Applications in Energy, Drug and Food*,
https://doi.org/10.1007/978-3-319-99602-8_7

important (Suri et al. 2007). Adsorption or covalent linking of the drug to the surface of nanocarriers or encapsulation of the drug in the nanocarriers is possible. Covalent linking is above the other ways due to the ability to control specific amount of drug molecules attach to the nanocarriers. Cell-specific targeting can be done either by active or passive mechanism.

Active mechanism—The attraction of drug-nanocarriers conjugate to the affected site by using recognition ligands which are attached to the surface of conjugate antibodies, etc. It can also be achieved by physical stimuli manipulation (e.g., pH, temperature, etc.).

Passive mechanism—A result of enhanced vascular permeability and retention (EPR) in leaky tissues of tumors. The drugs will be released upon the arrival of the drug-nanocarrier conjugates at the affected tissues. A controlled drug release can be achieved by the changing physiological environment (e.g., pH, temperature, etc. or via enzymatic reaction) (Nevozhay et al. 2007).

Undesirable effects of nanoparticles depend on their size, shape, concentration, surface chemistry, administration route, reaction of the immunity system and residence time in the bloodstream. Factors that can affect the toxicity of nanoparticles should be considered. Toxicological studies are important when doing new drug delivery system formulations. However, their size can be a reference basis for generalization. Smaller particles have higher surface area. So they are more reactive, thus more toxic. Generally accepted, nanoparticles with less than 100 nm diameter have optimal pharmacokinetic characteristics which are suitable to be used in vivo. Smaller nanoparticles may be excreted via renal clearance and tissue leakage meanwhile the larger ones are removed through phagocytosis and via macrophage.

7.2.1 Liposomes

Reportedly, liposomes increase the drugs solubility and improve their pharmacokinetic properties in term of the rapid metabolism, lower side effects, and increase of anticancer activity (in vivo and in vitro) (Sunderland et al. 2006). Factors such as pH, osmotic gradient, composition of the liposome and the surrounding environment, affect the drug release from liposomes (Santos Giuberti et al. 2011). Liposomes interaction with cells can occur via adsorption, endocytosis, fusion and lipid transfer. Examples of drugs in liposomal formulations are anticancer drugs (Santos Giuberti et al. 2011), serotonin (neurotransmitter) (Afergan et al. 2008), antibiotics (Uner and Yener 2007), anti-inflammatory (Paavola et al. 2000), and antirheumatic drugs.

Reports on clinical outcomes and side effects—Photodynamic therapy (PDT) which is based on intense pulsed light (IPL) and spray (0.5% 5-aminolevulinic acid encapsulated in liposome), used in inflamed facial acne treatment (Yeung et al. 2011). Turkova et al. (2011) has performed the comparison between efficacy and safety of deoxycholate and lipid (liposomal) amphotericin B formulations (AMBF) in the treatment of invasive fungal disease in neonates. It was reported that the usage of deoxycholate amphotericin B is cheap, effective, and safe as a first-line therapy.

Safdar et al. (2010) has evaluated nephrotoxicity associated with amphotericin B lipid complex (ABLC) and liposomal amphotericin B (L-AmB) in patients receiving antifungal therapy and prophylaxis, and found no significant difference of nephrotoxicity for both drugs. Modified liposomes contain specific proteins or antigens, etc., that can be used to design drugs assigned to specific target tissue. Biswas et al. (2012) presented hydrazine-functionalized poly (ethylene glycol)—phosphatidylethanolamine (PEG-PE)-based amphiphilic polymer that operates various ligands which were reversible model ligands monoclonal antinucleosome antibody 2C5, antimyosin antibody 2G4, and glycoproteins concanavalin A (Con-A).

In the study of Kim et al. (2003), they investigated modified cationic liposomes either by PEG-grafting or PEG-adding as plasmid DNA transfection complexes. They tested on the toxicity, mitochondrial targeting and delivery efficacy of paclitaxel (PTX), the model drug. They concluded that TPP PEG-PE is a possible and safe drug delivery system.

7.2.2 Nanoparticles Based on Solid Lipids

Types of carrier systems based on solid lipid matrix- Solid lipid nanoparticles (SLN), nanostructured lipid carriers (NLC) and lipid drug conjugates (LDC) (Wissing et al. 2004). They have been exploited for dermal, peroral, parenteral, ocular, pulmonary, and rectal delivery. SLN are composed of solid lipids with major component of purified triglycerides. It provides good stability and tolerance, protecting drugs from degrading, and controlled drug release. However, few disadvantages were pointed out such as low capacity for loading, drug expulsion after crystallization, and high amount of water of the dispersions (Muller et al. 2002). NLC and LDC are modified lipids based nanoparticles. NLC are made of the mixture of solid lipids and liquid lipids. Three types: imperfect type, multiples type and amorphous type. NLC are exploited for dermal delivery. LDC is a soluble drug-lipid conjugate, linked by covalent bond or salt formation (Muller and Keck 2004).

7.2.3 Polymeric Nanoparticles

Polymeric nanoparticles (PNPs) are made of synthetic polymers, and may be classified as biodegradable e.g., poly (L-lactide) (PLA), poly glycolide (PGA), non-biodegradable such as polyurethane, based on behaviours shown in vivo. The PNPs are coated with nonionic surfactants for avoiding immunization and intermolecular interactions. Only after polymerization or during the process that the drugs can immobilized or encapsulated on PNPs surface. Drugs release are either through desorption, diffusion or by nanoparticles erosion in target tissue (Tong et al. 2011). Immobilizations of retinyl acetate (RA) on ethyl cellulose (EC) have been a stepping stone in the improvisation of aqueous stability and photostability.

Biodegradable thermo-responsive chitosan-*g*-poly (*N*-vinylcaprolactam)-biopolymer was used to deliver 5-fluoro-uracil to cancer cells. The hypothesized mechanism of the drug release is conformational changes following swelling during the transition of temperature known as lower critical solution temperature (LCST) in which significant release above LCST was shown in vitro. Results showed high toxicity to cancer cells but lower to normal ones (Rejinold et al. 2011). Reported by Kumari et al. (2010), minimal toxicity is associated with PLGA usage as drug delivery. Such nanoparticles has high biocompatibility rate with tissue and cells. Drug-biodegradable polymeric nanocarrier conjugates are stable in blood, non-toxic, non-thrombogenic, non-immunogenic, and non-inflammatory (Rieux et al. 2006).

7.2.4 Dendrimer Nanocarriers

The structure of dendrimers consists of a core, dendrons, and surface active groups. The core is an atom or molecule with dendrons (arms) attached to it. Factors that affect the usability of dendrimers are the selection of a core, monomers type and surface functional groups. The cytotoxicity depends on the material of the core and is greatly influenced by the nature of dendrimers surface (Caminade et al. 2005). Simultaneous interactions are possible between the surface functional groups with certain amount of receptors.

Encapsulation of drugs internally in dendrimers, chemically attached or physically adsorbed to the surface are ways to attach the drugs to dendrimers, and the choices depends on the properties of the drugs. Encapsulation is used when the drugs are toxic, low solubility and labile. Meanwhile, chemical bonding controls the number of covalent bonds and thus, control the number of drugs attached to the dendrimers surface (Singh et al. 2008). The selectivity of both methods can be enhanced by the attachment of targeting agents (e.g., folic acid) to the surface. The dendrimers surface provides attachment platform for specific ligands such as folic acid, PEG, antibodies, etc. (Wilczewska et al. 2012). Such attachment can improve the surface activity, biological and even physical properties of the dendrimers.

The structure and distribution of drugs or genes inside a common dendrimer known as poly (amido amide) (PAMAM) has been investigated. When compared with free cisplatin, PAMAM shows several advantages such as highly accumulated drug in solid tumors, low rate of drug release, and less toxicity effects in other organs (D'Emanuele and Attwood 2005). Other drug applications in PAMAM dendrimers are anticancer drugs, including doxorubicin, 5-FU, methotrexate, and anti-inflammatory drugs such as ibuprofen, indomethacin or piroxicam. Their toxicity is affected by their size and charge in which higher generation (G4-G8) PAMAM dendrimers exhibit toxicity due to higher density of cationic charge (Shah et al. 2011). Toxicity studies on the cationic and anionic dendrimers using Caco-2 has concluded anionic dendrimers has lower cytotoxicity compared to the other one (Kitchens et al. 2006). Destabilization of cell membranes and cell lysis has been observed after the introduction of positively charged dendrimers. Roberts et al.

(1996), has observed that cationic PAMAM caused a decrease in cell viability. They also tested on toxicity of cationic PAMAM Starburst® in mice and found no side effects occurred due to high concentration of low generation cationic dendrimers. G4 dendrimers that have amino terminal groups were found to be toxic, impairing the growth and development of embryos of zebrafish. Dendrimers with carboxylic acid functional groups exert no effects on zebrafish embryos. Dendrimers can modulate the release of cytokine and chemokine which is helpful in therapy and yet, cause major side effects (Duncan and Izzo 2005).

7.2.5 Silica Materials

The ones used in drug delivery system are classified as xerogels and mesoporous silica nanoparticles (MSNs) (Wei et al. 2010). Advantages are biocompatibility, high porosity, and functionalized easiness. They are often chosen for biological purposes. Silica xerogels has an amorphous structure that is highly porous and bigger surface area. The porous structure is influenced by the parameters of synthesis (Echeverria et al. 2010). Sol-gel technique is a common way to form silica xerogels filled with drugs. Modification in the synthesis conditions, such as reagents ratio, temperature, catalyst concentration and drying pressure, alters the xerogels properties in controlled drug release (Czarnobaj 2008). Example of drugs loaded using this technique are phenytoin, doxorubicin, cisplatin, nifedipine, diclofenac, metronidazole and heparin. The best known types of MSNs are MCM-41 and SBA-15 (Wei et al. 2010). The MSNs possess more homogenous structure, lower polydispersity and bigger surface area for adsorption of therapeutic agents (Di Pasqua et al. 2009).

7.3 Advantages of Nanocarriers for Drug Delivery

Without nanotechnology, the potential of a few therapeutics including nucleic acids and small molecules as a part of diseases treatment would not be proven. Polymeric and micelle-based-nanocarriers have hydrophobic core which are able to encapsulate poor water soluble drugs meanwhile, the hydrophilic surface helps to enhance water insoluble drugs delivery (Lukyanov and Torchilin 2004). Delivery of nanoparticulate drugs can improve the stability of payload in which the encapsulated drugs can be protected against enzymatic degradation (Mao et al. 2001).

Liposomal and virus-based nanocarriers can mimic natural environment of protein which allows protein to be stabilized and expressed naturally (Steinmetz 2010). Another main advantage is the ability of nanocarriers to specific targeting. Tumors-passive targeting can be achieved through the enhanced permeation and retention (EPR) effect. Vessels of tumors are often disorganized with enlarged gap junctions, thus promoting the permeation and retention of nanocarriers and subsequently, continuous releases of therapeutic substances (Kratz and Warnecke 2012). However, EPR effect is highly diversified between different tumours.

Delivered therapeutic drugs can often cause side effects on healthy tissues. However, with the formulation and design of active targeted nanocarriers, the release of therapeutics can be controlled sustainably. With the ability of multiple encapsulation and diversified therapeutic effects, would help to reduce not just the side effects, but the dosages too (Acharya and Sahoo 2011).

7.4 Nanoparticles Application (Healthcare/Medical)

7.4.1 Targeted Drug Delivery

The limitation of anticancer drugs is due to poor selectivity and poor diffusion. Patients undergoing chemotherapy for cancer are still facing life-threatening side effects due to the lack of drugs that are more specified for tumour cells. To overcome this problematic issue, cytotoxic drugs should be transported to specific sites of tumours. The concept of targeted drug delivery involves the modification of drugs or nanocarriers with specific ligand that can interact selectively with the marker present on the surface of the target cell. The crucial part of the concept is the identification of distinctive properties of the marker on the target cell surface which cannot be found or expressed in normal cells. This particular active targeting have been gaining attention and not just due to the limitations of conventional treatments, but through the discovery of new carcinogenic molecular targets and the clinical studies involving monoclonal antibodies (MAbs) and other molecules that target specific markers (Karra and Benita 2012).

Monoclonal antibodies (MAbs) have been used in targeting the surface of target epitopes. Their binding to tumour's specific marker has been covalently linked to specific targeted cancer drugs. So these mAb-immunoconjugates can be used to deliver drugs to specific targeted tumours. One example is the humanized anti-CD33 antibody-alicheamicin conjugate Mylotarg, used in acute myeloid leukemia treatment (Huang et al. 2009).

In 1981, targeting cancer using mAbs had been described by Milstein. Clinical demonstrations were done for antibody-based tissue targeting; using 17 FDA's approved mAbs. In 1997, mAb rituximab (Rituxan) was approved for the treatment of non-Hodgkin's lymphoma patients (Peer et al. 2007). One year later, Trastuzumab (Herceptin) was approved for the treatment of breast cancer. It was an anti-HER2 mAb binded to ErbB2 receptors (Albanell and Baselga 1999). Vascular endothelial growth factor (VEGF) was found to be the promising angiogenesis inhibitor, thus few approaches to targeting VEGF have been investigated. The best VEGF inhibition approach studied is the Bevacizumab (Avastin), which was approved as the first anti-angiogenic agent for the treatment of colorectal cancer in 2004 (Ferrara 2005). Over 200 antibody-based delivery systems are clinically studied. Broad studies have opened up to other options of targeting ligand tools such as aptamers, peptides, and growth factors, etc.

Since nanocarriers have higher ratio of surface area to their volume, there is possibility for targeting purposes, in achieving higher ligand density on the surface. Polymers and lipids are the common vectors used for drug delivery. A dozen of polymer drug conjugates have been clinically tested. Examples are anti-endothelial immunoconjugates, fusion proteins and caplostatin (First polymer-angiogenesis inhibitor conjugates) (Peer et al. 2007). Meanwhile, there are biodegradable polymers containing entrapped drugs such as Goserelin (Zoladex) and Leuprolide (Leupron Deprot), which are made of polylactideco glycolide entrapping luteinizing hormone releasing hormone, used for prostate cancer treatments. Carmustine (Gliadel) was used to treat brain cancer (Duncan 2006). There are also some polymer-drug conjugates and polymer-protein conjugates being developed for protein and gene delivery. The first practical use of polymer-protein conjugates as anticancer agents was in the early 1990s in which SMANCS and PEGylated proteins were introduced. Further studies into these polymer-protein conjugates were done. Polymer-drug conjugates were designed in late 1970s and studies involving HPMA copolymer conjugates and PGA conjugates (Duncan 2006).

7.4.2 Drug and Vaccine Delivery

Containment and prevention of diseases are depending on the key role of vaccines. Vaccine administration system is one of the obstacles involved. The delivery of the materials for vaccines administration should be in a good manner and should show high therapeutic effects. The enhancement of vaccine adjuvants in improving the quality of cellular and humoral immunity are possibly being swapped with the use of nanoparticles that improves the delivery of antigens to immune system. Some of the proved nanoformulations include MF59 (Novartis) (Wilson 2015). Other formulations using nanomaterials includes dendrimers, liposomes, micelle, etc. Nanoparticles as delivery tools benefited in ways that they can target specific tissues, dosage reduction as well as low toxicity effects, and improve the efficiency of drugs delivery. The size of nanoparticles is important for the drugs biodistribution in-vivo and influencing the cellular mechanism such as phagocytosis, endocytosis, and so on.

Nanoparticles too can facilitate the delivered antigens interaction with the Antigen Presenting Cells (APCs). Nanovaccines have received a lot of attentions due to their beneficial properties in overcoming the limitations of immunotherapy effects including low interactions with APCs, and instability of macromolecules. These immunomolecules can either be encapsulated within or conjugated on surface of polymeric nanoparticles. The usage of same nanoparticles with different surface charges in different studies has reported that nanoparticles with cationic groups were internalized more due to higher affinity for proteoglycans present on the cells surface (Gonzalez-Aramundiz et al. 2012).

Poly lactic-co-glycolic-acid is a biodegradable polymer that has been extensively studied as a delivery carrier as they can be packed as nanoparticles or microparticles according to the nature and formulations of delivery. Encapsulated antigens modifies physico-chemical properties of the nanodrugs delivery that it affects the results of assays on stability, surface charge, size and biodistribution cannot be extrapolated from one to another molecule using the same encapsulating particle. The properties of the antigen can be changed when it is encapsulated, and so the stability, functional structure and immunogenicity are in need of verification. Altogether, nanoparticles administration in lab animal testing using intraperitonial injection shows good protection against infections meanwhile oral administration shows less efficiency. One exception is the system that used chitosan or alginate as the DNA vaccine encapsulator. There is only one licensed vaccine commercialized in Canada (Ross et al. 2015).

Such important part in some studies is that the adjuvant effect of the nanodelivery system is almost potent as the loaded antigen itself and has been reported in mammals with the use of liposomes. Alum salts are the common immune adjuvants being used due to their "inflammasome" mechanism that leads to the increased of danger signals and subsequently, activate immunity. They still have few drawbacks in terms of inability to induce many cellular responses, degradation once being freeze-dried and multiple administration long term. These drawbacks have set researchers to find solutions and thus, emulsions may be the best suggested system to be used in immunotherapy. Few water-in-oil emulsions were developed and were able to formed depot at the site of injection, attracting immune cells; however the adversed reactions have maximized their application. Oil-in-Water (O/W) emulsions were used as the alternative adjuvants of vaccine and one example is the MF59™, the first O/W emulsion approved as influenza vaccine adjuvant in 1997 (Wilson 2015).

Liposome is also one of the proposed alternatives for the immunity responses stimulation. Due to their phospholipid bilayers structure, they have the capability to encapsulate antigens and deliver them to APCs, ease the cross transportation of antigens and stimulate immune responses. One unique benefit of these carriers is that there is possibility of immunostimulants co-encapsulation.

Virosomes are the most advanced liposomal structure developed as nanovaccine and one licensed example developed for influenza is the Inflexal V in which, two glycoproteins of influenza (Hemagglutinin (HA) and neuraminidase) are integrated onto the liposomes' surface either covalently or by electrostatic interactions. This increases the chance of capturing the antigens and APCs processing. The high immunotherapeutic effect of this component is due to the fusion of HA protein with the endosomal membrane and aiding in the escape of virosomes and thus, antigen will not be destructed. The antigen will be available for class I antigen presentation (Smith et al. 2013). Despite of all the promising potentials for nanovaccines development, intensive studies should be done to develop nanoformulations that can be applied in immunotherapy.

7.5 Conclusion

The modification or adaptation of nanotechnology either in therapeutic therapy or as diagnostic tools for diseases is the stepping stones for breakthrough of the technology. Nanocarriers in drug delivery system are designed to improve the therapeutical and pharmaceutical application of conventional drugs. Incorporation of drugs into nanocarriers can protect the drugs from enzymatic degradation as well as high possibility of specific targeting and controlled release. Nano-carrier-drug conjugates are more specific and effective. By accumulating drugs at the target sites, they can reduce toxicity and side effects in normal tissues and thus, lowering the drugs dosages.

However, there are challenges including developing protocols for toxicity testing, improving drug loading, transport and release, etc. that have to be met yet. A real breakthrough can be achieved solely through the painstaking performance in nanotherapy. Nanotherapeutic of diseases may contribute success or improvement in cancer treatment. Moreover, immobilization of such antibodies, growth factor or folic acid on the nanoparticles surfaces will improve the selectivity of drug carriers. Applications of nanoparticles such as in therapeutic treatments and diagnosis, in medicine are set to spread wider.

7.6 Future Perspectives

The nanotechnology of drug delivery offers considerable advances but major needs for drug carriers are still unmet. The keyword for controlled drug transportation is control, and can be achieved if there is flexibility in; modulation of size of drug carrier, stability, biocompatibility, and the rate of independency of drug delivery with the surrounding environment.

These properties are in need of improvement in order to be clinically used but considering the results of in vivo and in vitro studies, they are subject of interest. Throughout the development of drug delivery system up until now, tremendous improvement has been made. Yet, there are still challenges needed to be deal with such as scaling up the process of bringing therapeutic substances or drugs into the market, as well as developing multifunctional drug carriers that fulfil the biological and therapeutic requirements.

References

Acharya S, Sahoo SK (2011) PLGA nanoparticles containing various anticancer agents and tumour delivery by EPR effect. Adv Drug Deliv Rev 63:170–183

Afergan E, Epstein H, Dahan R, Koroukhov N, Rohekar K, Danenberg HD, Golomb G (2008) Delivery of serotonin to the brain by monocytes following phagocytosis of liposomes. J Control Release 132:84–90

Albanell J, Baselga J (1999) Trastuzumab, a humanized anti-HER2 monoclonal antibody, for the treatment of breast cancer. Drugs Today (Barc) 35:931–946

Biswas S, Dodwadkar NS, Deshpande PP, Torchilin VP (2012) Liposomes loaded with paclitaxel and modified with novel triphenylphosphonium-PEG-PE conjugate possess low toxicity, target mitochondria and demonstrate enhanced antitumor effects in vitro and in vivo. J Control Release 159(3):393–402

Caminade AM, Laurent R, Majoral JP (2005) Characterization of dendrimers. Adv Drug Deliv Rev 57:2130–2146

Czarnobaj K (2008) Preparation and characterization of silica xerogels as carriers for drugs. Drug Deliv 15:485–492

D'Emanuele A, Attwood D (2005) Dendrimer-drug interactions. Adv Drug Deliv Rev 57:2147–2162

Di Pasqua AJ, Wallner S, Kerwood DJ, Dabrowiak JC (2009) Adsorption of the Pt (II) anticancer drug carboplatin by mesoporous silica. Chem Biodivers 6:1343–1349

Duncan R (2006) Polymer conjugates as anticancer nanomedicines. Nat Rev 6:688–701

Duncan R, Izzo L (2005) Dendrimer biocompatibility and toxicity. Adv Drug Deliv Rev 57:2215–2237

Echeverria JC, Estella J, Barberia V, Musgo J, Garrido JJ (2010) Synthesis and characterization of ultra microporous silica xerogels. J Non-Cryst Solids 356:378–382

Ferrara N (2005) VEGF as a therapeutic target in cancer. Oncology 69(3):11–16

Gonzalez-Aramundiz JV, Cordeiro AS, Csaba N, Fuente MDL, Alonso MJ (2012) Nanovaccines: nanocarriers for antigen delivery. Biol Aujourdhui 206(4):249–261

Huang Y-F, Shangguan D, Liu H, Phillips JA, Zhang X, Chen Y, Tan W (2009) Molecular assembly of an aptamer-drug conjugate for targeted drug delivery to tumour cells. ChemBioChem 10:862–868

Karra N, Benita S (2012) The ligand nanoparticle conjugation approach for targeted cancer therapy. Curr Drug Metab 13:1–20

Kim JK, Choi SH, Kim CO, Park JS, Ahn WS, Kim CK (2003) Enhancement of polyethylene glycol (PEG)-modified cationic liposome-mediated gene deliveries: effects on serum stability and transfection efficiency. J Pharm Pharmacol 55:453–460

Kitchens KM, Kolhatkar RB, Swaan PW, Eddington ND, Ghandehari H (2006) Transport of poly (amidoamine) dendrimers across Caco-2 cell monolayers: Influence of size, charge and fluorescent labeling. Pharm Res 23(12):2818–2826

Kratz F, Warnecke A (2012) Finding the optimal balance: challenges of improving conventional cancer chemotherapy using suitable combinations with nano-sized drug delivery systems. J Control Release 164(2):221–235

Kumari A, Yadav SK, Yadav SC (2010) Biodegradable polymeric nanoparticles based drug delivery systems. Colloids Surf B: Biointerfaces 75:1–18

Lukyanov AN, Torchilin VP (2004) Micelles from lipid derivatives of water-soluble polymers as delivery systems for poorly soluble drugs. Adv Drug Deliv Rev 56:1273–1289

Mao HQ, Roy K, Troung-Le VL, Janes KA, Lin KY, Wang Y, Augst JT, Leong KW (2001) Chitosan-DNA nanoparticles as gene carriers: synthesis, characterization and transfection efficiency. J Control Release 70:399–421

Muller RH, Keck CM (2004) Challenges and solutions for the delivery of biotech drugs – a review of drug nanocrystal technology and lipid nanoparticles. J Biotechnol 113:151–170

Muller RH, Radtke M, Wissing SA (2002) Nanostructured lipid matrices for improved microencapsulation of drugs. Int J Pharm 242:121–128

Nevozhay D, Kanska U, Budzynska R, Boratynski J (2007) Current status of research on conjugates and related drug delivery systems in the treatment of cancer and other diseases. Postepy Hig Med Dosw 61:350–360

Paavola A, Kilpelainen I, Yliruusi J, Rosenberg P (2000) Controlled release injectable liposomal gel of ibuprofen for epidural analgesia. Int J Pharm 199(1):85–93

Peer D, Karp JM, Hong S, Farokhzad OC, Margalit R, Langer R (2007) Nanocarriers as an emerging platform for cancer therapy. Nat Nanotechnol 2(12):751–760

Rejinold NS, Chennazhi KP, Nair SV, Tamura H, Jayakumar R (2011) Biodegradable and thermo-sensitive chitosan-g-poly (N-vinylcaprolactam) nanoparticles as a 5-fluorouracil carrier. Carbohydr Polym 83:776–786

Rieux A, Fievez V, Garinot M, Schneider YJ, Préat V (2006) Nanoparticles as potential oral delivery systems of proteins and vaccines: a mechanistic approach. J Control Release 116:1–27

Roberts JC, Bhalgat MK, Zera RT (1996) Preliminary biological evaluation of polyamidoamine (PAMAM) Starburst dendrimers. J Biomed Mater Res 30:53–65

Ross KA, Brenza TM, Binnebose AM, Phanse Y, Kanthasamy AG, Gendelman HE, Salem AK, Bartholomay LC, Bellaire BH, Narasimhan B (2015) Nano-enabled delivery of diverse payloads across complex biological barriers. J Control Release 219:548–559

Safdar A, Ma J, Saliba F, Dupont B, Wingard JR, Hachem RY, Mattiuzzi GN, Chandrasekar PH, Kontoyiannis DP, Rolston KV, Walsh TJ, Champlin RE, Raad II (2010) Drug-induced nephrotoxicity caused by amphotericin B lipid complex and liposomal amphotericin B: a review and meta-analysis. Medicine 89(4):236–244

Santos Giuberti C, Oliveira Reis EC, Rocha TGR, Leite EA, Lacerda RG, Ramaldes GA, Oliveira MC (2011) Study of the pilot production process of long-circulating and pH-sensitive liposomes containing cisplatin. J Liposome Res 21(1):60–69

Shah N, Steptoe RJ, Parekh HS (2011) Low-generation asymmetric dendrimers exhibit minimal toxicity and effectively complex DANN. J Pept Sci 17:470–478

Singh P, Gupta U, Asthana A, Jain NK (2008) Folate and Folate-PEG-PAMAM dendrimers: synthesis, characterization, and targeted anticancer drug delivery potential in tumor bearing mice. Bioconjug Chem 19:2239–2252

Smith DM, Simon JK, Baker JR Jr (2013) Applications of nanotechnology for immunology. Nat Rev Immunol 13:592–605

Steinmetz NF (2010) Viral nanoparticles as platforms for next-generation therapeutics and imaging devices. Nanomedicine 6:634–641

Sunderland CJ, Stejert M, Talmadge JE, Derfus AM, Barry SE (2006) Targeted nanoparticles for detecting and treating cancer. Drug Dev Res 67:70–93

Suri SS, Fenniri H, Singh B (2007) Nanotechnology-based drug delivery systems. J Occup Med Toxicol 2:16

Tong Q, Li H, Li W, Chen H, Shu X, Lu X, Wang G (2011) In vitro and in vivo anti-tumor effects of gemcitabine loaded with a new drug delivery system. J Nanosci Nanotechnol 11:3651–3658

Turkova A, Roilides E, Sharland M (2011) Amphotericin B in neonates: deoxycholate or lipid formulation as first-line therapy—is there a 'right' choice? Curr Opin Infect Dis 24:163–171

Uner M, Yener G (2007) Importance of solid lipid nanoparticles (SLN) in various administration routes and future perspectives. Int J Nanomedicine 2(3):289–300

Wei L, Hu N, Zhang Y (2010) Synthesis of polymer-mesoporous silica nanocomposites. Materials 3:4066–4079

Wilczewska AZ, Niemirowicz K, Markiewicz KH, Car H (2012) Nanoparticles as drug delivery systems. Pharmacol Rep 64:1020–1037

Wilson A (2015) Nanovaccine delivery systems in vaccine formulations. SM Vaccine Vaccin J 1 (2):1008

Wissing SA, Kayser O, Muller RH (2004) Solid lipid nanoparticles for parenteral drug delivery. Adv Drug Deliv Rev 56:1257–1272

Yeung CK, Shek SY, Yu CS, Kono T, Chan HH (2011) Liposome-encapsulated 0.5% 5-aminolevulinic acid with intense pulsed light for the treatment of inflammatory facial acne: a pilot study. Dermatol Surg 37:450–459

Chapter 8
Targeted Therapeutic Nanoparticles for Cancer and Other Human Diseases

Rabiatul Basria S. M. N. Mydin, Wan Nordiana Rahman, Rosmazihana Mat Lazim, Amirah Mohd Gazzali, Nur Hazirah Mohd Azlan, and Said Moshawih

8.1 Introduction

Application of nanotechnology has been seen to grow exponentially in recent years. Currently, researchers have focused on usage of nanoparticles due to its wide variety potential in development of technology correspond to the demand of modern era. In biomedical field, nanoparticles have come to the forefront as targeted therapies either in cancer therapy or treatment of other human diseases (Brede and Labhasetwar 2013; Martín-Rapun et al. 2017).

Minute size nanoparticles which ranging from 1 to 100 nm in at least one dimension made them possible to be used as a drug delivery platform providing a promising tool in therapeutic approaches of human diseases. Furthermore, they possess unique properties; site specificity, wider surface area, chemically stable, ability to escape from multi-drug resistance, less toxicity which is safe for human use, and are able to cross biological barrier that limit drugs crossing (Urbanski et al. 2016; Banerjee et al. 2016; Khan et al. 2017; Lamprecht 2016; Mudshinge et al. 2011; Gelperina et al. 2005; Singh and Lillard 2009; Dikmen et al. 2011). These

R. B. S. M. N. Mydin (✉) · N. H. M. Azlan
Oncological and Radiological Sciences Cluster, Advanced Medical and Dental Institute, Universiti Sains Malaysia, Kepala Batas, Pulau Pinang, Malaysia
e-mail: rabiatulbasria@usm.my

W. N. Rahman · R. M. Lazim
Medical Radiation Programme, School of Health Sciences, Universiti Sains Malaysia (Health Campus), Kubang Kerian, Kelantan, Malaysia

A. Mohd Gazzali
School of Pharmaceutical Sciences, Universiti Sains Malaysia, Minden, Pulau Pinang, Malaysia

S. Moshawih
Jordan Center for Pharmaceutical Research, Amman, Jordan

S. Siddiquee et al. (eds.), *Nanotechnology: Applications in Energy, Drug and Food*,
https://doi.org/10.1007/978-3-319-99602-8_8

nanoparticles are engineered to combat cancer either through direct tumour cell targeting, or in combination with radiotherapy or phototherapy technologies.

The key goal of cancer therapeutics is to improve patient's survival rate and quality of life. The effectiveness of a therapeutic agent in cancer therapy is evaluated by its ability to suppress and kills tumour cell while sparing the healthy surrounding tissue unaffected (Sutradhar and Amin 2014; Brannon-Peppas and Blanchette 2004; Cho et al. 2008). Nanoparticles are designed to improved site specificity and internalization, so they enhanced the efficacy of treatment and suppress the possibility of the serious side effects that cancer patients often experience with current treatment modality (Cho et al. 2008; Steichen et al. 2013; Davis and Shin 2008).

This chapter discussed the emergent field of therapeutic nanoparticles which offers promising improvement in drug or gene delivery, radiotherapy and photodynamic therapy. Modified delivery systems allow improved drug delivery over traditional pH, microbe, or receptor dependent models, while antibody association allows for more advanced imaging modalities. Nanoparticles have potential clinical application could enhance the current treatment efficacy, specificity and reduced side effects.

8.2 Targeted Therapeutic Solid Lipid Nanoparticle for Cancer and Other Human Diseases

Solid lipid nanoparticles (SLN) have been widely studied for potential therapeutic applications in treating cancer other diseases related to central nervous system (CNS). The traditional nano-carriers such as emulsions, liposomes, and polymeric nanoparticles, were developed in the form of SLN that is sized in a submicron in diameter. Müller and Mehnert firstly introduced this colloidal system at the beginning of the 1990s, and it is simply composed of physiological lipids, dispersed in water in an aqueous surfactant solution. The main ingredients of this delivery system are lipids (usually include fatty acids, glycerides waxes, liquid and cationic lipids), surfactant, co-surfactant and active ingredients (drugs and nucleic acids), that is biodegradable and biocompatible with the human body (Mäder and Mehnert 2004). SLN has many advantages over the traditional nanoparticles in that it encapsulates the lipophilic and hydrophilic drugs, confers more stability, better-controlled release and it is less expensive and more feasible for scale-up sterile production.

8.2.1 Potential Application of SLN in Cancer Treatment

SLN have been studies extensively as promising drug delivery agent to improve the efficiency of chemotherapy. The hydrophobicity of SLN is formed from lipids; hence it can be manipulated in somehow to load the water-soluble and ionic drugs.

For instance, adding a lipophilic counter-ion to the charged drug molecules forms positive-negative ion pairs. This method increased the encapsulation efficiency for doxorubicin and verapamil from 35% to 80% (Wong et al. 2004). In the same context, SLN rate and content of drug release can be adjusted by enhancing some factors during the production process such as decreasing the concentration of surfactants, cooling the lipid emulsion rapidly and adding oils that are spatially different from solid lipids to decrease the "burst effect" of the NP. Additionally, the charged drugs can be partitioned much better by complexation with ionic polymers in polymer-lipid hybrid nanoparticles (PLN) and released in a controlled manner (Wong et al. 2006b). In order to avoid reticuloendothelial system (RES) clearance by phagocytic cells, SLN were coated with polymers such as polyethylene glycol or poloxamers. For example, doxorubicin and paclitaxel-loaded SLN were coated with PEG 2000 that made it bigger in diameters and circulating longer in the blood stream (Zara et al. 2002).

8.2.1.1 Advantages of Targeted SLN in Cancer Treatment

Targeted drug delivery by SLN improved the stability of loaded drugs, increases their half-lives, redefines their biodistribution and decreases multidrug resistance by cancer cells. Polymer decoration on the stealth SLN has added more circulation time and t½ compared to the non-stealth ones due to the ability to evade the reticuloendothelial system (Fundarò et al. 2000). Besides, blood, kidney, and spleen witnessed increased drug concentrations delivered by SLN, in contrast to that, heart was the organ to accumulate the lowest concentrations of the encapsulated drug compared to the drug solutions. This suggests lower side effects associated with anthracycline class of the cytotoxic agents such as doxorubicin and idarubicin, which are known for their cardiotoxicity, and thereby, a higher lifetime dose limitation of this class of medicines (Reddy et al. 2005; Yang et al. 1999).

In the same way, polymer-lipid hybrid nanoparticles (PLN)-loaded doxorubicin was assayed *in vitro* to kill multidrug-resistant (MDR) cancer cells and it was found that it is eight times more efficient than the free doxorubicin in preventing proliferation and survival of this kind of cells. This was because of the lipid nanoparticles are taken up by the cancer cell via endocytosis and, therefore, avoiding the P-glycoprotein drug efflux mechanism which is the most accepted mechanism that causes drug resistance to chemotherapy (Wong et al. 2006b). Furthermore, adding chemosensitizers to lipid–based NPs containing drugs can give tremendous advantages to overcome the P-glycoprotein efflux mechanism. For instance, when PLN was loaded with doxorubicin simultaneously with elacridar—a chemosensitizer—they showed a remarkable uptake and anticancer activity against MDR cancer cells, notably, this activity was much higher than the two single agents loaded separately into PLN and co-administered together (Wong et al. 2006a).

8.2.1.2 Tumor-Specific Targeting by SLN

Among the most important side effects that should be avoided in treating cancer with SLN is to reduce non-specific targeting for noncancerous cells. This can be achieved by SLN surface functionalization with targeting agents such as peptides, antibodies and polymers that are expressed on target cells more than non-affected cells. The overexpression of certain antigens and receptors on cancer cell surface such as folate, transferrin and chimeric antigen receptors (CDs) could become as a potential target decorated by ligands to avoid non-cancerous cell. Endocytosis by cancerous cells that happens right after ligand-receptor binding facilitates the delivery of a drug and/or nucleic acids inside the cells. Many surface modifications were applied to enable drug delivery to brain, liver and other tissues in the body for different diseases. Table 8.1 summarizes these modifications and molecule that can be grafted on the surface of solid lipid nanoparticles.

8.2.1.3 Gene Delivery by SLN

SLN based nano-carrier as gene delivery system can be considered a potential candidate for gene silencing that involves small interfering RNA (siRNA) and MicroRNA. Gene silencing is being increasingly used for the treatment of cancers by which gene expression is reduced by at least 70%. However, naked siRNA and MicroRNA cannot be injected directly into the blood stream, and this is due that, it has an anionic nature, which prevent passive uptake by cells. Moreover, plasma nucleases and immune responses degrade these molecules before they reach their target cells. Therefore, a nano-carrier is needed to protect genes and nucleic acids from degrading factors as well as to deliver it to its target cells (Xie et al. 2014).

Delivering nucleic acids such as siRNA into the nucleus of cancer cell leads to gene silencing that causes metastasis and subsequently, cancer cell cycle regulation. In a study that tests the feasibility of gene silencing, cationic nanoparticles were designed to actively target the angiogenesis vasculature during the remodeling process of the new tumor tissue. During this process, the endothelial cells express molecules that hasten the process of invasion and proliferation such as the integrin $\alpha v\beta 3$, which plays a critical role in endothelial cell for angiogenesis. A NP with cationic lipid polymerization was covalently coupled to a small organic $\alpha v\beta 3$ ligand with a gene encoding green fluorescence protein. It was found that the cationic nanoparticles could deliver genes selectively to angiogenic blood vessels in tumor-bearing mice. The therapeutic benefits in this study revealed that this kind of SLN, finally led to the apoptosis of the tumor-associated endothelium and a suppression of the primary and metastatic tumors (Hood et al. 2002).

On the other hand, MicroRNA (miRNA) is a class of small noncoding RNA bind to the 3′-untranslated region of target messenger RNA, leading to translational repression. It has been employed as a sensitizer to the cytotoxic agent paclitaxel for the treatment of breast cancer stem cells because noncoding regions of mRNA

Table 8.1 The classification and types of SLN surface modifications and their characteristics

Surface functionalization	Examples	Characteristics	References
Hydrophilic polymers	Brij 78, Brij 68, Poloxamer F68	• Non-ionic hydrophilic polymers with PEG chains • Decreased energy-dependent P-gp efflux as a resistance mechanism • Increased AUC and circulation time • Parenteral drug administration	Kaur et al. (2008), Ban et al. (2018)
Ligands designed to bind with receptors overexpressed on target site	Folic acid	• Folate receptor is highly expressed in numerous cancers especially epithelial origin • Folate demand of rapidly dividing cells	Chen et al. (2013), Kuo and Lee (2015)
	Low lipid lipoproteins (LDL)	• Overexpression of LDL receptors, e.g.: to target melanoma cells • Adsorption of (ApoE) and (ApoB) onto the surface of the nanoparticles coated with LDL facilitates endocytosis inside brain	Kim et al. (2007)
	Peptide ligands	• E.g. 1: L-protein is a surface antigen for hepatitis B virus can bind on liver cells • E.g. 2: vasoactive intestinal peptide is a neuropeptide that binds on receptors in the peripheral and central nervous system employed to facilitate BBB penetration by I.V. and nose-to-brain delivery	Mishra et al. (2010)
Antibodies grafted on SLN	(1) Immunoglobulin G (2) Anti-insulin receptor anti-body (3) Anti-epithelial growth factor receptor	• Highly specific by monoclonal antibodies (MAb) targeting agents and grafted on nanoparticles	Kuo and Shih-Huang (2014), Mishra et al. (2010), Kuo and Liang (2011)

(continued)

Table 8.1 (continued)

Surface functionalization	Examples	Characteristics	References
	(anti-EGFR) produced as monoclonal antibody	• E.g. 1: MAb targets insulin receptors on human brain endothelial cells employed in HIV infection of the brain • E.g. 2: Anti-EGFR enhances the transport of anti-cancer agents to glioblastoma cells and reduces the administered dosage	
Miscellaneous	Wheat germ agglutinin + Lactoferrin	• Cationic SLN increased the permeation of loaded drugs into brain tissue and gave a sustain release feature • Reduced transendothelial electrical resistance	Kuo and Wang (2017)

expression were altered in the case of this cancer. Liu et al. (2016) condensed the anionic miRNA with the cationic SLN to form a complex that has delivered miR-200c to the cancer cells and expressed it more than 11 times compared to miR-200c free solution. In the same study, the expression of class III β-tubulin was down regulated and the cellular cytotoxicity of paclitaxel-loaded NLC against breast cancer stem cells was enhanced significantly. Nevertheless, no in vivo study has confirmed such results as it is considered the mainstay of the effectiveness of separate gene-chemo therapy treatment by NPs.

8.2.2 Potential Therapeutic Advantages of SLN for Diseases Related to Central Nervous System

In treating diseases associated with central nervous systems (CNS), there is a limitation for drugs to pass through the blood brain barrier (BBB) in order to be delivered to the central nervous system. BBB key role is to provide all necessary nutrients to the brain with the avoidance of xenobiotics that are neurotoxic but are not in the bloodstream. In addition to that, it should protect the brain environment from ionic fluctuation that affects the homeostasis and perfect mental performance (Cecchelli et al. 2007). Molecules pass through the BBB cross either intercellularly by the so-called paracellular pathway, or intracellularly by the transcellular pathway. The tight junction between cells, trans-membrane proteins and unique endothelial cells that line cerebral capillaries control the paracellular diffusion of molecules, while the lipophilic nature of drugs or carriers and the affinity to p-glycoproteins efflux mechanism in addition to intra- and extra-cellular enzymes determine the diffusion through the transcellular pathway (Cacciatore et al. 2016).

8.2.2.1 SLN for Alzheimer Disease

Many in vivo and in vitro studies were carried out to investigate a drugs or phytochemical compounds delivery to the brain loaded into SLN and coated by different ligands in order to bypass the BBB. Chitosan the polysaccharide coating was used to protect the siRNA molecule that silences the BACE1 gene which is related the cleavage of β-amyloid precursor proteins and β-secretase that are responsible for amyloid generation in the brain of Alzheimer patient. Rivastigmine was loaded into SLN for the treatment of AD through intranasal delivery, showed a high safety profile for the nasal mucosa (Shah et al. 2015). In an in vivo study, polysorbate-80 was employed to decorate the surface of the SLNs that used to encapsulate the alkaloid piperine, which is known for its superior antioxidant properties but poor bioavailability. The study revealed a significant reduction in the neuronal oxidative stress and cholinergic degradation for the piperine-SLN compared to donepezil and free piperine solutions (Yusuf et al. 2013). Other modifications were applied on the surface of the SLN in order to facilitate the penetration to the monolayer of human brain microvascular endothelial cells and astrocytes to delivering the drugs required by active targeting. These modifications comprise the addition of tamoxifen, lactoferrin, and 5-HT-moduline. At present, these ligands were used for brain cancer drug delivery, however, it can be extrapolated in the future for other CNS-related diseases (Kuo and Cheng 2016; Kuo and Hong 2014).

8.2.2.2 SLN for Parkinson's Disease

The dopamine agonist Bromocriptine was encapsulated by NLP made of tristearin/miglyol and given i.p. for hemilesioned rats (Parkinson's induced animal model). It was found that NLP extended bromocriptine half-life for five more hours compared to the free solution (Esposito et al. 2012). Similarly, the oral administration of apomorphine—another dopamine agonist—was tested after being loaded into SLN and the bioavailability was increased 13-times, in addition to the enhancement of the antiparkinsonian effects due to the higher drug distribution in the striate nucleus (Tsai et al. 2011). On the other hand, three SLN formulations were prepared to deliver apomorphine or its prodrugs diisobutyryl apomorphine intravenously. All preparations were found to present high brain targeting and decreasing liver first pass metabolism (Hsu et al. 2010; Liu et al. 2012; Wen et al. 2012). Intranasal preparation was designed for Ropinirole, by triglyceride based SLN and was comparable to the oral marketed ones with decreased dose and dosing frequency in addition to sustain release best fitted to zero-order kinetic controlled by diffusion and erosion of lipid matrix (Pardeshi et al. 2013).

8.2.3 Future Prospective for SLN

Hydrophilic payloads can be encapsulated in the SLN by forming insoluble drug-lipid conjugates either by conjugating the drug salt with a lipophilic moiety such as fatty acids or by linking them with ether or ester through covalent bonding. On the other hand, SLN has disadvantages such as the high water content of dispersion (up to 99%), low drug loading capacity and in some cases drug expulsion during storage. Therefore, they were further developed in order to overcome those disadvantages, and subsequently increasing the drug payload and preventing drug expulsion. The multiplicity of SLN forms makes it suitable for different classes of anticancer agents with a wide range of structural formulas and physicochemical properties.

8.3 Targeted Superparamagnetic Iron Oxide Nanoparticles (SPIONS)

In radiotherapy, nanoparticles have potential application as radiosensitizer in enhancing the efficacy of the treatment. There are many types of the radiosensitizers that are grouped in metal-based nanoparticles, quantum dots, superparamagnetic iron oxides and non-metal based nanoparticles. Examples of metal-based nanoparticles are the gold nanoparticles (AuNP), Gadolinium based nanoparticles (Gd-NP), Titanium based (Ti-NP), Silver based nanoparticles (Ag-NP) and Hafnium-based nanoparticles (Hf-NP). The non-metal based nanoparticles can be classed as Silicon-based nanoparticles (Si-), Fullerene-based nanoparticles and chemotherapeutic entrapped nanoparticles (Kwatra et al. 2013). Superparamagnetic iron oxide nanoparticles (SPIONs) have been traditionally used for disease imaging and it is recently used for cellular-specific targeting, drug delivery and multi-modality imaging (Veiseh et al. 2010). SPIONs are magnetic nanoparticles that represent a class of non-invasive imaging agents that have been developed for magnetic resonance imaging (MRI) and currently the only types of nanoparticles that have been approved for clinical study (Mahmoudi et al. 2009). This iron oxide based nanostructure with proper surface architecture and conjugated targeting ligands or proteins have attracted a great deal of attention for targeted drug delivery applications in cancer therapy (Mahmoudi et al. 2010). Intriguingly, SPIONs are also highly biocompatible with negligible toxicity to healthy tissues which make them suitable for radiotherapy application (Kwatra et al. 2013).

8.3.1 Characteristics and Properties of SPIONS

SPIONs exist in variety of structures and characteristic. Ferrous and ferric iron oxides were presented in seven phases of crystalline. The most common were hematite, maghemite, magnetite and wustite. In examples, magnetite NPs have been used in cancer diagnosis and therapy meanwhile maghemite is used as MRI contrast agent (Martinez et al. 2009). SPIONs exhibit paramagnetism and diamagnetism properties and are classified by their response to an external magnetic field (Gupta and Gupta 2005). Iron oxide nanoparticles also show ferromagnetism properties in which it can be permanently magnetized. Among all types of nanoparticles, SPIONs are special in their superparamagnetic properties which allow them to be directed and localized to a particular organ using external magnetic force. Superparamagnetism is a phenomenon where the materials retain no remnant magnetization upon the removal of magnetic field (Rosen et al. 2012). SPIONs are useful as targeted drug delivery agents because it can be transported by magnetic field to desired site or tumour (Mahmoudi et al. 2009). The SPIONs will remains at the target for a certain period of time once the external magnetic field is removed (Mahmoudi et al. 2009; Gupta and Gupta 2005).

SPIONs are already been clinically applied as contrast agents in MRI to visualize many type of diseases and used to image liver tumour and metastases (Malekigorji et al. 2014). They are widely utilized as contrast agent in MR imaging due to their strong influence of T1 relaxation and T2 relaxation. The nanoparticles display very strong paramagnetic properties with very large susceptibility. This characteristic however depends on the particles size and coating. The behavior of SPIONs in magnetic field is highly dependent on their size in which their superparamagnetic behavior increases with the decreasing particles size (Gupta and Gupta 2005). Small particles below 15 nm, ferromagnetism is no longer observed and no permanent magnetization remains after removal of external magnetic field.

In term of biocompatibility, iron oxides are able to dissociate into iron and oxygen inside the body that makes it suitable for biological applications. SPIONs also can be safely eliminates in metabolic and oxygen transport systems (Gupta and Wells 2004). However, evaluation of the SPIONs toxicity must consider the mechanism of SPIONs system interaction with the whole human body during functional lifetime and how the components will affect the body during biodegradation processing. Naked SPIONs are quite unstable and even form bulk aggregates in biological fluids. Appropriate design of core and shell of SPIONs is extremely important for medical applications (Li et al. 2012). Plunoric F-127-coated iron oxide nanoparticles have been developed for application as MRI contrast agent and also anti-cancer therapies. These nanoparticles have relatively higher T2 relaxivity and capable to accommodate high payloads of multiple drugs. The drug released from SPIONs is effective in killing cancer cells. Studies also found out that SPIONs are able to inhibit the function of the receptor to induce apoptosis in glioma cells in vitro (Kievit and Zhang 2011).

The SPIONs are steadily used in biomedical research and magnetic nanoparticles with appropriate surface coatings are already used clinically as MRI contrast agent (Soenen et al. 2012). Nevertheless, the safety of SPIONs is still unclear whether it can be used with caution because of the variation in incubations conditions, SPIONs characteristics and cell types studied. The toxicity of SPIONs was found to be highly dependent on their physical properties such as size, shape and surface coating. The increment of the particles size increases the surface area and creates more reactive action towards surrounding biological components and affects the biocompatibility of the nanoparticles (Mahmoudi et al. 2008).

8.3.2 Multifunctional Targeted SPIONs for Radiotherapy

The study of dose enhancement using high atomic number (Z) material have been done for over 50 years starting by the seminal work on the dose enhancement effects at the interfaces between high and low Z materials (Spiers 1949). The studies stemmed from the concept of increase number of photoelectron due to increase photon interaction cross section in the presence of high Z materials that deposited the dose in the target. This study introduces the concept of dose enhancement in cancer radiotherapy using high Z materials by using iodine, the most commonly used X-rays contrast agent. Radiation dose enhancing effects of iodine have been observed on many studies conducted in vitro and in vivo using low energy X-ray (Matsudaira et al. 1980). Regressions of tumour growth in mice were observed when iodine contrast medium was injected directly into the tumour (Mello et al. 1983).

In these recent years, gold nanoparticles (AuNPs) are widely investigated for potential application in radiotherapy (Hainfeld et al. 2004; Rahman et al. 2009) The extreme small size, good biocompatibility and ease in chemical modification drawing interest to the AuNPs as particles of choice (Su et al. 2014). The application of AuNPs as dose enhancer or radiosensitizing agent is not only to enhance the effectiveness of radiation therapy but also compensate the insufficiency and limitation of current conventional treatment. The theoretical and experimental foundation of the dose enhancement effects based on AuNPs is well established and extensively optimized. The same principles could be applicable to other types of metallic nanoparticles but different mechanism could involve depending on the nanoparticle composition and surface coating. Among the particles of interest are SPIONs which have been the focus in nanomedicine study due to their numerous applications. The prime advantage of SPIONs in comparison to AuNPs is their cost effectiveness and magnetic properties. SPIONs are used in combination radiation applications as targeted therapy, drug delivery, MRI contrast agent and commonly coated with organic macromolecules such as poly (acrylic acid), Dextran and poly (ethyleneimine) or coatings such as silica, carbon or precious metals (such as gold or silver).

The unique properties of SPIONs as compare to other types of nanoparticles are their magnetic or superparamagnetic properties. These properties make them extremely useful to enhance the sensitivity of MRI and increase the contrast effects on MRI image. In addition to traditional application of SPIONs as MRI contrast agent, multi-modal imaging approaches have been explored to enhance the imaging sensitivity and accuracy. MRI-PET/SPECT dual mode imaging agent has been achieved via combination of SPIONs with radioisotope based imaging techniques. A combined MRI and PET/SPECT dual modal system can provide highly resolved tomographic image along with strong sensitivity (Shin et al. 2015). Accurate biological information such as in vivo distribution and pharmacokinetics could be obtained with this technique. For example, investigation on cancer xenograft labelled with SPIONs in nude mice reveal that the SPION distribution could be monitored using MRI (Wu et al. 2008). This particular characteristic allows their kinetic to be directed and control using external magnetic force.

SPIONs applications not only limited to diagnostic purpose but also have potential for therapeutic usage. SPIONs applications to treat diseases are normally in combination with ionizing or non-ionizing radiation. In a study conducted by Maier-Hauff et al. (2011), magnetite iron oxide (Fe3O4) nanoparticles with 12 nm diameter coated with aminosilane were administered to tumour and treated with thermothermal in combination with stereotactic radiotherapy (Maier-Hauff et al. 2011). The combination of thermotherapy and radiotherapy using magnetic nanoparticles leads to better outcome and survival compared to conventional therapies (Maier-Hauff et al. 2011).

Hyperthermia is form of therapy using heat to enhance the effect of radiation cancer treatment. The characteristic of hyperthermic effects make it a good complementary treatment to radiotherapy and enhance the therapeutic effectiveness of radiation cancer therapy. Cassim et al. (2009) have done a study using MTG-B murine breast adenocarcinoma cell to evaluate the therapeutic enhancement potential for SPIONs hyperthermia and radiation (Cassim et al. 2009). 5 mg/mL of iron concentration of 120 nm radius of Dextran-coated SPIONs were injected into the tumour and irradiated with 6 MeV photons with single dose of 15 Gy. From the experiment, the tumour regrowth delay achieved the increment approximately 50% using addition of SPIONs hyperthermia and 15 Gy radiation compared to 15 Gy radiation treatments alone. The combinations of SPIONs hyperthermia and external beam radiation have shown the tumour enhancement response in a mouse mammary adenocarcinoma.

Low linear energy transfer (LET) radiation such as X-rays can minimize undesired side effects for normal tissue cells. However, radiotherapy is always limited by its toxic effect on normal tissue and the intrinsic radioresistance of tumour cells. Regulations of tumour radiosensitivity are always induced by the balance between the DNA damage and repair processes. The coated SPIONs are nontoxic; hence it can be used as radiosensitizer in cancer radiation therapy (Klein et al. 2014). ROS not only induced DNA double strand breakage in the nucleus but also reduce the survival rates of the cells in high production of ROS in the mitochondria (Westman 2006). The ROS also increase the apoptosis of cancer cells as revealed by high proliferation rate and reduced the ability of DNA self-repair (Klein et al. 2014). The applications of SPIONs increase the production of free radical in cells that may cause more cellular death. Degradation of iron oxide into free ions in physiological environments contributes to generation of more free radical (Malekigorji et al. 2014).

In order to achieve more dose enhancement effects by increasing ROS production, SPIONs can be conjugated to anticancer drug or other therapeutic agents. Huang et al. (2013) have done a study using SPION with a novel anticancer drug, β-lapachone (β-lap) (Huang et al. 2013). The β-lap increase reactive oxygen species (ROS) stress in cancer cells. Pre-treatment of cells with 0.14 mM SPION-micelles followed by β-lap concentrations shows increment in cell lethality compared to the one without SPION-micelle pre-treatment. The SPION-micelles release iron ions into cancer cells which interact with hydrogen peroxide (H_2O_2) generated from β-lap. As a result of that, the ROS stress are increase in the cancer cells then enhanced the therapeutic index of β-lap. The increment of the ROS stress significantly increases the cell death.

Study on animal model or in vivo analysis on the SPIONs enhancement of tumour regression have been done using orthotopic rat glioma model. The rats firstly were examined using MRI to obtain the size and position of the tumour before irradiated with 45 MeV proton beam. The rats were irradiated at two single doses, 20 or 40 Gy. From the results, the group receiving SPIONs and proton beam showed 65–70% smaller tumour volume compared to the group receiving proton alone. The proton-impact SPIONs produced therapeutic enhancement compared to the proton irradiation alone (Seo et al. 2013).

The dose enhancing or radiosensitizing effects by SPIONs are highly dependent on the concentration of the particles in which similar to AuNPs, production of secondary radiation are increasing with concentration used. The CT26 cell lines irradiated with 45 MeV monoenergetic proton beams decrease in the cell viability with the increasing of nanoparticles concentration. However, the increasing of nanoparticles concentration will increase the cell cytotoxicity. This will contribute to the increasing of particle-induced x-ray emission (PIXE) effects from the increasing of cellular uptake of SPIONs (Kim et al. 2010).

Other than in-vitro and in vivo investigation, monte carlo simulation of various nanoparticles; Au, Ag, I and Fe2O3 was done to evaluate the dose enhancement effects induced by all this particles including iron oxide based nanostructures (Toossi et al. 2015). The simulation used 7, 18 and 30 mg/mL nanoparticles concentration irradiated at 8, 12 and 14 MeV electrons from medical linac. Results indicate the presence of dose enhancement with maximum electron dose enhancement was observed for 30 mg/mL of gold nanoparticles with 8 MeV electrons. The photon contaminations increase inside the tumor with the existence of the nanoparticles and hence increase tumor dose. Lower energy electron beams and high atomic number nanoparticles can be great benefits to use for electron dose enhancement in the electron mode of linac. The photons originated from the nanoparticles will increase the photon dose inside the tumor (Toossi et al. 2015).

8.3.3 Future Prospective for SPIONs

SPIONs are well known as contrast agents for magnetic resonance imaging and have shown potential as radiosensitizer in radiotherapy. Utilization of targeted SPIONs in both therapy and diagnostic application presented huge potential in the development of multifunctional theranostic agents that may hold the key to improved clinical outcomes of cancer treatment.

8.4 Targeted Nanoparticles for Photodynamic Therapy

8.4.1 Photodynamic Therapy (PDT)

Photodynamic therapy or more famously known as PDT is an anticancer theranostic approach which started to emerge in the late 1970s. Depending on the type of agents and methods used, this technique could serve as both a diagnostic tool and a therapeutic option for cancer eradication which is extraordinary and appealing to both researchers and clinicians for better patient outcome. As a diagnostic tool, the fluorescence capability of most photosensitizing agents has enabled the detection of tumors following light irradiation. As a therapeutic option, anticancer therapy by PDT is relatively non-invasive as compared to other treatment modalities because the induction of cell death is made possible only in the presence of three important components; a photosensitizer (PS), oxygen molecules and light at a specific wavelength.

PS are usually tetrapyrrolic compounds with heme-like structures, are beautifully pigmented depending on the presence or absence of metal in the center of the ring and are bulky molecules. PS are usually classified based on their generations and until now, there are four generations of PS. Photofrin®, the first-generation PS was the first agent granted approval by FDA in 1995 for treating esophageal cancer, 1998 for lung cancer treatment and in 2003 to treat Barrett's Esophagus. The second generations are composed of other molecules such as porphyrin, chlorin and phthalocyanines, among others.

However, until now the current available PS from the first and second generations have not showed a good selectivity towards cancerous cells. This has caused several drawbacks such as the need of administering higher PS doses to ensure sufficient PS accumulation in cancer cells to achieve adequate therapeutic outcomes. Administration of high doses leads to the issue of wastage especially because PS are costly materials and therefore, the development of tumor-specific PS is of great interest to PDT scientists.

Hence, research were directed towards two main approaches; (1) the generation of new PS with the hope that the newly developed molecule will accumulate in the targeted cells, and (2) the use of carriers such as nanoparticles (NP), liposomes and microspheres to enable site-specific delivery of the encapsulated or conjugated PS into the targeted cells (Rosenkranz et al. 2000). The latter is an example of third generation PS, in which the delivery of PS is designed to be specific and targeted with the help of carriers or certain targeting agents.

8.4.2 NP as Targeting Vehicle

The association of PS with nanoparticles has managed to show advantages in the improvement of overall PDT performance based on multiple research articles published to date. Since most PS are hydrophobic molecules, their encapsulation in nanostructures could potentially improve their water solubility, prevent aggregation, simplified the formulation step and most importantly enhance the delivery of encapsulated or conjugated PS into specific targeted cells. Briefly, the association of PS with nanostructures may have the following advantages (Rui et al. 2016):

1. Nanostructures have a high surface to volume ratios and hence a high drug loading.
2. The enhanced permeability and retention (EPR) effect associated with NP will increase the amount of delivered PS.
3. The small size of NP will enable the diffusion of the particles through cell junctions
4. The potential for targeting will increase the PS concentration at the desired site and subsequently will reduce the toxic effects on healthy cells.
5. The solubility, biodistribution, pharmacokinetics, cell uptake and specificity of the PS can be improved.
6. A constant therapeutic dose can be maintained at the site of action
7. The possibility to develop multifunctional nanoplatforms to carry multiple components which may include the PS and targeting agents.

Since the introduction of NP, there are numerous studies and reports in the literatures precising the application of NP made from different materials such as gold, silica, graphene and carbon nanotubes in PDT with excellent experimental results (Abrahamse and Hamblin 2016). Among the earliest publication on NP for PDT was an article by Bachor et al. (1991) which describe the use of polystyrene microspheres for the delivery of chlorin e6 to phagolysosomes which documented good PDT effect.

8.4.3 *Nanoparticles for PDT*

Many original research articles in the recent years reported on the production of NP as a targeting vehicle for PDT. In general, there are two types of materials used in the production of NP for PDT; organic and inorganic nanomaterials. This part will briefly introduce and describe examples of latest development of targeted NP in PDT.

8.4.3.1 Organic Nanomaterials for PDT

Organic-based NP are being explored extensively as a vehicle for drug delivery and targeting due to their biocompatibility characteristic. In PDT, there is no exception. Among the many different types of organic NP being explored for PDT, three will be discussed here-in which are liposomes, polymeric NP and hyaluronic acid NP.

In clinical practice, there is at least one liposomes-based formulation that is approved by FDA to be used in PDT—Visudyne®. This formulation is not prepared for cell-targeting; instead it is only indicated for local PDT application although in the literatures, there are multiple reports precising the potential of liposomes as carriers for targeted PDT. PEGylated liposomes with prolonged circulation time, immunoliposomes, tumor-specific, pH-sensitive, magnetoliposomes and cationic liposomes are among the examples of liposomal innovation for PDT reported to date. Researchers have also incorporated certain targeting agents such as folic acid on the surface of liposomes to further improve their targeting ability. These various techniques employed have shown to improve the dosing and targeting ability of liposomes besides reducing normal tissue damage and cutaneous sensitivity (Bovis et al. 2012).

The second type of organic NP is polymer-based NP. It can be divided into synthetic- and natural-based NP and both types are commonly used in PDT drug delivery research. A recent research by Lin et al. (2017) reported on the development of PS-encapsulated PLGA-chitosan NP with transferrin as the targeting agent. The system showed a good overall tumor inhibition rate (63% in 15 days) with fewer side effects on normal cells. Wennink et al. (2017) described a unique macrophages-targeting approach with polymeric NP from Ben-PCL-mPEG which encapsulates *m*-THPC. The PS was found to be accumulated in the atherosclerotic lesions in the blood vessels, paving a way for future application of PDT in the treatment of such lesion.

Hyaluronic acid (HA) is a naturally occurring material in human body composing of D-glucuronic acid and *N*-acetylglucosamine. The number of research describing the formation and utilization of HA in PDT is relatively low as compared to other nanomaterials. This may be due to the difficulty in synthesizing the NP albeit its effectiveness in ensuring tumor accumulation of the targeted drug. Among the most recent was described by Gao et al. (2017) in which they fabricated HA NP with encapsulation of chlorin e6. CD44 cell receptor was exploited as a target and they observed a satisfactory accumulation of chlorin e6 in tumor cells. Tumor eradication was also recorded following irradiation by light, with very low observable side effects.

8.4.3.2 Inorganic Nanomaterials

In a book chapter by Colombeau et al. (2016) entitled *Inorganic Nanoparticles for Photodynamic Therapy*, the authors have divided the inorganic NP into six groups—gold NP, silica NP, NP which can self-produced ROS, NP with other types of excitation (such as self-illuminating NP), iron oxide and theranostic NP and Mg-Al hydroxides. Among these, the most researched are gold and silica NP.

Through recent advancement, gold NP were designed with cellular penetration and internalization ability, intracellular delivery and subcellular localization of associated molecules (Lévy et al. 2010). In the targeted delivery of PS however, gold is less explored as compared to organic nanomaterials because gold NP are more useful as photothermal therapy agent (PTT) as compared to PDT. Its use in PDT are usually in conjugation with other materials, such as Rose Bengal as PDT agent (Fu et al. 2017) or peptides as targeting agent. Study by Wu et al. (2017) has developed gold NP with conjugated 5-aminolevulinic acid (5-ALA) on the surface, alongside a specific cell-penetrating peptide (CPP)—RRRRRRRR (R8). The CPP R8 will help the internalization of the nanoparticle and hence increases the delivery of the 5-ALA into the targeted cancer cells. This was successfully proven in their study and they have shown that the tumor volume in their test subject was significantly reduced as compared to the control group following PDT. Song et al. (2017) reported on the production of another type of gold NP, Au@Pt NP which could cause thermal-induced cell death upon exposure to laser light in addition to the common PDT effect which is through the production of ROS. The NP were decorated with folic acid molecules on the surface and this has helped to increase site-specific delivery of the NP.

Silica is another interesting material for NP production. There are two types of silica NP—mesoporous silica NP (MSN) and non-porous silica NP. Bharathiraja et al. (2017) described the production of silica NP with conjugated folic acid on the particles surface which has helped to direct the NP towards folate receptor positive MDA-MB-231 cells. In addition to the two types of inorganic NP mentioned above, there are many other materials that have been used to prepare NP for PDT, such as fullerene, titanium dioxide, graphene and zinc oxide (Youssef et al. 2017). These materials have several advantages on their application of PDT. Titanium dioxide (TiO_2) as an example is a wide-band gap semiconductor which is photoactive in the presence of UV light. The conjugation of PS on TiO_2 NP has allowed the activation of PDT by using only visible light, enabling future expansion of PDT application of TiO_2 NP.

8.4.4 *Future Prospective of Targeted Nanoparticles for Photodynamic Therapy*

The exploration of NP as a vehicle for PS delivery and targeting is expected to bloom further in the years to come. Based on current stage of research, PDT has its own potential to complement, if not function on its own as an effective anticancer treatment. The development of targeted NP formulations is believed to be able to improve the safety and efficacy of PDT and hence it might become an interesting therapeutic option for both physicians and patients in the future.

8.5 Conclusion and Outlooks

There is a lot more potential of nanoparticles in treatment of human diseases especially cancer to be discovered in order to achieve the maximum efficiency of current therapeutic modality and improving public health generally.

Acknowledgements The authors are thankful to the Ministry of Education (MOE) Malaysia for funding this work under Transdisciplinary Research Grant Scheme (TRGS) grant no. 6769003. The authors are very much grateful to Universiti Sains Malaysia (USM) for providing the necessary facilities to carry out the research work and financial support under USM-Short Term Research Grant (304/CIPPT/6315073).

References

Abrahamse H, Hamblin MR (2016) New photosensitizers for photodynamic therapy. Biochem J 473(4):347–364

Bachor R, Shea CR, Belmonte SJ, Hasan T (1991) Free and conjugated chlorin E6 in the photodynamic therapy of human bladder carcinoma cells. J Urol 146(6):1654–1658

Ban C, Jo M, Lim S, Choi YJ (2018) Control of the gastrointestinal digestion of solid lipid nanoparticles using PEGylated emulsifiers. Food Chem 239:442–452

Banerjee A, Qi J, Gogoi R, Wong J, Mitragotri S (2016) Role of nanoparticle size, shape and surface chemistry in oral drug delivery. J Control Release 238:176–185

Bharathiraja S, Moorthy MS, Manivasagan P, Seo H, Lee KD, Oh J (2017) Chlorin e6 conjugated silica nanoparticles for targeted and effective photodynamic therapy. Photodiagn Photodyn Ther 19:212–220

Bovis MJ, Woodhams JH, Loizidou M, Scheglmann D, Bown SG, MacRobert AJ (2012) Improved in vivo delivery of m-THPC via pegylated liposomes for use in photodynamic therapy. J Control Release 157(2):196–205

Brannon-Peppas L, Blanchette JO (2004) Nanoparticle and targeted systems for cancer therapy. Adv Drug Deliv Rev 56(11):1649–1659

Brede C, Labhasetwar V (2013) Applications of nanoparticles in the detection and treatment of kidney diseases. Adv Chronic Kidney Dis 20(6):454–465

Cacciatore I, Ciulla M, Fornasari E, Marinelli L, Di Stefano A (2016) Solid lipid nanoparticles as a drug delivery system for the treatment of neurodegenerative diseases. Expert Opin Drug Deliv 13(8):1121–1131

Cassim SM, Giustini AJ, Petryk AA, Strawbridge RA, Hoopes PJ (2009) Iron oxide hyperthermia and radiation cancer treatment. Proc SPIE Int Soc Opt Eng 181:71810O

Cecchelli R, Berezowski V, Lundquist S, Culot M, Renftel M, Dehouck MP, Fenart L (2007) Modelling of the blood–brain barrier in drug discovery and development. Nat Rev Drug Discov 6(8):650–661

Chen C, Ke J, Zhou XE, Yi W, Brunzelle JS, Li J et al (2013) Structural basis for molecular recognition of folic acid by folate receptors. Nature 500(7463):486–489

Cho K, Wang XU, Nie S, Shin DM (2008) Therapeutic nanoparticles for drug delivery in cancer. Clin Cancer Res 14(5):1310–1316

Colombeau L, Acherar S, Baros F, Arnoux P, Gazzali AM, Zaghdoudi K et al (2016) Inorganic nanoparticles for photodynamic therapy. In: Sortino S (ed) Light-responsive nanostructured systems for applications in nanomedicine. Springer International Publishing, Cham, pp 113–134

Davis ME, Shin DM (2008) Nanoparticle therapeutics: an emerging treatment modality for cancer. Nat Rev Drug Discov 7(9):771–782

Dikmen G, Genç L, Güney G (2011) Advantage and disadvantage in drug delivery systems. J Mater Sci Eng 5(4):468

Esposito E, Mariani P, Ravani L, Contado C, Volta M, Bido S et al (2012) Nanoparticulate lipid dispersions for bromocriptine delivery: characterization and in vivo study. Eur J Pharm Biopharm 80(2):306–314

Fu Y, Liu H, Ren Z, Li X, Huang J, Best S, Han G (2017) Luminescent CaTiO 3: Yb, Er nanofibers co-conjugated with Rose Bengal and gold nanorods for potential synergistic photodynamic/photothermal therapy. J Mater Chem B 5(26):5128–5136

Fundarò A, Cavalli R, Bargoni A, Vighetto D, Zara GP, Gasco MR (2000) Non-stealth and stealth solid lipid nanoparticles (SLN) carrying doxorubicin: pharmacokinetics and tissue distribution after iv administration to rats. Pharmacol Res 42(4):337–343

Gao S, Wang J, Tian R, Wang G, Zhang L, Li Y et al (2017) Construction and evaluation of a targeted hyaluronic acid nanoparticle/photosensitizer complex for cancer photodynamic therapy. ACS Appl Mater Interfaces 9(38):32509–32519

Gelperina S, Kisich K, Iseman MD, Heifets L (2005) The potential advantages of nanoparticle drug delivery systems in chemotherapy of tuberculosis. Am J Respir Crit Care Med 172 (12):1487–1490

Gupta AK, Gupta M (2005) Synthesis and surface engineering of iron oxide nanoparticles for biomedical applications. Biomaterials 26(18):3995–4021

Gupta AK, Wells S (2004) Surface-modified superparamagnetic nanoparticles for drug delivery: preparation, characterization, and cytotoxicity studies. IEEE Trans Nanobioscience 3(1):66–73

Hainfeld JF, Slatkin DN, Smilowitz HM (2004) The use of gold nanoparticles to enhance radiotherapy in mice. Phys Med Biol 49(18):N309

Hood JD, Bednarski M, Frausto R, Guccione S, Reisfeld RA, Xiang R, Cheresh DA (2002) Tumor regression by targeted gene delivery to the neovasculature. Science 296(5577):2404–2407

Hsu SH, Wen CJ, Al-Suwayeh SA, Chang HW, Yen TC, Fang JY (2010) Physicochemical characterization and in vivo bioluminescence imaging of nanostructured lipid carriers for targeting the brain: apomorphine as a model drug. Nanotechnology 21(40):405101

Huang G, Chen H, Dong Y, Luo X, Yu H, Moore Z et al (2013) Superparamagnetic iron oxide nanoparticles: amplifying ROS stress to improve anticancer drug efficacy. Theranostics 3 (2):116

Kaur IP, Bhandari R, Bhandari S, Kakkar V (2008) Potential of solid lipid nanoparticles in brain targeting. J Control Release 127(2):97–109

Khan I, Saeed K, Khan I (2017) Nanoparticles: properties, applications and toxicities. Arab J Chem In Press

Kievit FM, Zhang M (2011) Surface engineering of iron oxide nanoparticles for targeted cancer therapy. Acc Chem Res 44(10):853–862

Kim HR, Gil S, Andrieux K, Nicolas V, Appel M, Chacun H et al (2007) Low-density lipoprotein receptor-mediated endocytosis of PEGylated nanoparticles in rat brain endothelial cells. Cell Mol Life Sci 64(3):356–364

Kim JK, Seo SJ, Kim KH, Kim TJ, Chung MH, Kim KR, Yang TK (2010) Therapeutic application of metallic nanoparticles combined with particle-induced x-ray emission effect. Nanotechnology 21(42):425102

Klein S, Sommer A, Distel LV, Hazemann JL, Kröner W, Neuhuber W et al (2014) Superparamagnetic iron oxide nanoparticles as novel X-ray enhancer for low-dose radiation therapy. J Phys Chem B 118(23):6159–6166

Kuo YC, Cheng SJ (2016) Brain targeted delivery of carmustine using solid lipid nanoparticles modified with tamoxifen and lactoferrin for antitumor proliferation. Int J Pharm 499(1):10–19

Kuo YC, Hong TY (2014) Delivering etoposide to the brain using catanionic solid lipid nanoparticles with surface 5-HT-moduline. Int J Pharm 465(1):132–142

Kuo YC, Lee CH (2015) Inhibition against growth of glioblastoma multiforme in vitro using etoposide-loaded solid lipid nanoparticles with p-Aminophenyl-α-d-Manno-Pyranoside and folic acid. J Pharm Sci 104(5):1804–1814

Kuo YC, Liang CT (2011) Inhibition of human brain malignant glioblastoma cells using carmustine-loaded catanionic solid lipid nanoparticles with surface anti-epithelial growth factor receptor. Biomaterials 32(12):3340–3350

Kuo YC, Shih-Huang CY (2014) Solid lipid nanoparticles with surface antibody for targeting the brain and inhibiting lymphatic phagocytosis. J Taiwan Inst Chem Eng 45(4):1154–1163

Kuo YC, Wang IH (2017) Using catanionic solid lipid nanoparticles with wheat germ agglutinin and lactoferrin for targeted delivery of etoposide to glioblastoma multiforme. J Taiwan Inst Chem Eng 77:73–82

Kwatra D, Venugopal A, Anant S (2013) Nanoparticles in radiation therapy: a summary of various approaches to enhance radiosensitization in cancer. Transl Cancer Res 2(4):330–342

Lamprecht A (ed) (2016) Nanotherapeutics: drug delivery concepts in nanoscience. CRC Press, Boca Raton, FL

Lévy R, Shaheen U, Cesbron Y, See V (2010) Gold nanoparticles delivery in mammalian live cells: a critical review. Nano Rev 1(1):4889

Li L, Mak KY, Shi J, Koon HK, Leung CH, Wong CM et al (2012) Comparative in vitro cytotoxicity study on uncoated magnetic nanoparticles: effects on cell viability, cell morphology, and cellular uptake. J Nanosci Nanotechnol 12(12):9010–9017

Lin X, Yan SZ, Qi SS, Xu Q, Han SS, Guo LY et al (2017) Transferrin-modified nanoparticles for photodynamic therapy enhance the antitumor efficacy of Hypocrellin A. Front Pharmacol 8:815

Liu KS, Wen CJ, Yen TC, Sung KC, Ku MC, Wang JJ, Fang JY (2012) Combined strategies of apomorphine diester prodrugs and nanostructured lipid carriers for efficient brain targeting. Nanotechnology 23(9):095103

Liu J, Meng T, Yuan M, Wen L, Cheng B, Liu N et al (2016) MicroRNA-200c delivered by solid lipid nanoparticles enhances the effect of paclitaxel on breast cancer stem cell. Int J Nanomedicine 11:6713

Mäder K, Mehnert W (2004) 1—solid lipid nanoparticles—concepts, procedures, and physicochemical aspects. In: Nastruzzi C (ed) Lipospheres in drug targets and delivery: approaches, methods, and applications. CRC Press, Boca Raton, FL, pp 1–22

Mahmoudi M, Simchi A, Imani M, Milani AS, Stroeve P (2008) Optimal design and characterization of superparamagnetic iron oxide nanoparticles coated with polyvinyl alcohol for targeted delivery and imaging. J Phys Chem B 112(46):14470–14481

Mahmoudi M, Simchi A, Milani AS, Stroeve P (2009) Cell toxicity of superparamagnetic iron oxide nanoparticles. J Colloid Interface Sci 336(2):510–518

Mahmoudi M, Simchi A, Imani M, Shokrgozar MA, Milani AS, Häfeli UO, Stroeve P (2010) A new approach for the in vitro identification of the cytotoxicity of superparamagnetic iron oxide nanoparticles. Colloids Surf B: Biointerfaces 75(1):300–309

Maier-Hauff K, Ulrich F, Nestler D, Niehoff H, Wust P, Thiesen B et al (2011) Efficacy and safety of intratumoral thermotherapy using magnetic iron-oxide nanoparticles combined with external beam radiotherapy on patients with recurrent glioblastoma multiforme. J Neuro-Oncol 103 (2):317–324

Malekigorji M, Curtis ADM, Hoskins C (2014) The use of iron oxide nanoparticles for pancreatic cancer therapy. J Nanomed Res 1:1

Martinez AI, Garcia-Lobato MA, Perry DL (2009) Study of the properties of iron oxide nanostructures. Res Nanotechnol Devel 19:184–193

Martín-Rapun R, De Matteis L, Ambrosone A, Garcia-Embid S, Gutierrez L, De La Fuente M, J. (2017) Targeted nanoparticles for the treatment of Alzheimer's disease. Curr Pharm Des 23:1927–1952

Matsudaira H, Ueno AM, Furuno I (1980) Iodine contrast medium sensitizes cultured mammalian cells to X rays but not to γ rays. Radiat Res 84(1):144–148

Mello RS, Callisen H, Winter J, Kagan AR, Norman A (1983) Radiation dose enhancement in tumors with iodine. Med Phys 10(1):75–78

Mishra BBTS, Patel BB, Tiwari S (2010) Colloidal nanocarriers: a review on formulation technology, types and applications toward targeted drug delivery. Nanomedicine 6(1):9–24

Mudshinge SR, Deore AB, Patil S, Bhalgat CM (2011) Nanoparticles: emerging carriers for drug delivery. Saudi Pharm J 19:129–141

Pardeshi CV, Rajput PV, Belgamwar VS, Tekade AR, Surana SJ (2013) Novel surface modified solid lipid nanoparticles as intranasal carriers for ropinirole hydrochloride: application of factorial design approach. Drug Deliv 20(1):47–56

Rahman WN, Bishara N, Ackerly T, He CF, Jackson P, Wong C et al (2009) Enhancement of radiation effects by gold nanoparticles for superficial radiation therapy. Nanomedicine 5 (2):136–142

Reddy LH, Sharma RK, Chuttani K, Mishra AK, Murthy RSR (2005) Influence of administration route on tumor uptake and biodistribution of etoposide loaded solid lipid nanoparticles in Dalton's lymphoma tumor bearing mice. J Control Release 105(3):185–198

Rosen JE, Chan L, Shieh DB, Gu FX (2012) Iron oxide nanoparticles for targeted cancer imaging and diagnostics. Nanomedicine 8(3):275–290

Rosenkranz AA, Jans DA, Sobolev AS (2000) Targeted intracellular delivery of photosensitizers to enhance photodynamic efficiency. Immunol Cell Biol 78(4):452–464

Rui LL, Cao HL, Xue YD, Liu LC, Xu L, Gao Y, Zhang WA (2016) Functional organic nanoparticles for photodynamic therapy. Chin Chem Lett 27(8):1412–1420

Seo SJ, Jeon JK, Jeong EJ, Chang WS, Choi GH, Kim JK (2013) Enhancement of tumor regression by coulomb nanoradiator effect in proton treatment of iron-oxide nanoparticle-loaded orthotopic rat glioma model: implication of novel particle induced radiation therapy. J Cancer Ther 4(11):25

Shah B, Khunt D, Bhatt H, Misra M, Padh H (2015) Application of quality by design approach for intranasal delivery of rivastigmine loaded solid lipid nanoparticles: effect on formulation and characterization parameters. Eur J Pharm Sci 78:54–66

Shin TH, Choi Y, Kim S, Cheon J (2015) Recent advances in magnetic nanoparticle-based multi-modal imaging. Chem Soc Rev 44(14):4501–4516

Singh R, Lillard JW (2009) Nanoparticle-based targeted drug delivery. Exp Mol Pathol 86:215–223

Soenen SJ, De Cuyper M, De Smedt SC, Braeckmans K (2012) Investigating the toxic effects of iron oxide nanoparticles. Methods Enzymol 509:195–224

Song Y, Shi Q, Zhu C, Luo Y, Lu Q, Li H et al (2017) Mitochondrial-targeted multifunctional mesoporous Au@ Pt nanoparticles for dual-mode photodynamic and photothermal therapy of cancers. Nanoscale 9(41):15813–15824

Spiers FW (1949) The influence of energy absorption and electron range on dosage in irradiated bone. Br J Radiol 22(261):521–533

Steichen SD, Caldorera-Moore M, Peppas NA (2013) A review of current nanoparticle and targeting moieties for the delivery of cancer therapeutics. Eur J Pharm Sci 48:416–427

Su XY, Liu PD, Wu H, Gu N (2014) Enhancement of radiosensitization by metal-based nanoparticles in cancer radiation therapy. Cancer Biol Med 11(2):86

Sutradhar KB, Amin ML (2014) Nanotechnology in cancer drug delivery and selective targeting. ISRN Nanotechnol 2014:1–12

Toossi MTB, Ghorbani M, Sabet LS, Akbari F, Mehrpouyan M (2015) A Monte Carlo study on dose enhancement and photon contamination production by various nanoparticles in electron mode of a medical linac. Nukleonika 60(3):489–496

Tsai MJ, Huang YB, Wu PC, Fu YS, Kao YR, Fang JY, Tsai YH (2011) Oral apomorphine delivery from solid lipid nanoparticles with different monostearate emulsifiers: pharmacokinetic and behavioral evaluations. J Pharm Sci 100(2):547–557

Urbanski M, Mirzaei J, Sharma A, Hofmann D, Kitzerow H-S, Hegmann T (2016) Chemically and thermally stable, emissive carbon dots as viable alternatives to semiconductor quantum dots for emissive nematic liquid crystal–nanoparticle mixtures with lower threshold voltage. Liq Cryst 43:183–194

Veiseh O, Gunn JW, Zhang M (2010) Design and fabrication of magnetic nanoparticles for targeted drug delivery and imaging. Adv Drug Deliv Rev 62(3):284–304

Wen CJ, Zhang LW, Al-Suwayeh SA, Yen TC, Fang JY (2012) Theranostic liposomes loaded with quantum dots and apomorphine for brain targeting and bioimaging. Int J Nanomedicine 7:1599

Wennink JW, Liu Y, Mäkinen PI, Setaro F, de la Escosura A, Bourajjaj M et al (2017) Macrophage selective photodynamic therapy by meta-tetra (hydroxyphenyl) chlorin loaded polymeric micelles: a possible treatment for cardiovascular diseases. Eur J Pharm Sci 107:112–125

Westman JA (2006) Medical genetics for the modern clinician. Lippincott Williams & Wilkins, Philadelphia, PA

Wong HL, Bendayan R, Rauth AM, Wu XY (2004) Development of solid lipid nanoparticles containing ionically complexed chemotherapeutic drugs and chemosensitizers. J Pharm Sci 93 (8):1993–2008

Wong HL, Bendayan R, Rauth AM, Wu XY (2006a) Simultaneous delivery of doxorubicin and GG918 (Elacridar) by new polymer-lipid hybrid nanoparticles (PLN) for enhanced treatment of multidrug-resistant breast cancer. J Control Release 116(3):275–284

Wong HL, Rauth AM, Bendayan R, Manias JL, Ramaswamy M, Liu Z et al (2006b) A new polymer–lipid hybrid nanoparticle system increases cytotoxicity of doxorubicin against multidrug-resistant human breast cancer cells. Pharm Res 23(7):1574–1585

Wu W, He Q, Jiang C (2008) Magnetic iron oxide nanoparticles: synthesis and surface functionalization strategies. Nanoscale Res Lett 3(11):397

Wu J, Han H, Jin Q, Li Z, Li H, Ji J (2017) Design and proof of programmed 5-aminolevulinic acid prodrug nanocarriers for targeted photodynamic cancer therapy. ACS Appl Mater Interfaces 9 (17):14596–14605

Xie L, Jiang R, Zhu F, Liu H, Ouyang G (2014) Application of functionalized magnetic nanoparticles in sample preparation. Anal Bioanal Chem 406(2):377–399

Yang SC, Lu LF, Cai Y, Zhu JB, Liang BW, Yang CZ (1999) Body distribution in mice of intravenously injected camptothecin solid lipid nanoparticles and targeting effect on brain. J Control Release 59(3):299–307

Youssef Z, Vanderesse R, Colombeau L, Baros F, Roques-Carmes T, Frochot C et al (2017) The application of titanium dioxide, zinc oxide, fullerene, and graphene nanoparticles in photodynamic therapy. Cancer Nanotechnol 8(1):6

Yusuf M, Khan M, Khan RA, Ahmed B (2013) Preparation, characterization, in vivo and biochemical evaluation of brain targeted Piperine solid lipid nanoparticles in an experimentally induced Alzheimer's disease model. J Drug Target 21(3):300–311

Zara GP, Cavalli R, Bargoni A, Fundarò A, Vighetto D, Gasco MR (2002) Intravenous administration to rabbits of non-stealth and stealth doxorubicin-loaded solid lipid nanoparticles at increasing concentrations of stealth agent: pharmacokinetics and distribution of doxorubicin in brain and other tissues. J Drug Target 10(4):327–335

Chapter 9
Solid Lipid Nanoparticles: A Modern Approach for the Treatment of Neurodegenerative Diseases

Anisha A. D'Souza

9.1 Introduction

Neurodegeneration indicates degeneration of neuron, either leading to ataxia—impaired moment or dementia—impaired cognitive capabilities and memory (Goldsmith et al. 2014). If uncontrolled they lead to progressive neuron damage, aggravate the condition as well as causes bed-ridden situations. Neurodegenerative diseases mainly include Alzheimer's Disease (AD), Parkinson's Disease (PD), Huntington's disease (HD), Pick's disease, Lewy body dementia (LBD), Vascular Dementia (VD), Amyotrophic Lateral Sclerosis (ALS), Prion disease, Frontotemporal Dementia (FTD) and Spinocerebellar ataxia (Goyal et al. 2014; Spuch et al. 2012). By the age of 70 years, half of human population exhibit neurodegenerative and CNS disorders. Globally, 35.6 million of the population is suffering from dementia and would reach 115.4 million by 2050 (Goyal et al. 2014).

Neurodegenerative diseases affecting the central nervous system (CNS) are herculean task for medical treatment. The main challenge begins with deciphering the diagnosis, initiation and pathology of neurodegeneration. However they mostly are characterized by accumulation of atypical protein such as amyloid β peptides, amyloid precursor protein, paired helical filament of microtubule-associated protein, tau in frontotemporal dementia and tauopathies, α-synuclein in PD, filamentous prion proteins in Creutzfeldt-Jakob disease, etc. Another challenge is the site of treatment. The site of treatment being CNS, it poses many innate barriers of blood brain barrier (BBB) and the brain cerebrospinal fluid barrier (BCSFB). The basic function of these barriers is to prevent the influx of any pathogens and toxic agents to the CNS as well as to maintain homeostasis. Thus, sufficient concentration of drug is not attained in the CNS.

A. A. D'Souza (✉)
Formulation Development Laboratory, Pharmaceutical R&D, Piramal Enterprises Limited, Mumbai, India

S. Siddiquee et al. (eds.), *Nanotechnology: Applications in Energy, Drug and Food*,
https://doi.org/10.1007/978-3-319-99602-8_9

Nanotechnology has been successful throughout in harnessing impossible barriers with potential therapeutic efficacy (Cacciatore et al. 2016). Nanoparticles are more advantageous than any other approaches of targeting the brain. The smaller surface area permits easy permeation through BBB. High surface area to volume ratio changes the pharmacokinetic properties thereby altering the bio distribution of drug. The current chapter discusses the various neurodegenerative diseases, approaches to target the CNS with specific applications of solid lipid nanoparticles in these neurodegenerative diseases.

9.2 Neurodegenerative Diseases

Exact aetiology of neurodegeneration and neuroinflammation is not known. Excitotoxicity like hyperactivity of glutamate receptors and apoptosis are one of the contributors for degeneration. The p53 apoptotic pathway gets activated during stress and lead to programmed cell death (Goyal et al. 2014).

Alzheimer's disease (AD) is an age-related and progressive neurodegenerative disorder. Distinct neurological symptoms of dementia (memory loss), decrease in cognitive function, decline in learning, language orientation, thinking, behavioural complications are characteristic symptoms (Burla et al. 2014). Women are more prone than men (Misra et al. 2016). These symptoms are related to disrupted cholinergic symptoms resulting in memory loss. High concentrations of extracellular amyloid-β is accumulated which causes neurofibrillary tangles and development of senile plaques. Treatment is often with acetyl cholinesterase inhibitors to compensate for deficiency of choline (Saykin et al. 2004). Mild to moderate AD have been treated with USFDA (United States Food and Drugs Administration) approved three cholinesterase inhibitors—Donepezil, Rivastigmine and Galantamine (reversible competitive acetylcholine esterase inhibitor). Strategies targeting AD inhibit the production of amyloid-β peptide and try to increase its clearance. Two types of amyloid-β protein are known to be present, 1–40 and 1–42 residue protein with amyloid 1–42 being predominant. The monomer protein however aggregates and causes neurotoxicity (Spuch et al. 2012).

Parkinson's disease (PD) is yet another neurodegenerative disorder affecting 300 individuals among a million of people (Rajput 1992). This incidence of PD increases in late individuals aged 50 years (Samii et al. 2004). Mental disorders like psychosis or depression, bradykinesia, rigidity, hesitant speech and tremor are symptoms of PD. Formation of lewy bodies in neurons occurring due to lack of cellular dopamine level arises due to malfunction or degeneration of dopaminergic substantia nigra. The underlying cause for the same is yet to be known. Mitochondrial dysfunction, protein misfolding and oxidative stress also play a fundamental role (De Lau and Breteler 2006). Levodopa is the best treatment available for PD. Dopamine receptor agonists can also be used as supplemental therapy. Dopamine receptors are stimulated directly. Levodopa though can traverse the BBB, it fails to attain required concentration in CNS (Jankovic and Stacy 2007).

Huntington's disease (HD) is yet another fatal neurodegenerative disease resulting from inherited change of Huntington gene. Autosomal dominant mutation is seen on either of two copies of gene. Dysfunction of mitochondria seems to play an important role in the disease pathogenesis. Treatment for this disease includes those strategies which could attenuate mitochondrial impairments (Sandhir et al. 2014).

Frontotemporal Dementia (FTD) involves shrinkage of brain's temporal and frontal anterior lobes. Pick bodies i.e. abnormal protein-filled structures developed with neurons are affected with phosphorylation of Tau proteins present in neurofibrillary tangles. Other cellular inclusion proteins—TDP 43 and ubiquitin are also found. Behavioural changes like social functioning, change in personality, judgement, social functioning and speech issues are the two major clinical symptoms seen in FTD. Hence it is confused with psychiatric problem but it occurs in age span of 40–70. Individuals gain a strong desire to eat and as a result gain weight. No successful therapy has been achieved (Spuch et al. 2012).

After AD, Lewy Body Dementia (LBD) is another type of progressive dementia resulting in decline of mental abilities. LBD results in variable alertness, attention, hallucinations, visual spatial dysfunction, rigid muscles, tremors and slow moments. Neurological overlap between PD, AD and LBD is seen especially with respect to presence of Lewy neuritis and Lewy bodies. Additionally LBD causes variation in distribution of lewy body with abnormally accumulated alpha-synuclein (Spuch et al. 2012). Additionally in LBD cognitive function is degraded due to loss of cholinergic neurons while motor control degradation is seen due to loss of dopaminergic neurons. Accurate diagnosis of LBD is mistaken with early stages of VD or AD except that onset is rapid in LBD. Treatment of LBD involves cholinesterase inhibitors. However they are not effective therapy. Experiments to inhibit fibril formation are in progress.

In amyotrophic lateral sclerosis (ALS), it has been found the mutant Cu-Zn superoxide dismutase 1 exerts toxicity on motor neurons located in ventral horn of spinal cord resulting in locomotion defects (Li et al. 2014). The disorder causes muscle atrophy, progressive weakness and fast circulation (Spuch et al. 2012). Spinocerebellar ataxia is also an inherited brain disorder characterized by poor hands, leg, eye movement and speech coordination. More than 20 types of spinocerebellar ataxia have been diagnosed. Vascular dementia (VD) is also a type of dementia developed due to altered cellular and molecular events resulting from impaired blood flow to brain. The symptoms are more common after a major stroke where flow of blood is disrupted in portion of brain. Depending upon the brain areas deprived, functioning would be affected. Memory loss may not be a prominent feature (Spuch et al. 2012).

Prion disease transmissible spongiform encephalopathy involves structural modification of cellular prion protein into Scrapie prion protein—a pathological conformer. This is a unique condition where it can be infectious, inherited or sporadic. Currently there is no effective therapy for this. Progression of this disease is rapid with initial onset of death (Spuch et al. 2012). Neuroinflammation in all diseases is characterized by the glial cell riposte. They enhance lesion development, hyperphosphorylation of tau-proteins and neurite growth related disorders (Mendez-Huergo et al. 2014).

9.3 Physiology of CNS

9.3.1 Blood Brain Barrier and Blood Cerebrospinal Fluid Barrier

The brain is protected by Blood brain barrier (BBB) and blood cerebrospinal fluid barrier (BCSFB). The BBB controls the flow of molecules across the brain parenchyma while the BCSFB regulates across the Cerebrospinal fluid. The BBB constitutes specialized capillary endothelial cells held together with tight junctions on one side and foot processes of astrocytes on the other side. Astrocytes are glial cells responsible for maintaining homoeostasis. It is major uptake of serum ligands compared to BCSFB. Tight junctions are produced by interaction of transmembrane proteins like occludin, claudins and adhesion molecule-1. as well as cytoplasmic accessory proteins (Chen and Liu 2012). The BCSFB is similar to BBB except that there are choroid plexus cells with tight junctions as well as intracellular gaps and fenestrations around the micro vascular endothelium (Fig. 9.1). The BBB protects

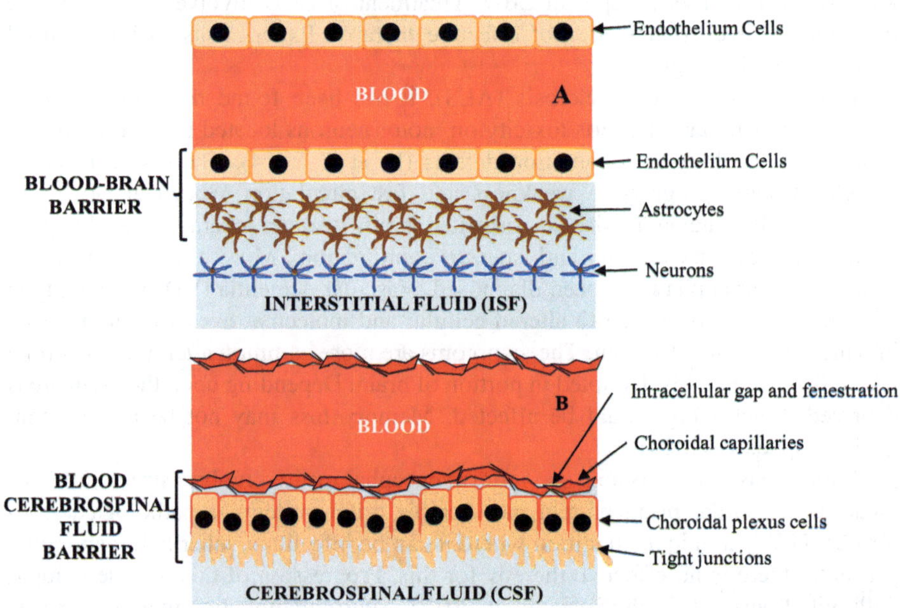

Fig. 9.1 Schematic representation of (**a**) Blood-Brain barrier and (**b**) Brain-cerebrospinal fluid barrier

from slightest fluctuation in ionic composition, flux of xenobiotic, provides essential nutrients and maintains homeostasis (Cecchelli et al. 2007).

9.3.2 Transport Mechanisms

Details of various transport mechanisms can be read from other reviews (Goyal et al. 2014). However, a brief of the same will be summarized below. Transportation can occur through transcellular, paracellular, carrier mediated, absorptive-mediated transcytosis and receptor mediated transport. The brain architecture however limits the transportation across the brain membrane. The tight junctions of BBB restrict paracellular diffusion of water soluble solutes.

Normally, molecules can gain access via transcellular route (active/passive or both) of capillary endothelial cells or choroid plexus cells to brain or cerebrospinal fluid or by transcytosis or solubilisation of endothelial cell membrane or though inhibition of P-glycoprotein (P-gp) efflux. However, it is dependent upon the characteristics of drug molecules. Lipophilic molecules i.e. lipid soluble with molecular weight of 400–600 Da can diffuse across the cells via transcellular pathway along concentration gradient. Molecules beyond this range such as glucose, nucleosides, amino acids, etc., would need receptor-mediated, carrier-mediated or absorptive-mediated transport (Cecchelli et al. 2007). The molecules should have the following properties for transcellular uptake (Pardridge 2010; Vytla et al. 2012)

1. Maximum activity is reported when log Po/w is near to 2.
2. Molecular weight less than 500 Da permits easy diffusion.
3. The number of hydrogen bonds should be less than 8–10 bonds. Each hydrogen or functional polar group decreases permeability by one unit (Clark 2003).
4. Cationic charged molecules are more permeable than anionic charged molecules

As a result delivery of drugs to the CNS is very difficult.

In carrier mediated transport, Carrier mediated proteins facilitate transport of water soluble molecules via active or facilitated diffusion (GLUT1, LT1, ABC transporters). ABC transporters are majorly responsible for transport of drugs. However the BBB also contains P-glycoprotein and multiple drug resistance proteins which remove the drug entering the brain. In adsorptive mediated transcytosis, electrostatic interactions across the membranes facilitate pinocytosis for internalization of drug molecules within cell (Chen and Liu 2012). Specific receptors are highly expressed across the membranes. Receptor mediated endocytosis is triggered when the ligands bind to the receptors and causes internalization of ligand (D'Souza and Devarajan 2015). Thus the ligand along with the nanoparticles is directed towards endosomes/lysosomes or transcytosis across membrane. For instance, coating of nanoparticles with polysorbate binds the nanoparticle to LDL (Low density lipoprotein) receptors of CNS (Lin et al. 2017).

9.4 Approaches to Harness the CNS

Intelligent approaches are required for delivery of drugs to the brain. Basically, there are two techniques to reach the CNS—invasive and non-invasive. Invasive approach indicates transportation of molecules through the BBB via application of physical stimuli to open up the tight junctions. Physical stimuli could include ultrasound, electromagnetic radiation or any surgery, etc. Chemical based stimuli are non-invasive and do not need any surgery.

9.4.1 Invasive Techniques

Not all Drug moieties and peptides have the ability to traverse the BBB (Wu et al. 2002). Administration of increasing doses for any drug targeting CNS would not be recommendable due to side effects associated with high volume of distribution. Hence, at times BBB is disrupted to facilitate direct delivery to brain tissue using intracerebral implants, intraventricular, interstitial or intracavitary delivery, biological tissue delivery, etc. (Wang et al. 2002).

9.5 Non-invasive Techniques

9.5.1 Chemical Approaches

These are chemical structure transformations carried out on drug molecule to improve its physicochemical property such that drug accumulates in CNS, For e.g. permeability, solubility, etc.

Lipophilic analogue or lipidization introduces lipid groups with a partition coefficient 1.5–2.5 to the polar functional groups of molecule. This improves the chances of permeability (Madrid et al. 1991). Lipidization however causes non-specificity of the drug with high volume of distribution as lipidization improves passive diffusion. Additionally, it alters the rate of cytochrome P-450 mediated oxidative metabolism. Pharmacokinetic and biopharmaceutical parameters are thereby significantly altered (Bodor and Buchwald 1999). Pro-drugs too are chemical modifications which improve lipophilicity of the drug and hence permeability. However, it undergoes chemical or metabolic modification which further frees up the active drug (Pardridge 1988). Apomorphine is only USFDA approved Dopamine D1/D2 agonist for treatment in Parkinson's disease. However undergoes rapid degradation by first pass effect with only 1.7% of bioavailability.

Chemical drug delivery system is similar to pro-drug mechanism but requires an activation step like activation by site-specific enzyme, receptor based or physicochemical enzymatic driven (Bodor and Buchwald 1999). Permeability of peptide molecules can be improved by molecular packaging. Molecular packaging not only improves passive diffusion, improves ability of drug from enzymes, it could also enhance targeting of drug by masking the groups which is a hurdle for BBB penetration and prevents undesirable attraction from peptidases (Dwibhashyam and Nagappa 2008).

9.5.2 Biological Approaches

Peptides with poor BBB penetration could be improved by tagging with a carrier peptide which facilitates absorptive-mediated or receptor mediated uptake by BBB triggered by endocytosis (Kang et al. 1994). The peptide molecule inadvertently is released in the interstitial space of brain, wherein the active peptide molecule is cleaved from the transport carrier peptide. For instance, gold nanoparticle conjugated transferrin led to brain accumulation of these nanoparticles dependent upon the optimal transferrin concentration. High concentration of transferrin however strongly adhered to the endothelial cells and failed transcytosis (Townsend et al. 2007).

Cell-penetrating peptide comprises a basic amino acid sequence responsible for their cationic behaviour (Vives et al. 1997). The cationic charge facilitates cell surface interaction and is receptor independent. Most commonly used TAT (trans-activating transcriptional activator) peptide have been linked to biomacromolecules such as human immunodeficiency virus type 1 (HIV-1), herpes simplex virus type-1 (Huwyler et al. 1996).The peptide forms reverse micelles and destabilize the cell membrane thereby making it more permeable (Derossi et al. 1996).

Viral vectors are commonly used for gene delivery by direct CNS administration (Kravcik et al. 1999). They are however associated with side effects of unnecessary immune response and mutagenicity or endogenous recombination of virus which may lead to oncogenesis (Kumar et al. 2009).

9.5.3 Colloidal Drug Delivery

Colloidal delivery systems range from nanoparticles, liposomes, dendrimers, emulsions, etc. Colloidal delivery systems are highly dependent upon surface affinity, particle size and systemic stability for targeting to the CNS (Bhaskar et al. 2010). Besides, the affinity of these nanoparticles could be further enhanced by conjugating to specific ligands. However, among the different colloidal delivery systems, nanoparticles especially solid lipid nanoparticles have been preferred over others due to their high in vivo stability and better physical protection from enzymes (Yoo et al. 2005). Also, the efflux of drug is minimized upon encapsulating within matrix.

Literature cites surface modification of nanoparticles for brain specific delivery using surfactant e.g. Tween 80, Poloxamer 188 or antibody like OX26 antibody targeting transferrin receptors (Kanwar et al. 2012). Receptor-mediated endocytosis has always been recommended for site-specific delivery. Receptor based nanoparticles like transferrin have been studied in cerebral malaria in delivery of malaria and cancer (Gupta et al. 2007), detailed targeting in neurodegenerative disease needs to be explored. Bioadhesive surfactants increase the adhesion of carriers and mask the efflux mechanisms while surfactants like Tween 80 and Poloxamer increases the brain permeation by inhibiting the P-glycoprotein efflux and solubilizes the lipids from the membrane of brain endothelial cells (Kulkarni and Feng 2011). They also trigger serum proteins and adsorb Apo lipoprotein B and/or apolipoprotein E mimicking low-density lipoproteins facilitating receptor-mediated endocytosis (Kaur et al. 2008). Recently vascular endothelial growth factor, lactoferrin, insulin-like growth factor, albumin, angiopep-2 have also been explored for BBB delivery (Farokhzad and Langer 2006).

The exact mechanism of nanoparticle uptake through brain is not yet known. However, the popular and probable accepted theories are;

1. High surface area to volume ratio can increase the adsorption of nanoparticles on endothelial cells which can increase the retention and movement across BBB.
2. Smaller size provides easy diffusion where leaky vasculatures are present which is more prominent in cancer than neurodegenerative diseases.
3. Excipients play an important role in nanoparticle uptake. Surfactants enhance penetration by membrane fluidization i.e. solubilization of membrane lipids.
4. Polysorbates and pluronics act as efflux transporter inhibitors while mucoadhesive chitosan, HPMC etc., enhance drug retention. Some nanoparticles can also play with tight junctions and open and permit transportation (Pathan et al. 2009).
5. Endocytosis and transcytosis can deliver drug directly to endothelial cells.

9.6 Solid Lipid Nanoparticles (SLN) Explored for BBB Delivery

SLN are structurally made up of solid hydrophobic lipid core with drug either dissolved or dispersed in matrix. They overcome the limitations of polymeric nanoparticles by avoiding use of organic solvents and toxic monomers and are easily scalable methods. Further lipid nanoparticles exhibit better transport efficiency with increased contact time with BBB. The biodegradable and biocompatible (prepared from physiologically accepted lipids) nature of lipids used in preparation of SLN makes them a safer bate for brain delivery (Kaur et al. 2008).

Under normal physiological conditions, hydrophobic surfaces of nanoparticles trigger interaction of the nanosystems with opsonins—plasma proteins. Additionally if the surface is negatively charged it activates the complement system followed by phagocytosis by macrophages of the reticuloendothelial system (Aggarwal et al. 2009). Systemic circulation time is thus drastically reduced for solid lipid nanoparticles (Moghimi et al. 1991). However, masking the surface with hydrophilic surfactants of polymers could prevent phagocytosis. PEGylation with Polyethylene glycol of molecular weights between 2000 and 5000 bypasses RES uptake (D'Souza and Shegokar 2016).

Some of the SLN widely explored for delivery of drugs in AD, PD and other neurodegenerative diseases is given in Tables 9.1–9.3.

Besides intravenous, intrathecal and oral routes of delivery, the intranasal route of delivery is another successive method for CNS delivery. This route offers the quickest delivery of drug through neural pathways (Thorne and Frey 2001). Again, drugs need to demonstrate high lipophilicity and low molecular weight with smaller particle size for enhanced uptake via nasal route (Sakane et al. 1994). Drug can be delivered to brain via the cellular processes of olfactory pathway. Once it enters the olfactory epithelium it enters the bulb through trigeminal and olfactory nerve distributing throughout the brain. The high surface area of nasal cavity provides faster action. Nanostructured lipid carriers of Valproic acid provided more neuro-protection from the electroshocks in a rat model. Further among intraperitoneal and intranasal, the intranasal route exhibited higher ratio of drug concentration in brain to plasma (Jogani et al. 2008). However, a combination of biological factors of nasal mucosa and molecular factors of drug is needed.

9.7 Conclusion

Neurodegenerative diseases are more distressful and devastating concerns for human. Physiological barriers of brain though provide protection, but are bottle neck for therapeutic effectiveness in CNS disorders. The focus of treatment using traditional techniques has now been drifted towards CNS targeting therapies. Though most of them are in experimental stages and still associated with concerns, however they provide a ray of hope for better clinical application. Some of the concerns are reproducibility in BBB opening strategies in complex environment. Besides, repeated opening of BBB also raises concerns. SLNs have been reported to have low drug loading which is almost <0.5% in case of hydrophilic drugs. Hence in such cases the purpose of solid lipid nanoparticles could be served in potent molecules. Recently to overcome this limitations, nanoparticles of lipid drug conjugates encapsulating 33% of diminazene diaceturate have been studied for brain targeting (Müller et al. 2002). Though reported to have high ability to traverse BBB, more data is needed to confirm the same. Tailored nanoparticles are promising carriers for delivery across BBB and BCSFB. They have opened up the ways for delivery of hydrophilic molecules and peptides.

However, there still exists lot of challenges for delivery.

Nevertheless, SLNs have demonstrated good potential in their ability to cross the BBB with reduced toxicity and high chances of controlling drug release. As also

Table 9.1 SLN's explored for delivery in treatment of Alzheimer's

Drug	Features	Reference
–	SLNs of phosphatidic acid or cardiolipin and poly (alkyl cyanoacrylate), Compritol 888, Lutrol F68 and phospholipon 80H prepared by melt method yielded anionic nanoparticles (−41 ± 1.60 mV before functionalization and −44 ± 1.78 mV after functionalization) and particle size of 250 ± 20 nm. Increased production of nitric acid however was seen after highest dose. Nanoparticles containing phosphatidic acid decreased viability at lower doses in cultured human umbilical vein endothelial cells (HUVEC) and macrophages (RAW264.7). They were non-toxic to blood cells	Orlando et al. (2013)
(−)-Epi gallocatechin-3-gallate	Epigallocatechin-3-gallate is well known for its anti-oxidant, up regulating ±-secretase non-amyloidogenic processing of precursor amyloid protein and thus preventing A2 plaque formation. Poor bioavailability makes poor efficacy of this polyphenol. SLNS of (−)-epigallocatechin-3-gallate prepared by co-solubilisation increased almost 91% of neuronal ±-secretase used in AD mouse model	Smith et al. (2010)
Curcumin	Curcumin is a potent antioxidant, anti-inflammatory and neuro-protection. However it exhibits poor bioavailability SLNs of compritol 888, polysorbate 80 and soy lecithin entrapped ~82% of curcumin and exhibited a particle size of ~135 nm. Curcumin nanoparticles showed ~98% recovery in membrane lipids and 73% of recovery in acetyl cholinesterase activity compared to 15% and 22% respectively observed with free curcumin in aluminium induced AD in mice. Thus aluminium induced alterations could be reversed	Kakkar and Kaur (2011)
Ferulic acid	Ferulic acid, an anti-inflammatory and antioxidant also develops resistance to A2 peptide toxicity in mice. SLNs of compritol 888 ATO with Epikuron™200 (surfactant) and taurocholate sodium (co-surfactant) entrapped ferulic acid up to 20% with a particle size of 96 nm and low polydispersity and suppressed oxidative stress i.e. decreased generation of reactive oxygen species, cytochrome c release, inhibited activation of caspases 9 and 3	Picone et al. (2009)
Galantamine hydrobromide	Nanoparticles were prepared by solvent emulsification diffusion technique improved the bioavailability of drug by 25%. A high entrapment of 78% and cationic particles of 772 nm was observed using Tween80. A release of 54% after 10 min followed by 80% after 1 h was observed in phosphate buffer	Cacciatore et al. (2016)

(continued)

Table 9.1 (continued)

Drug	Features	Reference
Galantamine hydrobromide	SLN of galantamine hydrobromide (Glycerylbehnate-Compritol, Pluronic F-127 and Tween 80) were of sizes 100 nm with an entrapment efficiency of 83%. The particles also showed an in vitro controlled release greater than 90% in 24 h. Significant memory restoration was seen in cognitive deficit rats with SLNs compared to plain drug with a two-fold increased bioavailability as well	Misra et al. (2016)
Neurotoxic A2 peptide	A2 neurotoxic peptide released as monomers in Alzheimer's disease undergoes aggregation to form oligomers, fibrils and plaques in brain. SLNs targeting A2 peptide aggregates ($A2_{1–42}$) has been suggested as a potential in AD treatment. Lipid was selected with the highest interaction with A2 1–42 (distearoylphosphatidic acid) and used for SLN preparation along with stearic acid, phospholipon 90G and taurocholate sodium salt by microemulsion. Particle size obtained was 76 nm with a zeta of −43.3 mV indicating anionic charge	Gobbi et al. (2010)
Piperine	Piperine is poorly water soluble compound and undergoes first pass metabolism. SLN of piperine (natural alkaloid) used as an antioxidant using glycerol monostearate and another glycerol monostearate and Tween-80 blocks anticholinesterase enzyme or increase cholinergics in brain. Polysorbate coating was possible after dipping SLNs in 1% polysorbate-80 for 1 h. The coated particles showed anionic values after coating and significant reduction in superoxide dismutase values and decreased cholinergic degradation and decreased amyloid plaques too	Yusuf et al. (2013)
Quercetin	NPs of Compritol 888 ATO containing glycerides of behenic acid and Tween as surfactant prepared by microemulsion technique gave a particle size lower than 200 nm, an entrapment of 85.73% and zeta potential of 21.05 mV. In vivo studies in aluminium chloride treated rats reversed the neurodegenerative effects with significant. Antioxidant potential enhanced in presence of SLN	Dhawan et al. (2011)
Resveratrol	Bioavailability of oral administration with SLN was eightfold higher. Adhesion to GI tract was obtained after coating SLN with N-trimethyl chitosan. Prolonged resveratrol release up to 120 h and is sensitive to light. But lipid prevented its degradation compared to plain solution after	Pandita et al. (2014); Ramalingam and Ko (2016)

Table 9.2 SLN's explored for delivery in treatment of Parkinson's

Drug	Feature	Reference
Alpha lipoic acid	Softisan entrapped 90% of lipoic acid with Miranol Ultra C32 as emulsifier. The formulation was stable for 3 months. This was suggested for used of lipoic acid as anti-aging for topical delivery	Souto et al. (2005)
Apomorphine	Apomorphine undergoes first-pass effect with only 2% of bioavailability. Apomorphine encapsulated in Tripalmitin, glycerylmonostearate and polyethylene glycol monostearate yielded particles of 155 nm and 63 nm respectively. Gastric medium caused particles fusion with increase in mean size due to high ionic strength. However polyethylene glycol monostearate nanoparticles were more stable in intestinal fluid. Bioavailability of SLN was 12-fold compared to 2.09% of plain drug after oral administration. Apomorphine was detected at high concentration in striatum and other parts of rat brain compared to aqueous solution upon oral administration	Tsai et al. (2011)
Bromocriptine (Dopamine agonist)	SLNs of Tricaprin/tristearin containing bromocriptine yielded an entrapment of 84% and a mean particle size of ~198 nm. Good stability was observed over period of 6 months. In vivo studies revealed rapid onset of release but sustained for a long term	Esposito et al. (2008)
Chrysin (inhibits inflammation)	Chrysin loaded SLN restored acetyl cholinesterase and lipid peroxidation in Amyloid β-injected rats. Memory retention was also reversed by SLN compared to free control	Vedagiri and Thangarajan (2016)
Idebenone	Idebenone encapsulated within cetylpalmitate and isoceteth-20/glyceryloleate and ceteth-20/glyceryloleate by phase-inversion temperature. Particle size of ceteth-20/glyceryloleate increased when 0.5–0.7% of drug was incorporated while 1.1% of drug reduced the particle size. The SLNs were able to inhibit Lactic dehydrogenase and ROS production in astrocytes cultures from rat cerebral cortex	Montenegro et al. (2011)
Levodopa	SLN of Levodopa sized 108 nm was prepared by microemulsion technology and exhibited stability at 4 °C for 2 months	Zhan et al. (2010)
Lipoyl-mematine	Lipoyl-mematine is pro-drug with poor water solubility. Drug-lipid ratio of 1:5 entrapped drug up to 88% and exhibited a particle size of 170 nm and a zeta of −33.8 mV. Stability was achieved in stimulated intestinal and gastric fluids. Lipoyl-mematine was released up to 20% in media with pH of 1.2 and 7.4 and enzymes, the formulation had no toxic effects	Laserra et al. (2015)
Luteolin	Luteolin reduces oxidation stress. Luteolin, glycerol monostearate, phosphatidyl choline and Tween-80 yielded particles sized 48 nm and zeta potential of −9.62 mV. In the first hour, 48% of Luteolin followed	Dang et al. (2014)

(continued)

Table 9.2 (continued)

Drug	Feature	Reference
	by 89% release over 48 h. Bioavailability improved upon incorporation in SLN	
Ropinirole hydrochloride	SLNs were prepared by emulsification-solvent diffusion using pluronic F-68 and stearylamine with particle size as low as 67 nm. Nasal mucosa showed no change in integrity. Comparable therapeutic activity was seen with marketed oral formulations. Intranasal delivery demonstrated good response	Pardeshi et al. (2013)

Table 9.3 SLN's explored for miscellaneous applications

Curcumin	Curcumin SLNs were prepared using Brij 78 and TPGS to inhibit P-glycoprotein efflux pump. The effective permeability coefficient for the SLNs with Brij was 1.3 times and those with Brij and TPGS were 1.4 times greater than plain SLNs in the in situ pass jejunum perfusion test. AUC for these SLNs were also 12.3 fold higher than Curcumin suspension	Ji et al. (2016)
Curcumin	SLNs of curcumin in rats induced with Huntington's disease revealed increase in cytochrome levels and mitochondrial complexes. Superoxide dismutase and Glutathione levels were also restored. A decrease in mitochondrial swelling, reactive oxygen species, lipid peroxidation, protein carbonyls was seen. Significant improvement in neuromotor coordination could be seen	Sandhir et al. (2014)
Riluzole	Riluzole is a potent neuro-protective agent useful in amyotrophic lateral sclerosis. SLN of Riluzole and Compritol S88 ATO, phosphatidyl choline and taurocholate sodium salt prepared by microemulsion technique were of size 88 nm. These SLNs developed clinical signs of experimental allergic encephalomyelitis much later than those with free riluzole upon treatment in rats induced with experimental allergic encephalomyelitis, a sign of multiple sclerosis. Significant amount of Riluzole was found in CNS after 16 h of administration	Bondi et al. (2010)
Rosmarinic acid	Encapsulated 62% of rosmarinic acid within glycerol monostearate tween 80 and lecithin to yield particles of 149.2 nm with an anionic zeta of −38.27 mV. These SLN could attenuate induced striatal oxidative stress and locomotors and motor co-ordination, beam walk in animal models of HD upon intranasal administration. The intranasal route was much more effective and site-specific than intravenous route	Bhatt et al. (2015)

aware, SLNs are known for the ease to provide large scale production. SLNs open up the avenues for efficacious delivery of drugs in individuals with neurodegenerative diseases. It is an ever challenging but interesting and promising field.

References

Aggarwal P, Hall JB, Mcleland CB et al (2009) Nanoparticle interaction with plasma proteins as it relates to particle biodistribution, biocompatibility and therapeutic efficacy. Adv Drug Deliv Rev 61:428–437

Bhaskar S, Tian F, Stoeger T et al (2010) Multifunctional Nanocarriers for diagnostics, drug delivery and targeted treatment across blood-brain barrier: perspectives on tracking and neuro-imaging. Part Fibre Toxicol 7:3

Bhatt R, Singh D, Prakash A et al (2015) Development, characterization and nasal delivery of rosmarinic acid-loaded solid lipid nanoparticles for the effective management of Huntington's disease. Drug Deliv 22:931–939

Bodor N, Buchwald P (1999) Recent advances in the brain targeting of neuropharmaceuticals by chemical delivery systems. Adv Drug Deliv Rev 36:229–254

Bondi ML, Craparo EF, Giammona G et al (2010) Brain-targeted solid lipid nanoparticles containing riluzole: preparation, characterization and biodistribution. Nanomedicine (Lond) 5:25–32

Burla C, Rego G, Nunes R (2014) Alzheimer, dementia and the living will: a proposal. Med Health Care Philos 17:389–395

Cacciatore I, Ciulla M, Fornasari E et al (2016) Solid lipid nanoparticles as a drug delivery system for the treatment of neurodegenerative diseases. Expert Opin Drug Deliv 13:1121–1131

Cecchelli R, Berezowski V, Lundquist S et al (2007) Modelling of the blood-brain barrier in drug discovery and development. Nat Rev Drug Discov 6:650–661

Chen Y, Liu L (2012) Modern methods for delivery of drugs across the blood–brain barrier. Adv Drug Deliv Rev 64:640–665

Clark DE (2003) In silico prediction of blood-brain barrier permeation. Drug Discov Today 8:927–933

D'Souza AA, Devarajan PV (2015) Asialoglycoprotein receptor mediated hepatocyte targeting - strategies and applications. J Control Release 203:126–139

D'Souza AA, Shegokar R (2016) Polyethylene glycol (PEG): a versatile polymer for pharmaceutical applications. Expert Opin Drug Deliv 13:1257–1275

Dang H, Meng MHW, Zhao H et al (2014) Luteolin-loaded solid lipid nanoparticles synthesis, characterization, & improvement of bioavailability, pharmacokinetics in vitro and vivo studies. J Nanopart Res 16:2347

De Lau LML, Breteler MMB (2006) Epidemiology of Parkinson's disease. Lancet Neurol 5:525–535

Derossi D, Calvet S, Trembleau A et al (1996) Cell internalization of the third helix of the Antennapedia homeodomain is receptor-independent. J Biol Chem 271:18188–18193

Dhawan S, Kapil R, Singh B (2011) Formulation development and systematic optimization of solid lipid nanoparticles of quercetin for improved brain delivery. J Pharm Pharmacol 63:342–351

Dwibhashyam V, Nagappa AN (2008) Strategies for enhanced drug delivery to the central nervous system. Indian J Pharm Sci 70:145–153

Esposito E, Fantin M, Marti M et al (2008) Solid lipid nanoparticles as delivery systems for bromocriptine. Pharm Res 25:1521–1530

Farokhzad OC, Langer R (2006) Nanomedicine: developing smarter therapeutic and diagnostic modalities. Adv Drug Deliv Rev 58:1456–1459

Gobbi M, Re F, Canovi M et al (2010) Lipid-based nanoparticles with high binding affinity for amyloid-beta1-42 peptide. Biomaterials 31:6519–6529

Goldsmith M, Abramovitz L, Peer D (2014) Precision nanomedicine in neurodegenerative diseases. ACS Nano 8:1958–1965

Goyal K, Koul V, Singh Y et al (2014) Targeted drug delivery to central nervous system (CNS) for the treatment of neurodegenerative disorders: trends and advances. Cent Nerv Syst Agents Med Chem 14:43–59

Gupta Y, Jain A, Jain SK (2007) Transferrin-conjugated solid lipid nanoparticles for enhanced delivery of quinine dihydrochloride to the brain. J Pharm Pharmacol 59:935–940

Huwyler J, Wu D, Pardridge WM (1996) Brain drug delivery of small molecules using immunoliposomes. Proc Natl Acad Sci 93:14164–14169

Jankovic J, Stacy M (2007) Medical management of levodopa-associated motor complications in patients with Parkinson's disease. CNS Drugs 21:677–692

Ji H, Tang J, Li M et al (2016) Curcumin-loaded solid lipid nanoparticles with Brij78 and TPGS improved in vivo oral bioavailability and in situ intestinal absorption of curcumin. Drug Deliv 23:459–470

Jogani V, Jinturkar K, Vyas T et al (2008) Recent patents review on intranasal administration for CNS drug delivery. Recent Pat Drug Deliv Formul 2:25–40

Kakkar V, Kaur IP (2011) Evaluating potential of curcumin loaded solid lipid nanoparticles in aluminium induced behavioural, biochemical and histopathological alterations in mice brain. Food Chem Toxicol 49:2906–2913

Kang YS, Bickel U, Pardridge WM (1994) Pharmacokinetics and saturable blood-brain barrier transport of biotin bound to a conjugate of avidin and a monoclonal antibody to the transferrin receptor. Drug Metab Dispos 22:99–105

Kanwar JR, Sriramoju B, Kanwar RK (2012) Neurological disorders and therapeutics targeted to surmount the blood-brain barrier. Int J Nanomedicine 7:3259–3278

Kaur IP, Bhandari R, Bhandari S et al (2008) Potential of solid lipid nanoparticles in brain targeting. J Control Release 127:97–109

Kravcik S, Gallicano K, Roth V et al (1999) Cerebrospinal fluid HIV RNA and drug levels with combination ritonavir and saquinavir. J Acquir Immune Defic Syndr 21:371–375

Kulkarni SA, Feng SS (2011) Effects of surface modification on delivery efficiency of biodegradable nanoparticles across the blood-brain barrier. Nanomedicine (Lond) 6:377–394

Kumar A, Singh TD, Singh SK et al (2009) Methods, potentials, and limitations of gene delivery to regenerate central nervous system cells. Biologics 3:245–256

Laserra S, Basit A, Sozio P et al (2015) Solid lipid nanoparticles loaded with lipoyl-memantine codrug: preparation and characterization. Int J Pharm 485:183–191

Li J, Li T, Zhang X et al (2014) Human superoxide dismutase 1 overexpression in motor neurons of Caenorhabditis elegans causes axon guidance defect and neurodegeneration. Neurobiol Aging 35:837–846

Lin CH, Chen CH, Lin ZC et al (2017) Recent advances in oral delivery of drugs and bioactive natural products using solid lipid nanoparticles as the carriers. J Food Drug Anal 25:219–234

Madrid Y, Langer LF, Brem H et al (1991) New directions in the delivery of drugs and other substances to the central nervous system. In: August JT, Anders MW, Murad F (eds) Advances in pharmacology. Academic Press, New York

Mendez-Huergo SP, Maller SM, Farez MF et al (2014) Integration of lectin-glycan recognition systems and immune cell networks in CNS inflammation. Cytokine Growth Factor Rev 25:247–255

Misra S, Chopra K, Sinha VR et al (2016) Galantamine-loaded solid-lipid nanoparticles for enhanced brain delivery: preparation, characterization, in vitro and in vivo evaluations. Drug Deliv 23:1434–1443

Moghimi SM, Porter CJ, Muir IS et al (1991) Non-phagocytic uptake of intravenously injected microspheres in rat spleen: influence of particle size and hydrophilic coating. Biochem Biophys Res Commun 177:861–866

Montenegro L, Campisi A, Sarpietro MG et al (2011) In vitro evaluation of idebenone-loaded solid lipid nanoparticles for drug delivery to the brain. Drug Dev Ind Pharm 37:737–746

Müller RH, Radtke M, Wissing SA (2002) Solid lipid nanoparticles (SLN) and nanostructured lipid carriers (NLC) in cosmetic and dermatological preparations. Adv Drug Deliv Rev 54:S131–S155

Orlando A, Re F, Sesana S et al (2013) Effect of nanoparticles binding β-amyloid peptide on nitric oxide production by cultured endothelial cells and macrophages. Int J Nanomedicine 8:1335–1347

Pandita D, Kumar S, Poonia N et al (2014) Solid lipid nanoparticles enhance oral bioavailability of resveratrol, a natural polyphenol. Food Res Int 62:1165–1174

Pardeshi CV, Rajput PV, Belgamwar VS et al (2013) Novel surface modified solid lipid nanoparticles as intranasal carriers for ropinirole hydrochloride: application of factorial design approach. Drug Deliv 20:47–56

Pardridge WM (1988) Recent advances in blood-brain barrier transport. Annu Rev Pharmacol Toxicol 28:25–39

Pardridge WM (2010) Biopharmaceutical drug targeting to the brain. J Drug Target 18:157–167

Pathan SA, Iqbal Z, Zaidi SM et al (2009) CNS drug delivery systems: novel approaches. Recent Pat Drug Deliv Formul 3:71–89

Picone P, Bondi ML, Montana G et al (2009) Ferulic acid inhibits oxidative stress and cell death induced by Ab oligomers: improved delivery by solid lipid nanoparticles. Free Radic Res 43:1133–1145

Rajput AH (1992) Frequency and cause of Parkinson's disease. Can J Neurol Sci 19:103–107

Ramalingam P, Ko YT (2016) Improved oral delivery of resveratrol from N-trimethyl chitosan-g-palmitic acid surface-modified solid lipid nanoparticles. Colloids Surf B Biointerfaces 139:52–61

Sakane T, Akizuki M, Yamashita S et al (1994) Direct drug transport from the rat nasal cavity to the cerebrospinal fluid: the relation to the dissociation of the drug. J Pharm Pharmacol 46:378–379

Samii A, Nutt JG, Ransom BR (2004) Parkinson's disease. Lancet 363:1783–1793

Sandhir R, Yadav A, Mehrotra A et al (2014) Curcumin nanoparticles attenuate neurochemical and neurobehavioral deficits in experimental model of Huntington's disease. NeuroMolecular Med 16:106–118

Saykin AJ, Wishart HA, Rabin LA et al (2004) Cholinergic enhancement of frontal lobe activity in mild cognitive impairment. Brain 127:1574–1583

Smith A, Giunta B, Bickford PC et al (2010) Nanolipidic particles improve the bioavailability and alpha-secretase inducing ability of epigallocatechin-3-gallate (EGCG) for the treatment of Alzheimer's disease. Int J Pharm 389:207–212

Souto EB, Muller RH, Gohla S (2005) A novel approach based on lipid nanoparticles (SLN) for topical delivery of alpha-lipoic acid. J Microencapsul 22:581–592

Spuch C, Saida O, Navarro C (2012) Advances in the treatment of neurodegenerative disorders employing nanoparticles. Recent Pat Drug Deliv Formul 6:2–18

Thorne RG, Frey WH 2nd (2001) Delivery of neurotrophic factors to the central nervous system: pharmacokinetic considerations. Clin Pharmacokinet 40:907–946

Townsend SA, Evrony GD, Gu FX et al (2007) Tetanus toxin C fragment-conjugated nanoparticles for targeted drug delivery to neurons. Biomaterials 28:5176–5184

Tsai MJ, Huang YB, Wu PC et al (2011) Oral apomorphine delivery from solid lipid nanoparticles with different monostearate emulsifiers: pharmacokinetic and behavioral evaluations. J Pharm Sci 100:547–557

Vedagiri A, Thangarajan S (2016) Mitigating effect of chrysin loaded solid lipid nanoparticles against Amyloid beta25-35 induced oxidative stress in rat hippocampal region: an efficient formulation approach for Alzheimer's disease. Neuropeptides 58:111–125

Vives E, Brodin P, Lebleu B (1997) A truncated HIV-1 Tat protein basic domain rapidly translocates through the plasma membrane and accumulates in the cell nucleus. J Biol Chem 272:16010–16017

Vytla D, Combs-Bachmann RE, Hussey AM et al (2012) Prodrug approaches to reduce hyperexcitation in the CNS. Adv Drug Deliv Rev 64:666–685

Wang PP, Frazier J, Brem H (2002) Local drug delivery to the brain. Adv Drug Deliv Rev 54:987–1013

Wu J, Yoon S-H, Wu W-M et al (2002) Synthesis and biological evaluations of brain-targeted chemical delivery systems of [Nva2]-TRH. J Pharm Pharmacol 54:945–950

Yoo JY, Kim JM, Seo KS et al (2005) Characterization of degradation behavior for PLGA in various pH condition by simple liquid chromatography method. Biomed Mater Eng 15:279–288

Yusuf M, Khan M, Khan RA et al (2013) Preparation, characterization, in vivo and biochemical evaluation of brain targeted Piperine solid lipid nanoparticles in an experimentally induced Alzheimer's disease model. J Drug Target 21:300–311

Zhan S, Hou D, Ping Q et al (2010) Preparation and entrapment efficiency determination of solid lipid nanoparticles loaded levodopa. Zhongguo Yiyuan Yaoxue Zazhi 30:1171–1175

Chapter 10
Nanoparticles in Nanomedicine Application: Lipid-Based Nanoparticles and Their Safety Concerns

Rabiatul Basria S. M. N. Mydin and Said Moshawih

10.1 The Toxicity Profile of Lipid-Based NPs

Lipid-based NPs and especially solid lipid nanoparticles (SLN) has an exceptional lower toxic effects for human granulocytes when compared to polyester particles by 10–20 folds, and lower cytotoxicity than poly (alkyl cyanoacrylate) (PACA) NPs. These relatively safe nanoparticles make them an attractive candidate as a drug delivery system. Lipid-based NPs toxicity is less related to lipid content or particle size but more linked to the surfactant employed. Table 10.1 shows examples for toxicities of lipid and surfactant ingredients. The size of SLN did not lead to macrophages activation or to cytokines' production. The cytotoxic effect is mainly due to the interaction of the positively charged SLN with the negatively charged cellular membrane (Blasi et al. 2007).

10.2 NPs Effect on Blood Coagulation System

10.2.1 Dendrimers

The *in vitro* cytotoxicity of the cationic dendrimers as well as other cationic organic NPs have been elucidated, while the neutral and anionic NPs are not (Malik et al. 2000). It has been found that G5 polypropylenimine dendrimers changed the

R. B. S. M. N. Mydin (✉)
Oncological and Radiological Sciences Cluster, Advanced Medical and Dental Institute, Universiti Sains Malaysia, Kepala Batas, Pulau Pinang, Malaysia
e-mail: rabiatulbasria@usm.my

S. Moshawih
Jordan Center for Pharmaceutical Research, Amman, Jordan

S. Siddiquee et al. (eds.), *Nanotechnology: Applications in Energy, Drug and Food*,
https://doi.org/10.1007/978-3-319-99602-8_10

Table 10.1 Toxicity induced by lipid-based NPs by lipid and surfactant ingredients

Lipid matrix	Toxicity type	Reference
Toxicity due to the lipid component		
Dimethyl stearoyl-ammonium bromide	Cell viability reduction in macrophages by 15.3% at concentration of (0.01%)	Schöler et al. (2000)
Cetyl palmitate	Cell viability reduction in macrophages by 61.8%	
Stearic acid	Cell viability reduction in macrophages by 2.2%	
Dynasan® and campritol® (both stabilized with lipoid S75)	Both are well tolerated *in vivo* and *in vitro* at a concentration range of (0.015–1.5%)	Müller et al. (1997)
SLN and polylactic/glycolic acid NPs	100% reduction of cells viability was observed with 0.5% polylactic/glycolic acid NPs (a polymeric NPs) *in vitro*, while SLN was well tolerated	
Toxicity due to stabilizer/surfactant		
Poloxamer 407[a]	Well tolerated *in vitro* up to 10% of concentration	Blasi et al. (2007)
Tween 80[a]	It showed a distinct reduction in cell viability *in vitro* at concentration ≥0.1% for bounded tween80 Free tween80 was found with high cytotoxicity Cytotoxicity was due to high cellular uptake Higher tween80 concentration reduced cellular uptake	
Poloxamer 184 Poloxamer 188 Poloxamer 235 Poloxamer 335	Poloxamers with lower molecular weight showed more decrease in cell viability *in vitro*	Müller et al. (1997)
Sodium dodecyl sulfate (SDS)	Has high toxic effect and decrease in cell viability SLN are well tolerated at concentration of 0.01% of SDS Free SDS is toxic even at a concentration of 0.00001%	

[a]Before incubation of cell lines with NPs, free surfactant was removed, and Campritol® was used as the SLN matrix

membrane permeability of the human umbilical vein endothelial cells. Moreover, the incubation time in addition to the surface charge employed in the *in vitro* studies lead to a perforation in the cell membrane (Stasko et al. 2007). On the other side, at dose levels higher than the maximum tolerated dose, G4 and G7 amine-terminated poly amido amine (PAMAM) lead to disseminated intravascular coagulation when given by i.v. route (Greish et al. 2012). Additionally, the cationic dendrimers with smaller particles induced platelet aggregation and leukocyte pro-coagulant activity *in vitro* (Dobrovolskaia et al. 2012).

10.2.2 Liposomes

Although liposomal NPs have less toxic effects compared to other NPs, but their surface charge plays a key role in blood coagulation toxicities. While the anionic liposomal NPs interact directly with platelets to inhibit agonist-induced aggregation, the cationic ones were found to act indirectly with plasma coagulation factors (Juliano et al. 1983). At the same time, the charged liposomes induced the abovementioned toxicities in addition to other transient thrombocytopenia, it has been noted that cationic liposomes interacted electrostatically with the negatively charged cell membrane of the endothelial cells in order to be uptaken and thus unload its drug and/or gene payloads inside it. This phenomenon can be exploited by employing liposomes with highly positive charge in order to deliver chemotherapeutic agents to cancer cells of endothelial origins (Ilinskaya and Dobrovolskaia 2013).

10.3 NPs Concerns on Macrophages

Although poly (alkyl cyanoacrylate) PACA toxicity profile as a polymeric colloidal drug delivery system is low, but more *in vivo* mutagenicity studies is still being investigated on the long-term use. In general, the toxicity of PACA is connected to its alkyl chain length, the longer the chain; the less toxic is NPs, as it correlates to the degradation process of the polymer, consequently, the monomer always shows higher toxicity (Graf et al. 2009). In an *in vitro* cytotoxicity study for PACA nanoparticles, polyisohexylcyanoacrylate NPs showed respiratory burst and activation of macrophages. On the other hand, polymethylcyanoacrylate nanoparticles in another study induced macrophages membrane perforation, while polyisobutylcyanoacrylate NPs did not. This can be attributed to the difference in the chain length of different PACA NPs in two combined mechanisms namely; PACA NPs degradation and adhesion (Blasi et al. 2007).

10.4 NPs Concerns on Cardiovascular System and Lungs

The inhaled nanoparticles settle in the peripheral lung depending on their sizes, as the larger NPs tend to be deposited more than the smaller ones and cross the pulmonary epithelium to reach eventually to the interstitium. The accumulation of inhaled NPs can cause inflammatory responses and potentiate oxidative stress which ultimately mediate endothelial dysfunction and atherosclerosis (Nadziejko et al. 2002). The process of systemic inflammation potentiates the thrombosis either by direct activation of platelets or by stimulating coagulation factors such as factor Xa and fibrinogen. Furthermore, the activation of platelets also increases the adhesion and aggregation and therefore, potentiates thrombosis. The activated platelets secrete more inflammatory and mitogenic molecules that alter the chemotactic and proteolytic properties of the endothelial cells and finally take part in the atherogenic formation (McGuinnes et al. 2010).

Surface charge of NPs affects their behavior and influences blood coagulation system in humans and living organisms dramatically. In a study to detect the oxidative stress caused by NPs on the cellular level and their interaction with platelet aggregation, three forms of polystyrene latex NPs (PLNP) were tried *in vitro* namely; unmodified, positively charged and negatively charged at a constant particle size (50 nm). It has been noted that surface charge modifications mediated the activation of platelet aggregation, in addition to that, the negatively and positively charged particles induced platelet-platelet aggregation, unlike the unmodified ones (McGuinnes et al. 2010). In another study the results indicated that only NPs with positive charge induced thrombogenesis in hamsters after an i.v. injection (Nemmar et al. 2003). Smaller NPs aggravated lung inflammatory condition related to bacterial infection due to the overexpression of pro-inflammatory mediators and triggered further oxidative stress (Inoue 2011).

10.5 Standard Guidelines/Reference for Toxicity Studies on NPs

Up to date, there are no standardized analytical and validation methods that are well established in detecting nanotoxicities of NPs. This kind of *in vitro* and *in vivo* studies are needed to be designed and validated by researchers individually until protocols are generated for preclinical assessment and put in practice to guarantee the safety of nanoparticles (Saiyed et al. 2011). Moreover, due to the multi-parts usually employed in nanocarriers in order to perform a multifunctional task, a rigorous assessment is required to measure the physicochemical properties, pyrogenic components, sterility, biodistribution and toxic potential for such systems. Table 10.2 summarizes *in vitro* and *in vivo* assays can govern parameters for assuring safety of NPs.

10.6 Future Prospective

Although the complex nature that lies in the interface between NPs and biological systems when tested *in vivo*, organic NPs are eligible to be used in a wider range looking to their relative low toxicity issues compared to inorganic nanocarriers. They are also characterized by the feasibility of scale-up production, application of sterile procedures and the biocompatability and biodegradability of their components. The incidents that are reported about systemic inflammation due to the inhaled NPs in

Table 10.2 Assays and instruments that check for parameters which produce organic and inorganic nanoparticles with less toxicity

Property	Parameters measured	Instrument/assay
Physicochemical properties	Particle size, shape, stability, particles size distribution, surface functionality, chemical composition	Scanning electron microscope, transition electron microscope, dynamic light scattering, photon correlation spectroscope
Sterility and pyrogenicity	The contamination by a viral, bacterial and/or fungal infection. Pyrogenicity tests detect fever causing substances (e.g. bacterial endotoxins)	Lumulus amebocyte lysate (LAL) assay, Rabbit pyrogen test
Biodistribution	*In vitro*: Cytotoxicity, immune cells induction, hemolysis, complement activation, thrombogenicity	Trypan blue exclusion test, lactate dehydrogenase, tetrazolium dye reduction, DNA staining, flow cytometry, NP phagocytosis assay (to check thrombogenicity), complement-activation assay (CH50), prothrombin time, activated partial thromboplastin time
	NP and drug localization in tissues	Energy dispersive X-ray test (EDX)
	In vivo: Absorption, distribution, metabolism, excretion, efficacy, hepatotoxicity and nephrotoxicity	Complete blood count, organ weight and histology, immunotoxicity, radiolabeled studies for NP and drug components, serum ALT, AST, LDH and blood urea nitrogen

addition to atherogenicity and thrombogenicity induced by i.v. administration of charged NPs are still scarce. On the other hand, the particle size and lipid/polymer employed in their formulation have a very little effect on activating the inflammatory cascade. Piling more toxicity and genotoxicity studies are needed so that to facilitate a more thorough perspective on the potential safety concerns of the organic nanostructures. Moreover, generating a new guidelines and protocols are imperative to standardize results from different studies.

Acknowledgements The authors are thankful to the Ministry of Education (MOE) Malaysia for funding this work under Transdisciplinary Research Grant Scheme (TRGS) grant no. 6769003 and Universiti Sains Malaysia (USM) for USM-Short Term Research Grant (304/CIPPT/6315073).

References

Blasi P, Giovagnoli S, Schoubben A, Ricci M, Rossi C (2007) Solid lipid nanoparticles for targeted brain drug delivery. Adv Drug Deliv Rev 59:454–477

Dobrovolskaia MA, Patri AK, Potter TM, Rodriguez JC, Hall JB, Mcneil SE (2012) Dendrimer-induced leukocyte procoagulant activity depends on particle size and surface charge. Nanomedicine 7:245–256

Graf A, McDowell A, Rades T (2009) Poly (alkycyanoacrylate) nanoparticles for enhanced delivery of therapeutics—is there real potential? Expert Opin Drug Deliv 6:371–387

Greish K, Thiagarajan G, Herd H, Price R, Bauer H, Hubbard D, Burckle A, Sadekar S, Yu T, Anwar A (2012) Size and surface charge significantly influence the toxicity of silica and dendritic nanoparticles. Nanotoxicology 6:713–723

Ilinskaya AN, Dobrovolskaia MA (2013) Nanoparticles and the blood coagulation system. Part II: safety concerns. Nanomedicine 8:969–981

Inoue KI (2011) Promoting effects of nanoparticles/materials on sensitive lung inflammatory diseases. Environ Health Prev Med 16:139–143

Juliano R, Hsu M, Peterson D, Regen S, Singh A (1983) Interactions of conventional or photopolymerized liposomes with platelets in vitro. Exp Cell Res 146:422–427

Malik N, Wiwattanapatapee R, Klopsch R, Lorenz K, Frey H, Weener J, Meijer E, Paulus W, Duncan R (2000) Dendrimers: relationship between structure and biocompatibility in vitro, and preliminary studies on the biodistribution of 125I-labelled polyamidoamine dendrimers *in vivo*. J Control Release 65:133–148

Mcguinnes C, Duffin R, Brown S, Mills NL, Megson IL, Macnee W, Johnston S, Lu SL, Tran L, Li R (2010) Surface derivatization state of polystyrene latex nanoparticles determines both their potency and their mechanism of causing human platelet aggregation in vitro. Toxicol Sci 119:359–368

Müller RH, Rühl D, Runge S, Schulze-Forster K, Mehnert W (1997) Cytotoxicity of solid lipid nanoparticles as a function of the lipid matrix and the surfactant. Pharm Res 14:458–462

Nadziejko, C., Fang, K., Chen, L., Cohen, B., Karpatkin, M. & Nadas, A. 2002. Effect of concentrated ambient particulate matter on blood coagulation parameters in rats. Research report (Health Effects Institute), 7–29; discussion 31–8

Nemmar A, Hoylaerts MF, Hoet PH, Vermylen J, Nemery B (2003) Size effect of intratracheally instilled particles on pulmonary inflammation and vascular thrombosis. Toxicol Appl Pharmacol 186:38–45

Saiyed M, Patel R, Patel S (2011) Toxicology perspective of nanopharmaceuticals: a critical review. Int J Pharm Sci Nanotechnol 4:1287–1295

Schöler N, Zimmermann E, Katzfey U, Hahn H, Müller R, Liesenfeld O (2000) Effect of solid lipid nanoparticles (SLN) on cytokine production and the viability of murine peritoneal macrophages. J Microencapsul 17:639–650

Stasko NA, Johnson CB, Schoenfisch MH, Johnson TA, Holmuhamedov EL (2007) Cytotoxicity of polypropylenimine dendrimer conjugates on cultured endothelial cells. Biomacromolecules 8:3853–3859

Chapter 11
Nanomaterials in Drug Delivery System

Nur Izzati Mohd Razali, Noor Syazwani Mohd Saufi, Raha Ahmad Raus, Wan Mohd Fazli Wan Nawawi, and Dayang Fredalina Basri

11.1 Introduction

Drugs have been used for a long time to improve health and extend lives. Traditionally, they are administered into body through various paths of administration such as oral, injection and topical. However, these conventional methods of delivering drug into body have some drawbacks. Some medicines may have difficulty to travel through blood circulatory as well as tissues and cells. Some might have degraded before it reaches the targeted site (Pires et al. 2009). Hence, drug delivery system, a combination of conventional methods and engineered technologies of administering pharmaceutical agents into humans and animals to maximize therapeutic effect is sought. Nanoparticles made by nanotechnologies might be the answer. The particles have been thoroughly investigated and found to be effective in assisting delivering drugs in various diseases. Their usage in cancer treatment is becoming a growing industry.

Nanoparticles are small size particles, from 10 to 1000 nm (Kreuter 1994) which expresses distinctive physiochemical and biological characteristics to benefit them as a drug carrier. Optimizations of their physiochemical and biological characteristics ease the cells to uptake the nanoparticles compared to bigger molecules (Svenson and Tomalia 2005). Nanoparticles are made to be biodegradable, able to stabilize readily undergoing change molecules such as proteins from deterioration and have capability to control drugs release (Singh and Lillard 2009). With the utilization of

N. I. M. Razali · N. S. M. Saufi · R. A. Raus (✉) · W. M. F. W. Nawawi
Department of Biotechnology Engineering, Faculty of Engineering, International Islamic University of Malaysia, Jalan Gombak, Selangor, Malaysia
e-mail: rahaar@iium.edu.my

D. F. Basri
Biomedical Science Programme, School of Diagnostic & Applied Health Sciences, Faculty of Health Sciences, Universiti Kebangsaan Malaysia, Kuala Lumpur, Malaysia

S. Siddiquee et al. (eds.), *Nanotechnology: Applications in Energy, Drug and Food*,
https://doi.org/10.1007/978-3-319-99602-8_11

nanoparticles, the release rate of drug to the target location is slowed and may release over duration of days to weeks (Panyam and Labhasetwar 2003). As the nanoparticles are biocompatible, it has the ability to coordinate with biological system without inducing immune response or any unfavourable feedbacks (Wilczewska et al. 2012).

By using nanoparticles, drugs can be delivered to the target treatment site avoiding side effects to the normal cells or tissues. A targeted drug delivery system by nanoparticles benefits in many ways, such as enhance the therapeutic effect, minimize the side effects of the drugs and damage to the tissues, and ensure that the drugs are being successfully delivered to the exact location (Pires et al. 2009). Targeted drug delivery by these particles can be attained either passively or actively. For passive targeting, the nanoparticles containing the drugs passively reach the target treatment site while for active targeting, the nanoparticles that carry the drugs are conjugated with tissue or cell-specific ligand (Lamprecht et al. 2001) of the targeted location. In cancer treatment, passive targeting take advantage of the condition of the tumor vessels which are leaky compared to normal tissue vessels. The gap junctions between endothelial cells of tumor vessels are enlarged allowing movement of nanoparticles into the immediate tumor region (Matsumura and Maeda 1986). In addition, tumor tissue lack of lymphatic drainage causing the drug to accumulate in the tumor tissue (Sinha et al. 2006). For active targeting in cancer treatment, the nanoparticles are conjugated with affinity ligands such as antibodies (Sudimack and Lee 2000), peptides (Arap et al. 1998) and small molecules (Leamon and Reddy 2004) that bind specifically to the receptor of tumor cells. The receptor should be highly concentrated on the tumor cells but not the normal cells.

In this article, specific nanomaterials instead of types of nanoparticles used in drug delivery system are discussed. The nanomaterials are liposome, chitosan, graphene oxide, carbon nanotube, nanodiamond and mesoporous silica. These nanomaterials are either already been applied in medical field or still in the clinical test stage and further investigations are needed.

11.2 Liposome

Liposome (also known as lipid vesicle), first found in 1960s (Bangham et al. 1965) is one of the well-known and widely used nanoparticle for drug delivery mechanism. Liposome primarily composed of phospholipid; which means it contains hydrophilic (water-loving) head and hydrophobic (water-hating) tail (Fig. 11.1). Since liposome is actually a sphere-like structure, the phospholipids help the entrapment of an aqueous liquid inside the hydrophobic membrane. For drug delivery, the aqueous region of liposome may carry crystallized drug and the hydrophobic region may bear hydrophobic-chemicals. Therefore, liposome can actually carry both hydrophobic and hydrophilic molecules (Cullis and Kruijff 1979).

In spite of this unique capability, liposome efficacy in carrying the drugs is limited by its chemical instability and physical instability. The phospholipid might

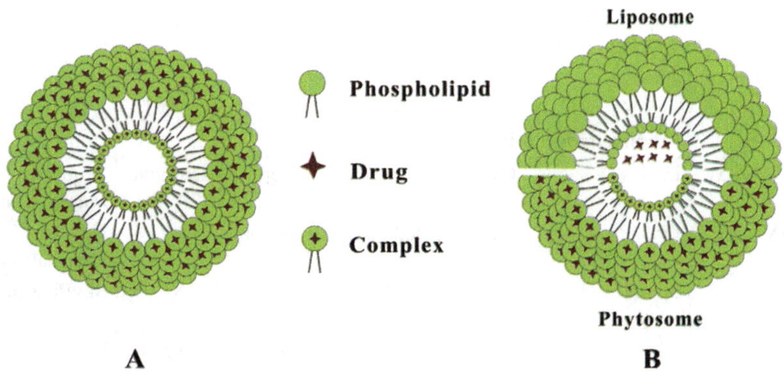

Fig 11.1 Schematic representation of liposomes (Source: Li et al. 2014)

be hydrolyzed thus cause the leakage of the drug before it reaches the targeted area (Li et al. 2014). Therefore, scientists are working to functionalize the liposomes by using polyethylene glycol (PEG). The PEGylated liposomes are capable to increase the circulating time thus it will reside longer in the human body in order to reach the targeted area (Klibanov et al. 1990). Through PEGylation, the size and molecular weight of the conjugated liposomes are increased therefore improving their pharmacokinetics and pharmacodynamics (Milla et al. 2012). The improvements made on the pharmacokinetics and pharmacodynamics include increasing water solubility, protection from enzymatic degradation, reducing real clearance and limiting the immunogenic and antigenic reactions (Milla et al. 2012). These features make liposomes as one of the widely used drug carrier in biomedical field. Some of the clinically approved liposomal drugs are listed in the Table 11.1.

Another alternative to increase the efficacy of liposome is by using the elastic liposomes (ELs). ELs are liposomes that has been improved in terms of their physicochemical and pharmacokinetics by mixing a suitable surface-active components (Cevc et al. 2008). This improved ELs are able to penetrate the skin without disintegration (El Maghraby et al. 2008) and are still in investigation for mechanisms of drug delivery due to significant difference outcomes in both *in vivo* and *in vitro* study (Hussain et al. 2017).

Liposomal drug carriers can also be synthesized to be targeted to a specific cell, tissues and receptor. The hydrophilic head of the lipids can be functionalized by adding active ligands (Vabbilisetty and Sun 2014). As reported by Vabbilisetty and Xue, two anchoring lipids used are Chol-PEG2000-TP and DSPE-PEG2000-TP. These active ligands will then ligate with the ligand at the targeted area and diffuse the carried drug. Therefore, functionalized the liposomes is very crucial and important in order for it to become novel drug carrier for the specific diseases.

Table 11.1 List of approved and administrated liposomes

Clinical products (approval year)	Administration	Indication	Company
Doxil® (1995)	i.v	Ovarian, breast cancer, Kaposi's sarcoma	Sequus Pharmaceuticals
DaunoXome® (1996)	i.v.	AIDS-related Kaposi's sarcoma	NeXstar Pharmaceuticals
Depocyt® (1999)	Spinal	Neoplastic meningitis	SkyPharma Inc.
Myocet® (2000)	i.v.	Combination therapy with cyclophosphamide in metastatic breast cancer	Elan Pharmaceuticals
Mepact® (2004)	i.v.	High-grade, resectable, non-metastatic osteosarcoma	Takeda Pharmaceutical Limited
Marqibo® (2012)	i.v	Acute lymphoblastic leukaemia	Talon Therapeutics, Inc.
Onivyde™ (2015)	i.v	Combination therapy with fluorouracil and leucovorin in metastatic adenocarcinoma of the pancreas	Merrimack Pharmaceuticals Inc.
Abelcet® (1995)	i.v	Invasive severe fungal infections	Sigma-Tau Pharmaceuticals
Ambisome® (1997)	i.v	Presumed fungal infections	Astellas Pharma
Amphotec® (1996)	i.v	Severe fungal infections	Ben Venue Laboratories Inc.
Visudyne® (2000)	i.v	Choroidal neovascularisation	Novartis

11.3 Chitosan

Chitosan is a co-polymer of glucosamine and N-acetylated glucosamine obtained after the deacetylation of chitin. The deacetylation process consists of boiling the chitin, which commonly extracted from crustacean's exoskeletons, in a concentrated alkaline solution. Chitosan differ than chitin by having higher percentage of deacetylated moieties across its polymer backbone and it is soluble in organic acids. Chitosan structure is similar to cellulose, except having amino group instead of hydroxyl group at C-2 of its monomer unit (Fig. 11.2). Over the last three decades, the study of its functionality and impressive potential applications in several fields has made chitosan as one of the promising renewable polymeric materials. Due to its biocompatibility, biodegradability, easy to formulate and various uses in applications associated with low toxicity, chitosan has become the topic of interest among the other polymeric materials for nanoparticulate drugs delivery (Nagpal et al. 2010).

Due to its smaller size, chitosan nanoparticles are capable to penetrate biological barriers and directly deliver the drugs to the lesion site efficiently. The amino and carboxyl groups in the chitosan molecules can form hydrogen bond with

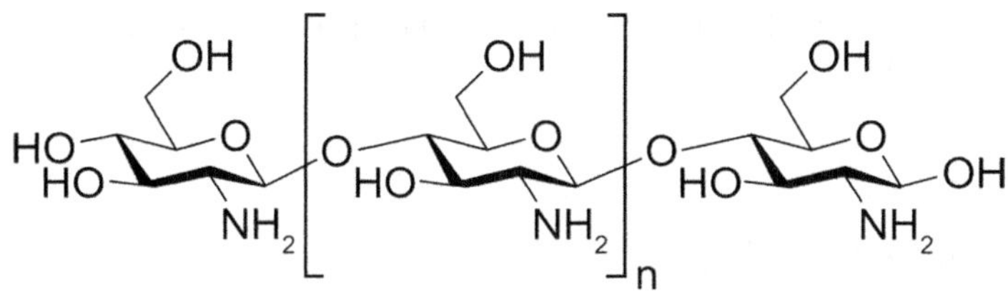

Fig. 11.2 Structure of chitosan (Wang et al. 2011)

glycoprotein in mucus, hence promoting adhesive effect. As mucoprotein in mucus is positively charged, chitosan and mucus are attracted to each other to extend the drug retention time which favour continuous drugs discharge and accessibility. In addition, protonation of chitosan's free amino group under acidic conditions renders chitosan to be polycationic, which in turn leads to gelation. This protonation also able to increase the paracellular permeability of peptide drugs across mucosal epithelia (Mansuri et al. 2016). *In vitro* testing shows that, controlled release of catechin is enhanced by chitosan's adhesivity and improved its bioavailability (Dudhani and Kosaraju 2010).

Apart from being an excellent mucoadhesive polymer for drug delivery, chitosan itself can stimulate immune system cytotoxic activity. Maeda and Kimura (2004) reported low molecular weight chitosan inhibit tumor growth in Sarcoma 180-bearing mice while Torzsas et al. (1996) discovered that chitosan contained in a diet could reduce the precancerous lesions development in colon cancer induced by azomethane compounds (Maeda and Kimura 2004; Torzsas et al. 1996). In drug delivery system, chitosan corrodes and degrades to release the drugs carried, leading to a clear sustained-released effect. Scientists are trying to improve in targeting and chitosan's bioavailability by doing some modifications characterized by pH sensitivity, targeting accuracy and thermosensitivity (Wang et al. 2011).

Studies on chitosan nanoparticles suggest it can transport and deliver various types of drugs (e.g. gene drugs, protein drugs and anticancer chemical drugs) which are unstable in the biological fluids and cannot readily diffused across the body system. Chitosan have been developed *in vitro* and *in vivo* for DNA and small interfering RNA (siRNA) delivery systems because of its cationic charge, mucoadhesive, and permeability enhancing properties (Mao et al. 2010). In chitosan nanoparticle/alginate drug encapsulation system, protein drug (i.e. insulin) was protected from being deteriorated by gastric fluid after complexing the drug with cationic β-cyclodextrin polymers (Zhang et al. 2010). Doxorubin (DOX), an anticancer drug has been grafted with carboxymethyl chitosan nanoparticles to suppress tumor-cell proliferation effectively as well as minimize the side effects of the drugs (Jeong et al. 2010).

The drug delivery systems mentioned above are only a few of chitosan's possible applications that being studied over the years. Other studied applications include controlled drug delivery, ocular drug delivery, brain targeting and many more

(Nagpal et al. 2010). Chitosan's physiochemical properties are the principle factor of its flexibility in those applications.

11.4 Graphene-Oxide

Graphene is a monolayer of sp^2-hybridized carbon atoms stacked in 2D honeycomb crystal lattice and its first appearance in 2004 has induced great enthusiasm among researchers (Fig. 11.3). Interest in graphene stems from the fact that it possess exceptional properties such as high Young's modulus, high fracture strength, high thermal conductivity, low electrical resistance for charge mobility, and large specific surface area (Geim 2009). As pure graphene is highly hydrophobic and poorly dispersible in water, hydrophilic versions of graphene such as graphene oxide (GO) is prepared. Although GO is soluble in water, it will aggregate should there any salt presence in solution as a result of charge screening effect (Liu et al. 2013). Therefore, surface modification of GO either by covalent or non-covalent binding of precursors is necessary step before it can be used as a drug carrier.

Defect on graphene lattice made it possible for GO surface to be covalently modified via nucleophilic substitution, electrophilic addition, condensation and addition (Liu et al. 2013). For example, amino and sulfonic group from side chains of water-soluble biocompatible polymers such as PEG (Liu et al. 2008), polyethyleneimine (PEI) (Zhang et al. 2011) chitosan and polyvinyl alcohol (Sahoo et al. 2011) has been successfully grafted on surfaces of GO sheets. Polymer-grafted GO nanoparticles increase the stability and biocompatibility as well as indirectly improve the water-insoluble anticancer drugs disintegration (Pan et al. 2012). However, during the intracellular drugs discharge from nanocarrier, diffusion barrier effect may happen caused by polymer coatings on the surface of GO sheets antagonistically influence the successive therapy. Alternatively, Wen and

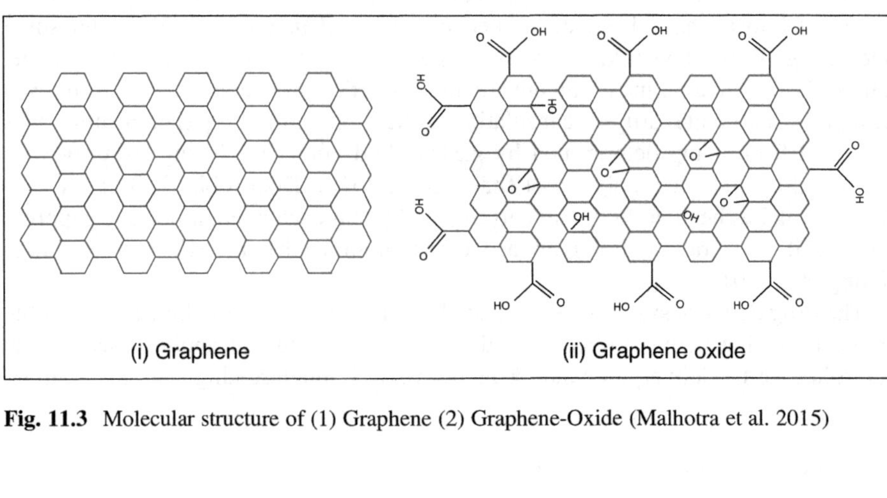

Fig. 11.3 Molecular structure of (1) Graphene (2) Graphene-Oxide (Malhotra et al. 2015)

colleagues developed a unique GO-based nanocarrier consisted of sheddable PEG shell attached to GO sheets surface (Wen et al. 2012).

Apart from covalent functionalization, GO surface can also be functionalized using non-covalent interaction with amphiphilic polymers for increasing stability, drug-loading capability and inducing anti-biofouling properties. Example of hydrophobic non-covalent functionalization between GO and Pluronic F127 polymer is reported by Hu et al. (2012), where the resulted composite was not only capable to encapsulate high amount of anticancer drug, doxorubicin (DOX), but also able to exhibits pH-controlled drug release behaviour (Hu et al. 2012). Yang et al. (2012), on the other hand exploiting cooperative noncovalent π–π interaction between DOX, GO, folic acid modified β-cyclodextrin, and adamante-grafted porphyrin for delivering DOX into folate-receptor-positive malignant cells. Although non-covalent functionalized GO may load smaller quantity of aromatic drugs as a result of conjugated areas on GO sheets, its polymer adsorption on GO surface is weaker than covalent binding, which result in less stable drug delivery system.

The application of GO and its derivative associated with anticancer drugs has attracted many scientists to further study about its flexibility in application. However, there still be a lot of challenges and long-term or short-term effect which need more explorations.

11.5 Carbon Nanotube (CNT)

The discovery of carbon nanotube (CNT) by Sumio Iijima in 1991 had sparked a revolution in carbon based nanomaterial in various fields. CNT can be grouped into single wall carbon nanotubes (SWCNTs) and multiple wall carbon nanotubes (MWCNTs) (Fig. 11.4) (Iijima 1991).

The way of graphene rolled into cylindrical form of CNT give rise to three possible structural models of SWCNTs: armchair, chiral or zigzag. Even though SWCNTs and MWCNTs have similar properties, the outer walls of the MWCNTs not only protect the inner CNT from chemical reaction with foreign substances but it also imparts high tensile strength properties to MWCNTs, in which SWCNTs do not exhibit (Eatemadi et al. 2014). Due to its unique one dimensional structure that promote cell membrane penetrability, CNTs have been explored as a transport vehicle for drug deliveries, be it for chemotherapy drugs, proteins or genes since 2004 (Liu et al. 2011). Another unique characteristic of SWCNT is that it can preferentially absorb and convert near infrared wavelength into thermal energy, and this properties has been exploited ever since for photothermal therapy against cancer cells (Moon et al. 2009).

However, CNT is not without drawback. Devoid of polar moieties, pristine CNTs tend to bundle up and typically insoluble in all types of solvents. In addition, the issues of toxicity surrounding the biomedical application of CNT are controversial. Poor solubility of CNTs can be resolved by appropriate surface functionalization and problem with CNT cytotoxicity that normally stem from heavy metal contamination

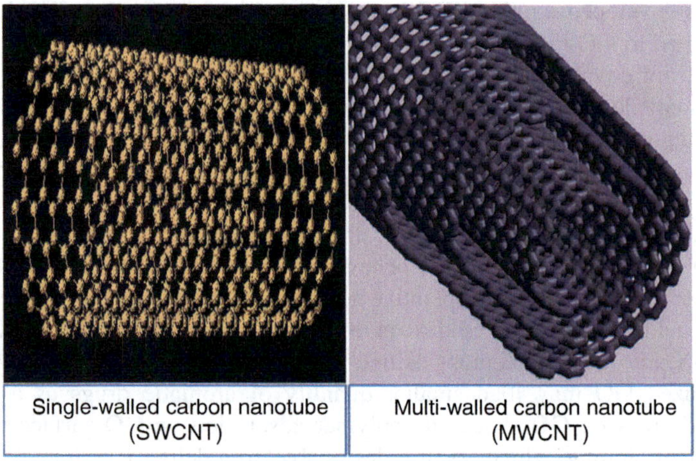

Fig. 11.4 Single-walled and multi-walled carbon nanotube (Dineshkumar et al. 2015)

during its synthesis can be rectified by purification (Wong et al. 2013). Water-soluble CNTs are biocompatible with bodily fluids and does not excrete any toxicity inside the body (Eatemadi et al. 2014).

According to Safari and Zarnegar (2014), functionalization steps can be directed into two, which are additional of reactions to the sidewalls and tips of CNTs and secondly, through oxidation that is followed by carboxyl-base coupling. The ability of CNTs to cross the cell membranes has increase the interest of researchers to make them as the drug carriers. The mechanism dominated by CNTs drug carriers is the desired drugs are attached by covalent or non-covalent bond (Safari and Zarnegar 2014). In the case of anti-cancer drugs, they are loaded into the CNTs and being accumulated at the cancer site due to the upgraded penetration capability (Dineshkumar et al. 2015).

Panchapakesan et al. (2005) reported the use of thermal-energy confinement on water filled-SWCNTs bundles as a precursor to nanobomb development. Conversion of near infrared radiation into thermal energy cause water molecules to evaporates and subsequently creates extreme pressure in SWCNTs before it explode in the vicinity of target cells (Panchapakesan et al. 2005). Drugs loaded into SWCNTs-based nanobomb have the prospective of exterminating human BT474 breast cancer cells without any toxicity (Dineshkumar et al. 2015). The usage of PEI-modified MWCNTs conjugated with fluorescein isothiocyanate (FI) and hyaluronic acid (HA) complexed with DOX shows a promising result as an efficient anticancer drug carrier. The MWCNT/PEI-FI-HA/DOX complexes can specifically target cancer cells and inhibit the growth of the cancer cells (Cao et al. 2015). CNTs functionalization increases their ability to enter the cell membranes via endocytosis pathway thus lead them to be one of the innovative *in vivo* drug carriers (Liu et al. 2011).

11.6 Nanodiamond

Nanodiamonds (NDs) have emerged as one of the versatile drug delivery carriers due to their exclusive combination of properties such as small in size, low cost in production, scalable production, negligible toxicity, chemical inertness of diamond core and rich chemistry of NDs surfaces. They are not only outstanding drug delivery carriers but also contributed in bone surgery as a biodegradable device, as tissues engineering scaffolds, helps in fighting virus and kill drug resistant microbes; and deliver genetic material into cell nucleus (Turcheniuk and Mochalin 2017). With a very small size approximately 5 nm or less in diameter (Fig. 11.5), NDs have the potential to penetrate even the smallest pore in the human body (Turcheniuk and Mochalin 2017). Based on the listed features above, no doubt NDs are definitely the most flexible nanoparticles for application in biomedical field.

Just like other nanoparticles involved in drug delivery system, NDs need to be functionalized in order to be the desired carrier for the desired drug-targeted delivery. Detonation is one of methods to functionalise the ND where explosion occur in which the speed of burning is higher than the speed of sound in the surroundings. However, since detonation NDs (DNDs) are achieved through explosion in steel blast chamber, the resulting DNDs might possess impurities (Turcheniuk and Mochalin 2017). Thus, the DNDs must go through post-synthesis purification before they can be used in biology and medicine application (Turcheniuk and Mochalin 2017). The post-synthesis purification methods include liquid phase oxidative treatment which will removes almost all graphitic and metals impurities (Pichot et al. 2008; Sushchev et al. 2008), gas phase oxidation which will remove metal impurities (Mitev et al. 2007; Petrov et al. 2007) and microwave-assisted liquid oxidation which the complete removal of metal can be achieved via the usage of EDTA complexes (Desai and Mitra 2013; Mitev et al. 2014).

To be the best nanoparticles in specific targeted therapy based on particular diagnostic tests or theranostics, NDs must have a well-defined surface chemistry (Turcheniuk and Mochalin 2017). To impose strategies for NDs chemical

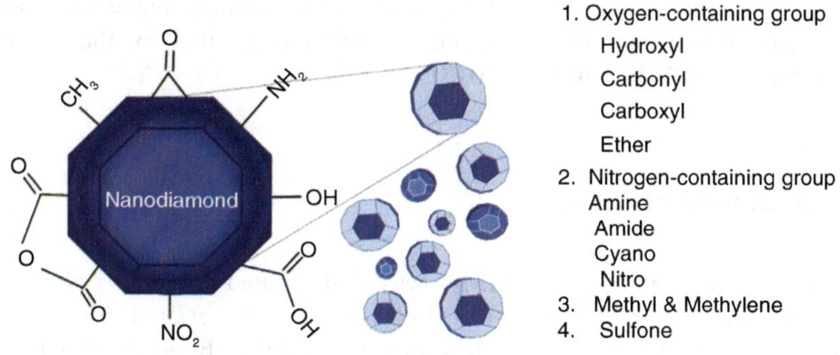

Fig. 11.5 The representation of nanodiamond with the functional groups presents of its surface (Source: Lim et al. 2016)

modification, the behavior of NDs in biological environment is to be determined before any modifications can be made (Turcheniuk and Mochalin 2017). Since DNDs are exposed to a vast array of functional groups (Fig. 11.5) like carboxyl, hydroxyl, epoxide, lactone, anhydride and alkenes (Krüger et al. 2006; Mochalin et al. 2012; Shenderova et al. 2006), it enables the usage of rich and well-known organic chemistry transformation to modify NDs surface chemistry according to desired environment (Turcheniuk and Mochalin 2017). The surface chemistry modification is aimed to increase NDs stability in biological media, to reduce toxicity and to create reactive sites for attachment of fluorescent dyes, gene vectors or other biologically active molecules (Turcheniuk and Mochalin 2017). Such surface chemistry modification can be achieved by graphitization for NDs with no surface functional group (Liang et al. 2011), by hydrogenated the NDs (Grall et al. 2015), air oxidation of NDs removes non-diamond carbon (Etzold et al. 2014; Osswald et al. 2008) and transfer surface functional groups carboxyls to NDs forming ND-COOH (Osswald et al. 2006). Polyethylene glycol modified NDs help ease the adsorptions of proteins, amino acids and salts on the NDs surface (Zareie et al. 2008).

Once the surface chemistry modifications have been achieved, the NDs can carry their duty to deliver specific drug to specific targeted area. Delivering doxorubicin (DOX) has been a typical treatment for many forms of cancer (Whitlow et al. 2017). However, the universal exposure of body to DOX results in lethal side effects such as alopecia, myelosuppression and cardiotoxicity. Chemoresistance of tumor cells unfortunately amplifies these side effects (Whitlow et al. 2017). In 2007, NDs and DOX (ND-DOX) have been investigated intensively and this could be the basis of a drug delivery platform (Whitlow et al. 2017). Based on Huang et al. (2007) findings in *in vivo* studies involving mouse, they reported that ND-DOX managed to bypass the drug efflux resistance mechanisms of mammary carcinoma and liver tumor thus increasing the half-life of DOX. ND-DOX also proved the retention enhancement of the drug over time when only traces of DOX still present after 7 days of administration of ND-DOX while free DOX was unable to be detected after 72 h of administration (Huang et al. 2007). Besides DOX, NDs have been studied together with epirubicin (Chen et al. 2009), paclitaxel (Liu et al. 2010), Purvalanol A and 4-hydroxytamoxifen (Chang et al. 2011). These chemotherapy agents are loaded by physical adsorptions and their released are often controlled by the pH of the surrounding media (Whitlow et al. 2017).

11.7 Mesoporous Silica

Mesoporous silica nanoparticle (MSN) was first created in early 1990s by two different research groups (Kresge et al. 1992; Yanagisawa et al. 1990). This leads to many studies in its potential applications especially in biomedical field. Later, other types of MSN that vary in pore size were developed. The distinguishable characteristics of MSNs are its uniformity in pore size and high pore volume that

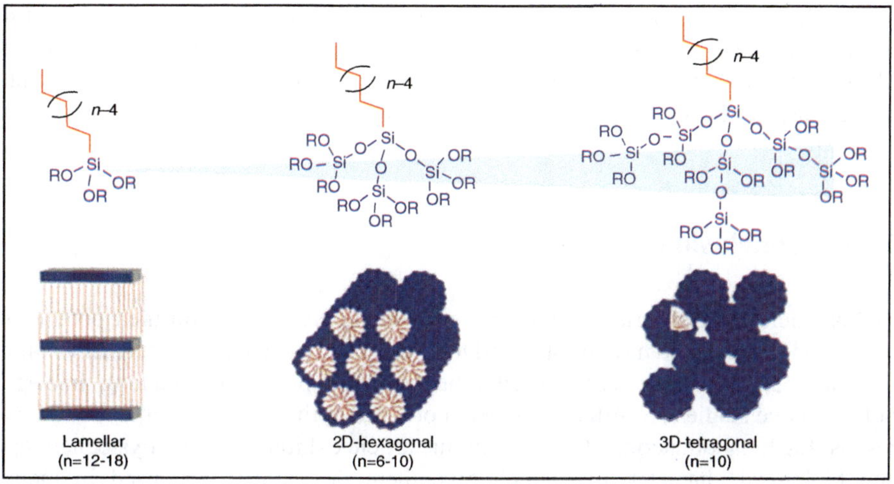

Fig. 11.6 Ideal structure of mesoporous silica (Source: Kuroda n.d.)

enhance the drugs carried capacity which can reduce the amount of carriers administered into body (Chen et al. 2014). In addition, MSN has high stability that help releasing of drugs in controlled manner unlike traditional liposome-based nanocarrier which has low stability in physiological conditions that may result the drugs to be released in explosive manner (Chen et al. 2014) (Fig. 11.6).

For the past decade, researchers have carried out numbers of *in vitro* evaluations of MSNs to test on their drug encapsulation, sustained drug released and therapeutic efficiency at different cell-level (Torney et al. 2007; Vallet-Regi et al. 2007). Throughout the years, MSNs have developed into three generations, MCM-41-type or SBA-15-type (Generation I) to nano-sized MSNs with steady spherical morphology, adjustable pore size and constituents (Generation II) and further integrated with multiple functionalities to cope with the requirements of cancer diagnosis and therapy (Generation III) (Chen et al. 2014).

The first reported application of MSNs as a drug carrier was in 2001 by Vallet-Regi who loaded an anti-inflammatory drug, ibuprofen into the mesopores of MCM-41 type and revealed that MCM-41 type was able to load high amount of drugs and extend the drug holding time (Vallet-Regi et al. 2001). In recent investigations, the scientists were further focused on MSNs' biocompatibility *in vivo* in terms of blood compatibility and circulation, biodegradability, biodistribution and more (Mamaeva et al. 2013). It is reported that MSNs' biocompatibility can be easily functionalized due to the plenty of silane surface chemistry available (Vallet-Regi et al. 2007). In recent finding, surface PEGylation of MSNs prolong the blood circulation time as well as increase the lung accumulation tumor of MSNs in passive-targeting drug delivery (Lu et al. 2010). This finding however is contradicted to what has been reported by He and colleagues saying that PEGylation reduces the lung accumulation of MSNs (He et al. 2011).

Despite of the fact that long term of *in vivo* biosafety evaluations are still inadequate and need further investigations, the present accessible information of MSNs reveal that they will be safe to be utilized as the anticancer drug carriers with appropriate drugs dosage.

11.8 Conclusion

In this article, nanomaterials that are currently being investigated for their potentials in biomedical field and have established to be drug carriers are elaborated. Liposome for example, has already been utilised in biomedical applications for a long time ago but extensive studies are still conducted in order to optimize its properties for better results. Each nanomaterials discussed in this article exhibits unique physiochemical and biological characteristics which give them flexibility to be used in many applications. Some exist as natural nanomaterials such as liposome, while some can be synthesized from certain compounds. Some nanomaterials can be functionalized to enhance their efficiency as drug carriers. However, not all synthetic nanomaterials are safe to be consumed by human. Therefore, further investigations and evaluations of each nanomaterial in term of long term effect *in vivo* must be carried out.

References

Arap W, Pasqualini R, Ruoslahti E (1998) Cancer treatment by targeted drug delivery to tumor vasculature in a mouse model. Science 279(5349):377–380

Bangham AD, Standish MM, Watkins JC (1965) Diffusion of univalent ions across the lamellae of swollen phospholipids. J Mol Biol 13(1):IN26–IN27. https://doi.org/10.1016/S0022-2836(65)80093-6

Cao X, Tao L, Wen S, Hou W, Shi X (2015) Hyaluronic acid-modified multiwalled carbon nanotubes for targeted delivery of doxorubicin into cancer cells. Carbohydr Res 405:70–77. https://doi.org/10.1016/j.carres.2014.06.030

Chang L-Y, Ōsawa E, Barnard AS (2011) Confirmation of the electrostatic self-assembly of nanodiamonds. Nanoscale 3(3):958. https://doi.org/10.1039/c0nr00883d

Cevc G, Mazgareanu S, Rother M, Vierl U (2008) Occlusion effect on transcutaneous NSAID delivery from conventional and carrier-based formulations. Int J Pharm 359(1–2):190–197. https://doi.org/10.1016/j.ijpharm.2008.04.005

Chen M, Pierstorff ED, Lam R, Li SY, Huang H, Osawa E, Ho D (2009) Nanodiamond-mediated delivery of water-insoluble therapeutics. ACS Nano 3(7):2016–2022. https://doi.org/10.1021/nn900480m

Chen Y, Chen H, Shi J (2014) Drug delivery/imaging multifunctionality of mesoporous silica-based composite nanostructures. Expert Opin Drug Deliv 11(6):917–930. https://doi.org/10.1517/17425247.2014.908181

Desai C, Mitra S (2013) Microwave induced carboxylation of nanodiamonds. Diamond Relat Mater 34:65–69. https://doi.org/10.1016/j.diamond.2013.02.005

Dineshkumar B, Krishnakumar K, Bhatt AR, Paul D, Cherian J, John A, Suresh S (2015) Single-walled and multi-walled carbon nanotubes based drug delivery system: cancer therapy: a review. Indian J Cancer 52(3):262–264. https://doi.org/10.4103/0019-509X.176720

Dudhani AR, Kosaraju SL (2010) Bioadhesive chitosan nanoparticles: preparation and characterization. Carbohydr Polym 81(2):243–251. https://doi.org/10.1016/j.carbpol.2010.02.026

Eatemadi A, Daraee H, Karimkhanloo H, Kouhi M, Zarghami N, Akbarzadeh A et al (2014) Carbon nanotubes: properties, synthesis, purification, and medical applications. Nanoscale Res Lett 9 (1):1–13. https://doi.org/10.1186/1556-276X-9-393

El Maghraby GM, Barry BW, Williams AC (2008) Liposomes and skin: from drug delivery to model membranes. Eur J Pharm Sci 34(4-5):203–222. https://doi.org/10.1016/j.ejps.2008.05.002

Etzold BJM, Neitzel I, Kett M, Strobl F, Mochalin VN, Gogotsi Y (2014) Layer-by-layer oxidation for decreasing the size of detonation nanodiamond. Chem Mater 26(11):3479–3484. https://doi.org/10.1021/cm500937r

Geim AK (2009) Graphene: status and prospects. Science 324(5934):1530–1534. https://doi.org/10.1126/science.1158877

Grall R, Girard H, Saad L, Petit T, Gesset C, Combis-Schlumberger M et al (2015) Impairing the radioresistance of cancer cells by hydrogenated nanodiamonds. Biomaterials 61:290–298. https://doi.org/10.1016/j.biomaterials.2015.05.034

He Q, Zhang Z, Gao F, Li Y, Shi J (2011) In vivo biodistribution and urinary excretion of mesoporous silica nanoparticles: effects of particle size and PEGylation. Small 7(2):271–280. https://doi.org/10.1002/smll.201001459

Hu H, Yu J, Li Y, Zhao J, Dong H (2012) Engineering of a novel pluronic F127/graphene nanohybrid for pH responsive drug delivery. J Biomed Mater Res A 100(1):141–148. https://doi.org/10.1002/jbm.a.33252

Huang H, Pierstorff E, Osawa E, Ho D (2007) Active nanodiamond hydrogels for chemotherapeutic delivery. Nano Lett 7(11):3305–3314

Hussain A, Singh S, Sharma D, Webster TJ, Shafaat K, Faruk A (2017) Elastic liposomes as novel carriers: recent advances in drug delivery. Int J Nanomedicine 12:5087–5108. https://doi.org/10.2147/IJN.S138267

Iijima S (1991) Helical microtubules of graphitic carbon. Nature 354(6348):56–58

Jeong YI, Jin SG, Kim IY, Pei J, Wen M, Jung TY et al (2010) Doxorubicin-incorporated nanoparticles composed of poly(ethylene glycol)-grafted carboxymethyl chitosan and antitumor activity against glioma cells in vitro. Colloids Surf B Biointerfaces 79(1):149–155. https://doi.org/10.1016/j.colsurfb.2010.03.037

Klibanov AL, Maruyama K, Torchilin VP, Huang L (1990) Amphipathic polyethyleneglycols effectively prolong the circulation time of liposomes. FEBS Lett 268(1):235–237. https://doi.org/10.1016/0014-5793(90)81016-H

Kresge C, Leonowicz ME, Roth WJ, Vartuli J, Beck JS (1992) Ordered mesoporous molecular-sieves synthesised by a liquid-crystal template mechanism. Nature 359(6397):710–712. https://doi.org/10.1038/359710a0

Krüger A, Liang Y, Jarre G, Stegk J (2006) Surface functionalisation of detonation diamond suitable for biological applications. J Mater Chem 16(24):2322–2328. https://doi.org/10.1039/B601325B

Kreuter J (1994) Nanoparticles. In: Swarbrick J, Boylan JC (eds) Encyclopaedia of pharmaceutical technology. New York, Marcel Dekker, pp 165–190

Kuroda K (n.d.) Mesoscale materials design utilizing layered materials. Retrieved from https://www.waseda-applchem.jp/en/faculty/kazuyuki-kuroda/

Lamprecht A, Ubrich N, Yamamoto H, Schäfer U, Takeuchi H, Maincent P, Kawashima Y, Lehr CM (2001) Biodegradable nanoparticles for targeted drug delivery in treatment of inflammatory bowel disease. J Pharmacol Exp Ther 299:775–781

Leamon CP, Reddy JA (2004) Folate-targeted chemotherapy. Adv Drug Deliv Rev 56:1127–1141

Li J, Wang X, Zhang T, Wang C, Huang Z, Luo X, Deng Y (2015) A review on phospholipids and their main applications in drug delivery systems. Asian J Pharm Sci 10(2):81–98. https://doi.org/10.1016/j.ajps.2014.09.004

Liang Y, Meinhardt T, Jarre G, Ozawa M, Vrdoljak P, Schöll A et al (2011) Deagglomeration and surface modification of thermally annealed nanoscale diamond. J Colloid Interface Sci 354 (1):23–30. https://doi.org/10.1016/j.jcis.2010.10.044

Lim DG, Prim RE, Kim KH, Kang E, Park K, Jeong SH (2016) Combinatorial nanodiamond in pharmaceutical and biomedical applications. Int J Pharm 514(1):41–51. https://doi.org/10.1016/j.ijpharm.2016.06.004

Liu Z, Jiao Y, Wang Y, Zhou C, Zhang Z (2008) Polysaccharides-based nanoparticles as drug delivery systems. Adv Drug Deliv Rev 60(15):1650–1662. https://doi.org/10.1016/j.addr.2008.09.001

Liu KK, Zheng WW, Wang CC, Chiu YC, Cheng CL, Lo YS et al (2010) Covalent linkage of nanodiamond-paclitaxel for drug delivery and cancer therapy. Nanotechnology 21(31):315106. https://doi.org/10.1088/0957-4484/21/31/315106

Liu Z, Robinson JT, Tabakman SM, Yang K, Dai H (2011) Carbon materials for drug delivery & cancer therapy. Mater Today 14(7–8):316–323. https://doi.org/10.1016/S1369-7021(11)70161-4

Liu J, Cui L, Losic D (2013) Graphene and graphene oxide as new nanocarriers for drug delivery applications. Acta Biomater 9(12):9243–9257. https://doi.org/10.1016/j.actbio.2013.08.016

Lu J, Liong M, Li Z, Zink JI, Tamanoi F (2010) Biocompatibility, biodistribution, and drug-delivery efficiency of mesoporous silica nanoparticles for cancer therapy in animals. Small 6 (16):1794–1805. https://doi.org/10.1002/smll.201000538

Maeda Y, Kimura Y (2004) Antitumor effects of various low-molecular-weight chitosans are due to increased natural killer activity of intestinal intraepithelial lymphocytes in sarcoma 180-bearing mice. J Nutr 134(4):945–950

Malhotra BD, Srivastava S, Augustine S (2015) Biosensors for food toxin detection: carbon nanotubes and graphene. MRS Online Proceedings Library Archive 1725

Mamaeva V, Sahlgren C, Lindén M (2013) Mesoporous silica nanoparticles in medicine—recent advances. Adv Drug Deliv Rev 65(5):689–702. https://doi.org/10.1016/j.addr.2012.07.018

Mansuri S, Kesharwani P, Jain K, Tekade RK, Jain NK (2016) Mucoadhesion: a promising approach in drug delivery system. React Funct Polym 100:151–172. https://doi.org/10.1016/j.reactfunctpolym.2016.01.011

Mao S, Sun W, Kissel T (2010) Chitosan-based formulations for delivery of DNA and siRNA. Adv Drug Deliv Rev 62(1):12–27. https://doi.org/10.1016/j.addr.2009.08.004

Matsumura Y, Maeda H (1986) A new concept for macromolecular therapeutics in cancer chemotherapy – mechanism of tumoritropic accumulation of proteins and the antitumor agent smancs. Cancer Res 46(12 Pt 1):6387–6392

Milla P, Dosio F, Cattel L (2012) PEGylation of proteins and liposomes: a powerful and flexible strategy to improve the drug delivery. Curr Drug Metab 13(1):105–119. https://doi.org/10.2174/138920012798356934

Mitev D, Dimitrova R, Spassova M, Minchev C, Stavrev S (2007) Surface peculiarities of detonation nanodiamonds in dependence of fabrication and purification methods. Diamond Relat Mater 16(4–7):776–780. https://doi.org/10.1016/j.diamond.2007.01.005

Mitev DP, Townsend AT, Paull B, Nesterenko PN (2014) Microwave-assisted purification of detonation nanodiamond. Diamond Relat Mater 48:37–46. https://doi.org/10.1016/j.diamond.2014.06.007

Mochalin VN, Shenderova O, Ho D, Gogotsi Y (2012) The properties and applications of nanodiamonds. Nat Nanotechnol 7(1):11–23. https://doi.org/10.1038/nnano.2011.209

Moon HK, Lee SH, Choi HC (2009) In vivo near-infrared mediated tumor destruction by photothermal effect of carbon nanotubes. ACS Nano 3(11):3707–3713. https://doi.org/10.1021/nn900904h

Nagpal K, Singh SK, Mishra DN (2010) Chitosan nanoparticles: a promising system in novel drug delivery. Chem Pharm Bull 58(11):1423–1430. https://doi.org/10.1248/cpb.58.1423

Osswald S, Yushin G, Mochalin V, Kucheyev SO, Gogotsi Y (2006) Control of sp2/sp3 carbon ratio and surface chemistry of nanodiamond powders by selective oxidation in air. J Am Chem Soc 128(35):11635–11642. https://doi.org/10.1021/ja063303n

Osswald S, Havel M, Mochalin V, Yushin G, Gogotsi Y (2008) Increase of nanodiamond crystal size by selective oxidation. Diamond Relat Mater 17(7–10):1122–1126. https://doi.org/10.1016/j.diamond.2008.01.102

Pan Y, Sahoo NG, Li L (2012) The application of graphene oxide in drug delivery. Expert Opin Drug Deliv 9(11):1365–1376. https://doi.org/10.1517/17425247.2012.729575

Panchapakesan B, Lu S, Sivakumar K, Teker K, Cesarone G, Wickstrom E (2005) Single-wall carbon nanotube nanobomb agents for killing breast cancer cells. Nanobiotechnology 1 (2):133–139. https://doi.org/10.1385/NBT:1:2:133

Panyam J, Labhasetwar V (2003) Biodegradable nanoparticles for drug and gene delivery to cells and tissue. Adv Drug Deliv Rev 55:329–347

Petrov I, Shenderova O, Grishko V, Grichko V, Tyler T, Cunningham G, McGuire G (2007) Detonation nanodiamonds simultaneously purified and modified by gas treatment. Diamond Relat Mater 16(12):2098–2103. https://doi.org/10.1016/j.diamond.2007.05.013

Pichot V, Comet M, Fousson E, Baras C, Senger A, Le Normand F, Spitzer D (2008) An efficient purification method for detonation nanodiamonds. Diamond Relat Mater 17(1):13–22. https://doi.org/10.1016/j.diamond.2007.09.011

Pires A, Fortuna A, Alves G, Falcão A (2009) Intranasal drug delivery: how, why and what for? J Pharm Pharm Sci 12(3):288–311. https://doi.org/10.18433/J3NC79

Cullis PR, De Kruijff B (1979) Lipid polymorphism and the functional roles of lipids in biological membranes. Biochim Biophys Acta 559(4):399e420

Safari J, Zarnegar Z (2014) Advanced drug delivery systems: nanotechnology of health design A review. J Saudi Chem Soc 18(2):85–99. https://doi.org/10.1016/j.jscs.2012.12.009

Sahoo NG, Bao H, Pan Y, Pal M, Kakran M, Cheng HKF et al (2011) Functionalized carbon nanomaterials as nanocarriers for loading and delivery of a poorly water-soluble anticancer drug: a comparative study. Chem Commun (Camb) 47(18):5235–5237. https://doi.org/10.1039/c1cc00075f

Shenderova O, Petrov I, Walsh J, Grichko V, Grishko V, Tyler T, Cunningham G (2006) Modification of detonation nanodiamonds by heat treatment in air. Diamond Relat Mater 15 (11–12):1799–1803. https://doi.org/10.1016/j.diamond.2006.08.032

Singh R, Lillard JW (2009) Nanoparticle-based targeted drug delivery. Exp Mol Pathol 86 (3):215–223. https://doi.org/10.1016/j.yexmp.2008.12.004

Sinha R, Kim GJ, Nie S, Shin DM (2006) Nanotechnology in cancer therapeutics: bioconjugated nanoparticles for drug delivery. Mol Cancer Ther 5(8):1909–1917

Sudimack J, Lee RJ (2000) Drug targeting via the folate receptor. Adv Drug Deliv Rev 41:147–162

Sushchev VG, Dolmatov VY, Marchukov VA, Veretennikova MV (2008) Fundamentals of chemical purification of detonation nanodiamond soot using nitric acid. J Superhard Mater 30 (5):297–304. https://doi.org/10.3103/S1063457608050031

Svenson S, Tomalia DA (2005) Dendrimers in biomedical applications—reflections on the field. Adv Drug Deliv Rev 57(15):2106–2129. https://doi.org/10.1016/j.addr.2005.09.018

Torney F, Trewyn BG, Lin VSY, Wang K (2007) Mesoporous silica nanoparticles deliver DNA and chemicals into plants. Nat Nanotechnol 2(5):295–300. https://doi.org/10.1038/nnano.2007.108

Torzsas TL, Kendall CWC, Sugano M, Iwamoto Y, Rao AV (1996) The influence of high and low molecular weight chitosan on colonic cell proliferation and aberrant crypt foci development in CF1 mice. Food Chem Toxicol 34(1):73–77. https://doi.org/10.1016/0278-6915(95)00083-6

Turcheniuk K, Mochalin VN (2017) Biomedical applications of nanodiamond (Review). Nanotechnology 28(25):252001. https://doi.org/10.1088/1361-6528/aa6ae4

Vabbilisetty P, Sun XL (2014) Liposome surface functionalization based on different anchoring lipids via Staudinger ligation. Org Biomol Chem 12(8):1237. https://doi.org/10.1039/c3ob41721b

Vallet-Regi M, Rámila A, Del Real RP, Pérez-Pariente J (2001) A new property of MCM-41: drug delivery system. Chem Mater 13(2):308–311. https://doi.org/10.1021/cm0011559

Vallet-Regi M, Balas F, Arcos D (2007) Mesoporous materials for drug delivery. Angew Chem Int Ed Engl 46(40):7548–7558. https://doi.org/10.1002/anie.200604488

Wang JJ, Zeng ZW, Xiao RZ, Xie T, Zhou GL, Zhan XR, Wang SL (2011) Recent advances of chitosan nanoparticles as drug carriers. Int J Nanomedicine 6:765–774

Wen H, Dong C, Dong H, Shen A, Xia W, Cai X, Song Y, Li X, Li Y, Shi D (2012) Engineered redox-responsive PEG detachment mechanism in PEGylated nano-graphene oxide for intracellular drug delivery. Small 8(5):760–769. https://doi.org/10.1002/smll.201101613

Whitlow J, Pacelli S, Paul A (2017) Multifunctional nanodiamonds in regenerative medicine: recent advances and future directions. J Control Release 261:62–86. https://doi.org/10.1016/j.jconrel.2017.05.033

Wilczewska A, Niemirowicz K, Markiewicz K, Car H (2012) [Review] Nanoparticles as drug delivery systems. Pharmacol Rep 64(5):1864–1882. https://doi.org/10.1016/S1734-1140(12)70901-5

Wong BS, Yoong SL, Jagusiak A, Panczyk T, Ho HK, Ang WH, Pastorin G (2013) Carbon nanotubes for delivery of small molecule drugs. Adv Drug Deliv Rev 65(15):1964–2015. https://doi.org/10.1016/j.addr.2013.08.005

Yang Y, Zhang YM, Chen Y, Zhao D, Chen JT, Liu Y (2012) Construction of a graphene oxide based noncovalent multiple nanosupramolecular assembly as a scaffold for drug delivery. Chem A Eur J 18(14):4208–4215. https://doi.org/10.1002/chem.201103445

Yanagisawa T, Shimizu T, Kuroda K, Kato C (1990) The preparation of alkyltrimethylammonium-kanemite complexes and their conversion to microporous materials. Bull Chem Soc Jpn 63(4):988–992. https://doi.org/10.1246/bcsj.63.988

Zareie HM, Boyer C, Bulmus V, Nateghi E, Davis TP (2008) Temperature-responsive self-assembled monolayers of oligo(ethylene glycol): control of biomolecular recognition. ACS Nano 2(4):757–765. https://doi.org/10.1021/nn800076h

Zhang L, Lu Z, Zhao Q, Huang J, Shen H, Zhang Z (2011) Enhanced chemotherapy efficacy by sequential delivery of siRNA and anticancer drugs using PEI-grafted graphene oxide. Small 7(4):460–464. https://doi.org/10.1002/smll.201001522

Zhang N, Li J, Jiang W, Ren C, Li J, Xin J, Li K (2010) Effective protection and controlled release of insulin by cationic β-cyclodextrin polymers from alginate/chitosan nanoparticles. Int J Pharm 393(1–2):213–219. https://doi.org/10.1016/j.ijpharm.2010.04.006

Chapter 12
Drug Discovery: A Biodiversity Perspective

Kholis A. Audah

12.1 Introduction

12.1.1 Current Health Issues

The complexity of health issues does not significantly reduce by the fast advancement of science and technology and the improvement of human civilization. As a matter of fact, there are more and more diseases, generative or degenerative, infectious or non infectious diseases affecting human's lives. Different factors have contributed to the current health issues which include but not limited to the increasing of human population and transmission, life styles, quality of foods, water and environment and the emerging and re-emerging microbial pathogens.

On the other hand, mankind keeps striving to find drugs not only for fighting the existing diseases but also current emerging ones. Furthermore, antibiotics or drugs that have been used to fight many diseases, constantly loose their effectiveness due to one reason or another. Human are always at a continuous endless battle between finding cure and diseases. Unfortunately, it is so obvious to deny that the speed of drug discovery is at a slower pace than the emerging of diseases. WHO reported that at least 30 new diseases emerged in 20 years period of time (World Health Report 1996). Therefore, it has become a necessity to search for alternatives to find different kinds of drug sources.

K. A. Audah (✉)
Department of Biomedical Engineering and Directorate of Academic Research and Community Services, Swiss German University, Tangerang, Indonesia
e-mail: kholis.audah@sgu.ac.id

S. Siddiquee et al. (eds.), *Nanotechnology: Applications in Energy, Drug and Food*,
https://doi.org/10.1007/978-3-319-99602-8_12

12.1.2 Conceptual Framework of Drug Discovery from Biodiversity

The concepts of antigen versus antibody, venom versus antidote somehow taught us that Mother Nature provided us the cures for every disease. It is just a matter of how to find the right drug for particular disease which is already available in the nature. By saying so, it is our duty to explore nature, in a responsible manner, to discover drugs for various diseases both existing and ones might emerge in the future. This work should be done collaboratively among individuals with different expertises even among countries that would contribute according their capacities. Some contribute in terms of facilities or expertise, while some others contribute in the form of natural resources. Some countries are simply blessed with very rich biodiversity either on the land (terrestrial) or in the ocean or even both. The most important thing to keep in mind is that all parties should work together for the sake of the betterment of mankind. So that none would think that they are neither more superior nor more important than others.

This chapter briefly described the power of Mother Nature as the abundant sources to find drugs for different kinds of illnesses include the challenges associated with the drug discovery process. Different findings reported by researchers had shown us that most if not all organisms in our planet from simplest form of lives to the most complex ones, from microbes to plants, either on the land or in the ocean possess various types of active compounds that potentially can be used as medicines. We, human being just need to search and discover these virtues provided by God the Almighty and to use them all for the sake of mankind's prosperity in the best possible manners.

Drug discovery through screening process utilizing natural products can become a solution of the slow and expensive drug discovery process using conventional way (Hassig et al. 2014; Li and Vederas 2009; Harvey 2008). In the United States, approximately 50% of drugs approved by Food and Drug Administration (FDA) originated from natural products or their derivatives (Newman and Cragg 2012). The key point of this drug discovery method is that the method requires the availability of a large number of extracts and or compounds for screening purposes. This can be possibly made by exploring the richness of biodiversity. Mother Nature has been known as long as human history as very rich sources for various types of human needs including as medicinal sources.

12.2 Drug Discovery from Natural Products

12.2.1 The History of Natural Products as Medicinal Sources

Since centuries ago, human had been relying on natural products, plants in particular, to fulfill their basic needs for food, cloth and housing. Plants can also be used as

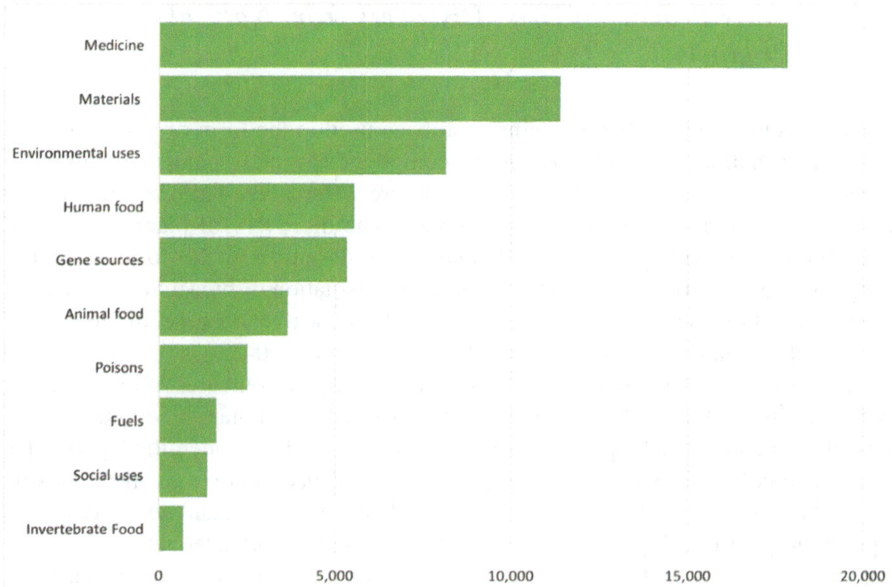

Fig. 12.1 Top documented uses of plant species. Graphic: Mongabay; Source: Royal Botanic Gardens Kew: State of the World's Plants 2017

poison on arrow for hunting, halusinogen in ritual, stimulant and medicines (Mann 2000). Plants had become very strong origin in traditional medicine practices in China, India and many other countries including Indonesia (Mittermeier et al. 2005). Long time before bioactive compounds were recognized pharmacologically, plants had been prescribed based on "similarity doctrine". For example, red color herbs were used to cure diseases related with blood and leaves with liver shape used to cure liver diseases (Sneader 2005). Traditional medicine practices generally use raw or processed materials according to traditions from generation to generation, such as well known boiled tea leaves or other herb formulas.

12.2.2 The Potential of Biodiversity as Medicinal Sources

There are about 391,000 species of vascular plants currently known to science, of which about 369,000 species (or 94%) are flowering plants, according to a report by the Royal Botanic Gardens, Kew, in the United Kingdom (Royal Botanical Garden Report 2017). Out of this number, only about 31,000 species have at least one documented use. These include uses for food, medicine, recreation, genes, poisons, animal feed, and building material (Fig. 12.1) (Royal Botanical Garden Report 2017). At least 28,187 plants species are currently recorded as being of medicinal use (State of the World's Plants 2017).

12.2.3 Early Process of Drug Discovery from Natural Products

In later development, herbal medicines had been used in the form of pure bioactive compounds (Salim et al. 2008). Bioactive compounds found in plants are known as secondary metabolites. Secondary metabolites are differentiated into several classes which are alkaloids, terpenoids and phenolics. The discovery of plants secondary metabolites used wider later as medicinal materials either in the form of native compounds or modified ones (Samuelsson 2004). Isolation of bioactive compounds eventually led to discovery of various drugs such as cocaine, codeine, digitoxin and quinine (Newman et al. 2000; Butler 2004; Samuelsson 2004).

Drug discovery processes from plants are involving different area of research and analytical methods. Such processes were initiated by botanist, ethnobotanist, ethnopharmacologist and plant ecologist who collected and identified particular plants. The collected plants were the plants that had been known as medicines or had not been known at all. Phytochemists then do the extraction and do screening by appropriate pharmacological tests and start isolation and characterization of bioactive compounds. Drug discovery processes were then developed to molecular level through determination and implementation of molecular target screening tests which were physiologically appropriate (Balunas and Kinghorn 2005). The combination of different scientific fields mentioned above had become different interdiscipline science which was named as pharmacognosy. During its development, pharmacognosy includes a larger natural product research, bioactive compounds from various sources including bacteria, fungi and marine organisms.

12.2.4 Significant Values of Medicinal Plants in Drug Discovery Process

In last decade, pharmaceutical industries started reducing their drug discovery program from natural products due to vast variety of chemical structure and stereocenter of natural products, so their synthetic form were not economical (Koehn and Carter 2005). Pharmaceutical industries are more interested in modelled molecules, combinatorial and other synthetic chemistry that can produce millions of compounds (Ganesan 2004; Tan 2004). However, utilization of plants or other natural products are still significant parts in drug discovery. Bioactive compounds from natural products contain so many proptotypes of new bioactive compounds which had been proven to be more relevant in relation with new drug discovery process (Koehn and Carter 2005). Drug materials from plants are not only used as new drug materials but also as drug candidates with elevated activity (Kramer and Cohen 2004).

Nowadays, most of pharmaceutical industries' roles related to natural products are taken over by small biotechnology industries which focus on identification of drug candidate compounds from natural product extracts and develop them to become drugs. Currently, there are many drug candidate compounds originated from plants are under clinical trial phases and introduced by the biotechnology industries, which some of them had become drugs that are commercially available. From the year 1981 to 2002, there were approximately 28% of new chemical compounds isolated from natural products or their derivatives. As many as 20% of them were chemical compounds analogues from natural products. Other researchers reported that in the United States, approximately 50% of drugs approved by Food and Drug Administration (FDA) originated from natural products or their derivatives (Newman and Cragg 2012). Natural products were also as initial point of synthesis of new synthetic compounds (Newman et al. 2003).

The nineteenth century was the beginning of the isolation of a number of other alkaloid compounds from medicinal plants, such as atropine (*Atropa belladonna*), caffeine (*Coffea arabica*), cocaine (*Erythroxylum coca*), efedrine (*Ephedra* sp.), morfine and codein (*Papaver somniferum*), pilocarpine (*Pilocarpus jaborandi Holmes*), physostigmine (*Physostigma venenosum*), quinine (*Cinchona cordifolia mutis* ex Humb), salicine (*Salix* sp.), theobromine (*Theobroma cacao*), theofilline (*Camelia sinensis*), and tubocurarine (*Chondrodendron tomentosum* Ruiz & Pav.). Up to now, isolation and characterization of new bioactive compounds from plants are still ongoing (Sneader 2005).

Artitir is a potential antimalarial compound and is as artemisinin derivative, sequiterpene lactone isolated from Artemisia annua L. (Asteraceae) (Graul 2001). Galantamine, drug for Alzheimer, discovered from lead ethnobotany and isolated for the first time from Galanthus woronowii Losinsk (Amaryllidaceae) in Russia in the year 1950 (Pirttila et al. 2004). Nitisinon was new drug from plant that works for rare generative disease, tyrosinemia, showing positive effect in lead structure. Nitisinon was a modification of mesotrione, a compound from Callistemon citrinus Stapf. (Myrtaceae) (Frantz and Smith 2003). Tiotropium was recently released in US market for curing chronic obstructive pulmonary disease (COPD). Tiotropium is an inhaled anticholinergic bronchodilator medicine, an atropine derivative isolated from Atropa Belladonna L. (Solanaceae) (Frantz 2005). Morphin-6-glucuronid is a morphine metabolite from Papaver somniferum L. (Papaveraceae) and was going to be used as morphine substitute due to much smaller side effect (Lotsch and Geisslinger 2001). Vinflunine was modified from Vinblastine from Catharantus roseus (L.) G. Don (Apocynaceae) as anticancer (Okouneva et al. 2003). Exatecan was camptothecin analogue from Camptotheca acuminata Decne. (Nyssaceae) and was being developed as anticancer (Cragg and Newman 2004). Calanolid was dipyranokumarine natural product isolated from Calophyllum lanigerum var. austrocoriaceum (Whitmore) p. F. Stevens (Clusiaceae), Malaysian tropical rain forest tree. Calanolid is an anti HIV drug with a unique work mechanism as reverse transcriptase non-nucleoside inhibitor of type 1 HIV and effective against HIV AZT-resistant strain (Yu et al. 2003).

In addition to direct use as medicinal materials, bioactive compounds from natural products can also be used as drug precursors, prototypes for modification of synthetic compounds or pharmacological markers. Out of 244 identified drug prototypes, 56 prototypes (23%) of them were secondary metabolites originated from plants (Sneader 2005). Bioactive compounds as drug precursors can be altered into preferred compounds through chemical modification process or fermentation methods (Salim et al. 2008).

By employing advance organic chemistry techniques, pharmacist can make analogue structure of natural products to obtain safer and more potential drugs. New produced compounds sometimes possess new pharmacological properties that can be categorized as derivative compounds. Podofillotoksin, camptotesine and guanidine were some examples of prototypes that their analogue had exactly the same pharmacological properties, while atropine was a prototype that its analogue had different pharmacological properties (Salim et al. 2008).

Bioactive compounds as pharmacological markers help researchers to understand work mechanism of intracellular signal transduction and biological mechanism related with certain diseases, so that better drug design can be achieved. For examples: Genisteine, an isoflavone naturally found in soybean (Glycine max Merr.) is an inhibitor of various protein tyrosine kinase (PTK), an important enzyme in intracellular signal transduction (Grynkiewicz et al. 2000). Different 12,13-diesther forbol also possess capacity to act as tumor promotor or activator protein kinase C (PKC) (Kazanietz 2005).

12.2.5 Challenges in Drug Discovery from Plants

Despite of many success stories of drug discovery from plants, following efforts remain facing many challenges. Pharmacognosist, phytochemist and natural product researchers still have to increase quality and quantity of compounds that enter drug development to be able to compete with chemical drug discovery business. Drug discovery process sometimes need about more than 10 years and spend about more than USD 800 million (Dickson and Gagnon 2004). Many of drug candidates considered failed after spending so much time and money. As a matter of fact, only one in 5000 drug candidate compounds which were entering clinical trials and accepted as drug. The steps mentioned above include identification, optimization and development of drug candidate compounds and then enter clinical trials (Balunas and Kinghorn 2005).

Those facts showed that drug discovery from plants requires longer time and more expensive compared to other drug discovery methods. Thus, many of pharmaceutical industries reduce their natural product research activities (Koehn and Carter 2005). To overcome such issues, it is important to develop a faster, better technology and or strategy, especially in increasing number of plant collection, screening methods, isolation of compounds and development of natural products and their derivatives (Do and Bernard 2004). Other challenge was limited availability of natural product extract, so it is not sufficient for optimization and development of drug candidate compounds. The challenge is also faced at clinical trials phases, so that natural products should be combined with synthetic compounds or chemicals.

Other technique is by developing extract library from natural products that combines natural products features with combinatorial chemistry (Butler 2004). Drug discovery through screening process utilizing natural products can become a solution for the slow and expensive drug discovery process using conventional way (Hassig et al. 2014; Li and Vederas 2009; Harvey 2008). The key point of this drug discovery method is that the method requires the availability of a large number of extracts and or compounds for screening purposes. This can be possibly made by exploring the richness of biodiversity. Mother Nature has been known as long as human history as very rich sources for various types of human needs including medicinal sources.

12.3 Extract Library

Extract library is a collection of extracts containing active compounds from natural products used for screening process of target biology. Although this looks simple, however development of extract library requires good understanding regarding modern paradigm of drug discovery process (Quinn 2012). High quality extract library leads the way in discovery of natural products that can be developed to become drug and early identification point for chemical compound optimization. Current drug discovery process is supported by HTS ability for larger chemical library (up to one million compounds), so that shorten the cycle of drug discovery project. Screening can be performed in 384 or 1536 formats of sample wells with volume of each sample approximately 2–20 μL in each well. Target-based screening and special technology platform is very suitable with current HTS. Unfortunately, this method applies only for pure synthetic compound library and not for crude extracts that contain hundreds of compounds (Quinn 2012).

The concept of extract library was firstly introduced when the High-Throughput Screening (HTS) concept was firstly recognized by pharmaceutical industries (Pereira and Williams 2007). Crude extract library has several benefits, such as cheap and easy preparations, minimum preparation time and possess high diversity level. However, crude extract library has some limitations such as natural physical form is not suitable for automated solution system (too concentrated), minor metabolite cannot be detected, requires longer time and continuous sample supply for isolation and identification of active compounds, the isolated compounds might have been known or non-targeted chemical compounds (Liu 2008).

Natural products pure compounds library was developed to cover the shortage of crude extract library and could become solution for drug discovery research from natural products suitable for high-throughput method (Bindseil et al. 2001). Similar to small synthetic molecules library, pure compounds extract library was designed with strategy adjusted to representative limitation of the desired chemical range (Brenk et al. 2008). Infrastructure and more sophisticated chemical informatics also need to be developed to determine potential extracts containing the desired compounds and to eliminate the undesired compounds. Potential minor components can be detected and included in screening process depend upon final detection or

Table 12.1 Top ten countries with their respective biodiversity in numbers

Country	Birds	Amphibians	Mammals	Reptiles	Fish	Vascular plants
Brazil	1753	1022	648	807	4521	56,215
Colombia	1826	771	442	601	2053	51,220
China	1240	411	551	478	3330	32,200
Indonesia	1615	347	670	728	4682	29,375
Mexico	1081	377	523	916	2602	26,071
South Africa	755	128	297	447	2059	23,420
Venezuela	1364	360	363	405	1709	21,073
United States	844	300	440	530	3067	19,473
India	1180	390	412	689	2465	18,664
Ecuador	1588	539	372	444	1098	19,362

Source: Plant data is from the World Conservation Monitoring Centre of the United Nations Environment Programme (UNEP-WCMC 2004). Species Data. Fish: Fishbase; Birds: Birdlife International; Amphibians: AmphibiaWeb; Mammals: IUCN; Reptiles: the Reptile Database (Butler, 2016)

isolation method used (Roy et al. 2010). Although some pure natural product compounds are already commercially available in affordable price, however majority of compounds need to be isolated individually or obtained from researchers through establishment of consortium. However, the cost, time and related resources with isolation and characterization of single compounds can be reduced.

Final objective of extract library is to obtain resource of different compounds for HTS evaluation (Dandapani et al. 2012). Compounds diversity is strongly correlated with biodiversity. Keep in mind that 17 countries with megabiodiversity in the world (Australia, Brazil, China, Colombia, Democratic Republic of Congo, Ecuador, India, Indonesia, Madagascar, Malaysia, Mexico, Papua New Guinea, Peru, Philippine, South Africa, United States of America and Venezuela) possess more than 70% of the entire biodiversity. Currently, there are approximately 391,000 species of vascular plants currently known to science, of which about 369,000 species (or 94%) are flowering plants, according to a report by the Royal Botanic Gardens, Kew in the United Kingdom (Royal Botanical Garden Report 2017). Out of this number, only about 31,000 species have at least one documented use. These include uses for food, medicine, recreation, genes, poisons, animal feed, and building material (Fig. 12.1) (Royal Botanical Garden Report 2017). In other words, with this large collection of plants available in the world, millions of extracts can be generated. The full list of countries with richest biodiversity in the world is shown in Table 12.1 (Royal Botanical Garden Report 2017).

Indonesia in particular has abundant natural products potential as drug and cosmetics materials both on land and in ocean. According to a report, Indonesia was listed as the big four of the richest country in the world in biodiversity both in the ocean and on the land as shown in Table 12.1 (UNEP 2004). One of the very popular collections of Indonesian herbal is Herbal collection managed by the Main Center for Research and Development of Plant Medicine and Traditional Medicine under the Ministry of Health of Indonesia. This sanctuary is located in Tawangmangu, Karanganyar, Central Java. It is the home of about 2000 Indonesian medicinal plants. Hundred of thousands of extracts can be generated from this single collection employing various solvents with different polarity.

The basic concept of this extract collection is that there is no single fraction discarded from every step of extraction processes. With the assumption that every extract has its own medicinal potential(s), either the one(s) generated from polar or non polar solvents. These collections can be utilized by different research institutions, universities or pharmaceutical industries for various research and development activities at different stages of research either basic, pre-clinical or clinical studies.

With the assumption, let say about 500 extracts or fractions can be collected from a single plant, there will be around one million extracts or fractions can be collected from 2000 medicinal plants available as in the sanctuary mentioned above (Audah 2015).

12.3.1 Sample Collection

Biota sample collection can be in the forms of bacterial culture, plant collection or marine biota. Some approaches in sample collection can be based upon their uses as traditional medicines or can be performed through random selection. Regardless of the basis of their collection, documentation, for examples place and time, and curation, for examples different parts of plants, are very important aspects for research continuation and downstream activities. Collection that is nationally programmed will ease the related research with conservation and understanding about the genetic resources. Species taxonomy identification is very important to increase the opportunity to discover new species that contain new compound(s) and to avoid the previously known one(s).

The latest biota sample collection strategy requires much less material compared to the requirement of the previous screening process. Development of screening technology, particularly the test with 384 and 1536 wells, approximately 200 mg extract is sufficient for screening several tests in HTS. With the development in spectroscopic structure elucidation, one milligram of compound is already sufficient, either for structure elucidation or biological profile for primary HTS test and some selectivity tests (Quinn 2012).

12.3.2 Extract Collection

Extracts can be collected from the whole biota or particular subsample according to the need for screening. Extraction from the whole biota ensures that all extracts are available for screening, isolation and chemical structure elucidation. However, degradation of compounds can occur to the extracts stored in a longer period of time. The need of time and solvent are larger compared to the extraction procedure of subsample of biota sample. Dried and ground plant sample (simplisia) can maintain the compound integrity. Moisture should be controlled to prevent sample damage. High moisture increases the chance of fungi and microorganisms to grow. Sample should be stored in each special container and barcoded.

Biota sample can be processed to become several form of extracts suitable for screening such as (1) crude extract which is extraction using organic or mixture of organic solvents; (2) prefractination crude extract library, which is crude extract that is fractionated using Solid Phase Extraction (SPE) or conventional liquid chromatography technique or combination of both; and (3) pure compound of natural products (Quinn 2012). Furthermore, extract collection requires representative storage room at every extract collection center. Ideally, this type of facility should be located at government institutions, universities or companies appointed as business partners that have the ability to do such big task. Extract collection in smaller numbers can be done in each research center or laboratories. One thing that needs to keep in mind is that all the procedures should be kept uniform (standardized). All these extract collections, either the small ones or the large ones should be well recorded in a database so that the database can be easily managed or organized, utilized and monitored. The database should be accessible to different parties with the permission of the authorized body.

12.3.3 Fractionation and Dereplication Handling

When an extract shows biological activity, fractionation is required to separate groups of compounds which have physical and chemical properties similarity such as solubility and acidity (Hughes et al. 2011). Every fraction is tested and active sample should be repeatedly fractionated to increase purity of the compounds. Test and fractionation process should be continuously performed until pure compounds responsible for particular biological activity obtained. However, repeated fractionation process should be prevented by dereplication process (Katiyar et al. 2012). Dereplication refers to the process of screening active compounds early in the development process to recognize and eliminate those compounds that have been studied in the past, thereby proactively decreasing the number of structures that will need to be fully elucidated and minimizing the amount of time spent on testing (Gaudêncio and Pereira 2015).

The isolated active compound should be purified and its molecular structure should be elucidated. If the active compound was commercially potential, the synthesis of its analog compound and structure-activity relationship (SAR) and quantitative structure-activity relationship (QSAR) studies were required (Guo 2017). However, if the natural source is commercially available to produce adequate active compound, so the synthesis that requires high cost production process and with the complexity of the stereochemistry of active compound can be avoided. For example, anticancer compound Vincristine, Etoposid and Taxol had already been synthesized, these compounds have chiral atom which is more economic when it is produced by extraction process from natural products compared to chemical synthesis (Demain and Vaishnav 2011). Plant conservation should also be taken into consideration, for example, isolation of Taxol anticancer drug from Yew Pacific plant's bark can cause extinction of the plant. So, nowadays Taxol can be obtained through semisynthetic process from Yew plant's leaf from Europe and America (Juyal et al. 2014).

12.3.4 Recollection

Recollection process should consider the conservation concept of species and habitat of collected subjects, because target compound research progressively will require large amount of material, it will give impact to environment. Therefore, it is important that initial collection is well documented. Previous sampling process should be equipped with GPS data, complete documentation and good taxonomy knowledge (WHO 2005).

In conducting biodiscovery research, it should be considered that if recollection of sample in large amount can not be done, target compound should be able to be chemically synthesized. Continuous supply of drug material from plants in adequate amount is very important in fulfilling market demand. Utilization of plant cell culture can be an alternative method for compounds in which their synthetic process is not economical and only available from plant in small amount. Plants accumulate secondary metabolites at certain stages of development. By manipulating environment condition and growth media, desired secondary metabolites can be obtained instead of directly from the entire plant. For example, paclitaxel was successfully produced from fermentation technology of plant cell (Ochoa-Villarreal et al. 2016).

Database formation when the time of sample collection is very important to be adjusted with biodiversity conservation concept and also to trace sample through HTS. Database collection includes taxonomy, time and location, collector either individual or institution and species availability. This will be very helpful in tracing and sample monitoring during research process for accessibility purpose and benefit sharing and recollection. This is also important to identify factors contributing to bioactivity, such as season, location and reproduction cycle stage (Atanasov et al. 2015).

12.3.5 Screening

In the past, drug discovery from plant bioactive compounds was a time consuming process. Even just for identifying structure of active compound, it took few weeks even years depend upon the complexity of the compound structure. Nowadays, the speeds of biological assay-based fractionations increase significantly due to the advancement of High Performance Liquid Chromatography (HPLC) tandemed with Mass Spectrometry (MS) or Liquid Chromatography-Mass Spectrometry

(LC-MS), Nuclear Magnetic Resonance (NMR) and High-Throughput Screening with robotic automation. Capillary NMR (cap-NMR) spectroscopy was a breakthrough in characterizing of rare compounds found in organisms (Schroeder dan Gronquist 2006).

Development of HTS technology accelerates extracts screening from plants. Nowadays, biological activity assay is not the limiting factor anymore in drug discovery process. As many as 100,000 samples can be tested in less than a week due to the advancement in data analysis system and the robotic automation usage (Butler 2005). However, plant extract library screening is still facing problem because of the compounds autofluorescence characteristic or the ability of UV light absorption that interferes reading results. To overcome this issue, pre-fractionation of extracts can be applied to reduce some of these problems. In general, HTS can also be equipped with computational screening method to identify and to avoid compounds that can possibly give false positive results (Walters and Namchuk 2003).

Researchers throughout the world are racing in finding new drugs for different illnesses, either the existing or the emerging ones. These include drugs for antibacterial and anticancer. Inappropriate use of antibiotics in some countries had caused antibiotic resistant to some strains, causing serious health problems. Nowadays, many antibiotics are not effective anymore against some infectious diseases caused by bacteria or any other microbes. Therefore, the effort in finding active compounds from natural products either in land or ocean becomes very important. The search also applied to find effective drugs for different types of cancer.

Various methods have been applied to evaluate or screen antibacterial activity in vitro from extracts or pure compounds. The most popular basic method is disk diffusion and broth dilution methods. Disk diffusion method has some advantages compared to other methods such as simplicity, relatively cheaper, and high reliability of the test results on some bacterial species and easiness of test results interpretation (measurement of zone of inhibition) (Balouiri et al. 2016).

In regards with cancer, research on anticancer from plants has become new attraction for researchers in the world. Many of natural products showed pharmacological potential that can be good starting point for discovery of anticancer drug. Vinblastin and Vincristin from plant *Catharantus roseus* has shown their effectiveness to cure cancer in human (Farnsworth and Soejarto 2009). Screening process for natural products with anticancer potential can be conducted using Brine Shrimp Lethality Test (BSLT). In the last 30 years, BSLT are commonly used for toxicity test of various natural products from plants (Mayorga et al. 2010).

BSLT is considered cheap and using less test materials. Since introduced for the first time, this in vivo test was proven to be representative as bioassay guidance for cytotoxic active fractionation and anticancer agent (Ahmed et al. 2010). In addition, other researches suggested that LC_{50} value obtained from BSLT correlately positive with the results of toxicity oral test in mice (Arslanyolu and Erdemgil 2006).

12.4 Indonesian Biodiversity in a Glimpse

Indonesia is a country that possesses more than 14,000 islands, located between Indian and Pacific Oceans. According to Fauna and Flora International (FFI), Indonesia is home of approximately 11% or more than 30,000 of the world's flowering plants and other biota both in land and marine with significant figures. This makes Indonesia become one of the richest country in the world in biodiversity.

One of potential plants as medicinal and cosmetics sources and widely spread along Indonesian coastline is Mangrove (Fig. 12.2) (UNEP-WCMC). Mangrove dan mangrove associates are very potential plants as medicinal sources (Bandaranayake 2002). Along roughly 90,000 km coastline, Indonesia is home of about 20 family with about hundreds species of mangroves and their associates. Indonesia has the largest mangrove forest or about 23% of total world mangrove forests (Giri et al. 2011). Figure 12.3 showed a mangrove conservation area in eastern coastline of Lampung, Indonesia which was previously an empty land. Today, this conservation area had attracted different species of birds and fishes that migrates from other places.

Since long time ago mangrove trees had been utilized either roots, branches, leaves, flowers and the fruits as food and medicinal sources. Qualitative phytochemistry analysis showed that leaf extract of Rhizophora stylosa and Avicenna marina contain flavonoid, terpenoid, alkaloid, flavonoid and glycosidic phenolic (Mouafi et al. 2014). Extracts of mangrove fruit also contain some bioactive compounds such as flavonoid, saponin and triterpenoid (Rohaeti et al. 2010).

Mangrove extracts had shown their activity against microbes or pathogen parasites in animals and plants (Batubara et al. 2010; Batubara and Mitsunaga 2013, Audah et al. 2018), including HIV (Rege and Chowdhary 2013) and Hepatitis-B virus (Yi et al. 2015). *Excoecaria agallocha* L can be used to reduce epilepsy,

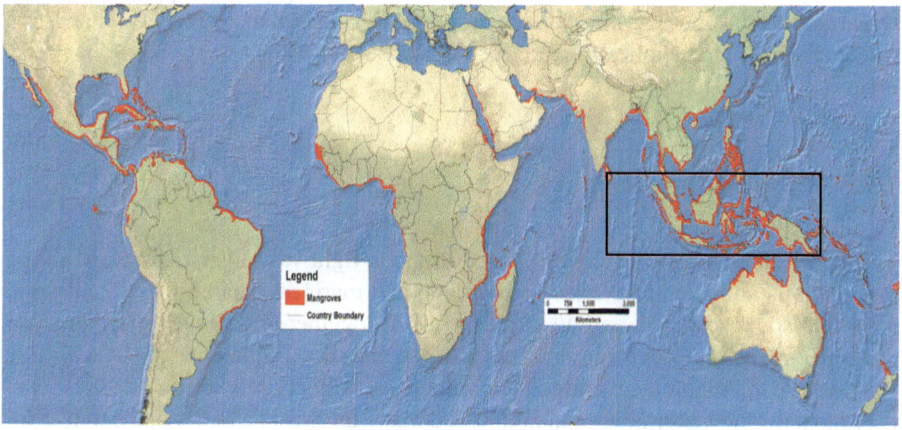

Fig. 12.2 Hot spots of world's mangrove. Inset is Indonesian mangrove hot spots (UNEP-WCMC)

Fig. 12.3 A mangrove conservation area in Eastern Coastline of Lampung, Indonesia (personal gallery)

conjunctivitis, dermatitis, leprosis, dermatitis, hematuria and toothache (Bandaranayake 2002). Mangrove extracts can also be used as antidiabetes (Gurudeeban et al. 2012), antinociceptive (Islam et al. 2012), antipyretic and antiinflammatory (Safari et al. 2016), anticancer (Singh and Kathiresan 2015), antiulcer (de-Faria et al. 2012).

Hibiscus tiliaceus can act as diuretic and laxative (Tambe and Bhambar 2016). According to Tanvira and Seenivasan (2014), some types of mangrove also can act as antivenom of snakebites. Mangrove extracts can also be used as mosquito's larvicide (Liem et al. 2013).

Up to now, mangrove research in Indonesia and worldwide is still very limited, especially for the purpose of drug discovery. This opens up opportunities for researches to start putting their efforts individually and collaboratively on mangrove research which also applies to mangrove's associates.

12.5 Concluding Remarks

Facing the challenges in discovering new drugs to fight vast diverse diseases, nature has given us abundant resources both in land and marine to search for. It's all up to us, scientists, researchers and all parties to work together to discover alternative for drugs to cure not only existing diseases but also ones that might emerge in the future. It's the time to complement and not to compete each other. People should share what they have for the betterment of mankind and make the world a better place to live for our generations in years to come.

References

Ahmed Y, Sohrab H, Al-Reza SM, Shahidulla Tareq F, Hasan CM, Sattar MA (2010) Antimicrobial and cytotoxic constituents from leaves of Sapium baccatum. Food Chem Toxicol 48:549–552

Arslanyolu M, Erdemgil FZ (2006) Evaluation of the antibacterial activity and toxicity of isolated arctiin from the seeds of Centaurea sclerolepis. J Fac Pharm 35:103–109

Atanasov AG et al (2015) Discovery and resupply of pharmacologically active plant-derived natural products: A review. Biotechnol Adv 33(2015):1582–1614

Audah KA (2015) Proceedings of the International conference on innovation, entrepreneurship and technology, 25–26 November, BSD City, Indonesia, ISSN: 2477-1538

Audah KA, Amsyir J, Almasyhur F, Hapsari AM, Sutanto H (2018) Development of extract library from Indonesian biodiversity: exploration of antibacterial activity of mangrove *Bruguiera cylindrica* leaf extracts. IOP Conf Ser Earth Environ Sci 130(1):012025

Balouiri M, Sadiki M, Ibnsouda SK (2016) Methods for in vitro evaluating antimicrobial activity: a review. J Pharm Anal 6(2):71–79

Balunas MJ, Kinghorn AD (2005) Drug discovery from medicinal plants. Life Sci 78 (2005):431–441

Bandaranayake WM (2002) Bioactive compounds and chemicals constituents of mangrove plants. Wet Ecol Manag 10:421–452

Batubara I, Mitsunaga T (2013) Use of Indonesian medicinal plant products against acne. Rev Agric Sci 1:11–30

Batubara I, Darusman LK, Mitsunaga T, Rahminiwati M, Djauhari E (2010) Potency of Indonesian medicinal plants as tyrosinase inhibitor and antioxidant agent. J Biol Sci 10(2):138144

Bindseil KU, Jakupovic J, Wolf D, Lavayre J, Leboul J, van der Pyl D (2001) Pure compound libraries: a new perspective for natural product based drug discovery. Drug Discov Today 6:840–847

Brenk R, Schipani A, James D, Krasowski A, Gilbert IH, Frearson J, Wyatt PG (2008) Lessons learnt from assembling screening libraries for drug discovery for neglected diseases. ChemMedChem 3(3):435–444

Butler MS (2004) The role of natural product chemistry in drug discovery. J Nat Prod 67 (12):2141–2153

Butler MS (2005) Natural products to drugs: natural product derived compounds in clinical trials. Nat Prod Rep 2005(22):162

Butler, R.A., 2016. The top 10 most biodiverse countries: What are the world's most biodiverse countries? https://news.mongabay.com/2016/05/top-10-biodiverse-countries/ Retreved from internet on April 30, 2018

Cragg GM, Newman DJ (2004) A tale of two tumor targets: topoisomerase I and tubulin. The Wall and Wani contribution to cancer chemotherapy. J Nat Prod 67(2):232–244

Dandapani S, Rosse G, Southall N, Salvino JM, Thomas CJ (2012) Selecting, acquiring, and using small molecule libraries for high-throughput screening. Curr Protoc Chem Biol 4:177–191. https://doi.org/10.1002/9780470559277.ch110252

Demain AL, Vaishnav P (2011) Natural products for cancer chemotherapy. J Microbial Biotechnol 4(6):687–699

Dickson M, Gagnon JP (2004) Key factors in the rising cost of new drug discovery and development. Nat Rev Drug Discov 3(5):417–429

Do QT, Bernard P (2004) Pharmacognosy and reverse pharmacognosy: a new concept for accelerating natural drug discovery. IDrugs 7(11):1017–1027

de-Faria FM et al (2012) Mechanisms of action underlying the gastric antiulcer activity of the *Rhizophora mangle* L. J Ethnopharmacol 139(1):234–243

Farnsworth NR, Soejarto DD (2009) Global importance of medicinal plants. In: Akerele O, Heywood V, Synge H (eds) Conservation of medicinal plants, 1st edn. Cambridge University Press, Cambridge, pp 25–52

Frantz S (2005) 2004 approvals: the demise of the blockbuster? Nat Rev Drug Discov 4(2):93–94
Frantz S, Smith A (2003) New drug approvals for 2002. Nat Rev Drug Discov 2(2):95–96
Ganesan A (2004) Natural products as a hunting ground for combinatorial chemistry. Curr Opin Biotechnol 15(6):584–590
Gaudêncio SP, Pereira F (2015) Dereplication: racing to speed up the natural products discovery process. Nat Prod Rep 32:779–810. https://doi.org/10.1039/C4NP00134F
Giri C et al (2011) Status and distribution of mangrove forests of the world using earth observation satellite data. Glob Ecol Biogeogr 20(1):154–159
Graul AI (2001) The year's new drugs. Drug News Perspect 14(1):12–31
Grynkiewicz G, Achmatowicz O, Pucko W (2000) Bioactive isoflavone—genistein; synthesis and prospective applications. Herba Polon 46:151–160
Guo Z (2017) The modification of natural products for medical use. Acta Pharm Sin B 7 (2):119–136
Gurudeeban S et al (2012) Antidiabetic effect of a black mangrove species *Aegiceras corniculatum* in alloxan-induced diabetic rats. J Adv Pharm Technol Res 3(1):52–56
Harvey AL (2008) Natural products in drug discovery. Drug Discov Today 13:894–901
Hassig CA et al (2014) Ultra-high-throughput screening of natural product extracts to identify proapoptotic inhibitors of Bcl-2 family proteins. J Biomol Screen 19(8):1201–1211
Hughes JP, Rees S, Kalindjian SB, Philpott KL (2011) Principles of early drug discovery. Br J Pharmacol 162:1239–1249
Islam MA et al (2012) Antinociceptive activity of methanolic extract of *Acanthus ilicifolius* Linn leaves. Turk J Pharm Sci 9(1):51–60
Juyal D, Thawani V, Thaledi S, Joshi M (2014) Ethnomedical properties of Taxus wallichiana Zucc. (Himalayan Yew). J Tradit Complement Med 4(3):159–161
Katiyar C, Gupta A, Kanjilal S, Katiyar S (2012) Drug discovery from plant sources: an integrated approach. Ayu 33(1):10–19
Kazanietz MG (2005) Targeting protein kinase C and "non-kinase" phorbol ester receptors: emerging concepts and therapeutic implications. Biochim Biophys Acta 1754:296
Koehn FE, Carter GT (2005) The evolving role of natural products in drug discovery. Nat Rev Drug Discov 4(3):206–220
Kramer R, Cohen D (2004) Functional genomics to new drug targets. Nat Rev Drug Discov 3 (11):965–972
Li JW, Vederas JC (2009) Drug discovery and natural products: end of an era or an endless frontier? Science 325(5937):161–165
Liem AF, Holle E, Gemnafle IY, Wakum DS (2013) Isolasi Senyawa Saponin dari Mangrove Tanjang (*Bruguiera gymnorrhiza*) dan Pemanfaatannya sebagai Pestisida Nabati pada Larva Nyamuk. Jurnal Biologi Papua 5(1):29–36
Liu Z (2008) Preparation of botanical samples for biomedical research. Endocr Metab Immune Disord Drug Targets 8(2):112–121
Lotsch J, Geisslinger G (2001) Morphine-6-glucuronide: an analgesic of the future? Clin Pharmacokinet 40(7):485–499
Mann J (2000) Murder, magic and medicine, 2nd edn. Oxford University Press, Oxford
Mayorga P, Pérez KR, Cruz SM, Cáceres A (2010) Comparison of bioassays using the anostracean crustaceans Artemia salina and Thamnocephalus platyurus for plant extract toxicity screening. Rev Bras Farmacogn 20:897–903
Mittermeier RA, Gil PR, Hoffman M, Pilgrim J, Brooks T, Mittermeier CG, Lamoreux J, da Fonseca GAB, Seligmann PA, Ford H (2005) Hotspots revisited: earth's biologically richest and most endangered terrestrial ecoregions. Conservation International, New York
Mouafi FE et al (2014) Phytochemical analysis and antibacterial activity of mangrove leaves (*Avicenna marina* and *Rhizophora stylosa*) against some pathogens. World Appl Sci J 29 (4):547–554
Newman DJ, Cragg GM (2012) Natural products as sources of new drugs over the 30 years from 1981 to 2010. J Nat Prod 75(3):311–335

Newman DJ, Cragg GM, Sneader KM (2000) The influence of natural products upon drug discovery. Nat Prod Rep 17(3):215–234

Newman DJ, Cragg GM, Snader KM (2003) Natural products as sources of new drugs over the period 1981–2002. J Nat Prod 66(7):1022–1037

Ochoa-Villarreal M, Howat S, Hong SM, Jang MO, Jin YW, Lee EK, Loake GJ (2016) Plant cell culture strategies for the production of natural products. BMB Rep 49(3):149–158

Okouneva T, Hill BT, Wilson L, Jordan MA (2003) The effects of vinflunine, vinorelbine, and vinblastine on centromere dynamics. Mol Cancer Ther 2(5):427–436

Pereira DA, Williams JA (2007) Origin and evolution of high throughput screening. Br J Pharmacol 152(1):53–61

Pirttila T, Wilcock G, Truyen L, Damaraju CV (2004) Long-term efficacy and safety of galantamine in patients with mild-to-moderate Alzheimer's disease: multicenter trial. Eur J Neurol 11 (11):734–741

Quinn RJ (2012) Basics and principles for building natural product-based libraries for HTS. In: Haian F (ed) Chemical genomics. Cambridge University Press, Cambridge

Rege AA, Chowdhary AS (2013) Evaluation of mangrove plants as putative HIV-protease inhibitors. Indian Drugs 50(7):41

Rohaeti, E et al. (2010) Potensi Ekstrak Rhizophora sp. Sebagai inhibitor tirosinase. Prosiding Semnas Sains III. IPB, Bogor, 13 November, pp 196–201

Roy A, McDonald PR, Sittampalam S, Chaguturu R (2010) Open access high throughput drug discovery in the public domain: a Mount Everest in the making. Curr Pharm Biotechnol 11 (7):764–778

Royal Botanical Garden Report (2017) State of the world's plants. Royal Botanic Gardens, Kew

Safari VZ et al (2016) Antipyretic, antiinflammatory and antinociceptive activities of aqueous bark extract of *Acacia nilotica* (L.) Delile in albino mice. J Pain Manag Med 2:113

Salim AA et al (2008) Drug discovery from plants. In: Ramawat KG, Mérillon JM (eds) Bioactive molecules and medicinal plants. Springer, Berlin. https://doi.org/10.1007/978-3-540-74603-4_1

Samuelsson G (2004) Drugs of Natural Origin, 5th edn. Apotekarsocieteten, Stockholm

Schroeder FC, Gronquist M (2006) Extending the scope of NMR spectroscopy with microcoil probes. Angew Chem Int Ed 45(43):7122–7131

Singh CR, Kathiresan K (2015) Effect of cigarette smoking on human health and promising remedy by mangroves. Asian Pacific Journal of Tropical Biomedicine 5(2):162–167

Sneader W (2005) Drug discovery: a history. Wiley, Chichester

Tambe VD, Bhambar RS (2016) Studies on diuretics and laxative activity of the *Hibiscus tiliaceus* Linn. bark extracts. Int J PharmTech Res 9(3):305–310

Tan DS (2004) Current progress in natural product-like libraries for discovery screening. Comb Chem High Throughput Screen 7(7):631–643

Tanvira P, Seenivasan R (2014) Targeting mangrove species as an alternative for snake bite envenomation therapy with special reference to phospholipase A2 inhibitory activity: a mini review. Res J Pharm Biol Chem Sci 5(2):1724–1731

Walters WP, Namchuk M (2003) Designing screens: how to make your hits a hit. Nat Rev Drug Discov 2:259–266

WHO (2005) WHO guidelines for sampling of pharmaceutical products and related materials. WHO Technical Report Series, No. 929

World Health Organization (1996) The World health report : 1996 : fighting disease, fostering development / report of the Director-General. Geneva : World Health Organization. http://www.who.int/iris/handle/10665/36848

Yi XX et al (2015) Four new cyclohexylideneacetonitrile derivatives from the hypocotyl of mangrove (*Bruguiera gymnorrhiza*). Molecules 20(8):14565–14575

Yu D, Suzuki M, Xie L, Morris-Natschke SL, Lee KH (2003) Recent progress in the development of coumarin derivatives as potent anti-HIV agents. Med Res Rev 23(3):322–345

Chapter 13
Nano TiO_2 for Biomedical Applications

Khairul Arifah Saharudin, Srimala Sreekantan, Rabiatul Basria S. M. N. Mydin, Siti Nor Qurratu Aini Abd Aziz, and G. Ambarasan Govindasamy

13.1 Fundamental Properties of TiO_2

Titanium dioxide (TiO_2) belongs to transition metal oxide groups. This oxide occurs in three different structures such as anatase (tetragonal), rutile (tetragonal) and brookite (orthorhombic) (Diebold 2003; Carp 2004). The fundamental structural units of TiO_2 crystal are TiO_6 octahedra; however, their arrangements are different from one phase to another. In rutile phase, TiO_6 octahedra are organized by sharing an edge along the c-axis to form chains, and then, corner-shared bonding character among chains leads to the formation of three-dimensional framework (Fig. 13.1a). In anatase, a framework with skewed chains is built using edge-shared bonding among all the TiO_6 octahedra as shown in Fig. 13.1b. Anatase is preferred over other polymorphs for photocatalytic applications because of its higher electron mobility and higher activity (Banerjee 2011). A greater extent of charge recombination in rutile is responsible for the generally lower photocatalytic activity of rutile in comparison to anatase. The basic units of TiO_2. This structure formed by edge-sharing TiO_2 octahedron. Brookite is more complicated structure and having large cell volume. It is the least dense compared to the other two forms and it is not often used for experimental investigations (Thompson and Yates 2006).

TiO_2 is an n-type semiconductor that has a valence band composed of 2p orbital O hybridized with 3d orbital of Ti and a conduction band formed by only 3d orbital of Ti (Park et al. 2002; Long and English 2009; Fitzpatrick et al. 2012). The relative

K. A. Saharudin · S. Sreekantan (✉) · S. N. Q. A. A. Aziz · G. A. Govindasamy
School of Materials & Mineral Resources Engineering, Engineering Campus, Universiti Sains Malaysia, Nibong Tebal, Pulau Pinang, Malaysia
e-mail: srimala@usm.my; rabiatulbasria@usm.my

R. B. S. M. N. Mydin
Oncological and Radiological Sciences Cluster, Advanced Medical and Dental Institute, Universiti Sains Malaysia, Kepala Batas, Pulau Pinang, Malaysia

S. Siddiquee et al. (eds.), *Nanotechnology: Applications in Energy, Drug and Food*,
https://doi.org/10.1007/978-3-319-99602-8_13

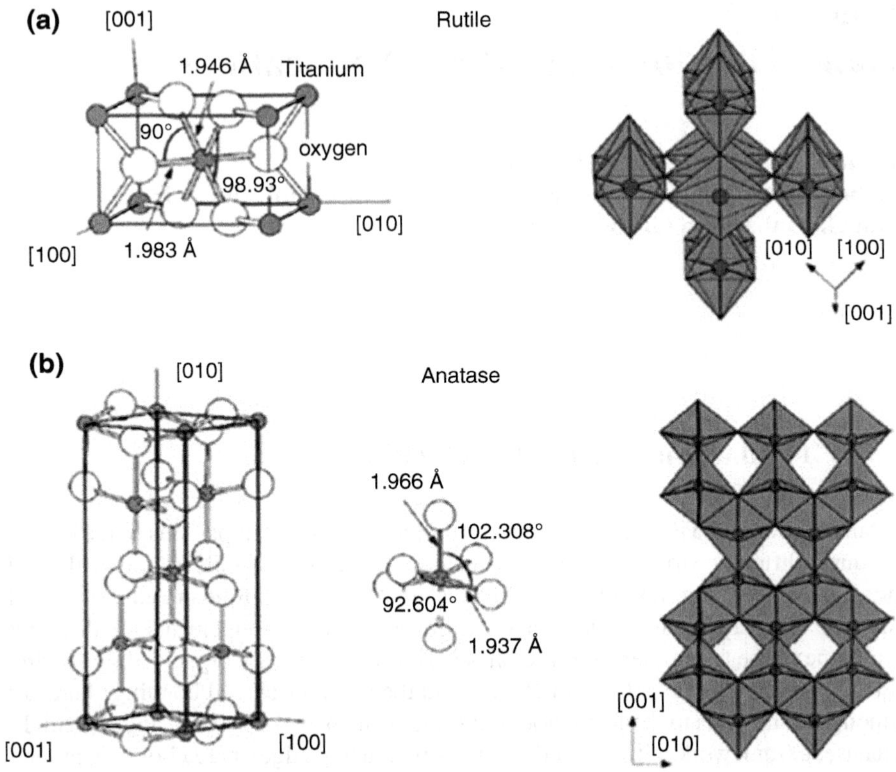

Fig. 13.1 Crystal structures of the (**a**) rutile and (**b**) anatase phases of TiO_2 (Reprinted with permission from Gupta SM, Tripathi M (2011) A review of TiO_2 nanoparticles. Chinese Science Bulletin 56 (16):1639–1657). Copyright © 2011, Springer Nature

Table 13.1 Band gap energy of metal oxide photocatalysts

Metal oxides	Band gap (eV)
TiO_2	3.2
ZnO	3.2
WO_3	2.8
a-Fe_2O_3	2.8

band gap position compared to redox potentials of variety organic reaction are of great importance for photocatalysis application. For example, to degrade organic compound by illumination with light, the redox potential of $H_2O/{}^{\cdot}OH$ couple (OH^- → ; OH + e^-; E0 = −2.8 eV) should be within the band gap of semiconductors (Hoffmann et al. 1995). TiO_2 fulfils these basic requirements relatively well compared to other metal oxides. Table 13.1 below summarized the metal oxides that could be used as photocatalyst.

To date, one dimensional (1D) structure of well-aligned TiO_2 nanotubes (NT) by anodization of Ti substrate under a specific set of environment conditions has been

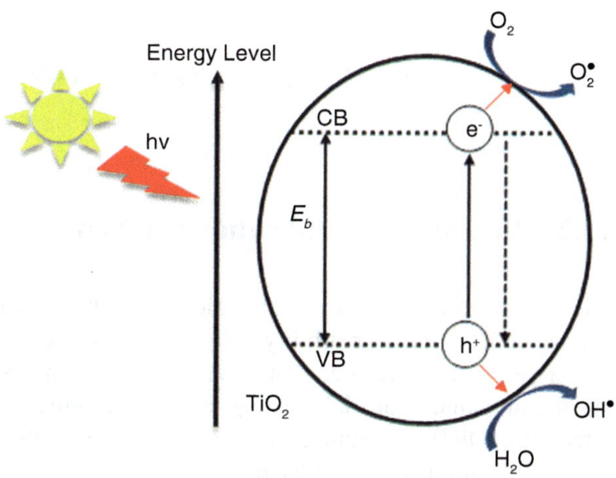

Fig. 13.2 Schematic illustration of main process in photocatalytic reaction

widely reported (Smith et al. 2009; Jiang et al. 2011). In recent report, TiO_2 NT is considered as one of the best forms of TiO_2 to enhance the photocatalytic activity (Gupta and Tripathi 2011). This is because of its tubular structure provides unique electronic properties such as, high electron mobility, low quantum confinement effect and high mechanical strength. In addition to that, the large surface area provided from the NT walls (inside and outside) and porosity structure of the hollow nanotube will facilitate light harvesting and provide more reaction interface (Abdelmoula et al. 2014).

Photocatalysis of TiO_2 involves three processes: the excitation, bulk diffusion and surface transfer of photoinduced charge carriers. Fig. 13.2 shows the three processes in a photocatalytic reaction. In general, the electrons in the conduction band (e^-_{CB}) and the holes in the valence band (h^+_{VB}) are generated (Smith et al. 2009) when TiO_2 surface is illuminated with light energy greater than the band gap energy of TiO_2 (Reaction 13.1). The h^+_{VB} can oxidize the organic molecules or react with OH^- or H_2O (Wang and Zhou 2011) which consequently produce $OH^{\cdot}$ (Reaction 13.2). The adsorbed OH ions on the NT also react with h^+_{VB} to form $OH^{\cdot}_{ad}$ (Reaction 13.3). Oxygen is usually supplied as electron acceptors (Jiang et al. 2011) and reacts with e^-_{CB} to produce $O_2^{\cdot-}$ (Reaction 13.4). Highly reactive oxidizing species (ROS), including $OH^{\cdot}$, $OH^{\cdot}_{ad}$, $O_2^{\cdot-}$ and h^+_{VB} eventually react with organic compound (R^+) resulting in the degradation of organic compounds (Reaction 13.5) (Smith et al. 2009).

$$TiO_2 + hv \rightarrow h^+_{VB} + e^-_{CB} \quad \text{(Reaction 13.1)}$$

$$h^+_{VB} + H_2O \rightarrow H^+ + OH^{\cdot} \quad \text{(Reaction 13.2)}$$

$$h^+_{VB} + OH^-_{ad} \rightarrow OH^{\cdot}_{ad} \quad \text{(Reaction 13.3)}$$

$$e^-_{CB} + O_2 \rightarrow O_2^{\cdot-} \quad \text{(Reaction 13.4)}$$

$$R^+ + \left(OH^{\cdot}, O_2^{\cdot-}, h^+_{VB}\right) \rightarrow \text{degradation products} \quad \text{(Reaction 13.5)}$$

13.2 Biomedical Applications of Nano TiO_2

The application of TNTs in biomedical field is a promising alternative because of their excellent biocompatibility, mechanical strength, and chemical resistivity. In recent years, there has been an extensive report on TNT use for dental and orthopaedics implant, drug delivery system and antimicrobial agents in device-related infections (DRI) for biomedical applications. Therefore, in the following sections those applications are elucidated.

13.2.1 Dental and Orthopaedics Implant

The repair of large bone defects due to injury or disease is one of the major problems in orthopedic and maxillofacial surgery (El-Ghannam 2005). Bone is a vascular and highly specialized form of connective tissue composed of ~20% collagen, ~70% bone mineral (mainly hydroxyapatite [HA]), and ~20% water, by weight (Wu et al. 2014). Conventional clinical treatments for bone repair and regeneration include autologous and allogeneic transplantations which have several limitations and complications, such as donor site morbidity and immunogenic response (Hoexter 2002). Replacement of tooth and bone with Ti and Ti alloys implants and plates is one of the most frequently applied material in orthopedic and dental implants for many years. Ti and its alloys, has been widely used in the biomedical field, due to their excellent properties that include a moderate elastic modulus of approximately 110 GPa, a good corrosion resistance and a low density (approximately 4700 kg m^{-3}) (Navarro et al. 2008). Titanium and its alloys are able to become tightly integrated into bone. This property significantly enhances the long-term behavior of the implanted devices, reducing the risks of loosening, infection dislocation. Presence of a natural oxide layer on titanium improves their bone-bonding ability. The oxidized surface is also believed to be responsible for Ti implants becoming osseointegrated *in vivo*, a process whereby bone is opposed to the implant without chronic inflammation and without an intervening fibrous capsule (Petersen 2014).

However, poor osseointegration between Ti implant and surrounding nature bone tissue to extend the lifetime of implant, still remain undetermined. Recent studies indicated development of nanotechnology to modify the TiO_2 nanoarchitectures on Ti and Ti Alloy implants to form highly ordered oxide nantube layer presented an ideal surface to resolve the osseointegration, thus induce great affinity for bone cell adhesion and differentiation. The diameters, lengths and shape of TiO_2 NT were

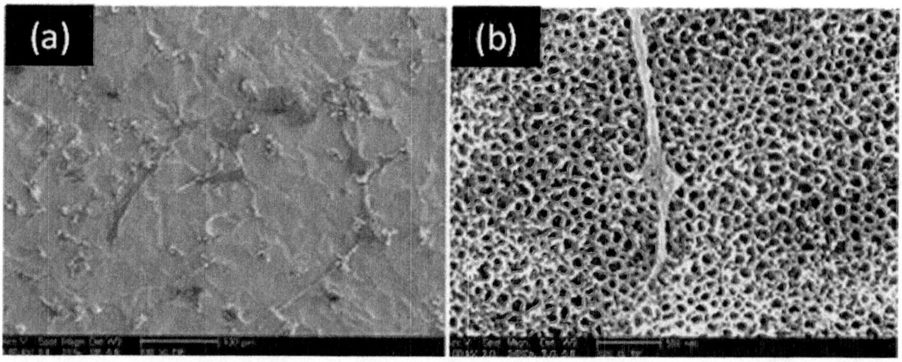

Fig. 13.3 FESEM images of marrow stromal cells on Ti and nanotubular surfaces for up to 7 days of culture. (**a**) Cells show spherical morphology on Ti and (**b**) high magnification FESEM on nanotubular surface shows that cell extensions are protruding into the nanotubular architecture (Reprinted with permission from Popat KC, Leoni L, Grimes CA, Desai TA (2007) Influence of engineered titania nanotubular surfaces on bone cells. Biomaterials 28 (21):3188–3197).

tunable in a controlled manner by optimizing electrochemical anodization parameters (Lai et al. 2011). Park demonstrated the effect of TiO_2 NT with different diameters could affect the proliferation and differentiation of mesenchymal with stem cells (MSCs) (Park et al. 2007, 2009). Diameters in the range of 15–100 nm are favourable for cell adhesion, spreading and growth of MSCs. Another report by Zhu et al. (2004) demonstrated the effects of topography and composition of TiO_2 NT surface on osteoblast (SaOS-2) response (Zhu et al. 2004). After anodization in an electrolyte composed of Ca-GP and CA, an increase of cell attachment and proliferation were observed. It is worth to note that their cell experiments showed an absence of cytotoxicity. The cells on the surfaces with micro-pores showed an irregular and polygonal growth and more lamellipodia. Also using Ca-GP and CA electrolyte, Li et al. (2004) reported an increased ALP activity of MG63 cell when the anodization voltage was above 300 V (Li et al. 2004).

In addition, the surface energy and surface roughness of NT would be essential factors for facilitating cells adhesion and proliferation. Popat et al. (2007) found that the MSCs may response differently to different surface structure. They observed an increased chondrocyte adhesion on TiO_2 NT surfaces compared with bare Ti after 7 days of culture. MSCs show a spherical morphology on bare Ti surface as shown in Fig. 13.3a. However, they show a filodopia like structure extending at certain region of the TiO_2 NT (Fig. 13.3b) (Popat et al. 2007). A study by Saharudin et al. (2013) showed that after 7 days of culture, the cell viability of PA6 cell (bone marrow stromal cell lines) grown on anodized Ti_6Al_4V substrate showed statistically higher ($p < 0\_001$) than those grown on flat Ti_6Al_4V alloy substrate (Saharudin et al. 2013). This suggests that topographical cues at nanoscale level promote cell adhesion and proliferation. The presence of Ca-ions also has been reported to be advantageous to cell growth (Maxian et al. 1994), and *in vivo* data show TiO_2 NT implant surfaces containing both Ca and P enhance bone apposition on the implant surface (Li et al. 2004; Fini et al. 1999; Ishizawa et al. 1995; Giavaresi et al. 2003a, b; Son et al. 2003; Baslé et al. 1993). Ishizawa et al. (1995) found strong bond bonding via push-out tests with HA (1–2 μm) coated TiO_2 NT after 8 weeks implantation into rabbits (Ishizawa et al. (1995). Giavaresi et al. (2003a, b) supported positive role of HA in accelerating bone ingrowth and bone mineralization (Giavaresi et al. 2003a, b). The results indicated that Ca and P concentrations were higher on the TiO_2 NT surface, which suggested that matrix deposition on the nanotubular surface enhance cell growth.

13.2.2 Drug Delivery System

Chemotherapy is a conventional cancer treatment that depends on the circulatory system to transport anticancer drugs to the tumor. Apart from the positive effect of the chemotherapy, this treatment also brings negative side effects such as non-specificity and toxicity of the drug, whereby the drugs not only attack cancerous cells but also healthy cells and organs. Therefore, targeted drug delivery is being developed as one alternative to chemotherapy treatment. Unlike conventional therapy, targeted drug deliveries are designed to direct the drug to a specific area and allow sufficient dosage of drug to be delivered at the tumor site. This reduces the side effects of the conventional treatment that attacks the healthy cells or organs.

In recent years, there has been a growing interest in using nano-systems as controlled and targeted drug delivery system or drug carrier for cancer cell. Generally, nanoparticles such as gold are coated with a biocompatible layer (polymers) to functionalize the nanoparticles. The anticancer drug can either be conjugated to the surface or encapsulated in the nanoparticles as shown in Fig. 13.4. When the drug/nanoparticle complex is administered, the complex is directed to the specific tumour site by an external magnetic field. Then, the drug is released by enzyme activity or by changes in pH, temperature or osmolality (Berry and Curtis 2003; Pankhurst et al. 2003; Tartaj et al. 2003; Chatterjee et al. 2014).

In targeted drug delivery, an ideal drug carrier should possess an appropriate pore size and a high surface area in order to accommodate the loading of the drug and to foster significant drug adsorption, respectively. Sufficiently, large and homogeneous pore network is required to control the uniform drug release kinetics. These requirements make mesoporous TiO_2 nanoparticles as an excellent candidate for controlled drug delivery systems. The capability to engineer the size and shape of TiO_2 conveys additional control of their chemical and electronic properties resulting in new opportunities for more efficient site-selective chemistry. In addition, TiO_2 nanoparticles with excellent photocatalytic property have made it popular as an excellent carrier for a light-induced drug delivery.

For example, TiO_2 nanoparticles has received considerable attention in efficient drug delivery systems as a carrier material for various drugs, such as doxorubicin (DOX) (Qin et al. 2011; Wu et al. 2011; McNamara and Tofail 2017; Du et al. 2015),

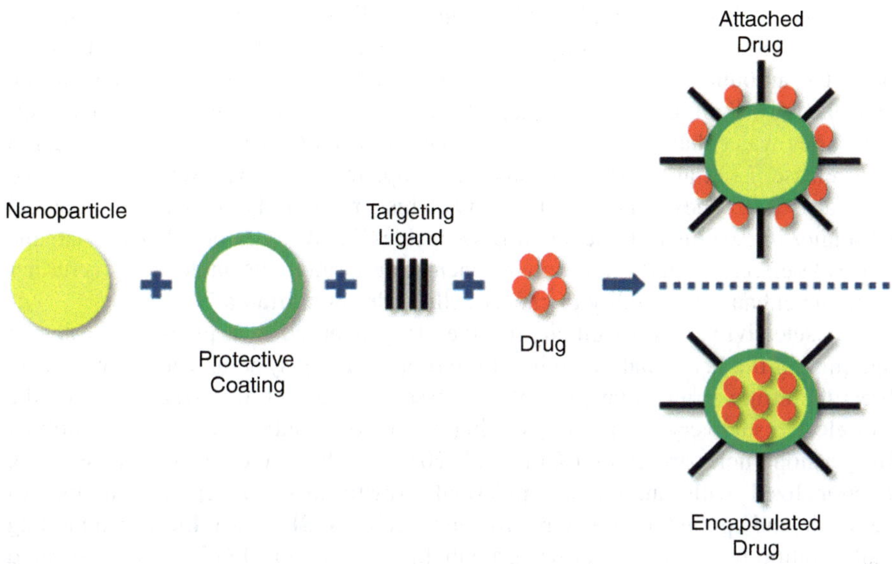

Fig. 13.4 Schematic of drug loading options in targeted drug delivery

daunorubicin (DNR) (Zhang et al. 2012; Wu et al. 2011) and paclitaxel (PTX) (Venkatasubbu et al. 2013). Qin et al. (2011) examined the cytotoxicity, cellular uptake and intercellular distribution of with and without DOX loaded TiO_2 nanoparticles using glioma (C6) cell line from rats (Qin et al. 2011). They found out that non-covalently conjugated (TiO_2/DOX) allow the main fraction of DOX inside the nuclei and exhibited greater cytotoxicity toward C6 glioma cancer cells than DOX alone. By controlling the pH environment to acidic nature, the anticancer activity of TiO_2/DOX nanocomposite was improved. McNamara and Tofail (2017) developed a pH sensitive DOX loaded TiO_2 nanoparticles for drug delivery to MDR breast cancer cells. It was found that at a pH environment of 4–5 allows the release of DOX from the surface of TiO_2 nanoparticles. The anticancer activity of TiO_2-DOX nanocomposite on multidrug resistant MCF-7/ADM breast cancer cell line and the drug sensitive MCF-7 breast cancer cell line, showed slightly higher anticancer activity than DOX by itself and 40% cytotoxic effect, respectively. Similarly, Zhang et al., developed DNR-loaded TiO_2 nanocomposites and examined their cytotoxicity, drug release behavior and cellular uptake (Zhang et al. 2012). The results showed that the release of DNR was accelerated when the pH of the environment decreased from 7.4 to 5. The cytotoxicity analysis carried out on leukaemia (K562) cell lines showed that DNR-TiO_2 nanocomposites were more toxic as compared to DNR alone. It shows that this DNR-TiO_2 nanocomposite has the potential to be used as pH controllable drug delivery system for leukaemia (K562) cell lines.

Another approach to control the release of drugs is via light-sensitive drug release systems. Yadav et al. (2017) developed a light responsive conjugated TiO_2

nanocorals/DOX (Yadav et al. 2017). Following light irradiation, TiO_2 nanocorals produce ROS and triggered drug release for killing of MCF7 cancer cell. Furthermore, the amount of drug released from the multimodal TiO_2 nanocorals can be controlled by light irradiation time. The anti-cancer effect of multimodal TiO_2 nanocorals was enhanced due to a combination of light-activated stimuli-triggered drug release. Li et al. (2009) prepared biocompatible TiO_2 whiskers photocatalyst (Li et al. 2009). They confirmed that TiO_2 whiskers not only enhance the potential anti-tumor intracellular efficiency of DNR inside SMMC-7721 but also alleviate the toxic side-effects of DNR in the environment. Furthermore, its photocatalysis further led to the enhanced mortality of cancer cells under light irradiation.

The selectivity and specificity of the drug released site plays key roles in determining the beneficial outcome of the drug delivery system. Therefore, how to direct the drug carrier to targeted cells is essential. Kim et al. (2012) explored the site-selective delivery of the drugs using monoclonal antibodies modified hollow TiO_2 nanoparticle structures (Kim et al. 2012). Hollow TiO_2 nanoparticles were functionalized with antihuman epidermal growth factor receptor 2 antibodies Herceptin and applied to human breast cancer cell line SK-BR-3. Enhanced binding of the antibody to the epidermal growth factor receptor (EGFR) leads to rapid internalization between the composites and the cell. The hollow particles were also loaded with the camptothecin anticancer drug. They are able to release an entire load of the drug in 24 h, producing efficient anticancer activity. Therefore, enhanced anticancer activity and reduced toxicity to normal cells.

13.2.3 Antimicrobial Agents in Device-Related Infections (DRI)

Device-related infections (DRIs) due to bacterial colonization and proliferation have become one of the major issues of short-term devices and implantable prosthesis (Vasilev et al. 2009). A wide range of bacteria, fungi and viral pathogens are responsible for such infections. Chronic infectious diseases arising from bacterial adhesion and colonization on the surfaces of medical devices were tremendously increased at an alarming rate in the last 20 years, mainly on immune compromised patients (Scorzoni et al. 2007). The direct contact of biomedical implant devices such as catheters and sutures with patients is a major concern because it causes cross contamination and casualties to immuno-compromised patients. Generally, the medical devices which implanted inside the human's vein would favor the adherence and breeding of microbial pathogens. The most common single isolate among patients is coagulase negative *Staphylococcus* (CoNS) (20.3%), followed by *Candida* spp. (17.5%), Gram-negative *bacilli* (49%), Gram-positive *cocci* (33%), and fungi (18%) (Wattal et al. 2014). Antibiotic resistance rates have been rising in the past two decades for all predominant organisms, including *S. aureus*, coagulase-negative *Staphylococci*, *enterococci* and Gram-negative pathogens (Lu et al. 2011). For this reason, a potential solution to prevent bacterial adhesion on the device surface is by antibacterial coating (Hetrick and Schoenfisch 2006).

Many attempts has been employed to reduce infection risk have been applied, such as implantation techniques, improvements in aseptic insertion and post-insertion care practices, and antibiotic-coated. However, the removal or eradication of bacteria is still limited. Generally, inorganic nanomaterials possess germicidal action, potentially serving a role in health service infection control (Leite and Padoveze 2012). Besides, it has been postulated that the various killing mechanisms of inorganic nanomaterials including reactive oxygen species (ROS) generated on the surface of particles, metal ions release, membrane dysfunction, particle internalization and cytolitic damage (Díaz-Visurraga et al. 2011) has capability of killing wide spectrum of microbes. For example, inorganic nanomaterials such as TiO_2, copper (Cu), silver (Ag), silver nitrate ($AgNO_3$), zinc oxide (ZnO) and copper (II) oxide (Cu_2O) is used in hygiene-sensitive areas to prevent infections. Among all these, inorganic metal oxides compounds incorporated polymer have gained priority in recent research due to unique properties such as strong antibacterial activity at low concentrations (Ren et al. 2009), stable in extreme conditions (Delgado et al. 2011; Rodriguez-Llamazares et al. 2012), non-toxic and some of them even contain mineral elements essential to the human body (Varghese et al. 2013; Fu et al. 2005; Sayılkan et al. 2009). This may be ascribed to the fact inorganic antimicrobial agent incorporated polymer only releases metal ions which is not harmful to human health for killing the microbes (Delgado et al. 2011). Inorganic polymer composites is accomplished via incorporation of antimicrobial agents directly into polymers (Azam et al. 2012; Gould et al. 2009), coating or adsorbing antimicrobials onto polymer surfaces (Delgado et al. 2011; Rodriguez-Llamazares et al. 2012), immobilisation of antimicrobials to polymers by ion or covalent linkages (Gabbay et al. 2006; Jašková et al. 2013) and the use of polymers that are inherently antimicrobial (Varghese et al. 2013).

Among various inorganic materials, TiO_2 nanoparticles is considered as an effective antimicrobial agents (Ravishankar Rai and Jamuna Bai 2011; Jašková et al. 2013). Antimicrobial efficacy of TiO_2 on *E. coli*, *S. choleraesuis*, *V. parahaemolyticus*, *L. monocytogenes*, *P. aeruginosa*, *S. aureus*, *D. actinidiae* and *P. expansum* have been reported (Emamifar 2011; Durairaj et al. 2014). Besides, the antibacterial activity of TiO_2 against Gram-positive and Gram-negative bacterial species is significantly greater ($p < 0.05$) in the presence of light than in the dark, and this difference was more pronounced for *B. subtilis* (Emamifar 2011). The greater bacterial elimination was found under light supports and it can be related to the photocatalytic ROS production. Light irradiated TiO_2 showed a 99.9% killing efficacy against all Gram-positive bacteria and Gram-negative bacteria (Durairaj et al. 2014). TiO_2 highly capable in killing bacteria such as *E. coli*, *S. aureus*, and viruses effectively under light irradiation when compared to other antimicrobial agents such as ZnO, MgO, CaO, CuO, Al_2O_3, Ag_2O, and CeO_2 (Raghupathi et al. 2011). Besides, certain study shows micron-sized TiO_2 particles and TiO_2 particles with mixture of rutile and anatase may cause DNA damage of bacterial cell (Karlsson et al. 2008, 2009). The micron-sized particles of TiO_2 (<5 μm) caused more DNA damage compared to the nanoparticles, which is likely explained by the crystal structures (Karlsson et al. 2009).

Maness and co-workers has reported the optimal dose of TiO_2 photocatalytic reactions on *E. coli* using light. For *E. coli* in a small concentration, when the TiO_2 concentration was increased from 0.1 to 0.5 mg/mL, a significant improvement in inactivation efficiency was observed (Maness et al. 1999). Many reported that the optimal TiO_2 concentration was 1 mg/ml (Seven et al. 2004; Amin et al. 2009). However, when the concentration lower than that, it does not kill the bacteria (Rahim et al. 2012), when the concentration higher than 1 mg/mL, they found a decrease in the inactivation efficiency (Kim et al. 2003; Maness et al. 1999). This is because a higher concentration of TiO_2 provides more surface sites, which results in lighter scattering and decreases the light penetration into the suspension. Rincon and Pulgarin have reported that at high concentration, the following terminal reactions may take place (Rincón and Pulgarin 2003).

$$^{\cdot}OH + {^{\cdot}OH} \rightarrow H_2O_2 \quad \text{(Reaction 13.6)}$$

$$H_2O_2 + {^{\cdot}OH} \rightarrow H_2O + HO_2^{\cdot} \quad \text{(Reaction 13.7)}$$

Since, $HO_2^{\cdot}$ is less reactive compared to $OH^{\cdot}$, it does not contribute to the oxidative process, which may also contribute to the diminution of the bacterial inactivation. The optimum concentration of TiO_2 is needed to inhibit the recombination of electron-hole pairs, so as to inactivate the viability of bacteria. Table 13.2 shows the optimum concentration to kill the bacteria.

13.3 Conclusion and Outlooks

TiO_2 photocatalyst has been widely used, due to its low toxicity, long-term photostability, strong oxidizing power, and ease of availability. The primary studies on biomedical applications of TiO_2 during the past decade, such as implant materials, drug delivery system and antimicrobial agents in DRIs have been summarized in this chapter. Owing to its biocompatibility, TiO_2 implants have been widely accepted to be used in orthopaedics and dental implant procedures. TiO_2 NT can be used as orthopedic and stent implants without limitations in their forms and shapes based on the remarkable properties of TiO_2 NT such as excellent biocompatibility and thermal and mechanical stability, thereby promoting bone cell adhesion, osseointegration, hydroxyapatite formation, differentiation and proliferation, and hemocompatibility. Frequently, TiO_2 is also considered as effective antimicrobial agents. It is confirmed that drug release from TiO_2 NT can be initiated by controlling pH, light and selectivity. These external stimulus strategies can be described as promising triggers for controlling drug release. Antibacterial surface functionalization represents the most effective way to reduce DRIs. The ROS induced by photoactivated TiO_2 can kill not only cancer cells, but also bacteria

Table 13.2 Optimum concentration of TiO_2 nanoparticles in killing various species of bacteria

Structure	Properties	Concentration TiO_2	Bacteria	Ref
TiO_2 anatase	• 2.95 m^2/g of surface area • 409 Å of pore size	1 mg TiO_2/mL	• *S. choleraesuis* subsp. • *V. parahaemolyticus*	Kim et al. (2003)
Degussa P25	NA	1 mg/mL	• *E. coli* • *Pseudomonas aeruginosa* • *S. aureus* • *Saccharomyces cerevisiae* • *Candida albicans*	Seven et al. (2004)
TiO_2-Ag	≈10 nm	1 mg/mL in sterilized water	• *E. coli*	Amin et al. (2009)
TiO_2 anatase	30 nm of particle size	0.1 g/250 mL	• *E. coli*	Swetha et al. (2010)
TiO_2 anatase	3–7 nm of particle size	• 1 g/L TiO_2 NP-optimum • 0.1 g/L is not enough • 2.5 g/L too high-not effective to inactivate	• *E. coli*	Rahim et al. (2012)
TiO_2 anatase and rutile	NA	• 3 and 7% Ag-TiO_2 at 60 mg/30 mL kill all the bacteria • TiO_2 at 80 mg/30 mL kill all the bacteria	• *S. aureus* • *E. coli*	Gupta et al. (2013)

and fungi. In particular, TiO_2 photocatalyst is easy to produce and use, efficient in catalysing reactions and has a wide spectrum of antimicrobial activity, since it is capable of killing a wide range of organisms including bacteria and endospores. Moreover, its biocompatibility is clinically proven, since a huge number of medical devices are made of Ti alloys.

There are still challenges that need to be tackled in order to make TiO_2 benefit human beings to the upmost in the near future. Although TiO_2 implants have been accepted to be used in orthopaedics and dentistry, many investigations are still confined *in vitro* and *in vivo*. In order for TiO_2 to be widely and successfully applied in clinic, a reasonable cooperation of researchers with the doctors, nurses and pharmacists must be conducted, while performing researches *in vitro* and *in vivo*. Recently, the safety issue of TiO_2 has led to growing concerns about the consequences of exposure of humans and the environment. It has been suggested that the ROS generation is owing to the cytotoxicity of TiO_2 nanoparticles. To evaluate the safety, human clinical trials after long-term toxicity assay and tolerability studies must be carried out and how to reduce the side effects should also be deeply concerned in the future. Testing of antibacterial coating efficacy is mainly conducted in simplified environments *in vitro* with no common procedures and clinical studies. Hence, the coating should be designed for specific applications, taking into account the bacteria involved, their attachment and growth mechanisms, and the biological environment envisaged for the application. Despite some limitation, there is a bright future that lies ahead of TiO_2 nanoparticles, which make a great contribution to the mankind.

Acknowledgements The authors are thankful to the Ministry of Education (MOE) Malaysia for funding this work under Transdisciplinary Research Grant Scheme (TRGS) grant no. 6769002. The authors are very much grateful to Universiti Sains Malaysia (USM) for providing the necessary facilities to carry out the research work and financial support under Research University (RU) grant no. 814281.

References

Abdelmoula M, Sokoloff J, Lu W-T, Close T, Menon L, Richter C (2014) Optical properties of titanium dioxide nanotube arrays. J Appl Phys 115(1):014306

Amin SA, Pazouki M, Hosseinnia A (2009) Synthesis of TiO_2-Ag nanocomposite with sol–gel method and investigation of its antibacterial activity against E. coli. Powder Technol 196 (3):241–245

Azam A, Ahmed AS, Oves M, Khan MS, Habib SS, Memic A (2012) Antimicrobial activity of metal oxide nanoparticles against Gram-positive and Gram-negative bacteria: a comparative study. Int J Nanomedicine 7:6003

Banerjee AN (2011) The design, fabrication, and photocatalytic utility of nanostructured semiconductors: focus on TiO_2-based nanostructures. Nanotechnol Sci Appl 4:35–65

Baslé MF, Chappard D, Grizon F, Filmon R, Delecrin J, Daculsi G, Rebel A (1993) Osteoclastic resorption of Ca-P biomaterials implanted in rabbit bone. Calcif Tissue Int 53(5):348–356

Berry CC, Curtis AS (2003) Functionalisation of magnetic nanoparticles for applications in biomedicine. J Phys D Appl Phys 36(13):R198

Carp O (2004) Photoinduced reactivity of titanium dioxide. Prog Solid State Chem 32(1–2):33–177

Chatterjee K, Sarkar S, Rao KJ, Paria S (2014) Core/shell nanoparticles in biomedical applications. Adv Colloid Interface Sci 209:8–39

Delgado K, Quijada R, Palma R, Palza H (2011) Polypropylene with embedded copper metal or copper oxide nanoparticles as a novel plastic antimicrobial agent. Lett Appl Microbiol 53 (1):50–54

Díaz-Visurraga J, Gutiérrez C, Von Plessing C, García A (2011) Metal nanostructures as antibacterial agents. In: Méndez-Vilas A (ed) Science and technology against microbial pathogens: research, development and evaluation. Formatex, Badajoz, pp 210–218

Diebold U (2003) The surface science of titanium dioxide. Surf Sci Rep 48(5):53–229

Du Y, Ren W, Li Y, Zhang Q, Zeng L, Chi C, Wu A, Tian J (2015) The enhanced chemotherapeutic effects of doxorubicin loaded PEG coated TiO_2 nanocarriers in an orthotopic breast tumor bearing mouse model. J Mater Chem B 3(8):1518–1528

Durairaj B, Xavier T, Muthu S (2014) Fungal generated titanium dioxide nanoparticles for UV protective and bacterial resistant fabrication. Int J Eng Sci Technol 6(9):621

El-Ghannam A (2005) Bone reconstruction: from bioceramics to tissue engineering. Expert Rev Med Devices 2(1):87–101

Emamifar A (2011) Applications of antimicrobial polymer nanocomposites in food packaging. In: Advances in nanocomposite technology. InTech. doi:https://doi.org/10.5772/18343

Fini M, Cigada A, Rondelli G, Chiesa R, Giardino R, Giavaresi G, Aldini NN, Torricelli P, Vicentini B (1999) In vitro and in vivo behaviour of Ca-and P-enriched anodized titanium. Biomaterials 20(17):1587–1594

Fitzpatrick P, Rowley A, Wright N, Bedel L (2012) Nanocatalysis for detoxification technologies. J Nanosci Nanotechnol 12(6):4911–4918

Fu G, Vary PS, Lin C-T (2005) Anatase TiO_2 nanocomposites for antimicrobial coatings. J Phys Chem B 109(18):8889–8898

Gabbay J, Borkow G, Mishal J, Magen E, Zatcoff R, Shemer-Avni Y (2006) Copper oxide impregnated textiles with potent biocidal activities. J Ind Text 35(4):323–335

Giavaresi G, Fini M, Cigada A, Chiesa R, Rondelli G, Rimondini L, Aldini NN, Martini L, Giardino R (2003a) Histomorphometric and microhardness assessments of sheep cortical bone surrounding titanium implants with different surface treatments. J Biomed Mater Res A 67(1):112–120

Giavaresi G, Fini M, Cigada A, Chiesa R, Rondelli G, Rimondini L, Torricelli P, Aldini NN, Giardino R (2003b) Mechanical and histomorphometric evaluations of titanium implants with different surface treatments inserted in sheep cortical bone. Biomaterials 24(9):1583–1594

Gould SW, Fielder MD, Kelly AF, Morgan M, Kenny J, Naughton DP (2009) The antimicrobial properties of copper surfaces against a range of important nosocomial pathogens. Ann Microbiol 59(1):151–156

Gupta SM, Tripathi M (2011) A review of TiO_2 nanoparticles. Chin Sci Bull 56(16):1639–1657

Gupta K, Singh R, Pandey A, Pandey A (2013) Photocatalytic antibacterial performance of TiO_2 and Ag-doped TiO_2 against S. aureus. P. aeruginosa and E. coli. Beilstein J Nanotechnol 4:345

Hetrick EM, Schoenfisch MH (2006) Reducing implant-related infections: active release strategies. Chem Soc Rev 35(9):780–789

Hoexter DL (2002) Bone regeneration graft materials. J Oral Implantol 28(6):290–294

Hoffmann MR, Martin ST, Choi W, Bahnemann DW (1995) Environmental applications of semiconductor photocatalysis. Chem Rev 95(1):69–96

Ishizawa H, Fujino M, Ogino M (1995) Mechanical and histological investigation of hydrothermally treated and untreated anodic titanium oxide films containing Ca and P. J Biomed Mater Res A 29(11):1459–1468

Jašková V, Hochmannová L, Vytřasová J (2013) TiO_2 and ZnO nanoparticles in photocatalytic and hygienic coatings. Int J Photoenergy 2013:795060

Jiang D, Zhou T, Sun Q, Yu Y, Shi G, Jin L (2011) Enhanced visible-light-induced photoelectrocatalytic degradation of methyl orange by CdS sensitized TiO_2 nanotube arrays electrode. Chin J Chem 29(11):2505–2510

Karlsson HL, Cronholm P, Gustafsson J, Moller L (2008) Copper oxide nanoparticles are highly toxic: a comparison between metal oxide nanoparticles and carbon nanotubes. Chem Res Toxicol 21(9):1726–1732

Karlsson HL, Gustafsson J, Cronholm P, Möller L (2009) Size-dependent toxicity of metal oxide particles—a comparison between nano-and micrometer size. Toxicol Lett 188(2):112–118

Kim B, Kim D, Cho D, Cho S (2003) Bactericidal effect of TiO2 photocatalyst on selected foodborne pathogenic bacteria. Chemosphere 52(1):277–281

Kim C, Kim S, Oh WK, Choi M, Jang J (2012) Efficient intracellular delivery of camptothecin by silica/titania hollow nanoparticles. Chem A Eur J 18(16):4902–4908

Lai M, Cai K, Zhao L, Chen X, Hou Y, Yang Z (2011) Surface functionalization of TiO_2 nanotubes with bone morphogenetic protein 2 and its synergistic effect on the differentiation of mesenchymal stem cells. Biomacromolecules 12(4):1097–1105

Leite GC, Padoveze MC (2012) Copper as an antimicrobial agent in healthcare: an integrative literature review. J Infect Control 1(2):33–36

Li L-H, Kong Y-M, Kim H-W, Kim Y-W, Kim H-E, Heo S-J, Koak J-Y (2004) Improved biological performance of Ti implants due to surface modification by micro-arc oxidation. Biomaterials 25(14):2867–2875

Li Q, Wang X, Lu X, Tian H, Jiang H, Lv G, Guo D, Wu C, Chen B (2009) The incorporation of daunorubicin in cancer cells through the use of titanium dioxide whiskers. Biomaterials 30 (27):4708–4715

Long R, English NJ (2009) Band gap engineering of (N,Ta)-codoped TiO_2: A first-principles calculation. Chem Phys Lett 478(4–6):175–179

Lu Y, Su C, Wang A, Liu H (2011) Hyphal development in Candida albicans requires two temporally linked changes in promoter chromatin for initiation and maintenance. PLoS Biol 9 (7):e1001105

Maness P-C, Smolinski S, Blake DM, Huang Z, Wolfrum EJ, Jacoby WA (1999) Bactericidal activity of photocatalytic TiO_2 reaction: toward an understanding of its killing mechanism. Appl Environ Microbiol 65(9):4094–4098

Maxian SH, Zawadsky JP, Dunn MG (1994) Effect of Ca/P coating resorption and surgical fit on the bone/implant interface. J Biomed Mater Res A 28(11):1311–1319

McNamara K, Tofail SA (2017) Nanoparticles in biomedical applications. Adv Phys X 2(1):54–88

Navarro M, Michiardi A, Castano O, Planell J (2008) Biomaterials in orthopaedics. J R Soc Interface 5(27):1137–1158

Pankhurst QA, Connolly J, Jones S, Dobson J (2003) Applications of magnetic nanoparticles in biomedicine. J Phys D Appl Phys 36(13):R167

Park M, Kwon S, Min B (2002) Electronic structures of doped anatase TiO_2: Ti1-xMxO_2 (M=Co, Mn, Fe, Ni). Phys Rev B 65(16):161201

Park J, Bauer S, von der Mark K, Schmuki P (2007) Nanosize and vitality: TiO_2 nanotube diameter directs cell fate. Nano Lett 7(6):1686–1691

Park J, Bauer S, Schlegel KA, Neukam FW, von der Mark K, Schmuki P (2009) TiO_2 nanotube surfaces: 15 nm—an optimal length scale of surface topography for cell adhesion and differentiation. Small 5(6):666–671

Petersen RC (2014) Titanium implant osseointegration problems with alternate solutions using epoxy/carbon-fiber-reinforced composite. Metals 4(4):549–569

Popat KC, Leoni L, Grimes CA, Desai TA (2007) Influence of engineered titania nanotubular surfaces on bone cells. Biomaterials 28(21):3188–3197

Qin Y, Sun L, Li X, Cao Q, Wang H, Tang X, Ye L (2011) Highly water-dispersible TiO_2 nanoparticles for doxorubicin delivery: effect of loading mode on therapeutic efficacy. J Mater Chem 21(44):18003–18010

Raghupathi KR, Koodali RT, Manna AC (2011) Size-dependent bacterial growth inhibition and mechanism of antibacterial activity of zinc oxide nanoparticles. Langmuir 27(7):4020–4028

Rahim S, Radiman S, Hamzah A (2012) Inactivation of Escherichia coli under fluorescent lamp using TiO_2 nanoparticles synthesized via sol gel method. Sains Malays 41(2):219–224

Ravishankar Rai V, Jamuna Bai A (2011) Nanoparticles and their potential application as antimicrobials. In: Mendez-Vilas A (ed) Science against microbial pathogens: communicating current research and technological advances. Formatex Research Center, Badajoz, pp 197–209

Ren G, Hu D, Cheng EW, Vargas-Reus MA, Reip P, Allaker RP (2009) Characterisation of copper oxide nanoparticles for antimicrobial applications. Int J Antimicrob Agents 33(6):587–590

Rincón AG, Pulgarin C (2003) Photocatalytical inactivation of E. coli: effect of (continuous–intermittent) light intensity and of (suspended–fixed) TiO_2 concentration. Appl Catal Environ 44(3):263–284

Rodriguez-Llamazares S, Mondaca M, Badilla C, Maldonado A (2012) PVC/copper oxide composites and their effect on bacterial adherence. J Chil Chem Soc 57(2):1163–1165

Saharudin KA, Sreekantan S, Abd Aziz S, Hazan R, Lai CW, Mydin R, Mat I (2013) Surface Modification and Bioactivity of Anodic Ti_6Al_4V Alloy. J Nanosci Nanotechnol 13 (3):1696–1705

Sayılkan F, Asiltürk M, Kiraz N, Burunkaya E, Arpaç E, Sayılkan H (2009) Photocatalytic antibacterial performance of Sn^{4+} doped TiO_2 thin films on glass substrate. J Hazard Mater 162(2):1309–1316

Scorzoni L, Benaducci T, Almeida AMF, Silva DHS, Bolzani VS, Gianinni MJSM (2007) The use of standard methodology for determination of antifungal activity of natural products against medical yeasts Candida sp. and Cryptococcus sp. Braz J Microbiol 38(3):391–397

Seven O, Dindar B, Aydemir S, Metin D, Ozinel M, Icli S (2004) Solar photocatalytic disinfection of a group of bacteria and fungi aqueous suspensions with TiO_2, ZnO and Sahara desert dust. J Photochem Photobiol A Chem 165(1):103–107

Smith YR, Kar A, Subramanian V (2009) Investigation of physicochemical parameters that influence photocatalytic degradation of methyl orange over TiO_2 nanotubes. Ind Eng Chem Res 48(23):10268–10276

Son WW, Zhu X, Hi S, Ong JL, Kim K (2003) In vivo histological response to anodized and anodized/hydrothermally treated titanium implants. J Biomed Mater Res B Appl Biomater 66 (2):520–525

Swetha S, Santhosh S, Geetha Balakrishna R (2010) Synthesis and comparative study of nano-TiO_2 over Degussa P-25 in disinfection of water. Photochem Photobiol 86(3):628–632

Tartaj P, del Puerto MM, Veintemillas-Verdaguer S, González-Carreño T, Serna CJ (2003) The preparation of magnetic nanoparticles for applications in biomedicine. J Phys D Appl Phys 36 (13):R182

Thompson TL, Yates JT Jr (2006) Surface science studies of the photoactivation of TiO_2-new photochemical processes. Chem Rev 106(10):4428–4453

Varghese S, ElFakhri SO, Sheel DW, Sheel P, Bolton FJE, Foster HA (2013) Antimicrobial activity of novel nanostructured Cu-SiO_2 coatings prepared by chemical vapour deposition against hospital related pathogens. AMB Express 3(1):53–60

Vasilev K, Cook J, Griesser HJ (2009) Antibacterial surfaces for biomedical devices. Expert Rev Med Devices 6(5):553–567

Venkatasubbu GD, Ramasamy S, Reddy GP, Kumar J (2013) In vitro and In vivo anticancer activity of surface modified paclitaxel attached hydroxyapatite and titanium dioxide nanoparticles. Biomed Microdevices 15(4):711–726

Wang S, Zhou S (2011) Photodegradation of methyl orange by photocatalyst of CNTs/P-$TiO(_2)$ under UV and visible-light irradiation. J Hazard Mater 185(1):77–85

Wattal C, Raveendran R, Goel N, Oberoi JK, Rao BK (2014) Ecology of blood stream infection and antibiotic resistance in intensive care unit at a tertiary care hospital in North India. Braz J Infect Dis 18(3):245–251

Wu KC-W, Yamauchi Y, Hong C-Y, Yang Y-H, Liang Y-H, Funatsu T, Tsunoda M (2011) Biocompatible, surface functionalized mesoporous titania nanoparticles for intracellular imaging and anticancer drug delivery. Chem Commun 47(18):5232–5234

Wu S, Liu X, Yeung KW, Liu C, Yang X (2014) Biomimetic porous scaffolds for bone tissue engineering. Mater Sci Eng R Rep 80:1–36

Yadav HM, Thorat ND, Yallapu MM, Tofail SA, Kim J-S (2017) Functional TiO_2 nanocoral architecture for light-activated cancer chemotherapy. J Mater Chem B 5(7):1461–1470

Zhang H, Wang C, Chen B, Wang X (2012) Daunorubicin-TiO_2 nanocomposites as a "smart" pH-responsive drug delivery system. Int J Nanomedicine 7:235–242

Zhu X, Chen J, Scheideler L, Reichl R, Geis-Gerstorfer J (2004) Effects of topography and composition of titanium surface oxides on osteoblast responses. Biomaterials 25 (18):4087–4103

Chapter 14
Nanotechnology: Recent Trends in Food Safety, Quality and Market Analysis

Zamri Nurfatihah and Shafiquzzaman Siddiquee

14.1 Introduction

Nanotechnology define as the technology that involve in manipulation of matter at nano scale with the size of less than 100 nm and up to 1 nm size (Cushen et al. 2012). It has different nano forms; namely nanoparticles, nanotubes, fullerenes, nanowhiskers, nanosheet and nanofibres that are constituents of nanomaterials and nanoparticles. According to EC Cosmetics Regulation ((EC = European Commission) No 1223/2009), nanomaterial is usually an "insoluble or bio-persistent and intentionally manufactured material with one or more external dimensions, or an internal structure, on the scale from 1 to 100 nanometers". Nanotubes define as nanomaterials that have cylindrical structure formation whereas fullerenes have spherical molecular formation and nanofibers have a ratio of minimum 3:1 of length to diameter in nano range (Hoet et al. 2004; O'Brien and Cummins 2008). Nanowhiskers are fine fibres in the ranged of 5–20 nm in cross-section and few micrometers in lengths (Pandey et al. 2008) and nanosheets are one-dimensional order of material in nano range (Kumar et al. 2009).

Nanotechnology is the recent trend advancement technology that has widely applied in various fields such as food industries, medicine, electronics, communication, and energy production (Sozer and Kokini 2009). Researchers have developed a rapid tool of biosensor that can evaluate the activities of soil microorganisms. Biosensor is able to predict the probable existence of soil disease by estimating the relative activity of microbes in the soil through the difference in oxygen consumption in their respiration. It also analyze the moisture range in agriculture soil using wireless nanotechnology which consist of micromachine microelectromechanical system (MEMS) cantilever beams enveloped with a water-sensitive nanopolymer

Z. Nurfatihah · S. Siddiquee (✉)
Biotechnology Research Institute, Universiti Malaysia Sabah, Kota Kinabalu, Sabah, Malaysia
e-mail: shafiqpab@ums.edu.my

S. Siddiquee et al. (eds.), *Nanotechnology: Applications in Energy, Drug and Food*,
https://doi.org/10.1007/978-3-319-99602-8_14

and optical sensor that can be detected and quantified the presence of humic acid and pesticides in water (Baruah and Dutta 2009).

Nanotechnology has been widely used in medicine field to improve the quality of health care. Application of fluorescent markers for diagnostic and screening purposes are ameliorated by the use of nanotechnology. Florescence nanoparticles are used to observe metabolism and disease conditions, targeting different tissues for selective imaging and superparamagnetic nanoparticles that are used as magnetic resonance (MR) contrast agent for imaging specific target molecules. Besides, with the improvement of nanomedicine, the constructing an artificial binding sites are able to recognize specific protein that are made easier as DNA strands can be differentiated and characterized using nanopores. Nanoparticles facilitate in manufacturing a unique drug delivery system as the size of macroparticles help the drug move across the membrane layer easily, and the encapsulation of the drug protect them from degradation or denaturation in the area with extreme pH (Emerich and Thanos 2003).

Another example of the application of nanotechnology is in energy source sector. In photovoltaic (PV) solar technology which converts the light directly into electrical current, nanostructured materials increase their productiveness whilst lowering the manufacturing and electricity expenses greatly. In fuel cell, nanotechnology give an advantage to hydrogen production, storage and transformation into electricity as it is a more efficacious catalyst for water splitting (Serrano et al. 2009).

Food and beverage sector is a substantial industry. Large-scale food companies are frequently improving the production, food safety and quality, thus research and development projects are completed diligently (Cushen et al. 2012). The breakthrough in nanotechnology lead to significant enhancement in food safety and food quality aspects such as improvement in food security, increasing shelf life, microorganism or pesticide detection, better flavor and nutrient delivery, and serving functional foods. This chapter briefly discussed on the current application of nanotechnology in food industry.

14.2 Food Safety

14.2.1 Food Packaging

The basic features of food packaging materials are used such as mechanical, thermal and optical characteristics as well as protecting the food contamination from microbial and chemical, preventing the loss and gain of moisture, hinder the loss of other volatile compounds such as flavors and act as a barricade against the oxygen, carbon dioxide and water vapor (Rhim et al. 2013). Hence, a lot of innovations researches have been conducted for the improvement of the current trend of food packaging with the uses of nanotechnology. One of the improvements of food packaging is introducing particularly nanomaterials into polymer matrix to achieve a nanocomposite (Cushen et al. 2012). Food Contact Materials (FCM) have been

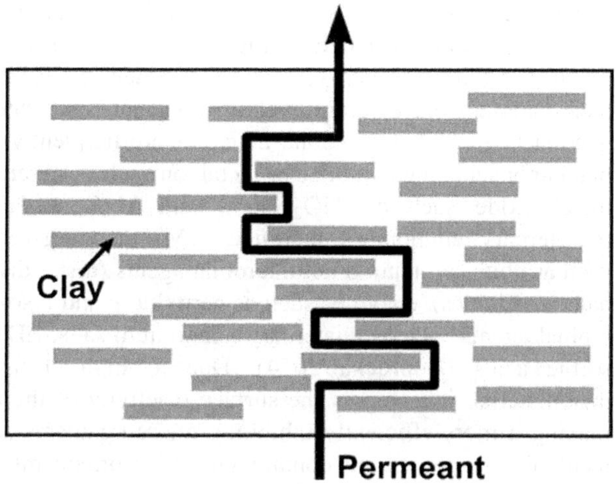

Fig. 14.1 Complex course of a permeant in a clay nanocomposite (Adapted from Adame and Beall 2009; de Azeredo 2009)

developed by adding the nanoscale mixture to enhance the flexibility, gas barrier function, moisture balance and temperature regulation to overcome the problem exists in conventional polymers. The larger polymer combined with nanoscale filler form a more difficult path for gas to travel through the matrix thus making it more efficient (Cushen et al. 2012).

Clay montmorillonite is one of the materials used as nanoscale filler in the manufacturing of nanocomposite. It has composed of stacked silicate sheet with plate-like arrangement and high aspect ratio of length to thickness. The existence of polymer formulations increased the strength of the barriers as they can create an irregularity in the diffusive path for the penetrant molecule (Fig. 14.1). Many studies have been conducted on montmorillonite, it can be integrated effectively with composite materials and the benefit of the composite films that come with high aspect ratio of the material (Rhim and Ng 2007). Laponite is new filler that is being studied for its competency to be incorporated in nanocomposites; it has higher aspect ratio compared to montmorillonite (Chung et al. 2010). The incorporation of laponite filler will elongate the path of diffusion of gas so it has to move around the obstructive filler hence inhibiting the gas exchange through the packaging material (Adame and Beall 2009; de Azeredo 2009; Sinha Ray and Okamoto 2003).

Bionanocomposites commonly used as biodegradable components and inclusion of nanomaterial fillers to overcome the drawback using traditional packaging materials (Sorrentino et al. 2007). Besides protecting the food, they enveloped and extended the shelf life of the product with the incorporation of nanofillers, the uses of bionanocomposites cut down the usage of plastics as packaging materials, reducing the use of fossil fuel in production, and biodegradable, thus making bionanocomposite is a more environmentally friendly packaging (Sozer and Kokini 2009).

Packaging that able to exhibit or release nanoscale antimicrobial compounds, antioxidants and/or flavours to extend the shelf life or sensory attribute of a food that is called an active packaging. However, the inclusion of active compounds in the packaging is more beneficial compared to the one design to migrate as latter treated as food additive under regulation (Cushen et al. 2012). Packaging that use nanomaterials have antimicrobial characteristic categorized as antimicrobial nanopackaging. Antimicrobial materials are frequently used to develop antimicrobial nanopackaging comprise of metal ion such as silver, copper, gold and platinum, metal oxide such as TiO_2, ZnO and MgO, organically modified nanoclay (quarternary ammonium modified MMT and Ag-zeolite), natural biopolymers such as chitosan, natural antimicrobial agents (nisin, thymol, carvacrol, isothiocyanate, antibiotics), enzymes such as peroxidase and lysozyme, and synthetic antimicrobial agents such as quaternary ammonium salts, EDTA, propionic, benzoic and sorbic acids (Dobrucka 2014). Due to high surface-to-volume ratio of the nanomaterial, it increases the surface reactivity of the antimicrobial agent. Hence making it more efficiently inhibiting microorganisms and decreasing the development of post-processing contaminant microorganisms which lead to prolong the product shelf life, improve food quality & safety and most importantly producing less food waste (Dobrucka 2014).

Silver and silver ions are efficient antimicrobial agents, it has large surface area of nanosilver particles and also easier for the particles to contact with fungi and bacteria. Nanosilver particles inhibit the reproduction of bacteria by binding to microbial DNA and inactivate the bacteria by binding to the sulphydryl groups of the metabolic enzymes in electron transport chain of bacteria (Maki and Tambyah 2001; Thomas et al. 2007). Due to the potent antimicrobial activity with high stability, the silver nanoparticles are administered in the antimicrobial packaging films (Emamifar et al. 2011; Llorens et al. 2012). Chitosan, a natural biopolymer, is an antimicrobial against a variety of microorganism (Entsar et al. 2003; Wu et al. 2005), because of the larger numbers of free amino acid are responsible for the antimicrobial activity when protonated, chitosan shows a better antimicrobial activity as compared to chitin (Rabea et al. 2003). In addition, zinc oxide nanoparticle has antimicrobial characteristic and it can be employed as functional fillers and used extensively as UV absorbers in cosmetics, coating materials, pigments, and barrier enhancer in some of industrial products (Kumar and Singh 2008; Li et al. 2009; Yu et al. 2004). Zinc oxide is inexpensive, non-toxicities and biocompatible in comparison to other inorganic antimicrobial compound (Paisoonsin et al. 2013).

14.2.2 Food Preservation

Currently nanotechnology is open up a new opportunity for biosensors in food technology. There are several types of nanosensor developed based on its application. The main purpose of nanosensor is to reduce the detection time of pathogens from days to hours or minutes (Bhattacharya et al. 2007). Nanosensor could be planted into packaging material to detect chemical discharged during food spoilage by work as 'electronic tongue' or 'noses' (Sozer and Kokini 2009). One type of nanosensor available is based on microfluidics devices (Baeummer 2004), it could be effectively detected of pathogens in real-time with high sensitivity. Microfluidic sensors' have ability to identify the compounds of interest in small amount of sample, rapidly and cost effective so it could apply in medical, biological and chemical analysis (Vo-Dinh et al. 2001; Mabeck and Malliaras 2006).

Aside from that, devices that used nanoelectromechanical systems (NEMS) technology have been used in food-analysis sector. These systems incorporated moving parts varying from nano- to milli-meter scale which can regulate the storage environment and determine the deadline that product must be sold (Sozer and Kokini 2009). Detection of trans-fat content in food can be conducted using a digital transform spectrometer (DTS) (Polychromix; Wilmington, MA, USA) which used microelectromechanical system technology (Ritter 2005). NEMS include advanced transducers for specific identification of chemical and biochemical signal. Advantages of the uses of micro- and nano-technologies (MNTs) for food safety and quality are rapid response, low costs, and capable of monitoring and detecting any deviation in packaging and storage condition (Canel et al. 2006).

Another innovation of biosensor is nanocantilever. Nanocantilevers detection is basically dependent to their capability of detecting biological-binding interaction such as between enzyme and substrate, cofactor and receptor and ligand, and antigen and antibody, using physical and/or electromechanical signaling (Hall 2002). The tiny pieces of silicon-based materials that made up nanocantilevers have their ability to identify proteins and detect pathogenic bacteria and viruses (Kumar 2006). These sensors have been proven successfully in detection of toxins, contaminant chemicals and antibiotic remnant in food products and molecular interaction (Ramirez Frometa 2006). Recently, nanocantilever device have been developed by a European Union-funded project to detect pathogens in food and water based on the sensing of ligand-receptor interaction. The detection of pathogen is depending on their vibration ability with different frequencies based on the biomass of the pathogenic organisms (Jain 2008).

14.3 Food Quality

In food industry, food quality is an essential part as it is related to the consumers' acceptability. In order to achieve a great organoleptic quality of a product, it depends on the proper control of raw material and processing (Hawthorn 1967). Nanotechnology applications in food quality assist in monitoring the quality parameters of the raw materials, during processing and the final product (Bücking et al. 2017).

Organic nanomaterials can be formed some specific function such as to encapsulated nutrient to enhance bioavailability, to cover the undesirable odour or taste, improve taste, texture or consistency of the food (Cushen et al. 2012). For example, nanoemulsion is one of the uses of nanotechnology in food processing technology. Nanoemulsion can define as emulsion with the diameters of the dispersed droplets of 500 nm or less (Ravichandran 2010). The small droplets of the emulsion contribute the unique rheological and textural properties of nanoemulsions which impart them transparent and acceptable to the touch (Sonneville-Aubrun et al. 2004). The smaller the size of the droplets, the less likely for the emulsion to breakdown and separate (Cushen et al. 2012). These unique properties make them preferable in food and cosmetic industry. Products that adopted nanoemulsion are fat nanostructured mayonnaise, spread and ice cream (Chaudhry et al. 2008).

Another current practice of nanotechnology in food is nanoencapsulation by immobilizing the nutrients, food ingredients or additives in a tailored carrier (Cushen et al. 2012). Nanoencapsulation of food ingredients and additives gives protective barriers against harsh environment, cover the flavor and taste, enhanced bioavailability, controlled release and improved dispersion in aqueous system for water-soluble food ingredients and additives (Chaudhry et al. 2008; Mozafari et al. 2006). Carriers are different based on its functions and need. The materials used as carrier include zein nanomaterials for encapsulating flavor compounds and α-lactalbumin nanotubes from milk protein as vehicle for vitamins and enzymes. These carriers are treated as food grade as their origin is zein and milk protein (Cushen et al. 2012). Example of the food that benefitted by encapsulation is the fortified bread with microencapsulated fish oil for health benefits (Cushen et al. 2012).

Nanolaminate is a very thin food grade film (1–100 nm/layer) that consists of two or more layers of nanometer dimensions material that is physically or chemically bonded dimensions. The primary advantage of nanoleminates is formed an edible film. These edible films are used in large variety of foods such as meats, fruits, vegetables, baked goods, fries, chocolate, and candies (Morillon et al. 2002; Cagri et al. 2004; Cha and Chinnan 2004; Rhim 2004). These films protect foods products against moisture, gases, and lipid, and they can enhance the textural properties of foods and act as agent that carry the flavors, colors, antioxidants, nutrients, and antimicrobials. These nanolaminates are produced from polysaccharides, proteins, and lipids. Polysaccharide- and protein-based nanolaminates provide good barrier against oxygen and carbon dioxide, however they are not a good barricade against moisture. On contrary, lipid-based films are a good barrier against moisture but a weak barrier to gasses and they have feeble mechanical durability (Park 1999).

Along with the innovation, microbial biosensors have been developed for food quality assessment. A biosensor is an analytical instrument that translate the response with analytes into a measurable signal which is proportional to the concentration of analytes using both biological recognition element with a signal transducer (Mohanty and Kougianos 2006; Lei et al. 2006; D'Souza 2001; Chauhan et al. 2004). Microbial biosensor uses microorganisms which comprise various enzymes as the bioelement and these enzymes can generate a response to the analytes selectively and specifically (Dai and Choi 2013). Microbial biosensors have advantageous to the food industries as to ensure the product quality and when process controls are needed; it is a rapid and affordable method (Mello and Kubota 2002). These sensors are used to control the materials in order to regulate the fermentation process. Microbial biosensors have used for sensitive determination of ethanol because ethanol is greatly crucial and necessary in different fermentation process (Dai and Choi 2013). However, biosensors can also detect other compound including caffeine and mycotoxins.

Valach et al. (2009) developed a new microbial amperometric biosensor for the quantification of ethanol in flow injection analysis and a linear response to ethanol in the range of 10 μM–1.5 mM in 3 min. Babu et al. (2007) developed an amperometric biosensor by immobilizing *Pseudomonas alcaligenes* MTCC 5264 on a cellophane membrane for the identification of caffeine and achieved a linear response to caffeine in the range of 0.1–1 mg/mL in 3 min. In addition, contaminants can also be detected using microbial biosensors. Zearalenone family mycotoxins which can cause mycotoxicoses are a familiar contaminant in milk (Bryden 2007). Genetically modified *Saccharomyces cerevisiae* used as the bioelement of the sensors for the detection of Zearalenone family mycotoxins in milk to ensure the quality of the milk (Valimaa et al. 2010).

14.4 Market Analysis

Food industry has driven by innovation, competitiveness and most importantly profitability. Hence, the rapid development of new technologies provided of products with enhanced flavor, texture, taste, extended shelf lives, improved safety, traceability and competitive costs. Although nanotechnology is not a common practice in small and medium food enterprises, larger multinational food and beverages companies such as Kraft, Nestlé, Heinz, PepsiCola and Unilever (ObservatoryNano 2010). It has estimated that about 400 food companies are involved in nanotechnology research and development; however the figure is expected to reach more than 1000 companies in 2018. As reported by the market research consultancy Cientifica in 2007, there are 464 different entries associated with nanotechnologies application in food and food contact packaging and the nanotechnology incorporated food products are worth approximately US$5.8 billion (Ehnert 2017). In Japan, the market for nanotechnologies incorporated food is approximately one billion yen (US$11 million) in 2005 and predicted to increase rapidly to 150 billion yen (US$1.65 trillion) in 2020 (Science and Technology Committee 2009). Ergo, nanotechnology in food is growing rapidly into a global, multi-billion industries.

Nevertheless, it is necessary to note that the market can be affected by considerable larger regulatory uncertainty, negative consumer recognition, and investment fear. The negative aspects of these effects can build a higher barrier for nanotechnology industries. Due to the science and the practical approach on methods to determine the potential health risk of nanotechnology applications trying to forge ahead with the development of nanotechnology applications, it causes regulatory uncertainty. As the functionality and nanotechnology application is developed to evaluate the effect of the new technology in exposure and toxicity testing have to be executed. In a consequence of the delay in establishing for methods, regulatory has difficulties in analyzing the risks which then lead to irrational fear of unknown element we are considering to use in food (Canady 2017).

Besides, the immense gap between innovative concept of nanotechnology and market acceptance within the new technology is due to the profuse research funding for innovation in nanotechnology applications without considering the uprising drawback. Food sector is especially susceptible to consumer mistrust and the biggest safety concern for nanotechnology applications and its utilization in food by public opinion polls albeit this perception can very much differ based on perceived benefit. As the understanding of regulators, consumers, and investor of those risks can be different from the understanding of experts. There is confusion in the definition drafts and variations across regulation and boundaries for nanotechnology applications in product development strategy and their effects on risk communication that have to understand. Because of the consumer and investor doubt of the regulatory scope of novel matter, they are likely to dismiss the early adopters of nanotechnology (Canady 2017).

14.5 Conclusion

The inclusion of nanotechnologies in food, packaging and food technology is an ongoing research and many of food and beverages firms are actively exploring the potential application of nanotechnology in food sector. Promising results and applications are successfully obtained and applied in the areas of food packaging, food safety and food quality.

Nanotechnology can offer exclusive advantages on food products and packaging in many ways. Nanotechnology-based techniques and devices such as nanoencapsulation, nanoemulsions and nanobiosensor are being developed to aid in the preparation of food and to ensure the food quality control by providing a precise and inexpensive analysis method. The inclusion of nanomaterials into food packaging is expected to improve the characteristic of the packaging, enhance the barrier properties of the packaging materials and help in reducing waste and the usage of scarce raw materials.

However, the food industry is more conservative than other sectors where nanotechnology can be applied. Consumers are more aware and cautious about what's in their food, whether the ingredient will bring any harm or any adverse effect to health. Thus, the corporation of nanotechnology which considered new and unknown might cause disapproval and rejection. Further studies should be conducted on the potential risk of nanomaterials in food to abate the wariness of consumers and regulatory issues, must be addressed before industry utilization. The law maker should issue a law and regulation regarding the application of nanotechnology in food sector to ensure the safety of the incorporated food and to prevent any misused of nanotechnology. On that account, this breakthrough can grow steadily, multidisciplinary and multiple stakeholder development strategies are essential to resolve the issues of novelty and regulatory scope for specific products so that recognition and trust can be cultivated.

Acknowledgement This work was supported by grants from the Ministry of Education Malaysia, Fundamental Research Grant Scheme (FRGS) (No. FRGS/1/2014/SG05/UMS/02/4).

References

Adame D, Beall GW (2009) Direct measurement of the constrained polymer region in polyamide/clay nanocomposites and the implications for gas diffusion. Appl Clay Sci 42:545–552

Babu VRS, Patra S, Karanth NG, Kumar MA, Thakur MS (2007) Development of a biosensor for caffeine. Anal Chim Acta 582(2):329–334

Baeummer A (2004) Nanosensors identify pathogens in food. Food Technol 58:51–55

Baruah S, Dutta J (2009) Nanotechnology applications in pollution sensing and degradation in agriculture: a review. Environ Chem Lett 7:191–204

Bhattacharya S, Jang J, Yang L, Akin D, Bashir R (2007) Biomems and nanotechnology-based approaches for rapid detection of biological entities. J Rapid Methods Auto Microb 15:1–32

Bryden WL (2007) Mycotoxins in the food chain: human health implications. Asia Pac J Clin Nutr 16(1):95–101

Bücking M, Hengse A, Grüger H, Schulte H (2017) Smart systems for food quality and safety. In: Monique AVA, de Voorde MV (eds) Nanotechnology in agriculture and food science. Wiley-VCH, Oxford

Cagri A, Ustunol Z, Ryser ET (2004) Antimicrobial edible films and coatings. J Food Prot 67:833–848

Canady R (2017) Potential benefits and market drivers for nanotechnology in the food sector. In: Chaudhry Q, Castle L, Watkins R (eds) Nanotechnologies in food. The Royal Society of Chemistry, London

Canel C et al. (2006) Micro and nanotechnologies for food safety and quality applications. In: MNE'06 Micro-and nano-engineering, 5C-3INV Microsystems and their fabrication 2 proceedings, Barcelona, Spain, 17–20 Sept 2006

Cha DS, Chinnan MS (2004) Biopolymer-based antimicrobial packaging: review. Crit Rev Food Sci Nutr 44:223–237

Chaudhry Q, Scotter M, Blackburn J, Ross B, Boxall A, Castle L et al (2008) Applications and implications of nanotechnologies for the food sector. Food Addit Contam Part A Chem Anal Control Expo Risk Assess 25(3):241–258

Chauhan S, Rai V, Singh HB (2004) Biosensors. Resonance 9:33–44

Chung YL, Ansari S, Estevez L, Hayrapetyan S, Giannelis EP, Lai HM (2010) Preparation and properties of biodegradable starch-clay nanocomposites. Carbohydr Polym 79(2):391–396

Cushen M, Kerry J, Morris M, Cruz-Romero M, Cummins E (2012) Nanotechnologies in the food industry—recent developments, risk and regulation. Trends Food Sci Technol 24:30–46

D'Souza SF (2001) Microbial biosensors. Biosens Bioelectron 16(6):337–353

Dai C, Choi S (2013) Technology and applications of microbial biosensor. Open J Appl Biosens 2:83–93

de Azeredo HMC (2009) Nanocomposites for food packaging applications. Food Res Int 42:1240–1253

Dobrucka R (2014) Application of nanotechnology in food packaging. J Microbiol Biotechnol Food Sci 3:353–359

Ehnert T (2017) The EU and nanotechnologies: a critical analysis. Hart, Oxford

Emamifar A, Kadivar M, Shahedi N, Soleimanian-Zad S (2011) Effect of nanocomposite packaging containing Ag and ZnO on inactivation of *Lactobacillus plantarum* in orange juice. Food Control 22(3):408–413

Emerich DF, Thanos CG (2003) Nanotechnology and medicine. Expert Opin Biol Ther 3 (4):655–663

Entsar IR, Badawy MET, Stevens CV, Smagghe G, Walter S (2003) Chitosan as antimicrobial agent: application and mode of action. Biomacromolecules 4(6):1457–1465

Hall RH (2002) Biosensor technologies for detecting microbiological food borne hazards. Microbes Infect 4:425–432

Hawthorn J (1967) The organization of Quality Control. In: Herschedoerfer SM (ed) Quality control in food industry, vol 1. Academic, London

Hoet P, Bruske-Hohlfeld I, Salata O (2004) Nanoparticles—known and unknown health risks. J Nanobiotechnol 2(1):12

Jain KK (2008) The handbook of nanomedicine. Springer, New York

Kumar AP, Depan D, Tomer NS, Singh RP (2009) Nanoscale particles for polymer degradation and stabilization—trends and future perspectives. Prog Polym Sci 34(6):479–515

Kumar AP, Singh RP (2008) Biocomposites of cellulose reinforced starch: Improvement of properties by photo-induced crosslinking. Bioresour Technol 99(18):8803–8809

Kumar CSSR (2006) Nanomaterials for biosensors. Wiley-VCH, Weinheim

Lei Y, Chen W, Mulchandani A (2006) Microbial biosensors. Anal Chim Acta 568:200–210

Li X, Xing Y, Jiang Y, Ding Y, Li W (2009) Antimicrobial activities of ZnO powder coated PVC film to inactivate food pathogens. Int J Food Sci Technol 44(11):2161–2168

Llorens A, Lloret E, Picouet PA, Trbojevich R, Fernandez A (2012) Metallic based micro and nanocomposites in food contact materials and active food packaging. Trends Food Sci Technol 24:19–29

Mabeck JT, Malliaras GG (2006) Chemical and biological sensors based on organic thin-film transistors. Anal Bioanal Chem 384(2):343–353

Maki DG, Tambyah PA (2001) Engineering out the risk for infection with urinary catheters. Emerg Infect Dis 7(2):342–347

Mello LDL, Kubota T (2002) Review of the use of bio-sensors as analytical tools in the food and drink industries. Food Chem 77(2):237–256

Mohanty SP, Kougianos E (2006) Biosensosrs: a tutorial review. IEEE Potentials 25(2):35–40

Morillon V, Debeaufort F, Blond G, Capelle M, Voilley A (2002) Factors affecting the moisture permeability of lipid-based edible films: a review. Crit Rev Food Sci Nutr 42:67–89

Mozafari RM, Flanagan J, Matia-Merino L, Awati A, Omri A, Suntres EZ et al (2006) Recent trends in the lipid-based nanoencapsulation of antioxidants and their role in foods. J Sci Food Agric 86:2038–2045

O'Brien N, Cummins E (2008) Recent developments in nanotechnology and risk assessment strategies for addressing public and environmental health concerns. Hum Ecol Risk Assess 14 (3):568–592

ObservatoryNano (2010) Agrifood sector: report on economic impact of nanotechnologies (FP7 Report, April 2010) 9

Paisoonsin S, Pornsunthorntawee O, Rujiravanit R (2013) Preparation and characterization of ZnO-deposited DBD plasma-treated PP packaging film with antibacterial activities. Appl Surf Sci 273:824–835

Pandey JK, Lee JW, Chu WS, Kim CS, Ahn SH (2008) Cellulose nano whiskers from grass of Korea. Macromol Res 16(5):396–398

Park HJ (1999) Development of advanced edible coatings for fruits. Trends Food Sci Technol 10:254–260

Rabea EI, Badawy ME, Stevens CV, Smagghe G, Steurbaut W (2003) Chitosan as antimicrobial agent: applications and mode of action. Biomacromolecules 4(6):1457–1465

Ramirez Frometa N (2006) Cantilever biosensors. Biotechnol Apl 23:320–323

Ravichandran R (2010) Nanotechnology applications in food and food processing: Innovative green approaches, opportunities and uncertainties for global market. Int J Green Nanotechnol Phys Chem 1:72–96

Rhim JW (2004) Increase in water vapor barrier property of biopolymer-based edible films and coatings by compositing with lipid materials. Food Sci Biotechnol 13:528–535

Rhim JW, Ng PKW (2007) Natural biopolymer-based nanocomposite films for packaging applications. Crit Rev Food Sci Nutr 47(4):411–433

Rhim JW, Park HM, Ha CS (2013) Bio-nanocomposites for food packaging applications. Prog Polym Sci 38(10):1629–1652

Ritter SK (2005) An eye on food. Chem Eng News 83:28–34

Science and Technology Committee (2009) Great Britain Parliament, House of Lords, Nanotechnologies and Food: Report. Authority of House of Lord, London

Serrano E, Rus G, García-Martínez J (2009) Nanotechnology for sustainable energy. Renew Sustain Energy Rev 13:2373–2384

Sinha Ray S, Okamoto M (2003) Polymer/layered silicate nanocomposites: a review from preparation to processing. Prog Polym Sci 28(11):1539–1641

Sonneville-Aubrun O, Simonnet JT, L'Alloret F (2004) Nanoemulsions: a new vehicle for skincare products. Adv Colloid Interface Sci 108-109:145–149

Sorrentino A et al (2007) Potential perspectives of bionanocomposites for food packaging applications. Trends Food Sci Technol 18:84–95

Sozer N, Kokini JL (2009) Nanotechnology and its applications in the food industry. Trends Biotechnol 27:82–89

Thomas V, Yallapu MM, Sreedhar B, Bajpai SK (2007) A versatile strategy to fabricate hydrogel-silver nanocomposites and investigation of their antimicrobial activity. J Colloid Interface Sci 315(1):389–395

Valach M, Katrlik J, Sturdik E, Gemeiner P (2009) Ethanol *Gluconobacter* biosensor designed for flow injection analysis: application in ethanol fermentation off-line monitoring. Sens Actuators B 138(2):581–586

Valimaa A, Kivisto AT, Leskinen PI, Karp MT (2010) A novel biosensor for the detection of zearalenone family mycotoxins in milk. J Microbiol Methods 80(1):44–48

Vo-Dinh T, Cullum BM, Stokes DL (2001) Nanosensors and biochips: frontiers in biomolecular diagnostics. Sens Actuators B Chem 74:2–11

Wu T, Zivanovic S, Draughon FA, Conway WS, Sams CE (2005) Physicochemical properties and bioactivity of fungal chitin and chitosan. J Agric Food Chem 53(10):3888–3894

Yu D, Cai R, Liu Z (2004) Studies on the photodegradation of Rhodamine dyes on nanometer-sized zinc oxide. Spectrochim Acta A Mol Biomol Spectrosc 60(7):1617–1624

Chapter 15
Nanotechnology Applications in Food: Opportunities and Challenges in Food Industry

Afiqah Silon Ummi and Shafiquzzaman Siddiquee

15.1 Introduction

The food and beverage sectors are global multi trillion dollar industry. All the larger food companies are constantly finding ways to enhance the efficiency of production, food safety and also food appearances. Todate continue research and development (R&D) are conducted to the main purposes of gaining competitive advantage and wide market share. As one of the industries where competition is intense and new inventions is vital, nanotechnologies have emerged as a potential assist to improve in the production of quality food with functionalized properties (Cushen et al. 2012). The opportunities of nanotechnology applications in food increased day by day, the challenges that need to be faced; it also cannot be ignored. The identification of emerging opportunities and challenges for the food industry are discussed in below.

15.2 Opportunities of Nanotechnology in Food Industry

Food related applications of nanotechnology mostly offer a wide range of benefit to the consumer and also the opportunities for the future of nanotechnology in food industry. Nanotechnology is expected to impacts in the diverse areas of food science so it will benefit both the food industry and consumers. There are numbers of potential benefits arising from the application of nanotechnology in food, which make it of real relevance for developing countries, as well as for developed nations.

For examples, this technology is used to increase the safety and quality of foods. Other application is the nano-derived innovations of food packaging that is main aim

A. S. Ummi · S. Siddiquee (✉)
Biotechnology Research Institute, Universiti Malaysia Sabah, Kota Kinabalu, Sabah, Malaysia
e-mail: shafiqpab@ums.edu.my

S. Siddiquee et al. (eds.), *Nanotechnology: Applications in Energy, Drug and Food*,
https://doi.org/10.1007/978-3-319-99602-8_15

to improve food safety and reduce food wastes (Arora and Padua 2010). Nanomaterials at the other side could be developed to enhance the delivery of nutrients and the pesticides to crops in which some experts predict might help developing countries (Chun 2009). Others are involving nano-formulations where it can improve the uptake, absorption and bioavailability of nutrients and supplements in the body as compared to bulk equivalents (Chaudry and Castle 2011).

15.2.1 Nanocapsules and Nanocarriers as Delivery System

Encapsulation can be defined as a technique to entrap or capture one substance within another substance which in real case; active agents within a carrier material. The active agent that is encapsulated also well-known as core material, fill, internal phase, or payload phase while the other name for the carrier material is coating, membrane, shell, wall material, external phase, or matrix (Zuidam and Shimoni 2010). Materials used for design of protective shell of encapsulate in food products or processes should be food-grade, biodegradable and able to perform a barricade between the core material and its surroundings.

15.2.1.1 Method of Encapsulation

Technology used for encapsulation process including spray-drying, fluid bed coating, spray-chilling/cooling, melt injection, melt extrusion, emulsification, coacervation, co-extrusion, inclusion complexation, liposome entrapment, rapid expansion of supercritical fluid (RESS), and freeze- or vacuum drying (Zuidam and Shimoni 2010). Among all methods, spray drying is the oldest and most widely used technique for encapsulation in food industry due to its flexibility as well as more economical. This method emits high quality of particles with size less than 40 μm which is useful featured from sensorial and textural characteristics of final products. However, the saying goes "nothing is perfect", same goes to this method. It does have its own flaws like non-uniform condition in the drying chamber, complexity of the equipment and inability to always control particle size.

15.2.1.2 Current Application of Encapsulation in Food

The most significant reasons for food encapsulation are provided to improve stability in final products and during food processing. For instance, the presence of live microorganisms and bioactive food components are known as probiotic bacteria have a very great benefit to health if they are present in sufficient amounts. However, this microorganism is very sensitive towards many things such as pH, mechanical stress, transport conditions, and digestive enzymes in the stomach. They need to survive the food process, storage, and food intake before can be deemed as useful and beneficial. Therefore, encapsulations can help them to survive not only their bioavailability but most importantly their functionality (Kim et al. 1996). Besides, encapsulation also help in lessens the evaporation and degradation of volatile actives like aroma which usually contains mixture of volatile and odorous organic molecules. Food manufactures have serious concerned about the preservation of aromatic additives since the flavors is usually costly (Milanovic et al. 2010). It helps to cover a food compound like aroma with a protective wall material and protected against evaporation, chemical reactions or even migration in a food (Madene et al. 2006).

Another application of encapsulation can be used to mask or cover the unpleasant feelings during eating, such as bitter taste and astringency of polyphenols and other compounds that yield high antioxidant activities. Encapsulated polyphenols instead of free compounds can lead to the improvements of the stability and bioavailability of the compounds in vivo and in vitro. This technology also allows additional nutrients to be added into food and beverage products without altering its flavor or quality. The delivery of particular ingredients and additive to a targeted site within the body is also possible, thus providing consumers with additional health benefits (Handford et al. 2014). Some food can be derived to be used further for encapsulation purposes such as alcoholic lecithin and sodium caseinate. These two food-derived substances are reduced to nanoscale before added to chitosan for use as nano capsules in the food industry (Sato et al. 2011).

Furthermore, when nanoencapsulation has applied in food additives and supplements, it can increase disperse ability of fat soluble additives in food products, improve food tastes, enable hygienic food storage, less use of fat, salt, sugar and preservatives. It also improves the bioavailability of nutrients and supplements. At present, the available examples including vitamins, antioxidants, colors, flavors and preservatives. Other technology developed for use in food products are nanosized carrier systems for nutrients and supplements. The nanocarrier system is used for taste making of certain ingredients and additives, or to protect them from degradation during processing (Chaudry and Castle 2011).

15.2.2 Nanopackaging

Main role of food packaging is to protect the food from its surrounding. Something covered is much more appealing to consumer compare to exposed one. Other function of packaging is providing nutritional information for consumers, extend food shelf life, and increase the quality of food. Over $860 million sales worldwide are produced by nanopackaging in 2006 and are likely to be a $30 billion market within the next 10 years (Coles et al. 2003). Usage of nanoparticles can help to improve the mechanical and heat resistance properties of food packaging. It leads to the increasing of shelf life, through by affecting gas or water vapor permeability. For instance, ordinary polymers are impermeable towards gases or water vapor, but polymer silicate nanocomposites have the ability to improve gas barrier, mechanical strength, and the heat resistance properties of food packaging (Holley 2005). It also offers new lightweight but much stronger food packaging materials that can keep secure food products during transportation, fresh for longer during storage and safe from microbial pathogens (Chaudry and Castle 2011).

One example of packaging material is composed of potato starch and calcium carbonate. This foam can be used to replace polystyrene used for fast food since it is thermal stabile and biodegradable too. Innovative packaging that integrated with antimicrobial properties are possible in prolong the shelf life of food products. These applications propose massive potential to the food industry by allowing the reduction in food waste, as well as a better quality and safer food supply (Handford et al. 2014). Furthermore, food contained in nanopackaging with antimicrobial ability may need less refrigeration during transport and it will be able to stay fresh over long journey (Rhim et al. 2013)

Nanolaminates consist of two (2) or more layer of nanomaterials (physically or chemically bonded to each other). It used as edible coatings and films as well as foaming which are applied in the food industry such as fruits, vegetables, meats, chocolate, candies, bakery products, and also French fries. These edible coatings act as barriers to moisture, lipid, or gases and increase the textural properties of foods or applied as carriers of functional agents including colors, flavors, antioxidants, nutrients and antimicrobials (Morillon et al. 2002; Rhim 2004). To date, nanolaminates are used as coating materials on food surface because of their intensely thin nature that make them very fragile (Kotov 2003).

Another nanomaterials used for packaging is nanoclay. Nanoclay is composed of fine-grained minerals of naturally occurring aluminium silicate having a layered sheet like geometry with sheet thicknesses <100 nm. One type of nanoclay is nanoscale montmorillinite (MMT) used in polymers with the purpose to improve the gas barrier properties, but it turns out the compound also has the ability to improve polymeric strength, heat resistance and thermal stability. At present, both industry and academia have high interested in nanoclays for used in food packaging to overcome these long standing issues (Majeed et al. 2013).

15.2.3 Nanotubes

Nanotubes are buckballs that have been bond two sides with additional atom groups added in the characteristic hexagon shape to form a hollow carbon tube (Scott 2005). Due to its thermal resistance and a strong and flexible structure, it applied in industrial food processing equipment. For example, carbon nanotubes are applied for crystallization of proteins and building of bioreactors and biosensors (Huang et al. 2000). Another application of nanotubes membranes can be used in food systems for the analytical purposes like molecular detection (enzymes, antibodies, proteins and DNA) as well as membrane separation of biomolecules (proteins, peptides, vitamins or minerals) (Lee and Martin 2002).

15.2.4 Nanosensors

Biosensor can be used for detecting gases, pathogens or toxins in package foods. However, using nanobiosensor revealed to be used not only for detection of pathogens in processing plants but also alerting consumers, procedures and distributors on the food safety status (Cheng et al. 2006). It may help in reducing foodborne illnesses in consumers (Handford et al. 2014). Other uses of nanosensor as mention by Rivas et al. (2006) is an electrochemical glucose biosensor that nanofabricates by layer-by-layer self-assembly of polyelectrolyte for detection and quantification of glucose. Liposome nanovesicle also employed for the purpose of detecting peanut allergenic proteins in chocolate (Wen et al. 2005). Similar, universal protein G-liposomal nanovesicles and an immounomagnetic bead sandwich assay are able to detect *E. coli* O157:H7, *Salmonella* spp. and *Listeria monocytohenes* simultaneously (Chen et al. 2006). Furthermore, zinc oxide nanowires are used in gas sensor for electronic nose. Electronic nose is a device designed for identifying different types of odors. Not only limited that, nanosensors also developed for detection of microorganisms and have the potential ability to differentiate the presence of pathogenic from benign microorganisms (Brody 2010). In a retail outlet, nanosensor labels on packaging allow the consumer to know if the products have been mistreated like subjected to temperature abuse during transport or if it contains an unsafe level of bacteria. When the consumer has finished with the product, waste packaging could be placed in a compost bin where it would degrade into eco-friendly constituents over a short period of time (Hannon et al. 2015)

15.2.5 Emerging Opportunities in the Agri-Food Sector

Nanotechnology is foreseen as a new technology that can be applied to assist the progress of the next development stage for genetically modified crops, input to animal production and fisheries, chemical pesticides and precision farming techniques. Precision farming is including as one of the most important techniques used for advancing mass production of crop by supervising environmental variables and applying appropriate action (Chen and Yada 2011).

The applications of nanotechnology in animal production improved efficacy and nutrition of animal feeds (e.g. fortified with nanosupplements, antimicrobial additives, detoxifying nanomaterials) and nanobiosensors for animal disease diagnostic. There is limited number of commercially available products integrating nanosized additives that specifically made for animal feed. The examples are mycotoxin binders such as nanoclay to protect animals from mycotoxicosis (Handford et al. 2014).

15.2.6 Others

Chaudry and Castle (2011) have overviewed the opportunities and challenges of nanotechnologies applications in food for developing countries. They found that the new advancement emerging from nanotechnology in food sector benefitted developing countries. The opportunities included offering more hygienic food/feed processing which lead to better food and feed safety, food quality and also reducing the risks of food borne illnesses as well. However, nanotechnologies can help to improve nutritional value of foods, it does contribute to having a more healthy and nutritious food.

Cushen et al. (2012) reviewed recent developments in nanotechnologies in the food sector, both in terms of the food matrix and in food-related industries such as food packaging and opportunities of nanotechnology. One of the applications is nanotechnology derived food ingredients. Organic components found naturally in food such as carbohydrate, protein and fat that can differ in size from large polymers to simpler molecules in the nano-range. These organic nanomaterials can be synthesized for significant intention such as the encapsulation of nutrients to enhance taste, texture and consistency of foodstuffs or else can be used to mask an undesirable taste or odor. However, the science behind the production of nano-derived food ingredients is still at early stage but it shows more promising with the anticipation of improving product functionality without ignoring the quality and safety aspect of the product.

A second application reviewed is emulsion stability. The small droplet size gives nanoemulsions distinct rheological and textural properties which contribute to the transparent and more favorable to the touch (Sonneville-Auburn et al. 2004). By applying nanoemulsions in food products, less fat will be used without a compromise in the creaminess properties, thus offering a healthier option to the consumer. Some of the products are involved low fat mayonnaise, spreads and also ice creams. At the same time, the use of nanoemulsions may able to decrease the need for certain stabilizers in a product. This application of nanoemulsions has a bright future in transforming the production of spreads and mayonnaise regardless it still in development stages.

Another nanotechnology application is nutraceuticals at the nanoscale. Nutraceutical is a compound like bioactive protein that usually present in functional food to improve health benefit to user in addition to the nutrition that the food itself offers. Nanomaterials can be used as bioactive in functional foods (Chau et al. 2007). By reducing the particle size of bioactive compounds to desire nanosize, it may improve the availability and delivery properties, and also solubility and biological activity of the compound. In addition, the applications of nanotechnologies are able to improve the stability of such micronutrients during processing, storage and distribution (Chen et al. 2006). Despite all the benefits, maintaining nutraceuticals in a stable state throughout the production process is undeniably challenging. For future reference, production of nutraceuticals at the nanoscale will be able to increase the stability throughout the processing chain hence it will be ultimately giving a great contribution to the consumer.

Nanosilver or nanoclay products are also being developed to enhance water filtration (Gruere et al. 2011). Surface functionalized nanomaterials are being developed that may add a certain functionality to food or packaging products. Current examples include the use of organically-modified nanoclays in food packaging applications. Due to the possible convergence of nanotechnologies with other technologies (e.g. biotechnology), the development of new functionalized nanomaterials is likely to grow in the future. Apart from, nanosilver is incorporated to destroy food pathogens and bacteria owing to its intense antimicrobial activity and low toxicity to mammalian cells and tissues. It considered as an important solution to overcome the growing problem of antibacterial resistance in the future (Alfadul and Elneshwy 2010).

It is well known that fresh-cut fruits will tend to undergo browning effect causing by the conversion of phenolic compounds into dark colored pigments in the presence of O_2. This is a very undesirable effect that can affect the marketing purposes. It has been reported that nano zinc oxide-coated active packaging are suitable as a viable alternative to common technologies for improving the shelf-life properties of "Fuji" apples as a fresh-cut product (He and Hwang 2016).

15.3 Challenges of Nanotechnology in Food Industry

Despite all the encouraging benefits and opportunities, nanotechnologies are still challenges that need to be faced as the current level of applications in food and related sectors globally. It becomes phenomenon due to its recently advantages, knowledge about the potential toxicological effects and impacts of nanotechnology have been receiving little attention but rapidly growing. This is due to a few numbers of researches that are still at R&D or near-market stages.

15.3.1 Technology Push and Market Pull

Major challenges that faced by the food sector is the ability to maintain its growth and move towards for commercialization. Every products or technologies are released to the market, it behind have countless number of researches by the expert researchers. In research, sometimes can take years to reach a conclusion, it does affect the investors due to the equipment cost as well as the uncertainty of results. Besides, the most companies are involved supporting technology-push rather than market-pull innovations, which is possible to cause challenges and risk in the long term. So a new invention technology should be entering the market by proper research and development (R&D), production as well as sales, however, it is completed without considering whether it is satisfies a user need or not (Gruere 2012)

15.3.2 Potential Risk to Health and Environment

While nano-sized material grants lots of potential benefit in food sector, its potential risk also manage to spark an interest not only to researcher but also to the public especially the consumer. In the food area, the questions arise on health risk and greater exposure on each stage of consumption (digestion and absorption by the intestinal tract), and also the effect and the chances of accumulation into the body. One of the concerns is the possibility of some insoluble and bio persistent nanoparticles (known as "hard" nanomaterial) to exposed to consumer *via* consumption of food and beverages. The matter of concern is when the nanoparticles get inside the body; it may cross biological barriers and somehow reach those parts of the body which should be supposedly protected from entry of particulate material (Chaudry and Castle 2011). As mentioned by Cushen et al. (2012), exposure to nanomaterials application in food can pass via three main routes; dermal contact, inhalation and ingestion. Routes via ingestion is become main concern in food industry. Even though there are no concrete results obtained from this kind of exposure due to lack of data. It is still worth noting that there may be risks to the consumer in the form of migrating particulate nanomaterials from food packaging into the foods.

Simon et al. (2008) showed that the rate of migration of a system increase with decreasing polymer dynamic viscosity and nanoparticle size. This indicates that there is a probability for migration of nanoparticulate materials from packaging material and the urge for quantitative risk assessment to be conducted. Nanoparticles of clay or nanoclay are highly reactive due to their larger surface area. This arise concern on its possibility to form more toxic of clay particles during use or production. Despite numerous of food packaging materials containing nanoparticles are already commercially available in few countries, the migration studies are currently limited. Several migration experimental and modelling studies have been designed so far, hence it main propose the likelihood of nanoparticles migration from polymer packaging to be either very low or nil, thus do not pose any significant risk to the consumer (Chaudry and Castle 2011).

Study conducted by Avella et al. (2005) proposed that there are insignificant mineral increases in vegetable packaged in a nanocomposite that derived from a biopolymer with integrated montmorillonite (nanoclay). There is a small amount of Silicon detected in the food samples, considering it as the main component of nanoclay. Even though it is tiny amount but it stills a proof of migration that can happen from packaging to foods. Since the transport mechanism of montmorillonite is remained an open question, considering the fact that the presence of nanoparticles in food could result distinctly from the release of the whole particles or from the post-migration assembly of ionic forms. Apart from the nature of food (solid/liquid, fatty/aqueous/acid, etc.) and conditions (time, temperature), the emanation of new processing technologies such as high-pressure, pulsed electric fields and advanced heating can influence the potential release of montmorillonite (Mauricio-Iglesias et al. 2010).

Huang et al. (2011) conducted four different kinds of foods simulating solutions that put into a commercially available food storage bag containing silver (Ag) nanoparticles. The result showed that there are indeed migrations of Ag nanoparticles from the polyethylene bags into the food simulating solutions. The main worst part is that the amount of migration increased with storage time and temperature. This result may have negative impacts on the progress of such packaging materials.

Following the trends in packaging industry nowadays, they use biodegradable packaging due to its positive effect on the environment. However, their intention to applying nanotechnologies for producing bionanocomposites packaging raises the issues of incidental environmental contamination as a result of nanomaterials being released following polymer degradation. To overcome this issue, it is better for the industry to conduct an ecotoxicity test in order to determine the risks caused by nanomaterials to the environment. The results should be revealed in an understandable manner on packaging to enable the consumer to make an informed choice.

The purpose of applying nanoingredients in foods and health supplement has improved the functionality and/or nutritional value of a food. However, there is still probability of such nanoparticles to forming compounds with other food materials; either interaction with each other or remain still in a free state while in the alimentary canal. The mechanism is how it will affect absorption remaining unknown; however this should alert the responsible committee to revise the Recommended Daily Allowances (RDA) for the increase use of such nano forms.

The toxicology and safety aspects of nanoparticles are reviewed by several researchers (Bouwmeester et al. 2009; Bradley et al. 2011; Cushen et al. 2012) mentioned that the association of nanomaterials into food may cause an entire new array of risks for the consumers. The applications of nanotechnology in food sector related to human health is the increased number of toxicological effects of nanoparticles at smaller concentrations causing by greater surface area, increased toxicity owing to improved bioavailability, higher access to the human body, compromised immune system response, and possible longer pathological effects (Momin et al. 2013).

15.3.3 Public Perception of Nanotechnology

Public perception of the various applications of nanotechnologies is a major factor that determines the commercial success of certain field. Same goes to its application in food industry. Thus consumers' attitudes are particularly sensitive when it comes to the foods and beverages they consume. Truly as the saying goes by, 'knowledge is power', a comparison of two U.S. surveys from 2004 (Scheufele and Lewenstein 2005) and 2007 (Scheufele et al. 2007) proven that respondents with higher education have a better knowledge about the issue of nanotechnology compared to the least educated respondents. The gaps in knowledge lead to several perceptions from the public, either good or bad. These gaps are mainly related to knowledge of the applications as well as consumer and environmental safety, which are impeding regulation and market update, and thus require further research.

For instance, Siegrist et al. (2008) have conducted the experiment on potential public responses among 337 respondents to perceived risks and perceived benefits of different nanotechnology foods and nanotechnology food packaging. The results suggested that participants perceived nanotechnology foods differently than nanotechnology packaging applications. Applications are not consumed by humans or animals which then referred as nano-outside are chosen as the least risky but high benefit compared to the applications that are directly consumed by humans or animals (nano-inside). The examples of nano-outside applications are UV-protection packaging and antibacterial food container while for nano-inside applications including nutritional supplement capsules and bread with higher nutritional value. In addition, the results indicated that respondents for whom naturalness of food is important perceived more risks associated with nanotechnology compared with respondents for whom naturalness of foods is less important. Another factor that affects the acceptability of nanotechnology application in food is trust which in this case, trusts in the food industry and retailers as well as trust in food science and consumer organizations.

Lacking of a clear and uniform definition of nanomaterials and nanotechnologies is lead to the existence of scientific gaps in knowledge. Efforts are currently underway to implement a more comprehensive international definition that can be understood by everyone. In addition, young age should be improved the awareness of nanotechnology, the authorities should include nano-eduacation programmes commencing at the school level too (Handford et al. 2014)

Without clear information, peoples may get the wrong perception about nanotechnology. This become one of the challenges issue to be used at maximum level first, they must be well received by consumer. Todate, media is playing a very important role as primary sources to deliver clear information to the public. It is including acknowledge of both benefits and risks of nanotechnology in food application. The idea is to notify the audiences of this technology not only does the benefits outweigh the risks, but at the same time informing the risks acceptable too. For novel issues like food nanotechnology, media can play an important role in shaping the awareness and mental associations that underlie public opinion (Cushen et al. 2012).

15.3.4 Requirements for a Risk Assessment

In order to determine toxicity profiles, nanoparticles in specific size, shape, solubility, reactivity, properties of the substance in a non-nano form and other physicochemical parameters must take into account. The toxicological properties are differ among particulate nanomaterials, and then a risk assessment should be conducted on a case by case basis (Munro et al. 2009).

The Scientific Committee under European Food Safety Authority (EFSA) has identified two specific hurdles in performing risk assessments on nanomaterials which are difficulty in characterizing, detecting and measuring nanomaterials and insufficient information on toxicology data (EFSA 2009). The fundamental requirement needed for designing risk assessment is the characterization of nanomaterials. Without a precise description on certain material, there is no way to quantify the risk (Cushen et al. 2012). For a risk assessment to be done on any exposure route related to the food industry, more results for the ingestion of nanomaterials must be reported. To reassure the consumer on the safety of using nanotechnology application in their food, standard analytical methods need to be established to detect the presence of nanomaterials in foods and also in food contact materials (FCMs).

A main issue that brought up in collective reports, it is the low scientific knowledge on key factors that needed for risk assessment, such as the issue of bioaccumulation, toxicity of nanoparticles, oral exposure or even the lack of study on risks by ingestion (Bouwmeester et al. 2007). Numerous studies has been done *in vitro* or *in vivo* but without oral exposure and lot of toxicological studies are conducted without following proper protocols provided, lead to their questionable results (Reich 2011).

15.4 Conclusion

Nanotechnology offers promising future in the agriculture and food industry where it has massive useful applications, but not to forget there is a lack of awareness and uncertainty about the balance between potential benefits on side and potential risks and possible long-term side effect on the other side. All the emerging opportunities of nanotechnology application in food industry not only benefitted to certain countries but also to the whole world. This technology is really helpful especially for developing countries. However, not everyone can adapt to changes so easily. Same goes to the idea applying nanotechnology in food sector. The gaps in scientific knowledge really affect the consumer perception on whether to accept or reject this new technology.

In the future, research aim to ensure that a thorough nanoscale hazard and risk analysis is conducted before nanotechnology exploited by industry and use by consumers. Nano should not be commercialized and marketed in food and agriculture until nano-specific safety guidelines and laws have been issued, nano-specific standard specifications have been agreed, and nano-specific regulatory issues for food and environment technology have established with public input to decision-making. Better consumer awareness on nanoscience and nanotechnologies is equally vital.

Acknowledgement This work was supported by grants from the Ministry of Education Malaysia, Fundamental Research Grant Scheme (FRGS) (No. FRGS/1/2014/SG05/UMS/02/4).

References

Alfadul SM, Elneshwy AA (2010) Use of nanotechnology in food processing, packaging and safety—review. Afr J Food Agric Nutr Dev 10(6):2719–2739

Arora A, Padua GW (2010) Review—nanocomposites in food packaging. J Food Sci 75(1):43–49

Avella M, De-Vlieger JJ, Errico ME, Fischer S, Vacca P, Volpe MG et al (2005) Biodegradable starch/nanocomposite films for food packaging applications. Food Chem 93(3):467–474

Bouwmeester H, Dekkers S, Noordam M, Hagens W, Bulder A, de Heer C, ten Voorde S, Wijnhoven S, Sips A et al (2007) Health impact of nanotechnologies in food production. Wageningen, RIKILT-Institute of Food Safety

Bouwmeester H, Dekkers S, Noordam MY, Hagens WI, Bulder AS, de Heer C et al (2009) Review of health safety aspects of nanotechnologies in food production. Regul Toxicol Pharm 53 (1):52–62

Bradley EL, Castle L, Chaudry Q et al (2011) Applications of nanomaterials in food packaging with a consideration of opportunities for developing countries. Trends Food Sci Technol 22 (11):604–610

Brody AL (2010) Packaging and the science of tiny. Food Technol 64(2):74–76

Chau CF, Wu SH, Yen GC et al (2007) The development of regulations for food nanotechnology. Trends Food Sci Technol 18(5):269–280

Chaudry Q, Castle L (2011) Food applications of nanotechnologies: an overview of opportunities and challenges for developing countries. Trends Food Sci Technol 22:595–603

Chen H, Weiss J, Shahidi F et al (2006) Nanotechnology in nutraceuticals and functional foods. J Food Tech. 60(3):30–36

Chen H, Yada R (2011) Nanotechnologies in agriculture: new tools for sustainable development. Trends Food Sci Technol 22(11):585–594

Cheng MMC, Cuda G, Bunimovich YL, Gaspari M, Heath JR, Hill HD, Mirkin CA, Nijdam AJ, Terracciano R, Thundat T, Ferrari M et al (2006) Nanotechnologies for biomolecular detection and medical diagnostics. Curr Opin Chem Biol 10:11–19

Chun AL (2009) Will the public swallow nano food? Nat Nanotechnol 4(12):790–791

Coles R, McDowell D, Kirwan MJ et al (eds) (2003) Food packaging technology. Blackwell, Oxford, p 346

Cushen M, Kerry J, Morris M, Cruz-Romero M, Cummins E et al (2012) Nanotechnologies in the food industry—recent developments, risks and regulation. Trends Food Sci Technol 24:30–46

European Food Safety Authority (EFSA) (2009) The potential risks arising from nanoscience and nanotechnologies on food and feed safety. Scientific Opinion of the Scientific Committee

Gruere G, Narrod C, Abbott L et al (2011) Agriculture, food and water nanotechnologies for the poor: opportunities, constraints and role of the CGIAR: IFPRI Discussion Paper 1064. International Food Policy Research Institute, Washington, DC

Gruere GP (2012) Implications of nanotechnology growth in food and agriculture in OECD countries. Food Policy 37:191–198

Handford CE, Dean M, Henchion M, Spence M, Elliott CT, Campbell K et al (2014) Implications of nanotechnology for the agri-food industry: opportunities, benefits and risks. Trends Food Sci Technol 40:226–241

Hannon JC, Kerry J, Cruz-Romero M, Morris M, Cummins E et al (2015) Advances and challenges for the use of engineered nanoparticles in food contact materials. Trends Food Sci Technol 43:43–62

He X, Hwang HM (2016) Nanotechnology in food science: functionality, applicability and safety assessment. J Food Drug Anal 24:671–681

Holley C (2005) Nanotechnology and packaging. Secure protection for the future. Verpack Rundsch 56:53–56

Huang W, Taylor S, Fu K, Lin Y, Zhang D, Hanks TW, Rao AM, Sun YP et al (2000) Attaching proteins to carbon nanotubes via diimideactivated amidation. Nano Lett 2:311–314

Huang Y, Chen S, Bing X, Gao C, Wang T, Yuan B et al (2011) Nanosilver migrated into food-simulating solutions from commercially available food fresh containers. Packag Tech Sci 24:291–297

Kim KI, Yoon YH, Baek YJ et al (1996) Effects of rehydration media and immobilization in Ca-alginate on the survival of Lactobacillus casei and Bifidobacterium bifidum. Korean J Dairy Sci 18:193–198

Kotov NA (2003) Layer-by-layer assembly of nanoparticles and nanocolloids: intermolecular interactions structure and materials perspective. Invited review to be published. In: Decher G, Schlenoff JB (eds) Thin films-polyeletrolyte multilayers and related multicomposites. Weinheim, Wiley-VCH, pp 207–243

Lee SB, Martin CR (2002) Electromodulated molecular transport in gold nanotube membranes. J Am Chem Soc 124:11850–11851

Madene A, Jacquot M, Scher J, Desobry S et al (2006) Aroma encapsulation and controlled release—a review. Int J Food Sci Technol 41:1–21

Majeed K, Jawaid M, Hassan A, Abu Bakar A, Abdul Khalil HPS, Salema AA et al (2013) Potential materials for food packaging from nanoclay/natural fibres filled hybrid composites. Mater Des 46:391–410

Mauricio-Iglesias M, Peyron S et al (2010) Wheat gluten nanocomposite films as food-contact materials: migration tests and impact of a novel food stabilization technology (high pressure). J Appl Polym Sci 116:2526–2535

Milanovic J, Manojlovic V, Levic S, Rajic N, Nedovic V, Bugarski B et al (2010) Microencapsulation of flavors in Carnauba wax. Sensors 10:901–912

Momin JK, Jayakumar C, Prajapati JB et al (2013) Potential of nanotechnology in functional foods. Emir J Food Agric 25(1):10–19

Morillon V, Debeaufort F, Blond G, Capelle M, Voilley A et al (2002) Factors affecting the moisture permeability of lipid-based edible films: A review. Crit Rev Food Sci Nutr 42:67–89

Munro IC, Haighton LA, Lynch BS, Tafazoli S et al (2009) Technological challenges of addressing new and more complex migrating products from novel food packaging materials. Food Addit Contam 26(12):1534–1546

Reich ES (2011) Nano rules fall foul of data gap. Nat 480(7376):160–161

Rivas GA, Miscoria SA, Desbrieres J, Berrera GD et al (2006) New biosensing platforms based on the layer-by-layer self-assembling polyelectrolytes on Nafion/carbon nanotubes-coated glassy carbon electrodes. Talanta 71(1):270–275

Rhim JW (2004) Increase in water vapor barrier property of biopolymerbased edible films and coatings by compositing with lipid materials. J Food Sci Biotechnol 13(4):528–535

Rhim JW, Park HM, Ha CS (2013) Bio-nanocomposites for food packaging applications. Prog Polym Sci 38(10-11):1629–1652

Sato ACK, Quintas MA, Vincente AA, Cunha RL (2011) Food grade nanoparticles obtained from natural source ingredients. ICEF 11th International Congress on Engineering and Food

Scheufele DA, Lewenstein BV (2005) The public and nanotechnology. How citizens make sense of emerging technologies. J Nanopart Res 7(6):659–667

Scheufele DA, Corley EA, Dunwoody S, Shih T, Hillback E, Guston DH et al (2007) Scientist worry about some risks more than the public. Nat Nanotechnol 2(12):732–734

Scott NR (2005) Nanotechnology and animal health. Rev Sci Tech Off Int Epiz 24(1):425–432

Siegrist M, Stampfli N, Kastenholz H, Keller C et al (2008) Perceived risks and perceived benefits of different nanotechnology foods and nanotechnology food packaging. Appetite 51:283–290

Simon P, Chaudry Q, Bakos D et al (2008) Migration of engineered nanoparticles from polymer packaging to food physicochemical view. J Food Nutr Res 47(3):105–113

Sonneville-Auburn O, Simonnet JT, L'Alloret F et al (2004) Nanoemulsions: a new vehicle for skincare products. Adv Colloid Interface Sci 108:145–149

Wen HW, Borejsza-Wysocki W, DeCory TR, Baeumner AJ, Durst RA et al (2005) A novel extraction method for peanut allergenic proteins in chocolate and their detection by liposome-based lateral flow assay. Eur Food Res Technol 221:564–569

Zuidam NJ, Shimoni E (2010) Overview of microencapsulates for use in food products or processes and methods to make them. In: Zuidam NJ, Nedovic V (eds) Encapsulation technologies for active food ingredients and food processing. Springer, New York. https://doi.org/10.1007/978-1-4419-1008-0_2

Chapter 16
Improvement of Food Packaging Based on Functional Nanomaterial

Bambang Kuswandi and Mehran Moradi

16.1 Introduction

Nanotechnology consists the development and fabrication as well as characterization of materials, or structures in length size of a nanometer (around 1–100 nm). If a macro-size particle is down-sized into nanometer size, the producing nanoparticle creates chemical and physical characteristics that significantly change their properties from the original macro-size particles. Hence, it looks like unbelievable that nano-structures could not only be developed and yet also has tremendous applications in many areas with essential implications on a human for a better life (Sozer and Kokini 2009; Ravichandran 2010). This technology offers tremendous improvements for a better life in many ways. It can be developed to create better products, wealth, health, and quality of life while at the same time and due to its nanometer-size, it could also reduce the impact on the environmental. Besides a tremendous development in many areas, nanotechnology in food packaging is still an infant. However, since the various of nanomaterial developed with functional properties can be used to improve materials for food packaging, hence even in its infancy, the nanomaterials are being developed and applied increasingly in the packaging industry for food nowadays.

Food packaging is those products that are manufactured with any kind of materials in which food is protected, packed, manipulated, distributed, transported and identified via the supply chain, from farm to consumers (Kuswandi 2017). These

B. Kuswandi (✉)
Chemo and Biosensors Group, Faculty of Pharmacy, University of Jember, Jember, East Java, Indonesia
e-mail: b_kuswandi.farmasi@unej.ac.id

M. Moradi
Department of Food Hygiene and Quality Control, Faculty of Veterinary Medicine, Urmia University, Urmia, West Azarbaijan, Iran

S. Siddiquee et al. (eds.), *Nanotechnology: Applications in Energy, Drug and Food*,
https://doi.org/10.1007/978-3-319-99602-8_16

main packaging functions are mandatory for any type of food packaging, however, based on the food types that have been packed and preserved, a broad range of characteristics such as barrier properties, mechanical, and thermal, are also required. Since the wide range of nanomaterials with their functional properties can be used to improve materials of packaging, then, nanomaterials are intensively being applied in the industry of food packaging. It has been presented that approximately 500 foods with nanotechnology in food packaging are predicted to be in marketed products, while in the next decade, this technology is estimated to be employed in 25% of the fabrication of all packaging in food (Reynolds 2007). In nano-packaging for food, functional nanomaterial could also be developed to have properties which could modulate the release antioxidants, antimicrobials, flavors, and enzymes as well as nutraceuticals to improve shelf life and food overall quality (Cha and Chinnan 2004). Even more, novel and innovative materials-based nanotechnology for packaging were in line with some films for antimicrobial in order to extend the food shelf life, for example, products of dairy, which have already been commercially available (El Amin 2005).

The most advantages of nanotechnology for the food industry is the introducing of novel food packaging technology as reported in the literature (Dasgupta et al. 2015), where it has been described that applications of the nanotechnology in agriculture and food is one of the fastest growing fields in the study of nanotechnology. This includes current study trends in processing and packaging of food, functional food, and quality control as well delivery system for nutraceutical products. In this area, many researchers and innovators in universities or industries are developing novel methods, techniques, procedures, and products that used direct application of nanotechnology in the agro-food field (Dasgupta et al. 2015). For instance, using nanotechnology in food packaging, industries have been producing nanomaterials for food packaging, which are purposed to extend food shelf life and increasing safety of packaged food. In the industry, packaging of food and its monitoring, as well as their improvements, are a major concern that relates to research and development using nanotechnology (Brody 2003). Besides improved packaging materials using nanomaterials, in food packaging, the leading innovative development is active and intelligent packaging, since it could increase safety and quality of food as well as expand the food shelf life (Kuswandi et al. 2011, Kuswandi 2017). In the latter case, many organizations and companies are developing food packaging with the ability to alert if contaminant presence in food, via responses to an environmental conditions change using intelligent functions, and self-repair tears and leaks via an active function in packaging.

Nanotechnology was employed to make nanoparticles as a film that could increase the certain gases permeability via the films in order to refuse unpleased gas, such as CO_2 or oxygen, which could shorten the foods shelf life. Furthermore, by employing nano-film, it is also studying that this film could serve as barrier protection and unwanted gases, such as ethylene or oxygen from food spoilage (Duncan 2011; Silvestre et al. 2011). Currently, many pharmacy and food scientists are studying to use nanomaterial with the target to increase drug delivery, vitamins or micronutrients in the dairy products by producing edible nanoparticles as nano-capsules which able to deliver their contents in the human body at the targeted organs on demand. Hence, it will improve significantly on human health, by reducing risks of stroke, heart attack, cancer and neurodegenerative diseases (Koo et al. 2005; Yan and Gilbert 2004). Furthermore, nanoparticles may also be used to introduce multiple functionalities, such as color, odors and taste, and even more active as a container for the functions of drugs or fungicides with controlled release (Lee et al. 2003). Even though it is promising for tremendous results could be achieved, the way to success in the development of real multiple functionalities of nano packaging still need long development to go (Kuswandi 2017).

Commonly, in food packaging industries, the most widely used materials are plastic polymers which are non-biodegradable materials. Consequently, these non-degradable plastics represent a serious threat to human and the environment (Kirwan and Strawbridge 2003). Hence, the biomaterials use in food packaging, such as edible and biodegradable films from regenerable materials are critically important (Tharanathan 2003). Since biomaterials could for some reason solve the serious problem of waste by reducing packaging materials, or even could also prolong the shelf life as well as extending safety and quality of food. In this regard, by proper biomaterials selection and packaging technologies, it could able to maintain the product freshness and safety during the distribution, storage, as well as consumption period (Stewart et al. 2002; Kuswandi et al. 2011, Kuswandi 2017). However, nowadays the bio-materials use in food packaging is still very limited. Since the natural polymers still have disadvantages to be used as food packagings, such as weak mechanical and poor barrier properties. Hence, the biomaterials were often composted or blended with other synthetic polymers or chemically modified materials with the target to extend use as food packaging materials (Petersen et al. 1999).

Similar to plastic packaging, biomaterials packaging must also provide several major functions as packaging, such as food protection, preserve overall sensorial characteristics and safety, as well as provide information or data to users (Robertson 1993). Further nanotechnology that could be used to reduce waste of the food packaging is based on nanocomposites materials. The use of nanocomposites could also increase the application of edible and biodegradable film and coating in food packaging (Lagaron et al. 2005; Sinha Ray and Bousmina 2005). It also will support the freshness and extend the food shelf life (Vermeiren et al. 1999). The way to go for successful usage of bio-nanocomposites is still required an accurate and comprehensive study (Sorrentino et al. 2007).

Besides a terrific improvement in this area, it is still in its infancy in the nanotechnology spectrum. This chapter focussed to address this knowledge gap by presenting a current advance review in the development of food packaging based on functional nanomaterial, targeting particularly on the development of nano-packaging that could fulfill user preferences and authority bodies as given in Fig. 16.1. Covered issues include nanomaterial in food packaging, physical improved packaging for increasing mechanical strength, barrier properties, flexibility and stability, biochemical improved packaging for biocompatible, biodegradable, low-waste and eco-friendly, improved packaging with active functions, such as

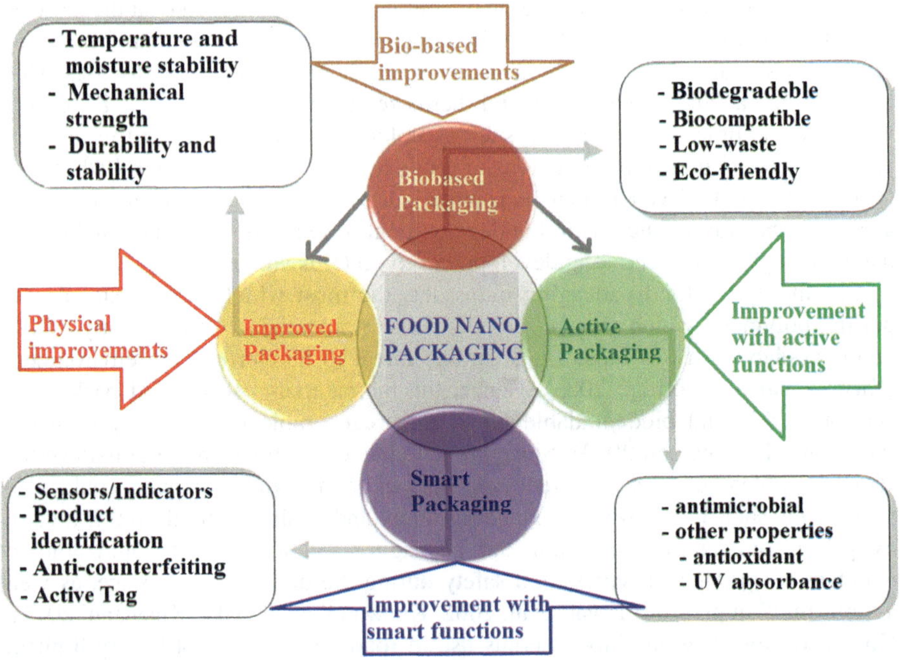

Fig. 16.1 Principles of improvement of food packaging based on functional nanomaterial that can be divided into several improvements including: physic-chemical improved packaging, such as temperature and moisture stability, mechanical strength, durability and flexibility; bio-based improved packaging for biodegradable, biocompatible, low-waste and eco-friendly; improved packaging with active functions, such as antimicrobial and other properties, e.g. antioxidant, UV absorbance etc.; and improved packaging with smart or intelligent functions such as nanosensors for the detection of food relevant analytes (gasses and small organic molecules), active stage and product identification and anticounterfeiting (Adapted from Kuswandi 2017)

oxygen scavenging and antimicrobials, improved packaging with smart or intelligent functions, e.g. nanosensors for freshness, contaminants and monitoring of food packaging integrity or conditions. Concerns about human health and environment related to these various biomaterials of food packaging with functional nanoparticles are discussed briefly. The chapter ended with a short overview of the future trend of nano-packaging for food.

16.1.1 Food Packaging Based on Nanomaterial

Generally, the application of nanomaterial used in food packaging can be classified into main two classes: (1) nano-object materials and (2) nanostructured materials, as given Fig. 16.2. In the nano-object materials, commonly they employed as a filler or nano-reinforcement that incorporates the application of nanoparticles, nanoplate,

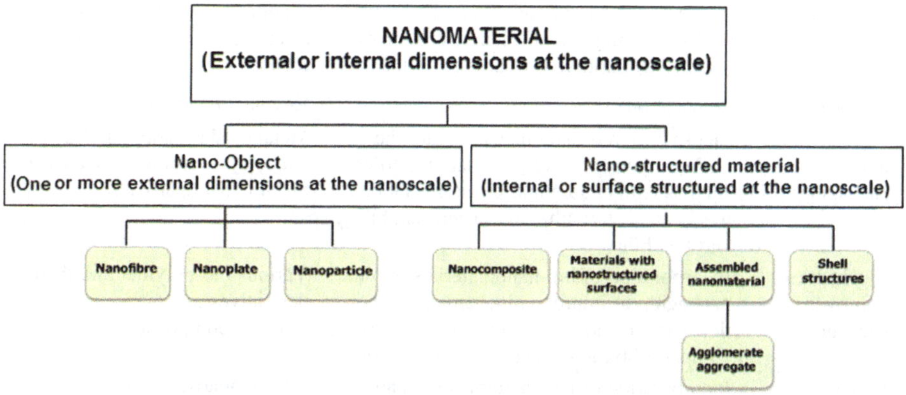

Fig. 16.2 The terminology of functional nanomaterial used in the food packaging: (a) nano-object and (b) nano-structured material. Nano-object consist of nanofibre, nanoplate, and nanoparticle to increase reinforcements, while nano-structured material consists of the nanocomposite, materials with nanostructured surfaces, assembled nanomaterial and shell structure

and nanofibers, e.g. nanoclays, metal oxides nanoparticles, carbon nanotubes and other fillers (metallic nanoparticles). In the case of the nanostructured materials, the nanomaterials are incorporated into a matrix of the polymer by dispersing it in the form of nanocomposites.

In the case of the nanomaterial used for the improvement of food packaging, the approaches can be classified mainly into four types of improvements: (1) packaging with physical improvement, i.e. gas barrier properties that pertaining to unwanted gases such as oxygen, and carbon dioxide, as well as ultraviolet rays. Moreover it is also improved in strengthness, stiffness, heat resistance, and dimensional stability; (2) biochemical improved packaging, i.e. biodegradation, edible, biocompatibility and low-waste as well as eco-friendly (3) improved packaging with active functions, i.e. antimicrobial or other functional compounds, such as UV absorbance, antioxidant, that could also improve the packaged food, including freshness, taste, and odor as well as self-life; (4) smart/intelligent functions improved packaging, e.g. nanosensors for detection of freshness, oxygen, and pathogen. These major used of nanomaterials in packaging as food nano-packaging also became trends in the development of nanotechnology in food packaging currently (Ranjan et al. 2014, Kuswandi 2017). In this context, the improvement of food packaging using nanomaterial mainly related with nano-reinforcements, bio-nanocomposites, nanocomposites and also in the case of active packaging, while in the case of smart packaging, the nanomaterials used as nanosensors. Table 16.1 shows a highlight of the promising applications of nanotechnology for improvements in packaging.

In physical improved packaging case, the nanoparticles incorporation in the packaging polymeric matrixes increases the physical characteristics of food packaging. For example, strength, barrier properties, heat resistance, stiffness, and dimensional stability of packaging materials can be increased by the addition of

Table 16.1 Improvement of food packaging based on functional nanomaterial can be classified as (a) physical improved packaging; (b) biochemical improved packaging, (c) improved packaging with active functions and (d) improved packaging with smart function

Applications	Descriptions	Key nanomaterials
Physical improved packaging	Incorporation of nanomaterials into the packaging to improve physical properties, such as temperature and moisture stability, mechanical strength, gas barrier, durability, and flexibility	Metal oxides nanoparticles, nanoclays, carbon nanotubes and metallic nanoparticles
Biochemical improved packaging	Incorporation of bio-nanomaterials into the packaging to improve biochemical properties, such as biodegradation, edible, biocompatibility, low-waste and eco-friendly	Various bio-polymer and their derivatives as bio-nanocomposites
Active packaging	Incorporation of nanomaterials with antimicrobial or other properties (e.g. antioxidant, UV absorbance) with intentional release into- and the consequent effect on the packaged food in term of taste, freshness and shelf life	Ag nanoparticles, gold nanoparticles, metal oxides nanoparticles
Smart packaging	Incorporation of nanosensors to monitor and report on the condition of the food (e.g. oxygen indicators, freshness indicators and pathogen)	High variability of nanosensors

nano-clays or SiO_2 nanoparticles. While in the biochemical improved packaging, the bio-nanomaterials incorporation into the packaging materials could increase biochemical properties, such as biodegradation, edible, biocompatibility, low-waste and eco-friendly. In the improved packaging case with active functions, nanoparticles were integrated into the packaging materials to act as active agents, for example, nanoparticles are being created especially for antimicrobial, where gold, silver, and metal oxide nanoparticles are the well-known nanoparticles to be developed as an antimicrobial agent. Here, AgNPs (silver nanoparticles) mainly applied as antimicrobials in much marketed food packaging. In addition, they can also be used as other active function, such as removers of ethylene in food packaging. While in improved packaging case with intelligent or smart functions, nanoparticles could act as a sensitive part of packaging materials to communicate and give information regarding the food packaged conditions. Commonly, this sensitive part is called nanosensors that are capable to give respond toward internal or external change to communicate, inform, and detect the food product in order to assay food safety and quality. The current innovation in nanomaterial for nanosensor as a smart function in food packaging include nanosensors for freshness, oxygen, carbon dioxide and even for the pathogen detection.

16.1.2 Physical Improved Packaging

In the improvement of food packaging by functional nanomaterial, the basic goal is to increase mechanical and physical characteristics of packaging. This goal can be achieved by mixing nanomaterial or nanostructure into the polymer matrix used for food packaging to increase its mechanical and physical characteristics, such as gas barrier, temperature and humidity resistance properties, mechanical strength and flexibility. Various nanocomposites or nanoparticles with reinforced polymers and nanostructures have been created that commonly consist of nanoparticles up to 5% (w/w), such as composites of clay nanoparticles which improve barrier characteristics (80–90% reduction) in the manufacturing for beer bottles, carbonated drinks and edible oils as well as films (Brody 2007; Chaudhry et al. 2008). The products, in which the nanocomposite used in the food direct contact have been approved by USFDA (the United States Food and Drug Administration) (Sozer and Kokini 2009) and also used in Europe and Asia.

16.1.2.1 Nano-coatings

Nano-coasting is a most common nanotechnology method used in food packaging. Coating of food can be in the form of thin layer or film of edible material which is placed or covered on food in order to make a mass transfer barrier (Guilbert et al. 1997). The film coatings could act as a barrier for gas, moisture, and lipid. Edible coating films are commonly formed and applied to the food directly both with a liquid film forming solution addition or using molten compounds (Baldwin et al. 1996). Materials used as edible coating films can be classified into three categories, i.e. hydrocolloids (polysaccharides with water-soluble properties), proteins and lipids. The suitable hydrocolloids are derivatives of cellulose, starch, pectin, chitosan and alginate (El Ghaouth et al. 1991). In addition, there are many edible coatings have also been made from lipid compounds that are originated from vegetable and animal fats. These lipid compounds include fatty acids, acylglycerols, and waxes. Lipids based coating biomaterials have suitable properties as a coating solution for adding gloss to confectionery products with an excellent moisture barrier. While waxes are typically applied for coating vegetables and fruits in order to delay respiration and reduce the moisture loss (Avena-Bustillos et al. 1997).

Currently, food grade coating or edible coatings are extensively used on various food products, e.g. meats, fruits, vegetables, cheese, chocolate, candies, and French fries as well as bakeries (Morillon et al. 2002; Cagri et al. 2004; Rhim 2004). Even though, not many types of research, have been presented in the incorporation of nanomaterial into coating films for improvement of their physicochemical characteristics. For example, to reduce diffusion of oxygen, pectins have been added with clay montmorillonite (Mangiacapra et al. 2006). Gelatin and montmorillonite have been prepared as a nanocomposite and employed for the considerable enhancement of the physical characteristics (Zheng et al. 2002). Moreover, layered nanocomposites of chitosan have also been produced to increase its stability (Darder et al. 2003).

Beside the limited data on related references, there is enough proof to support the advantages of inorganic nanofiller materials that improve physical properties, such as control of sugars, acids, texture, color, and flavor, while increasing its stability during shipping, distribution and storage, with an enhanced in appearance and retarded spoilage. In this case, nanoparticles can be employed as a carrier for antimicrobials and additives. In addition, it can also be applied as the additive stabilizer and the diffusion control efficiently into the food, and in the different regions, such as bulk vs. surface of a food. The diffusion control can be important for storage of food in long-term or for imparting specific suitable characteristics, e.g. flavor in a food product. In this case, a nano-coating with edible antibacterial properties prepared and used directly to bakery products by Sono-Tec Corporation, an American company (El Amin 2007).

16.1.2.2 Nano-laminates

Nanotechnology promises researchers including food scientist, to produce novel films for nano-laminate in many ways, with excellent physical properties suitable for the food. Typically, the nano-laminate films content of two or more layers with nano-layer which are linked to each other either physically or chemically. One of the most useful techniques in nano-laminate is using an LBL (layer by layer) deposition, where the surfaces are laminated with multiple nano-layers as interfacial films that can be made from different nanomaterials (Decher and Schlenoff 2003). Here, nano-laminate provides some benefits for the edible coatings and films preparation over conventional techniques that have many interesting features to be used in the food industries (Weiss et al. 2006).

A wide range of various adsorbing compounds could be employed to make the various layers properties, such as natural or biobased polyelectrolytes (polysaccharides, proteins), colloidal particles (vesicles, micelles, droplets), and charged lipids (phospholipids, surfactants). Hence, there is a possibility to open-up to integrate active compounds into the films, e.g. antimicrobials, antioxidants, anti-browning, enzymes, flavors, and odor. These active compounds could extend the quality and shelf life of coated or laminated food products. The basic of active agent properties of laminated films depend on the properties of the nanomaterial film used in the film preparation. Similarly with a nano-film coating, in the case of nano-laminated coatings, it could be made fully from edible or biobased ingredients, such as polysaccharides, proteins, and lipids. It can be prepared by employing simple techniques, e.g., dipping and washing. The film properties, such as thickness, structure, and composition, of the laminated multilayer, created surrounding the food could be managed in many ways, such as the number of total dipping steps, changing the adsorbing agents in the dipping solutions, the order where the food product is applied to the different dipping solutions, the solution and environmental conditions used, e.g., temperature, pH, ionic strength, and dielectric constant, etc.

16.1.2.3 Clay Nanoparticles and Nanocrystals

Clays with nano-size called nano-clays could be employed to improve physical characteristics of packaging materials, e.g. barrier properties by integrating and embedding of nano-clay inside the packaging materials. Typically used in nanocomposites, such as the layered silicates content of two-layers that is one nm thick with few microns length according to the type of nano-silicate used. In addition, the silicates content in polymer formulation layers could improve the diffusive path with the tortuosity for a molecule to penetrate through it (Fig. 16.3) that provides very good barrier characteristics (Bharadwaj et al. 2002; Cabedo et al. 2004; Mirzadeh and Kokabi 2007). In this case, the polymer chains interaction and silicate layers could produce three kinds of excellent nanoscale composites (Fig. 16.3). The nano-composites with intercalated is resulting due to the polymers chains penetration into the clay in its interlayer region that creating a multilayer with an ordered structure, as the results of various layers of polymer/inorganic at a repetition with a few nanometers distance as tactoid, phase-separated micro composites (Weiss et al. 2006). While exfoliated nanocomposites are due to extensive polymer penetration, where the clay layers are dispersed randomly and delaminated in the polymer matrix (Luduena et al. 2007). These nanocomposites have excellent physical characteristics for food packaging since it has the optimal interaction between polymer and clay (Alexandre et al. 2009; Adame and Beall 2009).

Another type of nano-clay is montmorillonite clay fillers as the most common developed nano-clay that consists of a clay layer with hydrated alumina-silicate. It is

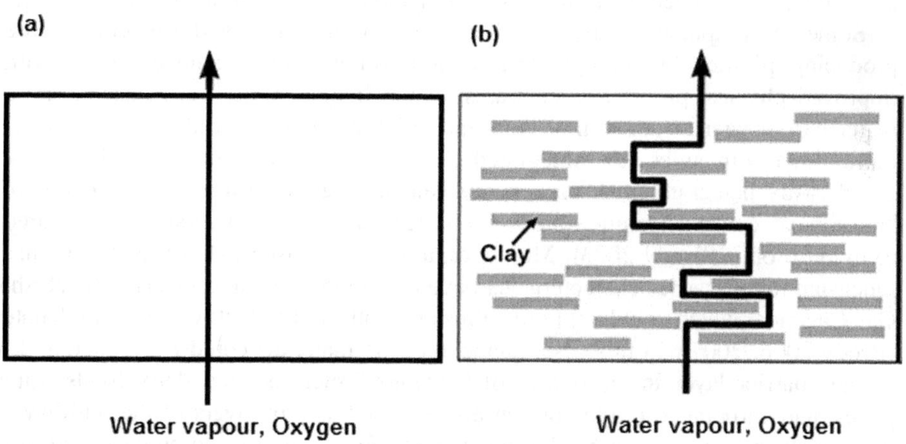

Fig. 16.3 Illustration of the "tortuous pathway" formed by incorporation of exfoliated clay nanoplatelets into a polymer matrix film. In a film composed only of polymer (**a**), diffusing gas molecules on average migrate via a pathway that is perpendicular to the film orientation. In a nanocomposite (**b**), diffusing molecules must navigate around impenetrable particles/platelets and through interfacial zones which have different permeability characteristics than those of the virgin polymer. The tortuous pathway increases the mean gas diffusion length and, thus, the shelf-life of spoilable foods (Adapted from Adame and Beall 2009)

containing an octahedral sheet with edge-shared, where aluminum hydroxide presence between two layers of silica tetrahedral (Weiss et al. 2006). The surfaces with the imbalance negative charges are compensated with the cations that are exchangeable, i.e. Na^+ and Ca^{2+}. The weak electrostatic forces, in this case, are related to the presence of linkage between the parallel layers (Tan et al. 2008). This type of clay is resulted by a medium negative surface charge that is an important parameter to determine the layer spacing equilibrium. Locally, the layer charges are not constant since it differs from one layer to another layer that should be calculated as an average value in the overall nanocrystal (Alexandre and Dubois 2000). The montmorillonite effective reinforcement filler results from its large aspect ratio (50–1000) as well as high surface area (Uyama et al. 2003).

In the case of nano-clays, they have been described mostly to increase several polymers mechanical strength (Yu et al. 2003; Avella et al. 2005; Chen and Evans 2005; Cyras et al. 2008; Jawahar and Balasubramanian 2006; Mangiacapra et al. 2006; Russo et al. 2007) and biopolymers that make it feasible to be used as packaging materials (Park et al. 2003; Petersson and Oksman 2006; Weiss et al. 2006; Xu et al. 2006; Dean et al. 2007; Marras et al. 2008). Other advantages have been described on the properties of various polymers due to integrating clay nanoparticles, such as improved glass transition (Cabedo et al. 2004; Petersson and Oksman 2006; Yu et al. 2003) and increased thermal degradation temperatures (Bertini et al. 2006; Cabedo et al. 2004; Cyras et al. 2008; Paul et al. 2003; Yu et al. 2003). In addition, the benefits of clay nanocomposite to be used as a packaging material, it provides enhanced shelf life, lightweight, heat resistant and shatterproof (Ravichandran 2010). However, the minor drawbacks of nano-clays integration on polymers are the observation in transparency reduction (Yu et al. 2003).

Some US companies, such as Southern Clay Products, Inc., and Nanocor Inc., are producing plastics by incorporating montmorillonite in a nanocomposite with improved physical properties, such as stronger, lighter, more heat-resistant, and a better barrier against gases, moisture, and volatiles (Moraru et al. 2003). Furthermore, other companies have integrated clays in nylon-6 as a fluid, which in turn, could easily penetrate small spaces between their layers. The nanocomposites of Nylon-6 can create a transmission rate of oxygen almost four times slower compared to nylon-6 only (Brody 2003). Moreover, a nanocomposite nylon MXD6 that has much-improved barrier characteristics have developed by Nanocor and Mitsubishi Gas Chemical which could apply for films and bottles of polyethylene terephthalate (Brody 2006, 2007). In this case, nanocomposite materials could be applied as an oxygen barrier layer in the bottles of beverages, such as beer, dairy foods, fruit juices, and carbonated drinks, or can also be used as thin layers of the multilayer films to extend a variety of food product shelf life, e.g. meats and processed meats, confectionery, cereals, cheese, and boil-in-bag foods (Brody 2007; Moraru et al. 2003).

16.1.3 Bio-Based Improved Packaging

Bio-based packaging or biochemical improved packaging could be applied to food packaging. This nanomaterial is biodegradable films that can be used for packaging of food to control the transfer of moisture or exchange of gas to increase shelf life, safety and maintain the nutritional ingredients and sensory quality (Siracusa et al. 2008). The bio-based packaging materials serve to be more eco-friendly compared to the plastic packaging. Similarly, bio-based materials provide a protection between a food and its surrounding environment, therefore avoiding food from deteriorated, such as microorganisms, gas conditions and relative ambient humidity. The excellent properties that benefiting from biodegradable films compared to other films are that the films are decomposing via the living microorganisms (Del Nobile et al. 2009). Generally, these materials serve to be pro eco-friendly as the decomposition of all products is completely occurred by nature, such as carbon dioxide, water, and biomass. The bio-based packaging nanomaterial, it does not or even less use of fossil fuels to create the nanomaterial, but uses more of the renewable resources, where upon disposal energy can be conserved via incineration.

Widely used polyester for food packaging is polyethylene terephthalate (PET), however, even it can be prepared from bio-based raw materials. It is currently manufactured using petroleum-derived materials. Polylactide (PLA) is biodegradable, bio-based polyester whereby incorporated with nanomaterial and could improve barrier or mechanical properties for application in food packaging. In addition, two new, bio-based polyesters are being marketed. Polyethylene furanoate (PEF) is made from 2,5-furan dicarboxylic acid (FDCA) and ethylene glycol derived from renewable resources, it is being produced by Synvina at its pilot plant in Geleen, the Netherlands. The polyester polytrimethylene furandicarboxylate (PTF) is made from bio-based furan dicarboxylic methyl ester (FDME) and bio-PDO™ (1,3-propanediol), it is being marketed in Chicago by ADM and DuPont. Both these polyesters are bio-based but not biodegradable and offer superior characteristics such as improved barrier properties and higher mechanical strength compared to ordinary plastics. They will challenge long-established material like a conventional PET, aluminum, and glass bottles. Even not eco-friendly, these bio-based materials could serve as the truly sustainable packaging materials available in the global market for the food industry.

Generally, biodegradable plastics that applied as bio-based packaging are polymer materials that at least require one step in the decomposition process by metabolism of the natural organism. In suitable situations, such as oxygen availability, temperature, and moisture, this biodegradation could lead to disintegration or fragmentation of the plastics without toxic or harmful effect on the environment (Chandra and Rustgi 1998). The biodegradable polymers plastics could be categorized based on their origin: (1) directly extracted polymers from biomass, e.g. polysaccharides, polynucleotides, polypeptides, and proteins, (2) fabricated polymers via bio-monomers chemical synthesis or mixed biomass and petrochemicals, e.g. polylactic acid or bio-polyester, (3) produced polymers by micro-organism

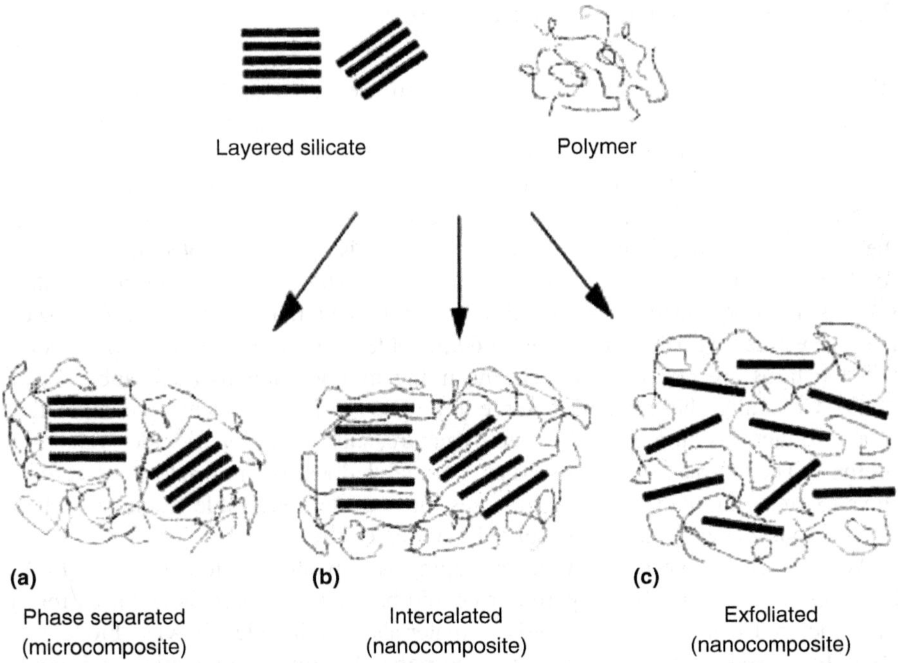

Fig. 16.4 Three types of composites when layered clays are incorporated with the polymeric matrix: (**a**) tactoid, phase-separated microcomposite; (**b**) intercalated nanocomposite and (**c**) exfoliated, polymer–clay nanocomposite (Courtesy of Alexandre and Dubois 2000)

or genetically modified bacteria, e.g. bacterial cellulose, polyhydroxybutyrate, xanthan, curdian, and pullan.

The ideal successful concept for the bio-based nanocomposite in synthetic polymers has developed new study on nanocomposites based on natural polymers or biodegradable materials for biochemical improved for applications in food packaging (Doi and Steinbuechel 2002; Kaplan 1998; Mohanty et al. 2005; Sorrentino et al. 2007; Steinbuchel 2003). The problems associated with biodegradable polymers are processing and performance as well its cost. Since in spite of their origin, "processing and performance" are the most common problem in the biodegradable polymer developments (Trznadel 1995; Scott 2000). The main problem arises with a biopolymer, such as low heat distortion temperature, brittleness, high gas and vapor permeability as well as poor resistance to protracted processing operations which have limited strongly their used as nanomaterials in food packaging. Hence, the used of nanotechnology to these bio-polymers could solve the problems and create the new opportunity for increasing both physical and chemical characteristics as well as at the same time its cost of efficiency.

Commonly bio-nanocomposites in packaging materials for food can be classified into three main types, where the clay layer is incorporated into a polymer, as presented in Fig. 16.4 (Alexandre and Dubois 2000). The nature of the compound employed, such as silicate layer, organic cation, and the polymeric matrix as well as the preparation method mostly dictate the types of composites formed. The micro composites are produced by the polymer chain that cannot intercalate into the silicate layer, where hence, make phase separated clay/polymer composites formation as given in Fig. 16.4a. The intercalated nano-composites are created if the inserted polymer chain into the clay layers is intercalated so that the interlayer spacing is extended with a well-defined spatial relationship layer that still bears to each other as given in Fig. 16.4b. The exfoliated nano-composites are created if the clay layers have already been separated completely, where the individual layers are dispersed in the matrix of polymer (Fig. 16.4c).

Since nanoparticles prepared using dispersion, thus the bio-nanocomposites can tremendously improve physical characteristics, e.g., as mechanical, barrier, and thermal as well as chemical characteristics, if they compared with the microscale composites and synthetic polymers. For instance, bio-nanocomposites greatly improve the barrier characteristics, as a result of the clay layers content which capable to reduce the molecular pathway that provides more tortuous the diffusive path (Bharadwaj 2001). Furthermore, the preparation and properties of various types of biodegradable nanocomposites polymer proving improved physicochemical characteristics that can be used for various food packaging applications (Sinha Ray and Bousmina 2005). Nowadays, the common developed bio-nanocomposites polymer desirable as a food packaging material is starch including its derivatives, such as polylactic acid, polyhydroxybutyrate, poly(butylene succinate), and aliphatic polyester. Hence, bio-nanocomposites polymer could be employed not only for biochemical improved packaging but also for improved packaging with active functions.

16.1.4 Starch-Based Nanomaterial

Starch is biopolymer that can be formed into bio-nanocomposites, since it is a promising natural material and renewable resources, as its cyclic productivity from various plants and extensive productivity suitable to the market demands as well as low cost (Gonera and Cornillon 2002; Smits et al. 1998). Starch can be used in many ways as a material for food packaging (Kim and Pometto 1994). With chemical modification or using plasticizers, starch can be created to make a film with suitable mechanical strength. For example, if it is prepared with an extruder using both mechanical and thermal energy, starch could be formed into a thermoplastic nanomaterial. In thermoplastic-based starch production, plasticizers are applied to lower hydrogen bonds in intra-molecular that make the product properties are stable. The main starch source for bioplastic polymers is corn; however more current development is studying the promising employment of starches based bioplastics that come from rice, oat, barley, wheat, potato, and soy sources.

Due to starch is hygroscopic; hence starch-based absorbent pads are typically applied as a promising alternative to a classical absorbent for a meat exudation

(Smith et al. 1995). The starch-based film can be used as materials for packaging of perishable foods (e.g. vegetables, fruits, and snacks and dry products). Even though, for these uses, mechanical efficiency, protection from moisture and oxygen are required. However, thermoplastic-based starch (TPS) often cannot fulfill all the requirements as packaging material. Due to its hydrophilicity properties, make its performance changes during processing and after it, as changes in the water content. In order to resolve this problem, the application of nano-filler clay as having been recommended as an excellent option.

The nano-clay as the promising filler has been applied to improve the TPS properties in food packaging materials (Chen and Evans 2005; De Carvalho et al. 2001; McGlashan and Halley 2003; Wilhelm et al. 2003; Yoon and Deng 2006). Films of clay/starch nanocomposite have been formed by montmorillonite nanoparticles dispersing through melt polymer processing techniques. It has been found that both the elongation and the tensile strength at break of TPS were improved by incorporation of sodium montmorillonite (>5%). In addition, the decomposition temperature was improved, and oppositely the relative diffusion coefficient of water vapor was reduced (Park et al. 2003). Here, mechanical properties also found to be improved by an enhance in the strength of tensile and modulus. Furthermore, the produced material conformity with real regulations on biodegradable materials according to European directives has been proved using the migration tests (Avella et al. 2005).

16.1.4.1 Polylactic Acid Based Nanomaterial

The polymers produced using classical chemical synthesis provides a varied range of biopolyesters, such as polylactic acid. It is a biodegradable plastic that synthesis from thermoplastic aliphatic polyester. It is biopolymer with the highest potential application to be used for a mass production of renewable materials for food packaging. Polylactic acid (PLA) is created using fermentation process from renewable resources by means of sugar corn and then followed with polymerization via ring-opening or with lactic acid polymerization via condensation. Within the degradable plastics, it is one of the most important polymers with biocompatible and biodegradable properties. Since, it represents excellent candidate to produce biodegradable and disposable material for packaging because it has excellent mechanical characteristics and capable to be processed in large scale (Murariu et al. 2008).

The PLA physicochemical properties are dependents on the two forms of the lactic acid (L or D) monomer ratio. L-PLA forms exhibit high crystallinity and high melting point, whereas a mixture of D- and L-PLA forms could produce an amorphous low-temperature polymer for the glass transition. Even though it is a bio-plastic with eco-friendly and good bio-compatibility, however, its slow degradation, low hardiness, and hydrophobicity, as well as low of reactive side-chain groups reduce its use as packaging material (Rasal et al. 2010). Then, the chemical modification is needed in order to improve its mechanical and physical properties and achieve customer needs. Furthermore, comparable to ordinary plastics, its mechanical properties are better with controlled surface characteristics, e.g. roughness, hydrophilicity, and reactive functionalities are the success story in the application of PLA as a packaging material. For example, PLA/clay nanocomposite materials for food packaging have been developed (Bandyopadhyay et al. 1999; Choi et al. 1997; Chang et al. 2003; Ogata et al. 1997; Pluta et al. 2002; Paul et al. 2003; Sinha Ray et al. 2002a, b, 2003).

The clays integration into PLA to create nano-composites of PLA/clay to enhance PLA's physic-chemical characteristics, for acceleration of its rate of degradation has been reported (Bandyopadhyay et al. 1999). The solvent mixtures casting between of PLA and organophilic clay in chloroform created the packaging materials with an improved crystallization tendency and Young's modulus (Ogata et al. 1997). On the other hand, the glass temperature transition improves slightly only with the addition of clay content. This could cause the structure of microcomposite despite the structure of nanocomposite. Due to this fact, a tactoids strong tendency formation was found in this case. The nanocomposites of PLA/layered silicate are made simply by melt extrusion, where it shows tremendous enhancement in physic-chemical characteristics in both solid and melt states compared to those with no clay (Sinha Ray et al. 2002a, b).

At the nano-scale level, mixing PLA and clays often improve remarkably physicochemical and functional characteristics compared with pure only PLA or ordinary composites (Okamoto et al. 2001). A various nano-composites blend of polylactic acid/silicate have been investigated, such as polylactic acid was blended with a mixture of montmorillonites and fluorohectorite clays, (Aguzzi et al. 2007; Oliva et al. 2007). In addition, nanocomposites that consist of a blend of the PLA and PLA/polycaprolactone were properly formed via melt-mixing using a modified kaolinite (Cabedo et al. 2006). Here, all nanocomposites produced have been shown a remarkable improvement in their properties, such as the mechanical, gas barrier, and thermal characteristics compared to the polymers or composites with no clay.

16.1.4.2 Polyhydroxybutyrate Based Nanomaterials

As an eco-friendly polymeric material, polyhydrobutyrate (PHB) has been extensively studied as the most well-known polyhydroxyalkanoate applied for packaging material. It is a polymer that can classify into the polyesters that are interesting for biodegradable and bio-based plastics (Frieder 2010). PHB is created by microorganisms (*Bacillus megaterium* or *Ralstonia eutrophus*) (Lenz and Marchessault 2005). This biopolyester is applied for a molecule energy storage in the cellular structure of microorganism. Since PHB has biochemical properties, such as biodegradability and biocompatibility, it could be used practically in food packaging industry (Weber 2000; Lenz and Marchessault 2005). PHB promises several benefits compared with ordinary plastics prepared from petrochemically derived materials in food packaging, since it exhibits compatibility with various perishable foods, e.g. fresh meat products, dairy products, ready meals, and beverages. Furthermore, it shows excellent biodegradability, and thus makes it shows better physical characteristics compared to PP (polypropylene) for applications in food packaging. It is non-toxic material. Even though, it is a crystalline polymer partially with a high degree of crystallinity and a high melting temperature, hence it is brittle which limits its application somewhat (Hankermeyer and Tjeerdema 1999). It has the poor low-impact strength that can be solved with hydroxyvalerate monomers addition into the polymer to make polyhydroxybutyrate-co-valerate (Liu et al. 2002), where this polymer could completely degrade into water and carbon dioxide in aerobic conditions (Lenz and Marchessault 2005).

There are numerous works presenting the application of PHB for the fabrication of bio-based materials, such as polymer/clay nanocomposites (Park et al. 2001; Liu et al. 2002; Maiti et al. 2003; Chen et al. 2004). Besides, the nanocomposite production based on polyhydrobutyrate looks impossible to create and only medium enhancements in material characteristics, however of polyhydrobutyrate has also been reported as bio-based packaging material (Maiti et al. 2003). Even though the products show poor mechanical characteristics with a poor extension at the break that reduces its applications in packaging. In addition, if the polyhydrobutyrate properties could be further improved, for example by small addition of nanoparticles of an eco-friendly material, this polymer would find a various number of used as food packaging material.

16.1.4.3 Polycaprolactone Based Nanomaterials

Linear polyester, such as polycaprolactone is created by either ring-opening of 2-methylene-1-3-dioxepane free radical polymerization or 3-caprolactone polymerization via ring-opening, employing various catalysts, e.g. anionic, cationic and coordination catalysts (Pitt 1990). Polycaprolactone is a polymer with a semicrystalline structure and displays a crystallinity at the high degree (around 50%). It exhibits low modulus and high elongation at break. Its commercial availability due to its physical properties, make it very interesting as a bio-based material for food packaging. Its applications, not only attractive for packaging but they are also attractive for the biomedical applications (Chandra and Rustgi 1998; Okada 2002; Nair and Laurencin 2007) as well as agricultural fields (Nakayama et al. 1997).

Since polycaprolactone has a low melting point, it is also necessary to be incorporated with other polymer materials (Ishiaku et al. 2002; Lee et al. 2002; Lim et al. 2002). For example, the incorporated of the fillers nanocomposites into polycaprolactone have been described, where the polycaprolactone/organically modified layered silicate have been developed as the nanocomposites that have better physical properties (Bharadwaj et al. 2002; Gorrasi et al. 2002, 2004; Tortora et al. 2002; Di et al. 2003).

16.1.5 Improved Packaging with Active Functions

In improved packaging with active functions, nanomaterials are employed in packaging to directly react with the contained food or its environment to ensure better protection and extend shelf-life of the food. For instance, AgNPs (silver nanoparticles) and silver nano-coatings could serve as an anti-microbial function, and also with other materials being applied as UV or oxygen scavengers. Nanosilver, nano-magnesium oxide, nano-copper oxide, and nano-titanium dioxide, as well as carbon nanotubes, are also reported as an antimicrobial with application in food packaging (Chaudhry et al. 2008; Doyle 2006; Miller and Senjen 2008). Typically, antimicrobial packaging as active packaging as oxygen absorbents has been produced and marketed by Kodak Company (Asadi and Mousavi 2006). Moreover, packaging with active oxygen scavenging employing enzymes between polyethylene films have been reported (Lopez-Rubio et al. 2006). Antimicrobial packaging can be developed to reduce and even completely inhibit microbial growth and ensure the safety of food (Brody 2007).

Food packaging with an antimicrobial agent is a type of active packaging in which antimicrobial substances incorporated into packaging material in order to ensure and extend microbial safety of food products. The technology has many advantages over conventional packaging (Lotfi et al. 2018). In this case, nanomaterial as antimicrobial agents can be coated, laminated, incorporated or immobilized onto a natural or synthetic polymer in order to reduce or inhibit the growth of the microorganism on packaged food. Antimicrobial function serves its application as a final hurdle in controlling the growth of unwanted and health threated microorganism particularly those causing surface contamination during a post-processing step. Several mechanisms of controlled release for antimicrobial substance have been investigated in order to extend an antimicrobial activity on the food surface. The antimicrobial activity retention during the storage of packaging polymers and selection of suitable substances that can withstand the extreme conditions of processing in polymer production is important that needs to be answered. Improvement of bio-based antimicrobial packaging and their successful commercialization would be a break-through step towards maintaining sustainability in food packaging applications.

16.1.5.1 Antimicrobial Function

The antimicrobial function in packaging typically involves incorporation of antimicrobial agents, such as silver nanoparticles or silver coatings into packaging materials. Antimicrobial films can decrease the spoilage and pathogenic microorganism growth. This film is especially interesting since its barrier properties and desirable structural integrity created by the used of nanomaterial, where the antimicrobial functions contributed by the antimicrobial agents of nanomaterial immobilized onto the film (Rhim and Ng 2007). In this case, nanomaterial in the film allows attaching more biological molecule copies that prove excellent efficiency (Luo and Stutzenberger 2008).

In antimicrobial functions, nanomaterials have been developed so that they can be employed as killing agents for microbe (Huang et al. 2005; Kumar and Munstedt 2005; Lin et al. 2005; Qi et al. 2004; Stoimenov et al. 2002), growth inhibitors (Cioffi et al. 2005) or even antibiotic carriers (Gu et al. 2003). The common typical and widely applied antimicrobial nanomaterials for food packaging are AgNPs that are popular for its excellent toxicity to many microorganisms, with low volatility and high-temperature stability (Kumar and Munstedt 2005). Film impregnated with AgNPs has been developed and their antimicrobial activity has been described in the literature.

The low content of silver nanoparticles in nanocomposites shows a better-improved activity against *E. coli* compared to micro composites using higher concentration of silver in polyamide 6/silver-nano- and microcomposites (Damm et al. 2007, 2008). Furthermore, ethylene absorbs and decomposes by silver nanoparticles (Hu and Fu 2003), that may increase fruits and vegetable shelf life. In addition, a film of polyethylene nanocomposite incorporated with AgNPs reduced the deterioration of a Chinese fruit, jujube (Li et al. 2009). A coating film impregnated with AgNPs was effectively reducing the growth of microbe and extended the shelf life of asparagus (An et al. 2008). AgNPs present in the film could increase modulus and matrix of a poly(vinyl) alcohol strength, and increased its thermal characteristics, extending its stability and antimicrobial activity (Mbhele et al. 2003). Moreover, the nanostructure of calcium silicate has also been applied as an adsorbent of Ag^+ from solution down to the 1 ppm (w/w) level (Johnston et al. 2008) as an antimicrobial film.

Other antimicrobial nanomaterials which have been applied in food packaging are TiO_2 (titanium dioxide), CNTs (carbon nanotubes), chitosan and nisin (Kuswandi 2017). TiO_2 is commonly applied as an agent for surface coatings of photocatalytic disinfectant (Fujishima et al. 2000). Photocatalysis of TiO_2 that creates the polyunsaturated phospholipids peroxidation in the microbial cell membranes (Maness et al. 1999), and it has been applied to inactivate several pathogenic bacteria related to food (Kim et al. 2003, 2005; Robertson et al. 2005). A packaging film consists of TiO_2 powder has been investigated due to its capability to decrease *E. coli* contamination on food surfaces, which indicating the film is promising to be applied for fresh-cut products (Chawengkijwanich and Hayata 2008).

Antibacterial properties of nanomaterial have also been found in carbon nanotubes. Direct contact with carbon nanotube aggregates was reported to have a toxic effect or even bacterocids on *E. coli*, this is due to the thin and long carbon nanotubes puncture microbial cells, which making damage irreversibly (Kang et al. 2007). Moreover, it is also found that CNTs are cytotoxic toward human cells, particularly if it in contact with the skin (Monteiro-Riviere et al. 2005) and lungs (Warheit et al. 2004) that would have a harmful impact on workers with CNTs directly during the processing stages compared to the users. Hence, it is compulsory to have knowledge regarding the effects of carbon nanotubes on human health, when it is ingested. The ingestion risk of incorporated particles into packaging material for food should be bear in mind since the migration into food during direct contact with them is possible.

Nisin as antimicrobial peptides can also be incorporated into LBL (layer by layer) structures to develop films that have antimicrobial activity (Haynie et al. 2006). Nisin creates pores in lipid bilayers and acts as an agent of depolarization on bacterial membranes (Sahl et al. 1987). Nano-films of the peptide with multilayer that intercalated with various peptides created at neutral pH to have oppositely charged that was more stable compared to the film of the peptide that used electrostatic interactions to stabilized (Li et al. 2006). Recently, nisin and chitosan coated on cellulose have been developed as a new antimicrobial film. Chitosan solution containing nisin (500 and 1000 μg/mL) via sol-gel method was developed and then coated on cellulose paper by dip coating technique (Divsalar et al. 2018). The incorporation of nisin into chitosan-cellulose as nanocomposite film caused a significant increase in antimicrobial activities on *Listeria monocytogenes.* Thus, the nanocomposite film of chitosan-cellulose containing nisin can be promising to be used for packaging in cheese (Divsalar et al. 2018).

Chitosan at the nanoscale has also been found to have antibacterial activity (Qi et al. 2004). One possible antimicrobial activity mechanism causes interactions between chitosan with positively charged and negatively charged cell membranes, which in turn, improving permeability of the membrane and making rupture eventually as well as intracellular material leakage. This is correlated with the study showed that both the engineered nanoparticles and the raw chitosan are not effective at above pH 6 that could caused by the absence of a protonated amino group (Qi et al. 2004). Very recently, nanostructured chitosan/monolaurin films were developed for its antimicrobial properties against *L. monocytogenes* in vitro and on ultrafiltered white cheese (Lotfi et al. 2018). Here, chitosan-zinc oxide mixtures containing monolaurin (0.5 and 1%) were prepared using the sol-gel method and the solution mixture was coated on cellulosic paper. The antimicrobial activity results found that the addition of monolaurin significantly increased the diameter of the zone of inhibition on *L. monocytogenes* as depicted in Fig. 16.5 (Lotfi et al. 2018).

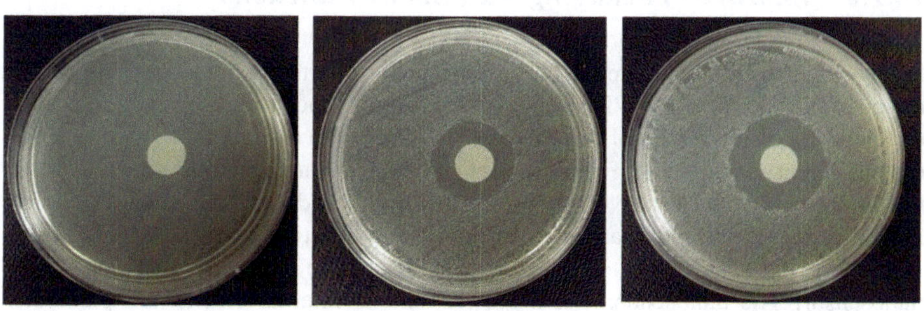

A B C

Fig. 16.5 Antimicrobial activity of cellulose-chitosan (CC) and monolaurin-incorporated CC film on *L. monocytogenes* according to disc surface spreading method. (**a**) CC film, (**b**) CC film: 0.5% mL and (**c**) CC film: 1% mL (Courtesy of Lotfi et al. 2018)

16.1.5.2 Oxygen Scavenging Film

Oxygen (O_2) causes degradation in many various foods directly or indirectly. For instance, direct reactions of oxidation cause browning in fruits or rancidity in vegetable oils. Food degradation causes by indirect reaction of O_2, such as spoilage of food due to aerobic microorganisms. Hence, the integration of O_2 scavengers into package material for food can reduce the very low level of O_2 that make it is useful for several uses, and will increase food shelf-life.

O_2 scavenger films were developed successfully by inserting TiO_2 (Titania nanoparticles) into different polymers (Xiao-e et al. 2004). Then, the films can be applied as packaging material for various oxygen-sensitive food products. Attention should be focused particularly on the photocatalytic activity of TiO_2 under UV radiation. Due to TiO_2 acts by a mechanism of photocatalytic, and its major disadvantage is the need of UVA light (Mills et al. 2006).

16.1.5.3 UV Absorbing Films

Nanocrystalline Titania (TiO_2) film is typically used nanomaterial as UV absorbing film. The TiO_2-coated films are effective when it irradiated to sunlight to inactivate fecal *E. coli* forms in water (Gelover et al. 2006). The visible light absorbance of TiO_2 improves by metal doping and improves its activity of photo-catalytic under UV radiation (Anpo et al. 2001). Moreover, it has been studied that doping TiO_2 film with AgNPs greatly increase inactivation of photo-catalytic bacteria (Page et al. 2007; Reddy et al. 2007). Hence, the combinations between in TiO_2/Ag + nanoparticles with PVC as nanocomposites were created an excellent antibacterial characteristic of the film (Cheng et al. 2006).

16.1.6 Improved Packaging with Smart Functions

In the improvement of food packaging with smart or intelligent functions, nanoparticles are employed for sensing chemical or biochemical or even microbial develops inside the food and its environment which was commonly known as smart or intelligent packaging. For instance, specific pathogens and specific gases nanosensors have been developed for food spoilage detection (Kuswandi et al. 2011). In this case, nanomaterials can be used as reactive materials in packaging to inform and communicate the condition inside the packaged product, called nanosensor. The nanosensor is able to respond to internal or external parameter changes to inform, communicate, and detect the food product with the goal to assure its quality and safety. Current studies in polymeric nanomaterials are to improve packaging with smart functions, such as freshness nanosensors, oxygen nanosensors, and nano-devices for food product traceability and identification.

Nanosensors integrated into food packaging are also developed to detect conditions (both the internally or externally) of various food products or their containers, via the supply chain. For instance, packaging can be sensitive to humidity or temperature change over a period of time and then serve related information of the food conditions via color development. For example, nanosensors integrated into plastic packaging could sense gases produced by spoiled food, where then packaging itself develop color as an alarm to customers. Therefore, nanosensors are capable to react to the changes in micro-environment inside the food package, such as humidity or temperature, oxygen exposure levels and etc., then, food deterioration or microbial level can be alert (Bouwmeester et al. 2009). Commonly, food producers predicted the date of expiration by calculating distribution, conditions of storage and display, particularly temperature where food exposed. Even though, these considered conditions are not always can be predicted, since food frequently are exposed to the abuse temperature, especially for food that needs a cold chain. Moreover, defects in the sealing of packaging systems could expose food to an unpredicted high level of oxygen. It could cause unwanted changes. When nanosensors equipped into food packaging, it could respond to certain chemical markers, pathogens and toxins in food, which can useful for real-time monitoring for food freshness status, and reduce the requirement for predicted expiration dates which are inaccurate (Liao et al. 2005).

Nano-biosensors have also been studied and marketed to sense spoilage and pathogenic microorganisms, contaminants, product tampering, to track ingredients or products during the processing chain (Nachay 2007). CNTs (carbon nanotubes) have been developed as nano-biosensors with several benefits compared to classical methods, e.g. high-performance liquid chromatography. Nano-biosensor based on CNTs is rapid with high-throughput screening capability, simple and low-cost; less power needed and can be recycled. Moreover, nano-biosensor based on MW-CNTs (multiwalled carbon nanotubes) has also been created to detect toxic proteins, microorganisms, and degradation of food including beverages (Nachay 2007). Moreover, they have also been integrated into packaging to develop color to alert the consumer, when food is starting to degrade, or contaminating with pathogens using e-noses (electronic nose) and e-tongues to smell or taste flavors and scents respectively (Asadi and Mousavi 2006; Joseph and Morrison 2006; Scrinis and Lyons 2007; Sozer and Kokini 2009). In real situations, many companies, such as British Airways, Nestlé, and MonoPrix Supermarkets, are typically employing nanosensors with color development (Pehanich 2006).

16.1.6.1 Freshness Nanosensors

Nanosensors for freshness indicator have been applied to the surface of packaging materials. Regarding this, several kinds of sensors for gases have been created that transduces chemical reaction on the surfaces of nanosensor into a signal correlated to the gas target. Electroactive conjugated polymers called conducting polymers can be prepared either by electrochemical oxidation or chemical reaction, are very attractive

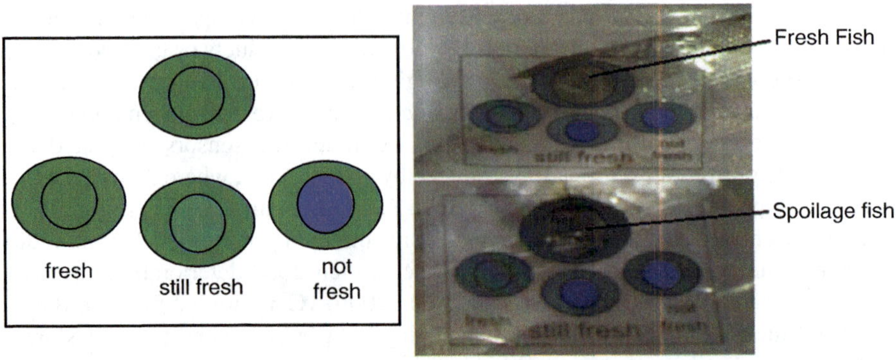

Fig. 16.6 Freshness sensor for smart packaging based on nanofibre of polyaniline. Left is sensor reference color change for detection of fish freshness, and right is sensor response towards fresh fish and spoilage fish (Courtesy of Kuswandi et al. 2012)

since their electrical, optical, and magnetic characteristics that are depended to the conjugated p electron in their backbones (Ahuja et al. 2007; Kuswandi et al. 2012). Polyaniline (PANI), polyacetylene or polypyrrole as conducting polymers have been used widely for smart packaging (Ahuja et al. 2007; Kuswandi et al. 2012). Conducting polymers that electrochemically polymerized have an excellent property to change between oxidized (doped) over reduced (undoped) states that are the basis of many applications for smart functions (Rajesh et al. 2004). On-package indicator fabricated with PANI film develops color change toward basic volatile amines produced during the period of fish spoilage (Fig. 16.6) (Kuswandi et al. 2012). Development of color, in terms of total color change of PANI, can be related to total volatile amine and microbial growth levels in the fish samples. This nanosensor is capable of the monitoring of fish spoilage in real-time at various constant or fluctuating temperatures.

Typically spoilage of food causes by microorganism's activities that creates gases can be sensed using nanocomposites of conducting polymer with metal oxides that could be employed for measurement or identification of microorganism growth based on their emitted gas which can also be used for freshness detection. Nanosensors based on conducting polymer nanocomposites contain conducting particles immobilized into an insulating polymer matrix. The color changes of the sensors create a pattern that correlates to the gas target (Arshak et al. 2007; Kuswandi et al. 2012). Nanosensors based on nanocomposites of conducting polymer consisting of carbon black with PANI have been prepared to determine and identify foodborne pathogens using a specific pattern corresponding to each microorganism. The bacteria, e.g. *Bacillus cereus*, *Vibrio parahemolyticus* and *Salmonella* spp. can be known based on the specific response pattern created by the nanosensors (Arshak et al. 2007). The freshness of chicken meat was determined using its smell of the metal oxide signal response, such as signal pattern from indium and tin oxide gas sensors were then further processed using a neural network (Galdikas et al. 2000). E- Tongues integrated into food packaging have also been reported

(Joseph and Morrison 2006). This sensor contains extremely sensitive of the nanosensors array of gases emitted by microorganisms that develop a color that indicates the food deterioration.

16.1.6.2 O_2 Nanosensors

During food storage, oxygen makes grow of aerobic microorganism. Therefore, it has become a concern in investigating non-toxic and irreversible oxygen nanosensors to make sure the oxygen absence in the food packaging systems, therefore food packaging needs to be vacuumed or degassed with nitrogen. A colorimetric oxygen nanosensor with UV-activated based on UVA light has been created that employs Titania (TiO_2) nanoparticles to photosensitize the methylene blue reduction by triethanolamine in a polymer encapsulated matric (Lee et al. 2005). Under UV irradiation, the nanosensor starts to bleach and change to colorless, till oxygen exposed to it, then it appears to blue color again. The color change rate is corresponding to the oxygen exposure level. Thin methylene blue/TiO_2 nanocomposite films deposited on glass using liquid phase deposition that has been used to oxides deposition into several substrates have been used as oxygen indicator in packaging systems for application in various foods that sensitive to oxygen (Gutierrez-Tauste et al. 2007).

SnO_2 based nano-crystal has been developed as a photosensitizer in an O_2 colorimetric nanosensor containing glycerol as an electron donor, methylene blue as a redox dye and an hydroxyethyl cellulose as encapsulating polymer (Mills and Hazafy 2009). When it is exposed to UVB light causes photobleaching or the nanosensor activation and methylene blue photoreduction by SnO_2 nanoparticles. The color of the nanosensor films will vary depend on to the O_2 exposure level, bleached or colorless when not exposed, and blue when underexposed.

16.1.6.3 Nano-devices for Tracking and Anti-counterfeiting

Some improved packagings integrated with smart functions have also been created as a tracking device for the safety of food or even for anti-counterfeiting purposes. Food Expert ID® have been produced by BioMerieux as a multi-detection test and used for nano surveillance that monitor food scares. pSiNutriam, the nanotech company have also been produced nano-based tracking technologies, such as an ingestible BioSilicon that could be incorporated in food packaging for monitoring purposes, pathogen detection as well as it could be safely consumed by consumers (Miller and Senjen 2008; Scrinis and Lyons 2007). Nano-barcodes have been produced by Oxonica Inc., (US Company) to be applied to each item or pellets that must be read using a modified microscope for anti-counterfeiting (Roberts 2007). Nanobarcodes® that already marketed produced by electroplating inert metals, e.g. gold, platinum, silver, or nickel, into templates, which determine the particle diameter, and then creating the striped nanorods from the templates (www.nanoplextech.com).

16.1.6.4 Nano-device for Active Tags and Traceability

Radiofrequency identification (RFID) as an active tag is given in Fig. 16.7. They are electronic information-based systems, where the RF (radio frequency) is used to transfer data from a tag that has been attached to a packaging to automatically trace and identify the object including food products. RFID has actually displayed some improvement to the barcodes or manual tracking systems. Moreover, they have a longer reading range and very strong that can operate under extreme condition, such as high temperatures and pressures. They can be detected at distances of over 100 m, and simultaneously they can be read as well (Abad et al. 2007, 2009). Nanotechnology offers nanosensor to incorporate cheap RFID tags called nano-enabled RFID tags. They are very small, flexible and even could be printed-out on the surface of the food labels. Thus, it enhances the versatility of the tags that enables low-cost production of them (www.thefreelibrary.com).

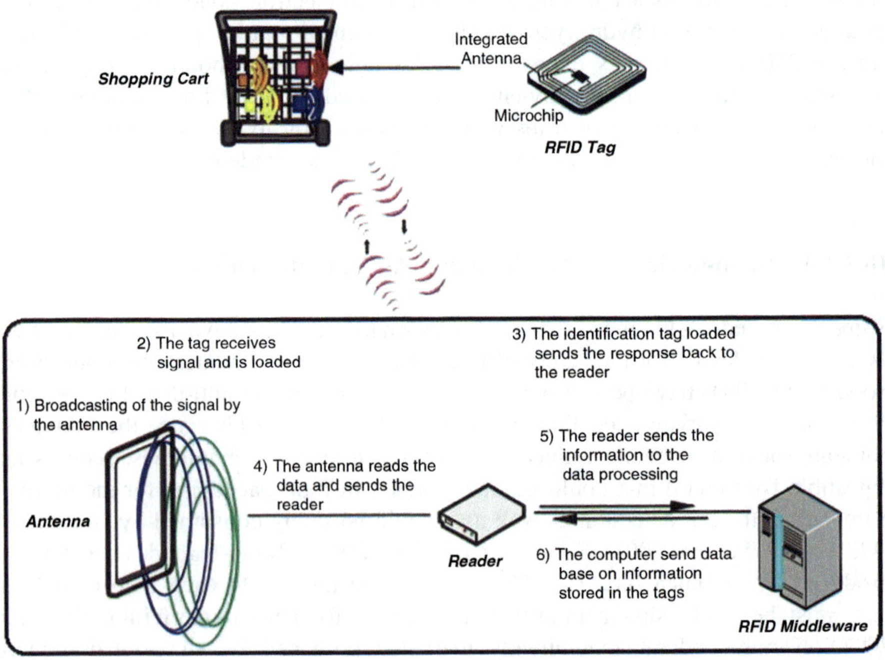

Fig. 16.7 Schematic representation of the RFID system that could employ nanodevice

16.1.7 Safety and Environmental Concerns

Functional nanomaterial that used in food packaging can also be classified as food-contact articles (FCAs) where they touch foods during all stages of production, transport, and preparation. FCAs are made of combinations of food contact materials (FCMs) like plastics, paper, and board, metals and glass, but also adhesives used in printing inks and coatings as well as functional nanomaterial used in improved packaging through many ways. During FCM and FCA manufacture, chemical starting substances are transformed into intended substances, but also non-intentionally added substances (NIAS). Safety must be determined for all food contact chemicals (FCCs), including nanomaterial that potentially migrates into foods and to which consumers can be exposed. Recently, awareness of the scientific challenges related to the risk assessment of FCCs has increased.

For safety and environmental concern purposes, it is necessary to study the potential migration of nanomaterials in packaging into food for the assessment of their potential hazard to human and environment. Currently, there are few works have been reported regarding the nanomaterial effects upon ingestion, or their interaction with food substances (Silvestre et al. 2011). Currently, the European legislation conducts an overall migration limit of 10 mg constituent per dm^2 surface area to all substances which can migrate from packaging materials to foodstuffs (Commission Regulation (EU) No. 10/2011). For example, a liter cubic packaging with 1 kg of food containing, this equates to a migration of 60 mg substance per kg of food. Even though, with the exception of a few materials, such as listed in Annex 1 of the legislation, risk assessment of nanomaterial has to be carried out on the basis a case-by-case (Silvestre et al. 2011; Commission Regulation (EU) No. 10/2011).

The silver migration of from the various kinds of nanocomposites into food stuff, such as migration analysis of the silver form (ions or particles) has been investigated (Echegoyen and Nerín 2013). They reported that in acidic food the highest level of silver migration could happen. In addition, heating was found to induce migration, where microwave heating increasing more migration than an old oven. It was recommended that silver migration could happen by two different mechanisms: the silver nanoparticles detachment from the composites, or and the silver ions undergo an oxidative dissolution. The silver and copper migration from antimicrobial nanocomposites, in food packaging has also been investigated (Cushen et al. 2014). It was found that the nanofiller percentage in the nanocomposites was one of the most important parameters that modulate the migration, more than temperature, contact time or particle size. A migration model of particles from food packaging into food has also been presented in this report. This model can be used as an excellent predictor for the migration level of nanosilver and less to nano copper migration into food, the model could promisingly benefit to industry, in term of reducing time and costs regarding the migration studies (Cushen et al. 2014).

Currently, the migration of an organo-modified clay polylactic acid nanocomposite and its toxicological profile to be applied as an FCM was studied. Migration investigations found that less than 10 mg per dm^2 of the nanocomposite migrated in water, under the experimental conditions (Maisanaba et al. 2014a). Moreover, the food stimulant analysis found that the metal levels measured were under the allowable concentration. The studies investigated the potential toxicity of migration extracts both in vitro and in vivo. Assessment of the potential the migration extracts cytotoxicity in vitro on two kinds of cells in the digestive system and their ability to increase DNA mutations were done and no evidence was reported regarding in vitro toxicity (Maisanaba et al. 2014a). Similarly, the rats that have been exposed for 90 days to the migration extracts in drinking water also did not show toxicity evidence, such as inflammation, oxidative stress (Maisanaba et al. 2014b).

These investigations inform that the possibility of nanomaterial migration from food contact materials into foodstuffs was high, particularly in the case of the potential migration rate associated with the nanofiller percentage present in the nanocomposites. There are needs is further investigation to study migration and toxicology to make sure the safety of nanotechnology applications in the food packaging industry. Thus, the improvement of food packaging with functional nanomaterial for safe and successful applications need to fits with at least three regulations, i.e. (a) food, (b) health and (c) environmental regulations as given in Fig. 16.8. The three regulations are required to assure that our society can take advantages from new developments of nanomaterial in food packaging, whilst a high standard for health protection, safety, and the environment concern are being implemented. In addition, these are required for the safety improved packaging with nanomaterial, with a special concern in the knowledge extension related to their toxicological effects, potential migration and exposure levels for both workers

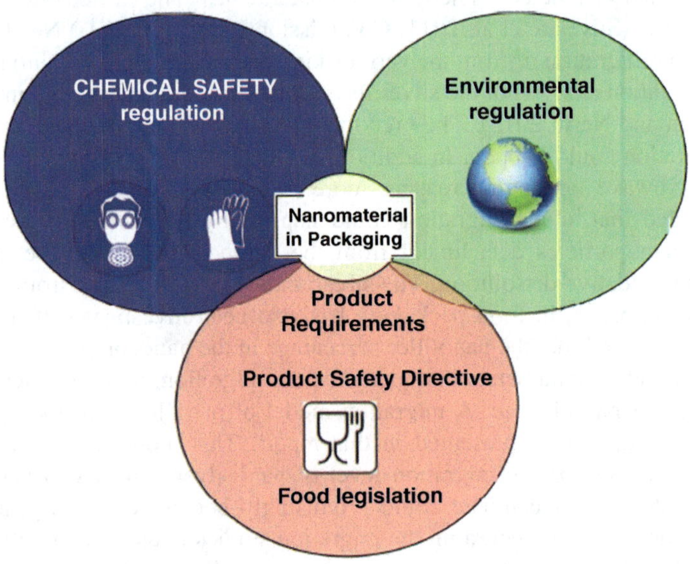

Fig. 16.8 Relevant regulation for improvement of food packaging based on functional nanomaterial, including food product legislation and product safety directive, chemical safety regulation, and environmental regulation (Adapted from Kuswandi 2017)

and consumers. Moreover, special attention on the effects of nanomaterial to human health that may cause a chronic effect must be applied. Furthermore, it is important for the food packaging authorities not only to control the quality of the packaged product but also to comply with the regulation that provides the consumer with clear information regarding its benefits and potential risks and environmental protection as well. If the three regulations have been compiled, food packaging using functional nanomaterial would play a predominant role in the production of safer, tastier, healthier, and more nutritious as well as in the same time eco-friendly food.

16.1.8 Conclusion and Future Prospective

Based on the current research results, it is tremendously show that nanotechnology offers various opportunities in improved food packaging based on functional nnanomaterials, ranging from biobased to smart packaging. The nanotechnology applications in food packaging give remarkable improvement in the packaging material properties, even in its infancy and still needs financial support for its research and development for the better understanding the role of nanotechnology use in food packaging materials both its benefits and draw-backs. The most benefits with nanotechnology used to improve packaging properties can be numerous, for example, it can give packaging materials with many functionalities, such as stronger, lighter, durability, enhance shelf-life, healthier, low-costs etc.

In order to generate new food packaging functions, nanotechnology promises to increase food properties as well, such as healthier, tastier, and more delicious as well as nutritious, when it packaged. Even though, currently many of these new improved packaging functions using nanomaterials are still in lab scale, and only targeted for products at high-value. Moreover, they can be applied to improve packaging where keeps food inside fresher for longer time, which in turn, extending food self-life and safety. The nanosensors integrated as intelligent packaging, could also serve users on the food state inside packaging with visual information. The nanosensors attach on the food packaging could alert users when a food inside is already spoilage. They can give warning before the food deteriorated or inform users the certain pathogen already growth in the foods. Furthermore, in the future, nanotechnology could change the entire packaging industry process and fabrication.

Besides, tremendous opportunities in food packaging applying nanomaterials, perhaps that conventional packaging will be changed with smart packaging that has multi-functionalities. The future of packaging materials based nanomaterials will fit with the needs for preservation of perishable foods and other products. By employing suitable nanomaterials, it could be able to make food packaging with better mechanical properties, such as stronger, better barrier and thermal properties. For instance, materials with nano-structure could avoid the bacteria and microorganisms invasion dedicated for safety of food. While, the nanosensors integrated in the packaging could use to alert user when food inside already spoilage and not safety to be consumed.

Regardless of how nanotechnology could improve food packaging could be commercialised, governed, or perceived by the consumers, it looks clear that the used of nanoscale material in packaging could continue to produce interesting and unforeseen food packaging. Even though, the threats arise with the toxic potentials and ethical issues, where the use of this nanotechnology should be wisely responded with more research and studies regarding these issues. Thus, the safe and successful applications of the nanotechnology will need continuous dialogue among the scientists and companies who manipulate them and the users who use them. If it runs successfully, then the fruitful of integration of nanomaterials into food packaging could play an important role in providing the world's food supply safer, tastier, healthier, and more nutritious, more delicious as well as eco-friendly.

Finally, authoritative bodies should write guideline dedicated to the criteria and followed by evaluating the food packaging safety regarding the use of nanomaterials with improved properties and novel functions. Novel techniques, methods, procedures and standardized test to investigate the nanomaterials effects upon ingestion, or the harmful potential interaction of nanomaterial-based packaging with food components are extremely required for the potential hazards evaluation relating to nanomaterials exposure to human. However, it is commonly predicted that nanotechnology-based packaging for food will be increasingly commercially available worldwide in the coming years forward.

Acknowledgments The authors gratefully thank DRPM, the Higher Education, Ministry of Science, Technology & Higher Education, the Republic of Indonesia for supporting this work via the Competency Grant.

References

Abad E, Zampolli S, Marco S (2007) Flexible tag microlab development: gas sensors integration in RFID flexible tags for food logistic. Sens Actuators B 127(1):2–7

Abad E, Palacio F, Nuin M, González de Zárate A, Juarros A, Gómez JM, Marco S (2009) RFID smart tag for traceability and cold chain monitoring of foods: demonstration in an intercontinental fresh fish logistic chain. J Food Eng 93(4):394–399

Adame D, Beall GW (2009) Direct measurement of the constrained polymer region in polyamide/clay nanocomposites and the implications for gas diffusion. Appl Clay Sci 42:545–552

Aguzzi C, Cerezo P, Viseras C, Caramella C (2007) Use of clays as drug delivery systems: possibilities and limitations. Appl Clay Sci 36:22–36

Ahuja T, Mir IA, Kumar D, Rajesh (2007) Biomolecular immobilization on conducting polymers for biosensing applications. Biomaterials 28:791–805

Alexandre M, Dubois P (2000) Polymer-layered silicate nanocomposites: preparation, properties and uses of a new class of materials. Mat Sci Eng R 28(1–2):1–63

Alexandre B, Langevin D, Mederic P, Aubry T, Couderc H, Nguyen QT et al (2009) Water barrier properties of polyamide 12/montmorillonite nanocomposite membranes: structure and volume fraction effects. J Membr Sci 328(1–2):186–204

An J, Zhang M, Wang S, Tang J (2008) Physical, chemical and microbiological changes in stored green asparagus spears as affected by coating of silver nanoparticles-PVP. Lebenson Wiss Technol 41(6):1100–1107

Anpo M, Kishiguchi S, Ichihashi Y, Takeuchi M, Yamashita H, Ikeue K et al (2001) The design and development of second-generation titanium oxide photocatalysts able to operate under visible light irradiation by applying a metal ion-implantation method. Res Chem Interm 27 (4–5):459–467

Arshak K, Adley C, Moore E, Cunniffe C, Campion M, Harris J (2007) Characterisation of polymer nanocomposite sensors for quantification of bacterial cultures. Sens Actuators B 126:226–231

Asadi G, M Mousavi (2006) Application of nanotechnology in food packaging. Available at http://iufost.edpsciences.org

Avella M, De Vlieger JJ, Errico ME, Fischer S, Vacca P, Volpe MG (2005) Biodegradable starch/clay nanocomposite films for food packaging applications. Food Chem 93:467–474

Avena-Bustillos RJ, Krochta JM, Saltveit ME (1997) Water vapour resistance of red delicious apples and celery sticks coated with edible caseinate-acetylated monoglyceride films. J Food Sci 62(2):51–354

Baldwin EA, Nisperos MO, Chen X, Hagenmaier RD (1996) Improving storage life of cut apples and potato with edible coating. Post Biol Technol 9(2):151–163

Bandyopadhyay S, Chen R, Giannelis EP (1999) Biodegradable organic–inorganic hybrids based on poly (L lactide). Polym Mater Sci Eng 81:159–160

Bouwmeester H, Dekkers S, Noordam MY, Hagens WI, Bulder AS, de Heer C et al (2009) Review of health safety aspects of nanotechnologies in food production. Regul Toxicol Pharmacol 53 (1):52–62

Bertini F, Canetti M, Audisio G, Costa G, Falqui L (2006) Characterization and thermal degradation of polypropylene–montmorillonite nanocomposites. Polym Degrad Stab 91:600–605

Bharadwaj RK (2001) Modeling the barrier properties of polymer-layered silicate nanocomposites. Macromolecules 34:9189–9192

Bharadwaj RK, Mehrabi AR, Hamilton C, Murga MF, Chavira A, Thompson AK (2002) Structure-property relationships in cross-linked polyestereclay nanocomposites. Polymer 43:3699–3705

Brody AL (2003) "Nano, nano" food packaging technology. Food Technol 57(1):52–54

Brody AL (2006) Nano and food packaging technologies converge. Food Technol 60(3):92–94

Brody AL (2007) Case studies on nanotechnologies for food packaging. Food Technol 07:102–107

Cabedo L, Gimenez E, Lagaron JM, Gavara R, Saura JJ (2004) Development of EVOH–kaolinite nanocomposites. Polymer 45(15):5233–5238

Cabedo L, Feijoo JL, Villanueva MP, Lagaron JM, Gimenez E (2006) Optimization of biodegradable nanocomposites based on a PLA/PCL blends for food packaging applications. Macromol Symp 233:191–197

Cagri A, Ustunol Z, Ryser ET (2004) Antimicrobial edible films and coatings. J Food Prot 67:833–848

Chawengkijwanich C, Hayata Y (2008) Development of TiO_2 powder-coated food packaging film and its ability to inactivate Escherichia coli in vitro and in actual tests. Int J Food Microbiol 123 (3):288–292

Cha D, Chinnan M (2004) Biopolymer-based antimicrobial packaging: a review. Crit Rev Food Sci Nutr 44:223–237

Chandra R, Rustgi R (1998) Biodegradable polymers. Prog Polym Sci 23:1273–1335

Chang J-H, Uk-An Y, Sur GS (2003) Poly (lactic acid) nanocomposites with various organoclays. I. Thermomechanical properties, morphology, and gas permeability. J Polym Sci B 41:94–103

Chaudhry Q, Scotter M, Blackburn J, Ross B, Boxall A, Castle L, Aitken R, Watkins R (2008) Applications and implications of nanotechnologies for the food sector. Food Additives Cont 25 (3):241–258

Chen GX, Hao GJ, Guo TY, Song MD, Zhang BH (2004) Crystallization kinetics of poly (3-hydroxybutyrate-co-3-hydroxyvalerate)/clay nanocomposites. J Appl Polym Sci 93:655–661

Chen B, Evans JRG (2005) Thermoplastic starch-clay nanocomposites and their characteristics. Carbohydr Polym 61(4):455–463

Cheng Q, Li C, Pavlinek V, Saha P, Wang H (2006) Surface-modified antibacterial TiO_2/Ag^+ nanoparticles: preparation and properties. Appl Surf Sci 252:4154–4160

Choi HJ, Kim JH, Kim J (1997) Mechanical spectroscopy studies on biodegradable synthetic and biosynthetic aliphatic polyesters. Macromol Symp 119:149–155

Cioffi N, Torsi L, Ditaranto N, Tantillo G, Ghibelli L, Sabbatini L et al (2005) Copper nanoparticle/polymer composites with antifungal and bacteriostatic properties. Chem Mater 17:5255–5262

Commission Regulation (EU) No. 10/2011 of 14 of January 2011 on plastic material and articles intended to come in contact with food. Official Journal of the European Union

Cushen M, Kerry J, Morris M, Cruz-Romero M, Cummins E (2014) Evaluation and simulation of silver and copper nanoparticle migration from polyethylene nanocomposites to food and an associated exposure assessment. J Agric Food Chem 62(6):1403–1411

Cyras VP, Manfredi LB, Ton-That MT, Vazquez A (2008) Physical and mechanical properties of thermoplastic starch/montmorillonite nanocomposite films. Carbohydr Polym 73:55–63

Darder M, Colila M, Ruiz-Hitky E (2003) Biopolymer-clay nanocomposites based on chitosan intercalated in montmorllonite. Chem Mater 15:3774–378

Damm C, Munstedt H, Rosch A (2007) Long-term antimicrobial polyamide 6/silver-nanocomposites. J Mater Sci 42(15):6067–6073

Damm C, Munstedt H, Rosch A (2008) The antimicrobial efficacy of polyamide 6/silver-nano- and microcomposites. Mater Chem Phys 108:61–66

Dasgupta N, Ranjan S, Mundekkad D, Ramalingam C, Shanker R, Kumar A (2015) Nanotechnology in agrofood: from field to plate. Food Res Int 69:381–400

Dean K, Yu L, Wu DY (2007) Preparation and characterization of mel-textruded thermoplastic starch/clay nanocomposites. Compos Sci Technol 67:413–421

Decher G, Schlenoff JB (2003) Multilayer thin films: sequential assembly of nanocomposite materials. Wiley-VCH, Weinheim, p 543

De Carvalho AJF, Curvelo AAS, Agnelli JAM (2001) A first insight on composites of thermoplastic starch and kaolin. Carbohydr Polym 45:189–194

Del Nobile MA, Conte A, Buonocore GG, Incoronato AL, Massaro A, Panza O (2009) Active packaging by extrusion processing of recyclable and biodegradable polymers. J Food Eng 93 (1):1–6

Di Y, Iannace S, Di Maio ED, Nicolais L (2003) Nanocomposites by melt intercalation based on polycaprolactone and organoclay. J Polym Sci B 41:670–678

Divsalar E, Tajik H, Moradi M, Forough M, Lotfi M, Kuswandi B (2018) Characterization of cellulosic paper coated with chitosan-zinc oxide nanocomposite containing nisin and its application in packaging of UF cheese. J Food Eng 111(2012):21–27

Doi Y, Steinbuechel A (2002) Polyesters, III, Applications and commercial products. In: Biopolymers, 4. Wiley-VCH, Weinheim

Doyle ME (2006) Nanotechnology: a brief literature review. Available at http://www.wisc.xv.edu/fri/briefs/FRIBrief_Nanotech_Lit_Rev.pdf

Duncan TV (2011) Applications of nanotechnology in food packaging and food safety: barrier materials, antimicrobials and sensors. J Colloid Interface Sci 363(1):1–24

Echegoyen Y, Nerín C (2013) Nanoparticle release from nano-silver antimicrobial food containers. Food Chem Toxicol 62:16–22

El Amin A (2005) Consumers and regulators push food packaging innovation. Available at http://foodproductiondaily.com/news/ng.asp?n=63704

El Amin A (2007) Nanoscale particles designed to block UV light. FoodProductionDaily.com Europe. 18 October. Available at http://foodproductiondaily.com/news/ng.asp?id=80676

El Ghaouth AE, Arul J, Ponnampalam R, Boulet M (1991) Use of chitosan coating to reduce water loss and maintain quality of cucumber and bell pepper fruits. J Food Process Preserv 15:359–368

Frieder W (2010) Lichtenthaler "Carbohydrates as organic raw materials". In: Ullmann's encyclopedia of industrial chemistry. Wiley-VCH, Weinheim. https://doi.org/10.1002/14356007

Fujishima A, Rao TN, Tryk DA (2000) Titanium dioxide photocatalysis. J Photochem Photobiol C Photochem Rev 1(1):1–21

Galdikas A, Mironas A, Senuliene V, Šetkus A, Zelenin D (2000) Response time based output of metal oxide gas sensors applied to evaluation of meat freshness with neural signal analysis. Sens Actuators B 69:258–265

Gelover S, Gomez LA, Reyes K, Leal MT (2006) A practical demonstration of water disinfection using TiO_2 films and sunlight. Water Res 40:3274–3280

Gonera A, Cornillon P (2002) Gelatinization of starch/gum/sugar system studied by using DSC, NMR and CSLM. Starch 54:508–516

Gorrasi G, Tortora M, Vittoria V, Galli G, Chiellini E (2002) Transport and mechanical properties of blends of poly (3-caprolactone) and a modified montmorillonite-poly (3-caprolactone) nanocomposite. J Polym Sci B 40:1118–1124

Gorrasi G, Tortora M, Vittoria V, Pollet E, Alexandre M, Dubois P (2004) Physical properties of poly (3-caprolactone) layered silicate nanocomposites prepared by controlled grafting polymerization. J Polym Sci B 42:1466–1475

Gu HW, Ho PL, Tong E, Wang L, Xu B (2003) Presenting vancomycin on nanoparticles to enhance antimicrobial activities. Nano Lett 3:1261–1263

Guilbert S, Cuq B, Gontard N (1997) Recent innovations inedible and/or biodegradable packaging materials. Food Additives Cont 14(6):741–751

Gutierrez-Tauste D, Domenech X, Casan-Pastor N, Ayllon JA (2007) Characterization of methylene blue/TiO_2 hybrid thin films prepared by the liquid phase deposition (LPD) method: application for fabrication of light-activated colorimetric oxygen indicators. J Photochem Photobiol A Chem 187:45–52

Hankermeyer CR, Tjeerdema RS (1999) Polyhydroxybutyrate: plastic made and degraded by microorganisms. Rev Environ Contam Toxicol 159:1–24

Haynie DT, Zhang L, Zhao W, Rudra JS (2006) Protein-inspired multilayer nanofilms: science, technology and medicine. Nanomed Nanotechnol Biol Med 2:150–157

Hu AW, Fu ZH (2003) Nanotechnology and its application in packaging and packaging machinery. Pack Eng 24:22–24

Huang L, Li DQ, Lin YJ, Wei M, Evans DG, Duan X (2005) Controllable preparation of nano-MgO and investigation of its bactericidal properties. J Inorg Biochem 99:986–993

Ishiaku US, Pang KW, Lee WS, Ishak ZAM (2002) Mechanical properties and enzymic degradation of thermoplastic and granular sago starch filled poly (3-caprolactone). Eur Polym J 38:393–401

Jawahar P, Balasubramanian M (2006) Preparation and properties of polyesterbased nanocomposite gel coat system. J Nanomater 4 [article ID 21656]

Johnston JH, Borrmann T, Rankin D, Cairns M, Grindrod JE, McFarlane A (2008) Nano-structured composite calcium silicate and some novel applications. Cur App Phys 8(3–4):504–507

Joseph T, Morrison M (2006) Nanotechnology in agriculture and food. www.nanoforum.org/nf06~modul~showmore~folder~99999~scid~377~.html?action=longview_publication

Kang S, Pinault M, Pfefferle LD, Elimelech M (2007) Single-walled carbon nanotubes exhibit strong antimicrobial activity. Langmuir 23:8670–8673

Kaplan DL (1998) Biopolymers from renewable resources. Springer, Berlin

Kim M, Pometto OR III (1994) Food packaging potential of some novel degradable starch-polyethylene plastics. J Food Protec 57:1007–1012

Kim B, Kim D, Cho D, Cho S (2003) Bactericidal effect of TiO_2 photocatalyst on selected food-borne pathogenic bacteria. Chemosphere 52(1):277–281

Kim TY, Lee YH, Park KH, Kim SJ, Cho SY (2005) A study of photocatalysis of TiO_2 coated onto chitosan beads and activated carbon. Res Chem Interm 31(4–6):343–358

Kirwan MJ, Strawbridge JW (2003) Plastics in food packaging. In: Coles R, McDowell D, Kirwan MJ (eds) Food packaging technology. Blackwell, London, pp 174–240

Koo OM, Rubinstein I, Onyuksel H (2005) Role of nanotechnology in targeted drug delivery and imaging: a concise review. Nanomed Nanotechnol Biol Med 1:193–212

Kumar R, Munstedt H (2005) Silver ion release from antimicrobial polyamide/silver composites. Biomaterials 26:2081–2088

Kuswandi B (2017) Environmental friendly food nano-packaging. Environ Chem Lett 15:205–221. https://doi.org/10.1007/s10311-017-0613-7

Kuswandi B, Wicaksono Y, Jayus AA, Heng LY, Ahmad M (2011) Smart packaging: sensors for monitoring of food quality and safety. Sens Inst Food Qual Safe 5:137–146

Kuswandi BJ, Restanty A, Abdullah A, Heng LY, Ahmad M (2012) A novel colorimetric food package label for fish spoilage based on polyaniline film. Food Cont 25:184

Lagaron JM, Cabedo L, Cava D, Feijoo JL, Gavara R, Gimenez E (2005) Improving packaged food quality and safety. Part 2: nanocomposites. Food Additives Cont 22(10):994–998

Lee CH, An DS, Park HJ, Lee DS (2003) Wide spectrum antimicrobial packaging materials incorporating nisin and chitosan in the coating. Pack Technol Sci 16:99–106

Lee SR, Park HM, Lim HL, Kang T, Li X, Cho WJ (2002) Microstructure, tensile properties, and biodegradability of aliphatic polyester/clay nanocomposites. Polymer 43:2495–2500

Lee SK, Sheridan M, Mills A (2005) Novel UV-activated colorimetric oxygen indicator. Chem Mater 17(10):2744–2751

Lenz RW, Marchessault RH (2005) Bacterial polyesters: biosynthesis, biodegradable plastics and biotechnology. Biomacromolecules 61:1–8

Li B, Rozas J, Haynie DT (2006) Structural stability of polypeptide nanofilms under extreme conditions. Biotechnol Prog 22:111–117

Li H, Li F, Wang L, Sheng J, Xin Z, Zhao L et al (2009) Effect of nano-packing on preservation quality of Chinese jujube (Ziziphus jujuba Mill. var. inermis (Bunge) Rehd). Food Chem 114 (2):547–552

Liao F, Chen C, Subramanian V (2005) Organic TFTs as gas sensors for electronic nose applications. Sens Actuators B 107(2):849–855

Lim ST, Hyun YH, Choi HJ, Jhon MS (2002) Synthetic biodegradable aliphatic polyester/montmorillonite nanocomposites. Chem Mater 14:1839–1844

Liu W, Yang H, Wang Z, Dong L, Liu J (2002) Effect of nucleating agents on the crystallization of poly(3-hydroxybutyratecohydroxy valerate). J Appl Polym Sci 86:2145–2152

Lin YJ, Li DQ, Wang G, Huang L, Duan X (2005) Preparation and bactericidal property of MgO nanoparticles on c-Al_2O_3. J Mater Sci Mater Med 16:53–56

Lopez-Rubio A, Gavara R, Lagaron JM (2006) Bioactive packaging: turning foods into healthier foods through biomaterials. Trends Food Sci Technol 17:567–575

Lotfi M, Tajik H, Moradi M, Forough M, Divsalar E, Kuswandi B (2018) Nanostructured chitosan/monolaurin film: preparation, characterization and antimicrobial activity against Listeria monocytogenes on ultrafiltered white cheese. Lebenson Wiss Technol 92:576–583. https://doi.org/10.1016/j.lwt.2018.03.020

Luduena LN, Alvarez VA, Vasquez A (2007) Processing and microstructure of PCL/clay nanocomposites. Mater Sci Eng A:121–129

Luo PG, Stutzenberger FJ (2008) Nanotechnology in the detection and control of microorganisms. In: Laskin AI, Sariaslani S, Gadd GM (eds) Advances in applied microbiology, vol 63. Elsevier, London, pp 145–181

Maness PC, Smolinski S, Blake DM, Huang Z, Wolfrum EJ, Jacoby WA (1999) Bactericidal activity of photocatalytic TiO_2 reaction: toward an understanding of its killing mechanism. Appl Environ Microbiol 65(9):4094–4098

Mangiacapra P, Gorrasi G, Sorrentino A, Vittoria V (2006) Biodegradable nanocomposites obtained by ball milling of pectin and montmorillonites. Carbohydr Polym 64:516–523

Maisanaba S, Pichardo S, Jordá-Beneyto M, Aucejo S, Cameán AM, Jos Á (2014a) Cytotoxicity and mutagenicity studies on migration extracts from nanocomposites with potential use in food packaging. Food Chem Toxicol 66:366–372

Maisanaba S, Gutiérrez-Praena D, Puerto M, Llana-Ruiz-Cabello M, Pichardo S, Moyano R, Blanco A, Jordá-Beneyto M, Jos A (2014b) In vivo toxicity evaluation of the migration extract of an organomodified clay-poly(lactic) acid nanocomposite. J Toxicol Environ Health A 77 (13):731–746

Maiti P, Batt CA, Giannelis EP (2003) Renewable plastics: synthesis and properties of PHB nanocomposites. Polym Mater Sci Eng 88:58–59

Marras SI, Kladi KP, Tsivintzelis I, Zuburtikudis I, Panayiotou C (2008) Biodegradable polymer nanocomposites: the role of nanoclays on the thermomechanical characteristics and the electrospun fibrous structure. Acta Biomater 4(3):756–765

Mbhele ZH, Salemane MG, van Sittert CGCE, Nedeljkovic JM, Djokovic V, Luyt AS (2003) Fabrication and characterization of silver–polyvinyl alcohol nanocomposites. Chem Mater 15 (26):5019–5024

McGlashan SA, Halley PJ (2003) Preparation and characterization of biodegradable starch-based nanocomposite materials. Poly Int 52:1767–1773

Miller G, Senjen R (2008). Out of the laboratory and on to our plates—nanotechnology in food and agriculture. Available at http://www.foeeurope.org/activities/nanotechnology/Documents/Nano_food_report.Pdf

Mills A, Doyle G, Peiro AM, Durrant J (2006) Demonstration of a novel, flexible, photocatalytic oxygen-scavenging polymer film. J Photochem Photobiol A Chem 177:328–331

Mills A, Hazafy D (2009) Nanocrystalline SnO_2-based, UVB-activated, colourimetric oxygen indicator. Sens Actuators B 136(2):344–349

Mirzadeh A, Kokabi M (2007) The effect of composition and draw-down ratio on morphology and oxygen permeability of polypropylene nanocomposite blown films. Eur Polym J 43 (9):3757–3765

Mohanty AK, Misra M, Drzal LT (2005) Natural fibers, biopolymers, and biocomposites. CRC, Boca Raton

Monteiro-Riviere NA, Nemanich RJ, Inman AO, Wang YY, Riviere JE (2005) Multi-walled carbon nanotube interactions with human epidermal keratinocytes. Toxicol Lett 155(13):377–384

Moraru CI, Panchapakesan CP, Huang Q, Takhistov P, Liu S, Kokini JL (2003) Nanotechnology: a new frontier in Food Science. Food Technol 57:24–29

Morillon V, Debeaufort F, Blond G, Capelle M, Voilley A (2002) Factors affecting the moisture permeability of lipid based edible films: a review. Crit Rev Food Sci Nutr 42:67–89

Murariu M, Ferreira AS, Pluta M, Bonnaud L, Alexandre M, Duboi P (2008) Polylactide (PLA)–CaSO4 composites toughened with low molecular weight and polymeric ester-like plasticizers and related performances. Eur Polym J 44:3842–3852

Nachay K (2007) Analyzing nanotechnology. Food Technol 61(1):34–36

Nair LS, Laurencin CT (2007) Biodegradable polymers as biomaterials. Prog Polym Sci 32:762–798

Nakayama A, Kawasaki N, Maeda Y, Arvanitoyannis I, Ariba S, Yamamoto N (1997) Study of biodegradability of poly (3-valerolactone-co-L-lactide). J Appl Polym Sci 66:741–748

Ogata N, Jimenez G, Kawai H, Ogihara T (1997) Structure and thermal/mechanical properties of poly (L-lactide)-clay blend. J Polym Sci B 35:389–396

Okada M (2002) Chemical syntheses of biodegradable polymers. Prog Polym Sci 27:87–133

Okamoto M, Morita S, Kim HY, Kotaka T, Tateyama H (2001) Dispersed structure change of smectic clay/poly(methyl methacrylate) nanocomposites by copolymerization with polar comonomer. Polymer 42:1201–1206

Oliva J, Paya P, Camara MA, Barba A (2007) Removal of famoxadone, fluquinconazole and trifloxystrobin residues in red wines: effects of clarification and filtration processes. J Environ Sci Health B 42:775–781

Page K, Palgrave RG, Parkin IP, Wilson M, Savin SLP, Chadwick AV (2007) Titania and silver–titania composite films on glass-potent antimicrobial coatings. J Mater Chem 17(1):95–104

Park SH, Choi HJ, Lim ST, Shin TK, Jhon MS (2001) Viscoelasticity of biodegradable polymer blends of poly(3-hydroxybutyrate) and poly(ethylene oxide). Polymer 42:5737–5742

Park HW, Lee WK, Park CY, Cho WJ, Ha CS (2003) Environmentally friendly polymer hybrids: Part I. Mechanical, thermal, and barrier properties of thermoplastic starch/clay nanocomposites. J Mater Sci 38:909–915

Paul M-A, Alexandre M, Degee P, Henrist C, Rulmont A, Dubois P (2003) New nanocomposite materials based on plasticized poly(L-lactide) and organo-modified montmorillonites: thermal and morphological study. Polymer 44:443–450

Pehanich M (2006) Small gains in processing, packaging. Food Proc 11:46–48

Petersson L, Oksman K (2006) Preparation and properties of biopolymer based nanocomposite films using microcrystalline cellulose. In: Oksman K, Sain M (eds) Cellulose nanocomposites, processing, characterization and properties. ACS symposium series 938. Oxford University Press, Oxford, pp 132–150

Petersen K, Nielsen PV, Bertelsen G, Lawther M, Olsen MB, Nilssonk NH et al (1999) Potential of biobased materials for food packaging. Trends Food Sci Technol 10:52–68

Pitt CG (1990) Poly-caprolactone and its copolymers. In: Chasin M, Langer R (eds) Biodegradable polymers as drug delivery systems. Marcel Dekker, New York, pp 71–120

Pluta M, Galeski A, Alexandre M, Paul M-A, Dubois P (2002) Polylactide/montmorillonite nanocomposites and microcomposites prepared by melt blending: structure and some physical properties. J Appl Polym Sci 86(6):1497–1506

Qi LF, Xu ZR, Jiang X, Hu C, Zou X (2004) Preparation and antibacterial activity of chitosan nanoparticles. Carbohydr Res 339:2693–2700

Rajesh, Takashima W, Kaneto K (2004) Amperometric phenol biosensor based on covalent immobilization of tyrosinase onto an electrochemically prepared novel copolymer poly(N-3-aminopropyl pyrrole-copolymer) film. Sens Actuators B 102:271–277

Ranjan S, Dasgupta N, Chakraborty AR, Samuel SM, Ramalingam C, Shanker R, Kumar A (2014) Nanoscience and nanotechnologies in food industries: opportunities and research trends. J Nanopart Res 16(6):2464. https://doi.org/10.1007/s11051-014-2464-5

Rasal RM, Janorkar AV, Hirt DE (2010) Poly(lactic acid) modifications. Prog Polym Sci 33:338–356

Ravichandran R (2010) Nanoparticles in drug delivery: potential green nanobiomedicine applications. Int J Nanotechnol Biomed 1:108–130

Reddy MP, Venugopal A, Subrahmanyam M (2007) Hydroxyapatite-supported Ag–TiO_2 as Escherichia coli disinfection photocatalyst. Water Res 41:379–386

Roberts, R. 2007. The role of nanotechnology in brand protection. Packaging Digest, January 2007. Available at www.packagingdigest.com/articles/200701/34.p

Robertson GL (ed) (1993) Food packaging: principles and practice. Marcel Dekker, New York

Robertson JMC, Robertson PKJ, Lawton LA (2005) A comparison of the effectiveness of TiO_2 photocatalysis and UVA photolysis for the destruction of three pathogenic micro-organisms. J Photochem Photobiol A Chem 175(1):51–56

Reynolds, G., (2007). FDA recommends nanotechnology research, but not labelling FoodProductionDailycom News. 26 July 2007. Available at www.foodproductiondailyusa.com/news/ng.asp?n=78574

Rhim JW (2004) Increase in water vapor barrier property of biopolymer-based edible films and coatings by compositing with lipid materials. J Food Sci Biotechnol 13:528–535

Rhim JW, Ng PKW (2007) Natural biopolymer-based nanocomposite films for packaging applications. Crit Rev Food Sci Nutr 47(4):411–433

Russo GM, Nicolais V, Di Maio L, Montesano S, Incarnato L (2007) Rheological and mechanical properties of nylon 6 nanocomposites submitted to reprocessing with single and twin-screw extruders. Polym Degrad Stab 92(10):1925–1933

Sahl HG, Kordel M, Benz R (1987) Voltage-dependent depolarization of bacterial membranes and artificial Iipid bilayers by the peptide antibiotic nisin. Arch Microbiol 149:120–124

Scott G (2000) Green polymers. Polym Degrad Stab 68:1–7

Scrinis G, Lyons K (2007) The emerging nanocorporate paradigm: nanotechnology and the transformation of nature, food and agrifood systems. Int J Sociol Food Agri 15(2):22–44

Silvestre C, Duraccio D, Sossio C (2011) Food packaging based on polymer nanomaterials. Prog Polym Sci 36(1):1766–1782

Sinha Ray S, Maiti P, Okamoto M, Yamada K, Ueda K (2002a) New polylactide/layered silicate nanocomposites. Preparation, characterization and properties. Macromolecules 35:3104–3110
Sinha Ray S, Yamada K, Okamoto M, Ogami A, Ueda K (2003) New polylactide/layered silicate nanocomposites. High-performance biodegradable materials. Chem Mater 15:1456–1465
Sinha Ray S, Yamada K, Okamoto M, Ueda K (2002b) New polylactide/layered silicate nanocomposite: a novel biodegradable material. Nano Lett 2:1093–1096
Sinha Ray S, Bousmina M (2005) Biodegradable polymers and their layered silicate nanocomposites: in greening the 21st century materials world. Prog Mater Sci 50:962–1079
Smith JP, Hoshino J, Abe Y (1995) Interactive packaging involving sachet technology. In: Rooney ML (ed) Active food packaging. Blackie Academic and Professional, Glasgow, pp 143–173
Smits ALM, Ruhnau FC, Vliegenthart JFG (1998) Ageing of starch based systems as observed by FT-IR and solid state NMR spectroscopy. Starch 50(11–12):478–483
Siracusa V, Rocculi P, Romani S, Dalla RM (2008) Biodegradable polymers for food packaging: a review. Trends Food Sci Technol 19:634–643
Stewart CM, Tompkin RB, Cole MB (2002) Food safety: new concepts for the new millennium. Innov Food Sci Emer Technol 3:105–112
Steinbuchel A (2003) General aspects and special applications, Biopolymers. Wiley-VCH, Weinheim
Stoimenov P, Klinger RL, Marchin GL, Klabunde KJ (2002) Metal oxide nanoparticles as bactericidal agents. Langmuir 18:6679–6686
Sorrentino A, Gorrasi G, Vittoria V (2007) Potential perspectives of bionanocomposites for food packaging applications. Trends Food Sci Technol 18(2):84–95
Sozer N, Kokini JL (2009) Nanotechnology and its applications in the food sector. Trends Biotechnol 27(2):82–89
Tan W, Zhang Y, Szeto YS, Liao L (2008) A novel method to prepare chitosan/montmorillonite nanocomposites in the presence of hydroxyl-aluminum olygomeric cations. Compos Sci Technol 68(14):2917–2921
Tharanathan RN (2003) Biodegradable films and composite coatings: past, present and future. Trends Food Sci Technol 14(3):71–78
Tortora M, Vittoria V, Galli G, Ritrovati S, Chiellini E (2002) Transport properties of modified montmorillonite-poly(3-caprolactone) nanocomposites. Macromol Mater Eng 287(4):243–249
Trznadel M (1995) Biodegradable polymer materials. Int Poly Sci Technol 22(12):58–65
Uyama H, Kuwabara M, Tsujimoto T, Nakano M, Usuki A, Kobayashi S (2003) Green nanocomposite from renewable resources: plant oil–clay hybrid materials. Chem Mater 15:2492–2494
Vermeiren L, Devlieghere F, Van Beest M, de Kruijf N, Debevere J (1999) Developments in the active packaging of foods. Trends Food Sci Technol 10:77–86
Warheit DB, Laurence BR, Reed KL, Roach DH, Reynolds GAM, Webb TR (2004) Comparative pulmonary toxicity assessment of single-wall carbon nanotubes in rats. Toxicol Sci 77:117–125
Weber CJ (ed) (2000) Biobased packaging materials for the food industry. (Food Biopack Project, EU Directorate 12). The Royal Veterinary and Agricultural University, Frederiksberg
Weiss J, Takhistov P, McClements DJ (2006) Functional materials in food nanotechnology. J Food Sci 71(9):107–116
Wilhelm HM, Sierakowski MR, Souza GP, Wypych F (2003) Starch film reinforced with mineral clay. Carbohydr Polym 52:101–110
Xiao-e L, Green ANM, Haque SA, Mills A, Durrant JR (2004) Light-driven oxygen scavenging by titania/polymer nanocomposite films. J Photochem Photobiol A Chem 162:253–259
Xu Y, Ren X, Hanna MA (2006) Chitosan/clay nanocomposite film preparation and characterization. J Appl Polym Sci 99(4):1684–1691
Yan SS, Gilbert JM (2004) Antimicrobial drug delivery in food animals and microbial food safety concerns: an overview of in vitro and in vivo factors potentially affecting the animal gut microflora. Adv Drug Deliv Rev 56:1497–1521

Yoon SY, Deng Y (2006) Clayestarch composites and their application in papermaking. J Appl Polym Sci 100(2):1032–1038

Yu YH, Lin CY, Yeh JM, Lin WH (2003) Preparation and properties of poly (vinyl alcohol)–clay nanocomposite materials. Polymer 44(12):3553–3560

Zheng JP, Li P, Ma YL, Yao KD (2002) Gelatine montmorillonite hybrid nanocomposite. I. Preparation and properties. J App Poly Sci 86:1189–1194.

Chapter 17
Applications of Polymeric Nanoparticles in Food Sector

Norizah Abdul Rahman

17.1 Polymeric Nanoparticles

Nanotechnology is a branch of manufacturing where dimensions on the order of a nanometer are important. Nanotechnology consists of science, engineering, and technology that conducted at the nanoscale, which is in the range of 1–100 nm. Applications of nanotechnology have been emphasized on the characterization, fabrication, and manipulation of nanostructures materials. Some unique features of matter develop when the size of the matter is in the nano range, and these new properties are that open up new opportunities to be explored. Rapid development of nanotechnology is expected to transform many areas including food science and food industry with increasing investment and market demand.

Polymeric nanoparticles generally are defined as solid colloidal particles with a normal size in the range of 10–1000 nm (Sabliov and Astete 2015). Polymeric nanoparticle structure consists of a polymeric particle incorporated with active component and usually surrounded with surfactant to stabilize the nanoparticles in a system (Sabliov and Astete 2015). The polymeric nanoparticles have two types of structures, which are nanospheres and nanocapsules. Nanospheres are matrix systems incorporated with bioactive compound are entrapped, chemically bound or absorbed to the polymer matrix. The bioactive compound is dispersing within the nanospheres polymers as shown in Fig. 17.1. Meanwhile, nanoencapsulates are a system which active compound is located in a core of nanoparticle that surrounded with a thin layer of polymer.

Polymer nanoparticles have attracted the interest of many research groups and have been utilized in an increasing number of fields during the last decades. The field

N. Abdul Rahman (✉)
Department of Chemistry, Faculty of Science, Universiti Putra Malaysia, Serdang, Selangor Darul Ehsan, Malaysia
e-mail: a_norizah@upm.edu.my

S. Siddiquee et al. (eds.), *Nanotechnology: Applications in Energy, Drug and Food*,
https://doi.org/10.1007/978-3-319-99602-8_17

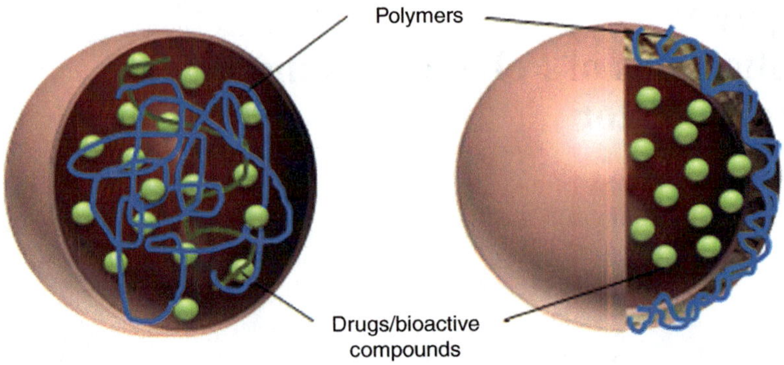

Fig. 17.1 Schematic representation of a nanosphere and a nanocapsule (Source: Reference with copyright permission)

of polymer nanoparticles is speedily growing and playing an essential role in a wide range of fields including electronics, sensors, medicine, biotechnology, environmental, and much more, during the past decades (Rao and Geckeler 2011). This trend is based on their excellent properties, which meet various applications and market needs. In the food industry for example, polymeric nanoparticles can be utilized to improve food quality such as extending shelf life, increase food safety, lower the cost, and enhance the nutritional benefits (Sozer and Kokini 2009). Different preparation technology could produce polymeric nanoparticles with different properties; thus, method preparation of polymeric nanoparticles has to be optimized depending on the particular application of interest.

In drug delivery application, polymeric nanoparticles prepared can be designed to provide protection of entrapped drugs, this may reduce side effects of the drug, overcome solubility limitation and stability of drug compound. Encapsulation of drug in the nanoparticle also could increase activity, and able to control the release of the drug. These indicate the importance of polymeric nanoparticles in disease treatment (Elzoghby et al. 2012). The same advantages also have the potential to transform the food industry by using the polymeric nanoparticles. In food industry, the polymeric nanoparticles can be formulated to encapsulate bioactive compound in the food and to control release the bioactive. The control release can stimulate interactions between the bioactive with the food materials for enhanced the flavor, texture, and also to increase nutritional of the food. The ultimate goal is the polymeric particles can be utilized as potential bioactive delivery systems for development of new food products with better quality, could prevent disease, and to improve the health of consumer.

Despite the very interesting results reported in literature, there is limitation of the use of polymeric nanoparticles in the food industry. The nanoparticles may cause health issues due to the host exposed repeatedly to nanoengineered particles for long periods of time (Sabliov and Astete 2015). Thus, in food application, it is crucial to use food grade materials in formulation of the polymeric nanoparticles, which is natural, biodegradable, biocompatible and nontoxic. A significant number of studies have done nanoparticle synthesis in the form of natural polymers that can be categorized mainly into two main groups which are protein (Jia et al. 2016) (e.g. albumin casein, zein, and β-lactoglobulin), and polysaccharides (Liu et al. 2008) such as alginic acid, carrageenan, xanthan gum, and chitosan. These natural polymers are replacing synthetic polymers that commonly used for polymeric nanoparticle preparation such as ξ-caprolactone (Muller et al. 2001), poly(lactic-*co*-glycolic acid) (PLGA) (Pool et al. 2012a) and poly(lactic acid) (PLA) (Zambaux et al. 1999).

17.2 Applications of Polymeric Nanoparticle in Food Sector

In food industry, it is important to incorporate substances that can address several limitations in food in processing and food formulation. The application polymeric nanoparticle could overcome the limitations by increase quality of food matrices, reduce nutritional losses during the food processing, add new flavor or texture to the food, offer release of aroma for longer period of time, enhance the stability, absorption and functionality of active compounds. These are some of the important application in which polymeric nanoparticles are foreseen to contribute in the future.

A number of techniques are available that can be used to produce polymer nanoparticles, such as solvent evaporation, salting-out, dialysis, supercritical fluid technology, micro-emulsion, mini-emulsion, surfactant-free emulsion, and interfacial polymerization (Rao and Geckeler 2011). The choice of technique preparation of polymeric nanoparticles is depending on a few factors, for example, particle size, particle size distribution, type of polymers, incorporated drug or bioactive compounds, type of application, etc. The details of methods available for the preparation of polymeric nanoparticle have been described in a number of reviews (Rao and Geckeler 2011; Sabliov and Astete 2015).

Several studies have applied polymeric nanoparticles to encapsulate many compounds, such as, phytonutrients (Sahoo et al. 2015), prebiotics (Zabot et al. 2016), bioactive compounds (López-Rubio and Lagaron 2012), antioxidants or some flavor enhancers (Li et al. 2014). Various types of polymers are used to prepare the polymeric nanoparticles including synthetic and natural polymers. However, for food application, it is important to use food-grade components for nanoparticle synthesis formulation. In this chapter, the focus is more on natural polymer based nanoparticles and its application in food sector.

17.2.1 Encapsulation of Bioactive Molecules of Polymeric Nanoparticles

In control release application, polymeric nanoparticle can be classified into two big categories, which are nanocapsules and nanosphere. Nanocapsule is when the systems in which the bioactive compound is confined to a cavity surrounded by a unique polymer membrane. It involved in the delivery of the desired component, or entrapment of the odor and unwanted components in the food and thereby resulting in the preservation of the food (Pradhan et al. 2015). In the biological system, nanocapsules carry the food supplements through the gastrointestinal tract and protect it from the harsh environment. This leads to increase the bioavailability of the substance. On the other hand, nanosphere is the matrix systems in which the bioactive compound is physically and uniformly dispersed. This is because these systems have very high surface areas, compounds may also be adsorbed on their surface (Sabliov and Astete 2015).

The advantages of the usage of polymeric nanoparticles are, it can increase the stability of any volatile compounds, and can be easily and cheaply fabricated in large quantities. Polymeric nanoparticles have significant advantages over traditional oral and intravenous methods of administration in terms of efficiency and effectiveness. It can deliver a higher concentration of the desired compound to the target. The choice of polymer and the ability to modify compounds release from polymeric nanoparticles have made them ideal candidates for many applications especially in food sectors.

17.2.2 Natural Polymers Used in the Formulation of Polymeric Nanoparticles

Two major classes of natural polymers have been extensively studied for the formulation of polymeric nanoparticles, which are natural carbohydrate and protein.

17.2.2.1 Natural Carbohydrates

Carbohydrates are molecules consisting of a basic unit, usually a six-membered ring, containing carbon and oxygen, either monosaccharides, oligosaccharides, or polysaccharides. Carbohydrates are found in a wide variety of foods. The important sources of carbohydrate are obtain directly from natural sources (known as natural carbohydrate) such as cereals (wheat, maize, rice), potatoes and fruits, and from processed carbohydrates (also known as refined carbohydrate) such as table sugar, bread, etc. Carbohydrates have various biological functions, such as energy storage, protein modification and regulation, and act as structural components of cell membranes.

Natural carbohydrates are commonly used for the preparation of polymeric nanoparticles. The role of these carbohydrates is to protect the bioactive compound or drug from the harsh conditions in the stomach. The structure of carbohydrate can be modified for optimal compatibility with the active components and also for the purpose of controlled release. The common polysaccharides used for the production of polymeric nanoparticles are alginates, chitosan, chitin, pectins, dextran, starch and inulin.

Alginates

Alginate or alginic acid is an anionic polysaccharide usually extracted from brown seaweed. Alginate is also an important component of the biofilms produced by the bacterium *Pseudomonas aeruginosa*. Its color ranges from white to yellowish-brown. It is sold in filamentous, granular or powdered forms. Due to its biocompatibility, low toxicity, relatively low cost, it has been extensively studied for many biomedical applications (Lee and Mooney 2012).

The preparation method of alginate nanoparticles has been discussed in details in review paper by Paques et al. (2014). Generally, two methods are primarily used to synthesize alginate nanoparticle. First is complexation, where alginate nano-aggregates can be formed from complex formation in an aqueous solution (Rajaonarivony et al. 1993) and alginate nanocapsule will form when the complex formation occur on the interface of an oil droplet (Lertsutthiwong et al. 2008; Fessi et al. 1989). The addition of calcium chloride is used for the complexation of alginate. Different degrees of cross-linking of alginic acid can be produced by varying the amounts of $CaCl_2$, which can be designed to control the release of the active component over time (Rastogi et al. 2007). Second is water-in-oil emulsification which forming alginate nanospheres. Machado et al. (2012) has reported the water-in-oil emulsions were produced from mixtures of the nonionic surfactant tetraethylene glycol monododecyl ether (C12E4), decane, and aqueous solutions of up to 2 wt% sodium alginate by means of the phase inversion temperature emulsification method. This method allows the preparation of finely dispersed emulsions without a large input of mechanical energy. With alginate concentrations of 1–2 wt % in the aqueous phase, emulsions showed good stability and narrow, monomodal distributions of droplets with radii <100 nm.

Chitosan

Chitosan is a linear polysaccharide composed of randomly distributed β-(1→4)-linked D-glucosamine (deacetylated unit) and *N*-acetyl-*D*-glucosamine (acetylated unit). Chitosan is the most important derivative of chitin. It is prepared by partial deacetylation of chitin under alkaline conditions. It has been widely used in food and bioengineering industries, including the encapsulation of active food ingredients, in enzyme immobilization, and as a carrier for controlled drug delivery, due to its significant biological and chemical properties such as biodegradability, biocompatibility, bioactivity, and polycationicity (Zhao et al. 2011).

A number of methods, such as the emulsion method (Fang et al. 2009), ionic gelation method (Ohya et al. 1994; Yokohama et al. 1998; Kataoka et al. 2000), reverse micellar method (Brunel et al. 2008), self-assembling method (Liu et al.

2005; Chen et al. 2003), etc., have been used to prepare chitosan nanoparticles. Jang and Lee (2008) investigated the stability and characteristics of vitamin C-loaded chitosan nanoparticles. The nanoparticles were prepared by ionic gelation of chitosan with tripolyphosphate (TPP). The results demonstrate the stability of chitosan nanoparticles for vitamin C during heat processing and suggest the possible use of vitamin C-loaded chitosan nanoparticles to enhance antioxidant effects because of the continuous release of vitamin C from chitosan nanoparticles. This indicated applicability of the system in food processing. Hu et al. (2008) investigated the process of fabricating chitosan nanoparticles to be used as carriers for delivering tea catechins. The release profile of tea catechins *in vitro* demonstrated that the controlled release of tea catechins using chitosan-TPP nanoparticles was achievable.

Starch

Starch is a low cost natural polymer has been studied for the past few decades due to its renewability and biodegradability. They were useful material because it is natural abundant polysaccharide produced as a storage polymer from plant (Nagasawa et al. 2004). Commonly, starch available in two major components and appears as a mixture of two glycosidic macromolecules which are very different in structure and properties. Linear amylose consisting of α-(1→4)-linked D-glucose, while amylopectin having the same backbone as amylose but with different myriads which α-(1→6)-linked branch points (Nagasawa et al. 2004; Ahmad et al. 1999).

In recent study by Saari et al. (2017), they investigated non-solvent precipitation to obtain nanosized starch. The starch particle size was found to be in the range 100–200 nm. The first step in such a method is to completely dissolve the starch granules (dissolution). New smaller particles were then re-created from the dissolved starch molecules by non-solvent precipitation. Starch granules are insoluble in cold water due to their inter-chain hydrogen bonding. During the dissolution process, starch granules disintegrate completely and the intermolecular and intramolecular hydrogen bonds are disrupted, thereby destroying the crystalline structure of the granule. In the study, non-solvent precipitation was carried out using ethanol at different starch solution-to-ethanol ratios, 1:1 and 1:10, and the starch solutions obtained with the two techniques; autoclave and microwave oven.

In another study, starch from Gadong tuber was used as a precursor material to synthesis starch nanoparticles by using acid hydrolysis method (Sisika et al. 2015). The nanoparticles were prepared by hydrolyzing the starch for seven days in sulfuric acid. Native Gadong starch was dispersed in an aqueous sulfuric acid solution and the dispersion was stirred for 7 days at 40 °C. On the seventh day, the suspension was centrifuged (20,000 rpm, 20 min) and washed using distilled water for several times until the suspension was neutralized. The starch particles were either round or irregular shape, with diameters ranging from 96–110 nm.

Pectins

Pectin is a structural heteropolysaccharide contained in the primary cell walls of terrestrial plants. It is mainly extracted from citrus fruits, and is used in food as a gelling agent, particularly in jams, jellies, and preserves. It is also used in dessert fillings, medicines, sweets, as a stabilizer in fruit juices and milk drinks, and as a source of dietary fiber. Arroyo-Maya and McClements (2015) prepared whey protein isolates and beet pectin composite nanoparticles using thermal processing and electrostatic complexation. The nanoparticles were encapsulated anthocyanin-rich extracts by the coacervation method. The nanoparticles had a diameter of less than 200 nm. The loading efficiency of anthocyanins was found 55%, and the nanoparticles displayed enhanced heat stability.

17.2.2.2 Protein

Proteins are natural food grade polymers and can be seen as a suitable material for the preparation of nanoparticles and microparticles. Protein offer several benefits, for example, their digestibility, price and capability to interact with a wide range of compounds and nutrients.

Zein

Zein is the major storage protein of corn and consists of about 45–50% of the protein in corn (Shukla and Cheryan 2001). Zein is a class of prolamine protein found in maize (corn). It is usually manufactured as a powder from corn gluten meal. Pure zein is odorless, tasteless, water-insoluble, and edible, and it has a variety of industrial especially in food industry. The ability of zein and its resins to form grease-proof coatings and their resistance to microbial attack have been of commercial interest. Zein also has potential applications include use in fiber, adhesive, coating, ceramic, ink, cosmetic, textile, chewing gum and biodegradable plastics (Anderson and Lamsa 2011). Zein has high thermal resistance and great oxygen barrier properties, which suitable candidate for the encapsulation of sensitive compounds to oxidation and/or temperature (Anderson and Lamsa 2011).

Zou et al. (2012) prepared cranberry procyanidins (CPs)-zein and blank zein nanoparticles by using a modified liquid-liquid dispersion method. The particle size of the CPs-zein nanoparticles increased from 392 to 447 nm with the increase of the CPs-to-zein mass ratios from 1:8 to 1:2. The loading efficiency of CPs in the CPs-zein nanoparticles decreased with an increase of CPs-to-zein mass ratios from 1:8 to 1:2, and ranged from 10 to 86%. In other study, Alqahtani et al. (2017) prepared the core-shell nanoparticles by using the phase separation method. This nanoparticle preparation is based on the differential solubility of zein and the milk protein. Zein is soluble in the hydroalcoholic solvent, but is insoluble in the aqueous phase. In contrast, β-casein and lactoferrin are insoluble in the hydroalcoholic phase.

The phase change leads to the colloidal aggregation of zein and self-assembly of milk proteins (β casein and lactoferrin) formed around the zein core. The size of the zein nanoparticle was less than 200 nm with a low polydispersity index.

Casein

Another interesting protein for micro- and nanoparticle design is casein. Casein is commonly found in mammalian milk, comprising 80% of the proteins in cow's milk and between 20 and 45% of the proteins in human milk (Kunz et al. 1990). Casein has a wide variety of uses, main ingredient of cheese production and also can be used as a food additive. Casein possesses many functional groups in its chemical structure and physicochemical properties that promote its functionality in drug delivery systems (Semo et al. 2007). Thus, casein-based nanoparticles have been proposed for delivering hydrophobic bioactives and drugs including vitamin D2 (Semo et al. 2007) and curcumin (Esmaili et al. 2011).

Penalva et al. (2015) synthesized casein nanoparticles for the oral delivery of folic acid. These nanoparticles were prepared by a coacervation process, stabilized with either lysine or arginine and, finally, dried by spray-drying. The resulting nanoparticles showed a mean size approximately 150 nm and a folic acid content of about 25 μg per mg nanoparticle. From the *in vitro* release studies, the results showed that casein nanoparticles acted as gastro-resistant devices, and thus, folic acid was only released under simulated intestinal conditions.

Albumins

Albumins are simple water-soluble proteins found in animal fluids and tissues and to a lesser extent plants. Albumins are nontoxic, non-immunogenic, biocompatible, biodegradable, easy to purify and soluble in water, their ease to scale up during manufacture compared to other drug delivery systems and thus make it potential candidate for nanoparticle preparation (Elzoghby et al. 2012).

Albumin-based nanoparticle as bioactive compound carrier are an attractive strategy, since a significant amount of compound can be incorporated into the particle matrix because of the different type of binding sites present in the albumin molecule structure. Albumin nanoparticles can be prepared by emulsion evaporation. The aqueous phase with dispersed albumin was added to an organic phase to obtain a water/oil (w/o) emulsion. The emulsion was homogenized with a high-speed homogenizer and the albumin was stabilized by heating (Yang et al. 2007; Jahanshahi and Babaei 2008), or by denaturation in the presence of a cross-linker such as formaldehyde or glutaraldehyde (Jahanshahi and Babaei 2008). Centrifugation was used to recover the albumin nanoparticles, which showed a size range from 500 to 900 nm as a function of the synthesis parameters (i.e., temperature, protein concentration, emulsification speed) (Yang et al. 2007).

17.2.3 The Encapsulated Compounds

17.2.3.1 Phenolic Compounds

Many plants which are rich in phenolic compounds possess several health benefits, such as preventing urinary tract infections, stomach ulcers, and dental diseases (Jia et al. 2016). When present in plants, these phenolic compounds are generally stable and bioactive. However, the compounds are prone to degradation after extraction process, and it is a challenge in foods processing. Encapsulation is a process where the compounds of interest are enclosed by one or a mixture of polymers. Thus, the polymers will protect the phenolic compounds from rapid degradation and helps to control the release of these compounds (Jia et al. 2016).

Flavonoids are polyphenolic compounds found in many plants and fruits that have been reported to exhibit certain bioactive properties that may be beneficial to human health e.g., antioxidant activity. Quercetin (3,3′,4′,5,7-pentahydroxy flavone) is a flavonoid that has been reported to have a particularly high antioxidant activity (Zhang et al. 2011). A delivery system containing polymeric (Eudragit) nanoparticles has been developed by Pool et al. (2012b) for encapsulation and controlled release of quercetin. The nanoparticles were fabricated using a solvent displacement method. The control release study showed quercetin remained trapped within the polymeric nanoparticles under acidic pH conditions, but was released under neutral conditions due to dissolution of the polymer matrix. The delivery systems may be suitable for encapsulating bioactive components within food, but releasing them within the small intestine where they can be adsorbed.

In other study conducted by Alqahtani et al. (2017), quercetin was loaded in poly (lactic-*co*-glycolic acid) (PLGA) nanoparticles was prepared by using three different methods, which including nanoprecipitation, salting out and single emulsion solvent evaporation. Among the three methods, the lowest particle size (~300 nm) with high entrapment efficiency (>90%) was obtained by the single-emulsion solvent evaporation method. The antimicrobial activity of the nanoparticles was evaluated, and the results showed that quercetin-loaded PLGA nanoparticles increased the duration and effectiveness of antimicrobial activity only against gram-positive bacteria.

The health benefits related with tea consumption have been attributed to the polyphenolic constituents, the flavan-3-ols, which known as catechins. Catechins showed promising results in preclinical studies as chemopreventive agents (Shirakami et al. 2012). However, the usage of catechin is limited, mostly due to its instability towards oxygen, change in pH, temperature and light, as well as their inefficient systemic delivery and low bioavailability (Sanna et al. 2015). The encapsulation of catechins into micro- and nanosystems is a useful approach to protect these bioactive compounds from undesirable effects of environmental conditions, thus, retaining the stability of the compounds until the time of consumption. Moreover, this approach provides carriers able to prevent the degradation during digestion, thus enhancing subsequent bioactivity and bioavailability and to promote a controlled release as well as targeted delivery.

In vitro release profiles of encapsulated polyphenols from nanoparticles were investigated by Sanna et al. (2015) in simulated gastrointestinal fluids. The antioxidant activity and stability of encapsulated extract were evaluated. The nanoparticles showed ability to control the polyphenols release where 20% of the polyphenols was released in simulated gastric medium, and 80% released after 5 h at pH 7.4. The results revealed that the white tea extract retained its antioxidant activity and the degradation of tea polyphenols was prevented by the nanoparticles. This is new opportunity for the utilization of white tea extract-loaded nanoparticles for nutraceutical applications.

17.2.3.2 Essential Oils

Essential oils are a concentrated liquid containing volatile, aromatic liquids with therapeutic activities that extracted from plants. They are commonly used for flavoring food and drink, cosmetics, perfumes, and pharmaceutical drugs. Essential oils such as turmeric oil and lemongrass oil are well-known in the pharmaceutical applications, and their active component can be used as antibacterial, antifungal, antioxidant, antimutagenic, and anticarcinogenic properties (Natrajan et al. 2015). However, similar to phenolic compounds, these essential oils are insoluble in water, unstable and volatile which hinder their use for new formulations. Encapsulation of the essential oil in nanoparticles is one of the ways to overcome these problems. Zein nanoparticles were applied for nanoencapsulation of essential oils by the liquid-liquid dispersion method (Wu et al. 2012). The study showed that encapsulation of essential oils in zein nanoparticles allows their dispersion in water, which greatly enhances their potential for use in food preservation without hindering their ability to scanvenge free radicals or to control *Escherichia coli* growth.

In other study, performed by Lertsutthiwong et al. (2009), turmeric oil-loaded chitosan-alginate nanocapsules were prepared. Turmeric oil is widely used in pharmaceutical and cosmetic applications because of its antibacterial, antifungal, antioxidant, and insect-repellent properties (Lertsutthiwong and Rojsitthisak 2011). However, turmeric oil is volatile, insoluble in water and unstable in certain environments, which causes difficulties with formulation of products. The preparation of nanocapsules for encapsulation of turmeric oil was performed using a simple procedure comprising of three steps: oil/water emulsification, gelification and solvent removal (Lertsutthiwong and Rojsitthisak 2011). The results show turmeric oil-loaded chitosan alginate nanocapsules were stable for up to 4 months at 4 and 25 °C (Lertsutthiwong et al. 2009). This is possibly due to strong ionic interactions between chitosan and alginate that lead to the formation of a polyelectrolyte complex, and resulting stabilization and decrease of the porosity of the alginate particles.

17.2.3.3 Flavor

Flavors are sensory impression of food, determined primarily from taste and smell to enhance the overall eating experience. This make flavors one of most important components in the food system. A food system is a natural encapsulation system that stabilizes nutrition, aroma, and flavor compounds in nano-/microstructures (Nakagawa 2014). During chewing and digestion, some parts of the encapsulated compounds are released from the food due to deformation of the food structure. To keep optimal flavor release during consumption is a big challenge for food engineers.

Nanoencapsulation techniques have widely been used to control flavor release and retention. The encapsulation helps to protect flavors from degradation, to stabilize and prolong shelf-life, and to enable controlled flavor release during consumption (Nakagawa 2014). A spray-drying technique is one of methods available for direct nanoparticle fabrication. The sizes of nanoparticle produced have effect on the encapsulations of flavor; the smaller particles (<38 μm) have lower surface oil content but poorer flavor retention than do larger ones (>63 μm) (Jafari et al. 2007).

17.2.3.4 Color Pigments

Foods from fruits and vegetables are naturally colored mainly by four groups of pigments: the green chlorophylls, the yellow-orange-red carotenoids, the red-blue-purple anthocyanins and the red betanin (He and Hwang 2016). Most of these natural pigments can be isolated from plant are contain hydrocarbon and low solubility in water. Thus, the application of these natural pigments in water-based food applications is limited due to their poor solubility in water. In addition, lack of stability of these natural compounds is a drawback. It would be of interest to the food industry to package these pigments in polymeric nanoparticles that would enable their dispersion in water, enhance its stability and prevent their degradation.

17.3 Solid-Phase Extraction Adsorbent for Chemical Residue in Food

A few studies have conducted on application of polymeric nanoparticle as adsorbent for separation of chemical residue in food. The first study was done by Zargar et al. (2016), they prepared zein nanoparticles as a dispersive solid-phase extraction (DSPE) adsorbent has been used for the preconcentration of dye additive, azorubine as a model compound in foodstuffs. The main advantages of use of zein are (1) zein biopolymer is a biocompatibility, biodegradability, and low toxicity biomaterials; (2) easy fabrication of zein nanoparticles; (3) it has relatively suitable stability; and

(4) fabrication of composites or conjugates formed between zein and other compounds as nanomaterials are relatively easy. The results demonstrated that the proposed method could be useful and practical in the future monitoring of azorubine in real samples.

17.4 Safety Concerns

The researches related to application of nanotechnology in the food sector are keeps increasing, and drive the expanding of the potential application of nanotechnology in food industry. Thus, the possibility of human contact to these nanomaterials also increased significantly either intentionally or unintentionally (Magnuson et al. 2011). However, only a few studies have focused on the potential toxicity of the presence of nanomaterials in foods, by analyzing food samples used in food additives/ingredients and food packaging (Amini et al. 2014).

Until now, our knowledge regarding the safety of used nanomaterials in food and nutrition industries is relatively very low. Bioavailability, biodistribution and the toxicity of these nanomaterials are very important to have deep understanding. Nanomaterials that used as food additives will have direct contact with internal human organs. The levels of exposure of the nanomaterials on human liable on their concentration in food and the quantity of that food consumed. Increasing uses of nanomaterial substances in foods have attracted major attention of public and government sectors (Jovanović 2015).

Nanoencapsulation could enter human's body through oral intake. For example, lipid nanoparticle was developed by using polyethylene glycol (PEG) with various chain lengths as emulsifiers to control their digestion fate in the gastrointestinal tract (Ban et al. 2018). However, the safety of nanoencapsulation remains uninvestigated and calls for further risk assessment (Borel and Sabliov 2014) particularly for long-term toxicity (Jovanović 2015; McCracken et al. 2013).

17.5 Conclusion

Significant progress has been made by researchers to study the application of polymeric nanoparticles in food sector. New materials, systems, processes, and formulations are being developed to solve problems and to create new potential application of the nanoparticles in the food industry. The focuses are more on delivery systems to improve stability, bioavailability, and functionality of the encapsulated bioactive compounds, and exploring new potential application of polymeric nanoparticles such as adsorbent for chemical residue in food. However, the questions are still remaining on the safety of these nanoparticles to human being.

References

Ahmad FB, Williams PA, Doublier JL, Durand S, Buleon A (1999) Physico-chemical characterisation of sago starch. Carbohydr Polym 38(4):361–370

Alqahtani MS, Islam MS, Podaralla S, Kaushik RS, Reineke J, Woyengo T, Perumal O (2017) Food protein based core-shell nanocarriers for oral drug delivery: effect of shell composition on in vitro and in vivo functional performance of zein nanocarriers. Mol Pharm 14:757–769

Anderson TJ, Lamsa BP (2011) Zein extraction from corn, corn products, and coproducts and modifications for various applications: a review. Cereal Chem 88(2):159–173

Arroyo-Maya IJ, McClements DJ (2015) Biopolymer nanoparticles as potential delivery systems for anthocyanins: fabrication and properties. Food Res Int 69:1–8

Amini SM, Gilaki M, Karchani M (2014) Safety of nanotechnology in food industries. Electron Physician 6(4):962–968

Ban C, Jo M, Lim S, Choi YJ (2018) Control of the gastrointestinal digestion of solid lipid nanoparticles using PEGylated emulsifiers. Food Chem 239:442–452

Borel T, Sabliov CM (2014) Nanodelivery of bioactive components for food applications: types of delivery systems, properties, and their effect on ADME profiles and toxicity of nanoparticles. Annu Rev Food Sci Technol 5:197–213

Brunel F, Varon L, David L, Domard A, Delair T (2008) A novel synthesis of chitosan nanoparticles in reverse emulsion. Langmuir 24(20):11370–11377

Chen XG, Lee CM, Park HJ (2003) O/W emulsification for the self-aggregation and nanoparticle formation of linoleic acids modified chitosan in the aqueous system. J Agric Food Chem 51:3135–3139

Elzoghby AO, Samy WM, Elgindy NA (2012) Albumin-based nanoparticles as potential controlled release drug delivery systems. J Control Release 157:168–182

Esmaili M, Ghaffari SM, Moosavi-Movahedi Z, Atri MS, Sharifizadeh A, Farhadi M et al (2011) Beta casein-micelle as a nano vehicle for solubility enhancement of curcumin; food industry application. Lebenson Wiss Technol 44(10):2166–2172

Fang H, Huang J, Ding L, Li M, Chen Z (2009) Preparation of magnetic chitosan nanoparticles and immobilization of laccase. J Wuhan Univ Technol Mater Sci Ed 24(1):42–47

He X, Hwang H-M (2016) Nanotechnology in food science: Functionality, applicability, and safety assessment. J Food Drug Anal 24(4):671–681

Hu B, Pan C, Sun Y, Hou Z, Ye H, Zeng X (2008) Optimization of fabrication parameters to produce chitosan tripolyphosphate nanoparticles for delivery of tea catechins. J Agric Food Chem 56(16):7451–7458

Fessi H, Puisieux F, Devissaguet JP, Ammoury N, Benita S (1989) Nanocapsule formation by interfacial polymer deposition following solvent displacement. Int J Pharm 55(1):R1–R4

Jafari SM, He Y, Bhandari B (2007) Role of powder particle size on the encapsulation efficiency of oils during spray drying. Drying Technol 25:1081–1089

Jahanshahi M, Babaei Z (2008) Protein nanoparticle: a unique system as drug delivery vehicles. Afr J Biotechnol 7:4926–4934

Jang K-I, Lee HG (2008) Stability of chitosan nanoparticles for l-ascorbic acid during heat treatment in aqueous solution. J Agric Food Chem 56(6):1936–1941

Jia Z, Dumont M-J, Orsat V (2016) Encapsulation of phenolic compounds present in plants using protein matrices. Food Biosci 15:87–104

Jovanović B (2015) Critical review of public health regulations of titanium dioxide, a human food additive. Integr Environ Assess Manag 11:10–20

Kataoka K, Matsumoto T, Tokohama M, Okano T, Sakurai Y (2000) Doxorubicin-loaded poly (ethylene glycol)-poly(b-benzyl-L-aspartate) copolymer micelles: their pharmaceutical characteristics and biological significance. J. Controlled Release 64:143–153

Kunz C, Lönnerdal B (1990) Casein and casein subunits in preterm milk, colostrum, and mature human milk. J Pediatr Gastroenterol Nutr 10(4):454–461

Lee KY, Mooney DJ (2012) Alginate: properties and biomedical applications. Prog Polym Sci 37:106–126

Lertsutthiwong P, Noomun K, Jongaroonngamsang N, Rojsitthisak P, Nimmannit U (2008) Preparation of alginate nanocapsules containing turmeric oil. Carbohydr Polym 74(2):209–214

Lertsutthiwong P, Rojsitthisak P, Nimmannit U (2009) Preparation of turmeric oil-loaded chitosan-alginate biopolymeric nanocapsules. Mater Sci Eng C 29:856–860

Lertsutthiwong P, Rojsitthisak P (2011) Chitosan-alginate nanocapsules for encapsulation of turmeric oil. Pharmazie 66:911–915

Li W, Li L, Wang Q, Pan Y, Wang T, Wang H, Song R, Deng H (2014) Antibacterial activity of nanofibrous mats coated with lysozyme-layered silicate composites via electrospraying. Carbohydr Polym 99:218–225

Liu CG, Chen XG, Park HJ (2005) Self assembled nanoparticles based on linoleic-acid modified chitosan: stability and adsorption of trypsin. Carbohydr Polym 62:293–298

Liu Z, Jiao Y, Wang Y, Zhou C, Zhang Z (2008) Polysaccharides-based nanoparticles as drug delivery systems. Adv Drug Deliv Rev 60:1650–1662

López-Rubio A, Lagaron JM (2012) Whey protein capsules obtained through electrospraying for the encapsulation of bioactives. Innov Food Sci Emerg Technol 13:200–206

Machado AHE, Lundberg D, Ribeiro AJ, Veiga FJ, Lindman B, Miguel MG et al (2012) Preparation of calcium alginate nanoparticles using water-in-oil (W/O) nanoemulsions. Langmuir 28:4131–4141

Magnuson BA, Jonaitis TS, Card JW (2011) A brief review of the occurrence, use, and safety of food-related nanomaterials. J Food Sci 76:R126–R133

McCracken C, Zane A, Knight DA, Dutta PK, Waldman WJ (2013) Minimal intestinal epithelial cell toxicity in response to short-and long-term food-relevant inorganic nanoparticle exposure. Chem Res Toxicol 26:1514–1525

Muller CR, Schaffazick SR, Pohlmann AR, de Lucca FL, Pesce da Silveira N, Dalla Costa T, Guterres SS (2001) Spray-dried diclofenac-loaded poly(ε-caprolactone) nanocapsules and nanospheres. Preparation and physicochemical characterization. Pharmazie 56(11):864–867

Nagasawa N, Yagi T, Kume T, Yoshii F (2004) Radiation cross-linking of carboxymethyl starch. Carbohydr Polym 58(2):109–113

Nakagawa K (2014) Nano- and microencapsulation of flavor in food systems (Chapter 10). In: Kwak H-S (ed) Nano- and microencapsulation for foods. Wiley, Oxford, pp 249–272

Natrajan D, Srinivasan S, Sundar K, Ravindran A (2015) Formulation of essential oil-loaded chitosan-alginate nanocapsules. J Food Drug Anal 23:560–568

Ohya Y, Shiratani M, Kobayashi H, Ouchi T (1994) Release behavior of 5-fluorouracil from chitosan-gel nanospheres immobilizing 5-fluorouracil coated with polysaccharides and their cell specific cytotoxicity. Pure Appl Chem 31:629–642

Paques JP, van der Linden E, van Rijn CJM, Sagis LMC (2014) Preparation methods of alginate nanoparticles. Adv Colloid Interface Sci 209:163–171

Penalva R, Esparza I, Agüeros M, Gonzalez-Navarro CJ, Gonzalez-Ferrero C, Irache JM (2015) Casein nanoparticles as carriers for the oral delivery of folic acid. Food Hydrocoll 44:399–406

Pool H, Quintanar D, de Dios FJ, Mano CM, Bechara JEH, Godínez LA, Mendoza S (2012a) Antioxidant effects of quercetin and catechin encapsulated into PLGA nanoparticles. J Nanomater 2012:145380

Pool H, Quintanar D, Figueroa JD, Bechara JE, McClements DJ, Mendoza S (2012b) Polymeric nanoparticles as oral delivery systems for encapsulation and release of polyphenolic compounds: Impact on quercetin antioxidant activity and bioaccessibility. Food Biophys 7:276–288

Pradhan N, Singh S, Ojha N, Shrivastava A, Barla A, Rai V, Bose S (2015) Facets of nanotechnology as seen in food processing, packaging, and preservation industry. Biomed Res Int 2015:365672

Rajaonarivony M, Vauthier C, Couarraze G, Puisieux F, Couvreur P (1993) Development of a new drug carrier made from alginate. J Pharm Sci 82:912–917

Rao JP, Geckeler KE (2011) Polymer nanoparticles: preparation techniques and size-control parameters. Prog Polym Sci 36:887–913

Rastogi R, Sultana Y, Aqil M, Ali A, Kumar S, Chuttani K, Mishra A (2007) Alginate microspheres of isoniazid for oral sustained drug delivery. Int J Pharm 334(1–2):71–77

Saari H, Fuentes C, Sjöö M, Rayner M, Wahlgren M (2017) Production of starch nanoparticles by dissolution and nonsolvent precipitation for use in food-grade Pickering emulsions. Carbohydr Polym 157:558–566

Sabliov CM, Astete CE (2015) Polymeric nanoparticles for food applications. In: Sabliov CM, Chen H, Yada RY (eds) Nanotechnology and functional foods: effective delivery of bioactive ingredients. Wiley, Chichester, p 272

Sahoo N, Sahoo RK, Biswas N, Guha A, Kuotsu K (2015) Recent advancement of gelatin nanoparticles in drug and vaccine delivery. Int J Biol Macromol 81:317–331

Sanna V, Lubinu G, Madau P, Pala N, Nurra S, Mariani A, Sechi M (2015) Polymeric nanoparticles encapsulating white tea extract for nutraceutical application. J Agric Food Chem 63 (7):2026–2032

Semo E, Kesselman E, Danino D, Livney YD (2007) Casein micelle as a natural nano-capsular vehicle for nutraceuticals. Food Hydrocoll 21(5):936–942

Shirakami Y, Shimizu M, Moriwaki H (2012) Cancer chemoprevention with green tea catechins: from bench to bed. Curr Drug Targets 13:1842–1857

Shukla R, Cheryan M (2001) Zein: the industrial protein from corn. Indust Crops Prod 13:171–192

Sisika R, Wan Ahmad WY, Fazry S, Mat Lazim A. Preparation and characterization of polymeric nanoparticles from Gadong starch, The 2015 UKM FST Postgraduate Colloquium, 050023

Sozer N, Kokini JL (2009) Nanotechnology and its applications in the food sector. Trends Biotechnol 27:82–89

Wu Y, Luo Y, Wang Q (2012) Antioxidant and antimicrobial properties of essential oils encapsulated in zein nanoparticles prepared by liquid-liquid dispersion method. Lebenson Wiss Technol 48(2):283–290

Yang L, Cui F, Cun D, Tao A, Shi K, Lin W (2007) Preparation, characterization and biodistribution of the lactone form of 10-hydroxycamptothecin (HCPT)-loaded bovine serum albumin (BSA) nanoparticles. Int J Pharm 340:163–172

Yokohama M, Fukushima S, Uehara R, Okamoto K, Kataoka K (1998) Characterization of physical entrapment and chemical conjugation of adriamycin in polymeric micelles and their design for in vivo delivery to a solid tumor. J. J Control Release 50:79–92

Zabot G, Silva EK, Azevedo VM, Meireles MAA (2016) Replacing modified starch by inulin as prebiotic encapsulant matrix of lipophilic bioactive compounds. Food Res Int 85:26–35

Zambaux MF, Bonneaux F, Gref R, Dellacherie E, Vigneron C (1999) Preparation and characterization of protein C-loaded PLA nanoparticles. J Control Release 60:179–188

Zargar B, Pourreza N, Bayat E, Hatamie A (2016) Zein bio-nanoparticles: a novel green nanopolymer as a dispersive solid-phase extraction adsorbent for separating and determining trace amounts of azorubine in different foodstuffs. RSC Adv 6:73096–73105

Zhang M, Swarts SG, Yin L, Liu C, Tian Y, Cao Y et al (2011) Antioxidant properties of quercetin. Adv Exp Med Biol 915:283–289

Zhao L-M, Shi L-E, Zhang Z-L, Chen J-M, Shi D-D, Yang J, Tang Z-X (2011) Preparation and application of chitosan nanoparticles and nanofibers. Braz J Chem Eng 28(3):353–362

Zou T, Li Z, Percival SS, Bonard S, Gu LW (2012) Fabrication, characterization, and cytotoxicity evaluation of cranberry procyanidins–zein nanoparticles. Food Hydrocoll 27(2):293–300

Chapter 18
Acetylcholinesterase (AChE) Biosensors for Determination of Carbamate Pesticides

Anwar Samsidar and Shafiquzzaman Siddiquee

18.1 Introduction

Pesticides are chemicals compounds practically used for controlling pests in agricultural sector, in homes and gardens, and in soil treatment and management. The globally escalation of the human population, the applications of pesticides have increased in the production and quality of agricultural products. Carbamate remains as the most important class of pesticides in terms of use. They are effectively eliminated pests/insects when applied extensively. However, the excessive and improper uses, it has created serious problem to both humans and environment. The uses approximately 0.1% of pesticides are reached the target pests, the remaining spreads in the non-targeted organisms including soil and other environmental matrices (Hart and Pimentel 2002).

Some pesticides may have dreadful toxic effects on non-target organisms because of their long half-lives, which cause them to persist in the environment (Smith and Gangolli 2002). Epidemiologists warn that pesticides currently available on the market may trigger cancer in non-target organisms, including human beings. Farmers and agricultural workers are the most at risk and suffer from unintentional pesticide poisoning every year (Alavanja et al. 2005; IARC 2003). The Codex Alimentarius Commission of the Food and Agriculture Organization (FAO) and the World Health Organization (WHO) have established the maximal residue limits (MRLs) for pesticides in a variety of foods. Because of the risks of acute toxicity and long-term damage from bioaccumulation in humans, rapid measurement of pesticide traces in environment, food and living things is highly needed (Verma and Bhardwaj 2015).

A. Samsidar · S. Siddiquee (✉)
Biotechnology Research Institute, Universiti Malaysia Sabah, Kota Kinabalu, Sabah, Malaysia
e-mail: shafiqpab@ums.edu.my

S. Siddiquee et al. (eds.), *Nanotechnology: Applications in Energy, Drug and Food*,
https://doi.org/10.1007/978-3-319-99602-8_18

Analytical methods, particularly chromatography and electrophoresis, are in use to detect carbamate residues at trace concentrations (Fernández-Ramos et al. 2014; Tegeler and El Rassi 2001), and paired with selective detection systems, so they are reliable and efficient (Sassolas et al. 2012). However, these techniques have the drawbacks of complex and time-consuming extraction procedures and the use of high volumes of organic solvents. In addition, expensive and sophisticated instruments, and skilled workers are required to conduct the pesticides analysis.

Biosensors are analytical devices in which biological elements integrated onto physical transducers to detect the target analytes (Guilbault et al. 2004). Biosensor advantages are rapid, simple, real-time, sensitive, and suitable for on-site analysis of target compounds (Badihi-Mossberg et al. 2007). Various types of analytes are used and merged with transducers, including enzymes, antibodies, microorganisms, whole cells, and DNA. Enzyme-based biosensors mostly developed for the detection of pesticide residues in matrices such as food, soil, ground, and surface water (Dyk and Pletschke 2011). The main aim of this book chapter is to discuss the available enzyme-based biosensors in the determination of carbamate pesticide residues and elaborate several enzyme immobilization techniques and carbamate toxicity effects.

18.2 Action and Toxicity of Carbamate Pesticides

Carbamates mostly applied as insecticides for crop protection against insects. Besides, carbamates are used such as fungicides on fruits, vegetables, ornamental plants, and seeds. The main advantages of carbamate are their high insecticidal action and low persistence on plants as they can degrade rapidly within weeks or months (IARC 1976). They act against a vast spectrum of pests by inhibiting cholinesterase enzymes, affecting the nervous system and causing neurotoxicity (Fukuto 1990). Carbamate metabolites may be more active than their parent substance; for instance, aldicarb sulfoxide, a metabolite of aldicarb, is a more effective cholinesterase inhibitor than the parent compound (Yang et al. 1996). Common carbamates like carbaryl, aldicarb, methomyl, oxamyl, primicarb and carbofuran are mainly accumulated in crops, fruits and leafy vegetables. Carbamate is known as endocrine disruptors, due to the intensive use of carbamates and lead to neurotoxicity and other negative human health effects and harmful to the environment (Schulte-Oehlmann et al. 2011).

Carbamates and Organophosphates (OP) have a similarity mechanism of action, as they are capable to inhibit cholinesterases. They are existed to each other in a way that OP irreversibly binds to the active sites of the enzyme while carbamates bind in a reversible way (Ivanov et al. 2000; Cesarino et al. 2012). Irreversible inhibition of the enzyme prevents the reuse of sensor devices which are linked to a reactivation agent, including oxime structures such as pyridine-2-aldoxime and 4-formylpyridinium bromide dioxime (Liu and Lin 2006; Du et al. 2007).

18.3 Enzyme Immobilization Techniques

Enzyme immobilizations become one of the most important aspects in the development of enzyme-based biosensors for determination of pesticides. Lower detection limits obtained in the developed biosensors using enzymes immobilized in solution (Ivanov et al. 2000). The proper enzyme immobilization techniques can influence the performance of the biosensor. Various types of immobilization techniques are available such as adsorption, entrapment, cross-linking and covalent coupling. Immobilization techniques are categorized into physical and chemical approaches as shown in Fig. 18.1 and their schematic illustrations are shown in Fig. 18.2.

18.3.1 Physical Adsorption

Adsorption is a simple and easy enzyme immobilization technique of biological elements onto transducers (Pohanka et al. 2007; Bonnet et al. 2003). In this immobilization process can provide high operational and storage stabilities. The working electrode surface causes simple deposition of enzymes through the formation of chemical bonds (Van der Waals forces, hydrogen bond or electrostatic interactions) between the amino acid residues of enzymes and chemical groups on the surface of the transducer (Ardao et al. 2011; Bayramoglu et al. 2010). AChE biosensor has developed by adsorption on screen-printed electrodes modified with multi-walled carbon nanotubes (MWCNTs). A small amount of AChE solution has placed onto the MWCNTs modified electrode surface and allowed to dry at room temperature after which the electrode rinsed twice to remove any unbound enzyme molecules (Joshi et al. 2005). AChE is physically involved in the adsorption with polyvinylpyrrolidone K 30 (Zou et al. 2006) or calcium carbonate-chitosan composite (Gong et al. 2009a, b). However, this immobilization technique has some

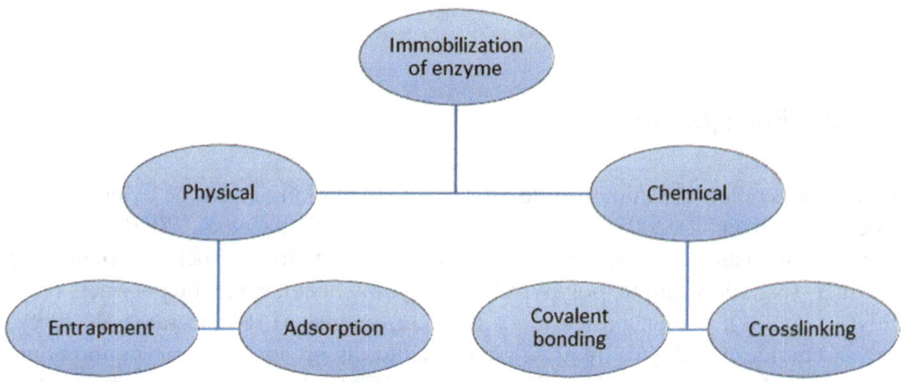

Fig. 18.1 Techniques available for enzyme immobilization in enzyme based biosensors

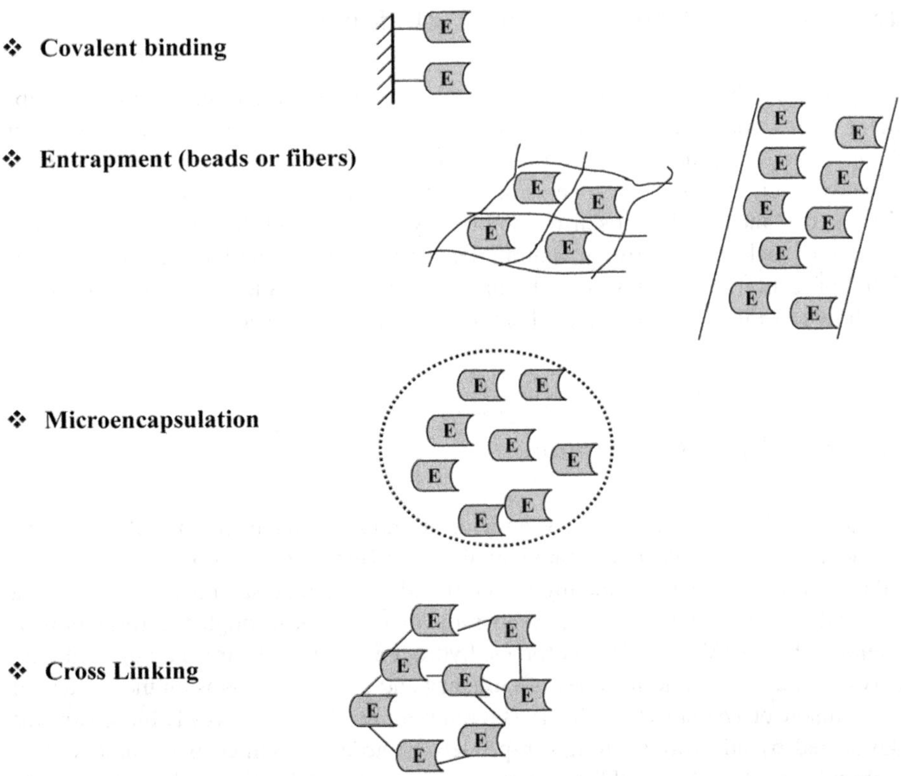

Fig. 18.2 Schematic diagram of the main enzyme immobilization techniques (Brena and Batista-Viera 2006)

limitations; enzymes are not tightly bound and desorption occur with changes in temperature, pH, and ionic strength. Hence, enzymes immobilizing by adsorption onto biosensors have showed in poor operational and storage stability (Sassolas et al. 2012).

18.3.2 Entrapment

Several researchers are immobilized enzymes onto working electrode by entrapping process (Zejli et al. 2008; Kuswandi et al. 2008; Sinha et al. 2010). Enzyme biomolecules can be entrapped in various types of matrices, such as conducting polymers (e.g., polyaniline, polypyrrole or polythiophene), synthetic polymers (e.g., polyacrylamide), carbon pastes (e.g., graphite, graphene) (Svancara et al. 2009), polysaccharide-based gels (e.g., alginate, chitosan or agarose), or in inorganic matrices such as silica or clay. The modified AChE electrodes have entrapped in sol-gel matrices and polymer or membrane lattices with AChE biomolecules (Gill and Ballesteros 2000; Kandimalla et al. 2006; Bucur et al. 2006). Application of polysaccharide-based gels as chitosan has some advantages due to their biocompatibility, ability to form biofilm and susceptibility to chemical modifications (Wang et al. 2008; Salam et al. 2011). Chitosan and other polymers naturally occur as poly (ethyleneglycol) or nafion can be used to avoid any cracking in the sol-gel process. Entrapment considered one of the most convenient methods for enzyme immobilization; however, application of this method has some limitation due to leach out of the bio-elements and disallows diffusion, which in turn reduces the performance of the biosensor (Sassolas et al. 2012).

18.3.3 Cross-linking

Enzyme immobilization through crosslinking is the most frequently used due to reach high stability in biosensors. This method commonly involves bifunctional reagents, such as glutaraldehyde. Several studies have successfully conducted the immobilization of AChE by crosslinking with glutaraldehyde vapor (Waibel et al. 2006; Arkhypova et al. 2004; Gogol et al. 2000). A part from glutaraldehyde, other bifunctional reagents like glyoxal or hexamethylenediamine has used to develop enzymatic biosensors (Sassolas et al. 2012). Bovine serum albumin (BSA) has used as functionally inert protein in conjunction with crosslinking reagents.

Amine et al. (2006) reported on crosslinking with glutaraldehyde attach enzymes to the transducer. The attachment of the enzymes onto the transducer caused by intermolecular links, stable covalent bonds that form between functional groups of AChE and transducer (Wong and Wong 1992; Margolin 1996; Lei et al. 2006). The AChE covalently attached to the transducer surfaces. Other researchers have reported that the enzymatic membrane is attached on the electrode surface with ChE, nafion® and glutaraldehyde layer (Suprun et al. 2003; Ivanov et al. 2003). Zheng et al. (2015) stated that the binding of AChE with glutaraldehyde at the ionic liquid-graphene-gel matrix *via* covalent linkage formed free $-NH_2$ groups of AChE and –CHO groups of glutaraldehyde. However, Nunes et al. (2004) addressed that when crosslinking is used, glutaraldehyde should be the last component in order to prevent deactivation of the enzymes.

18.3.4 Covalent Bonding

Covalent bonding is an appropriate alternative enzyme immobilization technique because of its excellent stability and more biomolecule activity (Shim et al. 2011; Shalini et al. 2014; Zucca and Sanjust 2014). Liu et al. (2015a, b) developed an amperometric carbaryl biosensor based on AChE for the analysis in real sample of cabbage. The LOD obtained was 1.7 nmol L^{-1} with good reproducibility and stability.

In electrochemical studies, different types of electrodes are mostly used such as graphite, carbon paste, gold, glassy carbon and so on. Enzyme biosensors are developed by using different nanomaterials such as ZrO_2 NPs-modified screen printed electrode (Lin et al. 2004), multi-walled carbon nanotubes (MWCNTs) modified glassy carbon electrode (GCE) (Lin et al. 2006), AuNPs-polypyrrole nanowires composite film modified GCE (Gong et al. 2009a, b), Au-MWCNTs modified GCE (Jha and Ramaprabhu 2010), and one-dimensional AuNPs onto GCE (YanRong et al. 2010) for determination of pesticides matrix. Vandeput et al. (2015) addressed a flow-through detector with gold disk immobilized AChE to a silver amperometric working electrode for characterization of several AChE inhibitors. Covalent based enzyme immobilization onto the gold electrode surface allowed the retention of enzymatic activity and facilitated the application of enzyme in a flow-through system for reversible inhibitors. AChE from electric eel is covalently immobilized onto a cysteamine modified gold disk adjacent to a silver disk-working electrode. Immobilization of enzymes by covalent bonding not only solves these problems but also promotes better bioelement activity and stability.

18.3.5 Other Immobilization Techniques

Adsorption, entrapment, crosslinking and covalent bonding are the most common techniques employed for the immobilization of enzymes in biosensors with high accuracy. Enzyme immobilizations are difficult to control due to the various chemical linkages, hydrophobic interactions, hydrogen bonding, and electrostatic attraction. The main forces involved in protein-particle interactions are often not well-organized (Meder et al. 2013). However, new enzyme immobilization techniques developed to overcome the limits of the existing immobilization techniques. Liu et al. (2011) developed a novel amperometric biosensor for detection of OP and carbamate pesticides. They immobilized the enzyme on a specific binding between boronic acid and glycosyl group. The results found that high enzyme activity levels retained after immobilization onto the electrode. Thus, a new enzyme immobilization method has successfully established based on the esterification between boronic acid groups and the glycoprotein of AChE for the development of carbamate biosensors.

18.4 Nanotechnology in Biosensor

Nanotechnology has been acknowedged to play an important role in the biosensors (Vo-Dinh et al. 2001; Haruyama 2003; Jain 2003). Nanotechnology has involved the application of materials, devices and systems typically with ranges smaller than 100 nm. Nanomaterials have proven several profitable aspects of their use in biosensor technology and displayed immense capacity for further development of biosensors. Nanomaterials characterized with unique features such as high surface per volume ratio, high reactivity, good electrical conductivity and high catalytic action (Kerman et al. 2008; Niemeyer 2001).

Nanomaterial-based biosensors developed from the association of applied material science, molecular engineering, and biotechnology. Nanomaterials have assisted to improve the sensitivity and specificity of biomolecule recognition from environment samples. The immobilized nanomaterials offer a novel interface for use as the sensing part of biosensors and enhance the sensitivity of the analytes. There are various classes of nanomaterials according to their chemical composition, organic or inorganic. Inorganic nanomaterials include metals, metal oxides, and quantum dots. While carbon nanotubes, fullerenes and graphene are classified as organic nanomaterials (Brownson and Banks 2010; Chen et al. 2013; Aragay et al. 2012; Altavilla and Ciliberto 2011; Valentini et al. 2004; Suarez-Martinez et al. 2012, Merkoçi 2013; Tamayo et al. 2013; Barberis et al. 2015; Ray 2015).

Zhao et al. (2015) have developed pesticide biosensors based on graphene and gold nanoparticles (AuNPs) for detection of carbamate (carbaryl). The developed methodology for pesticides detection is based on a direct electrodeposition of electrochemical reduced graphene oxide (ERGO)-gold nanoparticles (AuNPs)-ß-cyclodextrin (ß-CD) and Prussian blue-Chitosan (PB-CS) modified GCE. AChE enzyme is immobilized *via* adsorption in biosensor. Inhibition of both pesticides on AChE activity are analyzed which showed a LOD of 1.15 pg/mL.

Zhou et al. (2013) have developed AChE enzyme biosensor based on tin oxide nanoparticles (SnO_2 NPs), carboxylic graphene (CGR) and nafion (NF) modified GCE for detection of carbofuran. The formed nanocomposites on GCE showed excellent conductivity, catalysis and biocompatibility for adhesion of AChE. The LOD for carbofuran detected at 5×10^{-13} M. In a study conducted by Zhai et al. (2013) have incorporated chitosan, Prussian blue, MWCNTs and hollow gold nanospheres (HGNs) to detect carbofuran. The prepared hybrid films have increased electron transfer rate and enhanced electrochemical response of the developed AChE biosensor. The detection limit of carbofuran was calculated equal to 2.5 nM.

Silver nanoparticles (AgNPs) and CGR have implemented along with NF on the GCE in the development of AChE enzyme-based biosensor (Liu et al. 2014). The hybrid AgNPs/CGR/NF possessed excellent conductivity, catalytic activity and created a hydrophilic surface for adhesion of AChE. Addition of chitosan to immobilize the AChE is useful to keep the enzyme activities and the developed AChE biosensor showed high affinity to acetylthiocholine chloride (ATCl) substrate. Exposure to carbamate pesticide reduced the oxidation peak current with the increasing of pesticide concentration. This decrease in current signal correlated with the concentration of pesticide detected. Differential pulse voltammetry (DPV) has measured the peak currents and the LOD for carbamate (carbaryl) calculated to be 5.45×10^{-13} M. The recoveries test showed in the range of 93.1–105.6%, indicating with high accuracy and low matrix effect on the amperometric response.

In the detection of carbamates in fruit and vegetable samples, Kestwal et al. (2015) have developed AChE based biosensor with regards of fenugreek hydrogel-agarose matrix with AuNPs. The prepared hybrid nanomaterials with uniform distribution of AuNPs in the fenugreek hydrogel-agarose-AChE membrane has increased the enzyme retention efficiency and extended the shelf life of the enzyme. The developed AChE biosensor was capable to detect various carbamates such as carbofuran, oxamyl, methomyl and carbaryl, with respective LODs of 2, 21, 113 and 235 nM.

Using a sensing interface of gold electrode based on citrate-capped gold nanoparticles (AuNPs)/3-mercaptopropyl-trimethoxysilane (MPS) denoted as AuNPs/MPS/AuE, Song et al. (2016a, b) used the method to determine carbaryl pesticide. Atomic force microscopy (AFM) has used to study the surface topography and CV method is recorded to measure the inhibition of AChE by the carbaryl. The method showed sensitivity obtained of 32.04 $\mu Acm^{-2}/mM$ with LOD value for carbaryl equal to 1 nM.

Gold nanoparticles (AuNPs) provide a biocompatible microenvironment for biosensing material through immobilized biomolecules on electrode surfaces, which improves the sensitivity of biosensors (Liang et al. 2008; He et al. 2008). Du et al. (2008) have developed a gold nanoparticles-based electrochemical biosensor for determination of pesticides. AuNPs nanoparticles assembled on a sol–gel matrix and derived silicate network to immobilize AChE, which lead to a stable AChE biosensor for determination of pesticides. AChE biosensor was found to be high sensitive, specific, cost-effective, and lower detection limit (1.2×10^{-9} mol L^{-1}).

Carbon nanotubes (CNTs) have novel properties such as unique mechanical, physical, and chemical properties. In the field of nanoelectronics, biomedical engineering, biosensing, and bioanalysis, CNTs can play significant roles (Cui 2007); for example, polymer-CNTs composites have good electrical conductivity and good mechanical strength. Liu and Lin (2006) have fabricated an amperometric biosensor based on self-assembled AChE and CNTs with modified GCE for determination of OP pesticides. The electrocatalytic activity of CNTs highly improved the detection limit with high sensitivity, stability, reproducibility, and higher accuracy. Lower detection limit of paraoxon was found of 4×10^{-13} mol L^{-1} with a 6 minutes inhibition time. AChE and choline oxidase (CHO) based enzyme biosensors provide high sensitivity, wider range, and low detection limits for the analysis of OP compounds. The developed biosensor showed some advantages, such as simple preparation procedures, high sensitivity, and low cost. Nanotechnology offers many profitable aspects for biosensor technology in the pesticides detection. Table 18.1 summarizes the developed AChE based biosensor for determination of carbamate pesticides in agricultural crops.

18.5 Conclusion

Carbamate pesticides are mainly used in the agricultural field to enhance crop production. Due to their pest removal properties, various classes of pesticides have persisted in the environment over a long duration after reaching the maximum effectiveness. Apart from their recalcitrant structure and agriculture benefits,

Table 18.1 AChE enzyme based biosensors for determination of carbamate pesticides

Pesticide	Modifications	Samples	Detection limit	Reference
Carbofuran	AChE/PAMAM[a]-Au/CNTs/GCE	Onion Lettuce Cabbage	4×10^{-9} mol L^{-1}	Qu et al. (2010)
Carbaryl, Methomyl	GC/MWCNT/PANI[b]/AChE	Cabbage Broccoli Apple	1.4, 0.95 μmol L^{-1}	Cesarino et al. (2012)
Carbofuran	NF/AChE-CS/SnO_2NPs[c]-CGR-CS/GCE	Apple Cabbage	5×10^{-13} mol L^{-1}	Zhou et al. (2013)
Carbofuran	NF[d]/AChE/CS-PB-MWCNTs-HGNs/AuE	Cabbage Lettuce Leek Pakchoi	2.5 nmol L^{-1}	Zhai et al. (2013)
Carbofuran	NF/AChE-CS/NiONPs[e]-CGR[f]-NF/GCE	Apple Cabbage	5.0×10^{-13} mol L^{-1}	Yang et al. (2013)
Carbofuran, Aldicarb	AChE/CB[g]/AgNPs[h]/GCE	Peanut Grape	0.1, 10 nmol L^{-1}	Evtugyn et al. (2014)
Carbaryl	AChE-ChO/Au-MPA[i]-SAM	Watermelon juice Pomegranate juice	5.96 nmol L^{-1}	Hatefi-Mehriajrdi (2013)
Carbaryl	CS/AChE/PB-CS/ERGO[j]-AuNPs-β-CD[k]/GCE	Caraway Cabbage Rapeseed	1.15 pg mL^{-1}	Zhao et al. (2015)
Carbaryl	AChE-MWCNTs/GONRs[l]/GCE	Cabbages	1.7 nmol L^{-1}	Liu et al. (2015a)
Carbaryl	GO-IL[m]/GCE	Tomato Grape	0.02 μmol L^{-1}	Liu et al. (2015b)
Carbaryl	MWCNT–(PEI[n]/DNA)2/OPH/AChE	Apple	1 μmol L^{-1}	Zhang et al. 2015
Carbaryl	GA/AChE-IL-GR–Gel/GCE	Tomato juice	5.3×10^{-15} mol L^{-1}	Zheng et al. (2015)
Carbofuran Aldicarb	CB/P[5]A[o]	Peanut Beetroot	2×10^{-11}, 6×10^{-10} mol L^{-1}	Shamagsumova et al. (2015)
Formetanate hyrdocholirde (FMT)	CoPc[p]-fMWCNT/GCE	Mango Grape	9.7×10^{-8} mol dm^{-3}	Ribeiro et al. 2016
Carbaryl	AuNPs/MPS[q]/AuE	Fruit samples	1.0 nmol L^{-1}	Song et al. (2016)
Carbaryl	AChE-e-pGON[r]/GCE	Cabbage Spinach	0.15 ng mL^{-1}	Li et al. (2017)

Note: [a]Dendrimers polyamidoamine, [b]Polyaniline, [c]Tin oxide nanoparticles, [d]Nafion, [e]Nickel oxide nanoparticles, [f]Carboxylic graphene, [g]Carbon black, [h]Silver nanoparticles, [i]Mercaptopropionic acid, [j]Electrochemical reduced graphene oxide, [k]β-cyclodextrin, [l]Graphen oxide nanoribbons, [m]Ionic liquid, [n]Polyethyleneimine, [o]Pillar[5]arene, [p]Cobalt phthalocyanine, [q]Mercaptopropyl)-trimethoxysilane, [r]Electrochemically inducing porous graphene oxide nitrate carbamate pesticides are imposed acute toxicological effects on varying life forms. Their accumulation in living organisms may be harmful at higher concentrations. Thus, their prompt and accurate detection and analysis are crucial substances of concern. Analytical techniques as chromatographic techniques (HPLC, GC, etc.) are associated with various drawbacks of pesticides detection such as low sensitivity and efficiency, time-consuming, laborious, requiring expensive equipment and skilled labor, and so on. Enzyme inhibition-based biosensors offer a new alternative tool for the detection and determination of these neurotoxic compounds. These biosensors are high sensitivity, selectivity, rapidity, onsite handling and lower detection limits. The application of enzyme-based biosensor technology in the environment and agriculture relatively new compared to medical applications. Most commercialized biosensors devices are related to medical applications, only a few tailored for the monitoring of environment samples. Therefore, future research and development are required to produce reliable pesticide detection devices with lower detection limits, high stability, and onspot monitoring capability. To this end, researchers from all over the world are employed towards the development of biosensing techniques based on different variety of approaches, including investigating enzymes, exploring the latest nanomaterials, and applying new enzyme immobilization techniques for the detection of pesticides in the environment.

Acknowledgement This work was supported by grants from the Ministry of Education Malaysia, Fundamental Research Grant Scheme (FRGS) (No. FRGS/1/2014/SG05/UMS/02/4).

References

Alavanja MCR, Hoppin J, Kamel F (2005) Health effects of chronic pesticide exposure: cancer and neurotoxicity. Annu Rev Public Health 25:155–197

Altavilla C, Ciliberto E (2011) Inorganic Nanoparticles: Synthesis, Applications, and Perspectives, CRC Press, 33–68

Amine A, Mohammadi H, Bourais I, Palleschi G (2006) Enzyme inhibition-based biosensors for food safety and environmental monitoring. Biosens Bioelectron 21:1405–1423

Aragay G, Pino F, Merkoçi A (2012) Nanomaterials for sensing and destroying pesticides. Chem Rev 112(10):5317–5338

Ardao I, Alvaro G, Benaiges MD (2011) Reversible immobilization of rhamnulose-1-phosphate aldolase for biocatalysis: enzyme loading optimization and aldol addition kinetic modeling. Biochem Eng J 56:190–197

Arkhypova VN, Martelet C, Jaffrezic-Renault N, Chovelon JM, Elskaya AV, Soldtkin AP (2004) Potentiometric biosensor based on ISFETS and immobilized cholinesterases. Electroanalysis 16:1873–1882

Badihi-Mossberg M, Buchner V, Rishpon J (2007) Electrochemical biosensors for pollutants in the environment. Electroanalysis 19:2015–2028

Barberis A, Spissu Y, Fadda A, Azara E, Bazzu G, Marceddu S, Angioni A, Sanna D, Schirra M, Serra PA (2015) Simultaneous amperometric detection of ascorbic acid and antioxidant capacity in orange, blueberry and kiwi juice, by a telemetric system coupled with a fullerene-or nanotubes-modified ascorbate subtractive biosensor. Biosens Bioelectron 67: 214–223

Bayramoglu G, Metin AU, Altintas B, Arica MY (2010) Reversible immobilization of glucose oxidase on polyaniline grafted polyacrylonitrile conductive composite membrane. Bioresour Technol 101:6881–6887

Bonnet C, Andreescu S, Marty JL (2003) Adsorption: easy and efficient immobilisation of acetylcholinesterase on screen printed electrodes. Anal Chim Acta 481:209–211

Brena BM, Batista-Viera F (2006) Immobilization of Enzymes. In Methods in Biotechnology: Immobilization of Enzymes and Cells, Second Edition, 15–30. Totowa: Humana Press Inc.

Brownson DA, Banks CE (2010) Graphene electrochemistry: an overview of potential applications. Analyst 135(11): 2768–2778

Bucur B, Campas M, Prieto-Simon B, Andreescu S, Marty JL (2006) Enzymatic biosensors for screening carbamate insecticides: application to environmental and food monitoring. Ecol Chem Eng 13:339–348

Cesarino I, Moraes FC, Lanza MRV, Machado SAS (2012) Electrochemical detection of carbamate pesticides in fruit and vegetables with a biosensor based on acetylcholinesterase immobilized on a composite of polyaniline-carbon naotubes. Food Chem 135:873–879

Chen S, Yuan R, Chai Y, Hu F (2013) Electrochemical sensing of hydrogen peroxide using metal nanoparticles: a review. Microchim Acta 180(1-2):15–32

Cui D (2007) Advances and prospects on biomolecules functionalized carbon nanotubes. J Nanosci Nanotechnol 7:1298–1314

Du D, Huang X, Cai J, Zhang AD (2007) Comparison of pesticide sensitivity by electrochemical test based on acetylcholinesterase biosensor. Biosens Bioelectron 23:285–289

Du D, Wang M, Cai J, Tao Y, Tu H, Zhang A (2008) Immobilization of acetycholinesterase based on the controllable adsorption of carbon nanotubes onto an alkanethiol monolayer for carbaryl sensing. Analyst 133:1790–1795

Dyk JSV, Pletschke B (2011) Review on the use of enzymes for the detection of organochlorine, organophosphate and carbamate pesticides in the environment. Chemosphere 82:291–307

Evtugyn GA, Shamagsumova RV, Padnya PV, Stoikov II, Antipin IS (2014) Cholinesterase Sensor Based on Glassy Carbon Electrode Modified With Ag Nanoparticles Decorated With Macrocyclic Ligands. Talanta 127: 9–17

Fernández-Ramos C, Satinsky D, Smidova B, Solich P (2014) Analysis of trace organic compounds in environmental, food and biological matrices using large-volume sample injection in column-switching liquid chromatography. Trends Anal Chem 62:69–85

Fukuto TR (1990) Mechanism of action of organophosphorus and carbamate insecticides. Environ Health Perspect 87:245–254

Gill I, Ballesteros A (2000) Bioencapsulation within synthetic polymers (Part 1): sol-gel encapsulated biologicals. Tibtech 18:282–296

Gogol EV, Evtugyn GA, Marty JL, Budnikov H, Winter VG (2000) Amperometric biosensors based on nafion coated screen printed electrodes for the determination of cholinesterase inhibitors. Talanta 5:379–389

Gong J, Wang L, Zhang L (2009a) Electrochemical biosensing of methyl parathion pesticide based on acetylcholinesterase immobilized onto Au–polypyrrole interlaced network-like nanocomposite. Biosens Bioelectron 24:2285–2288

Gong J, Liu T, Song D, Zhang X, Zhang L (2009b) One step fabrication of three-dimensional porous calcium carbonate-chitosan composite film as the immobilization matrix of acetylcholinesterase and its biosensor on pesticides. Electrochem Commun 11:1873–1876

Guilbault GG, Pravda M, Kreuzer M (2004) Biosensors-42 years and counting. Anal Lett 37:14481–14496

Hart KA, Pimentel D (2002) Environmental and economic costs of pesticide use. In: Pimentel D (ed) Encyclopedia of pest management. Marcel Dekker, New York, pp 237–239

Haruyama T (2003) Micro- and nanobiotechnology for biosensing cellular responses. Adv Drug Deliv Rev 55:393–401

Hatefi-Mehriajrdi A (2013) Bienzyme self-assembled monolayer on gold electrode: an amperometric biosensor for carbaryl determination. Electrochim Acta 114:394–402

He X, Yuan R, Chai Y, Shi Y (2008) A sensitive amperometric immunosensor for carcinoembryonic antigen detection with porous nanogold film and Nano-Au/chitosan composite as immobilization matrix. J Biochem Biophys Methods 70:823–829

IARC (1976) Some carbamates, thiocarbamates and carbazides. International Agency for Research on Cancer, Lyon

IARC (2003) Monographs on the evaluation of carcinogenic risk to human. International Agency for Research on Cancer, vols. 5–53. International Agency for Research on Cancer, Lyon

Ivanov AN, Evtugyn GA, Gyurcsanyi RE, Toth K, Budnikov HC (2000) Comparative investigation of electrochemical cholinesterase biosensors for pesticide determination. Anal Chim Acta 404:55–65

Ivanov AN, Evtugyn G, Budnikov H, Ricci F, Moscone D, Palleschi G (2003) Cholinesterase sensors based on screen-printed electrodes for detection of orgaophoshorus and carbammic acid pesticides. Anal Bioanal Chem 377:624–631

Jain KK (2003) Nanodiagnostics: application of nanotechnology in molecular diagnostics. Expert Rev Mol Diagn 3:153–161

Jha N, Ramaprabhu S (2010) Development of Au nanoparticles dispersed carbon nanotube-based biosensor for the detection of paraoxon. Nanoscale 2:806–810

Joshi KA, Tang J, Haddon R, Wang J, Chen W, Mulchandani A (2005) A disposable biosensor for organophosphorus nerve agents based on carbon nanotubes modified thick film strip electrode. Electroanalysis 17:54–58

Kandimalla VB, Tripathi VS, Ju H (2006) Immobilization of biomolecules in sol-gels: biological and analytical applications: critical review. Anal Chem 36:73–106

Kerman K, Saito M, Tamiya E, Yamamura S, Takamura Y (2008) Nanomaterial-based electrochemical biosensors for medical applications. Trends Anal Chem 27:585–592

Kestwal RM, Bagal-Kestwal D, Chiang BH (2015) Fenugreek hydrogeleagarose composite entrapped gold nanoparticles for acetylcholinesterase based biosensor for carbamates detection. Anal Chim Acta 886:143–150

Kuswandi B, Fikriyah CI, Gani AA (2008) An optical fiber biosensor for chlorpyrifos using a single sol-gel film containing acetylcholinesterase and bromothymol blue. Talanta 74:613–618

Lei Y, Chen W, Mulchandani A (2006) Review microbial biosensors. Anal Chim Acta 568:200–210

Li Y, Shi L, Han G, Xiao Y, Zhou W (2017) Electrochemical biosensing of carbaryl based on acetylcholinesterase immobilized onto electrochemically inducing porous graphene oxide network. Sens Actuators B 238:945–953

Liang LZ, Qi JS, Mu WJ, Chen ZG (2008) Biomolecules/gold nanowires-doped sol-gel film for label-free electrochemical immunoassay of testosterone. J Biochem Biophys Methods 70:1156–1162

Lin S, Liu CC, Chou TC (2004) Amperometric acetylcholine sensor catalyzed by nickel anode electrode. Biosens Bioelectron 20:9–14

Lin TJ, Huang KT, Liu CY (2006) Determination of organophosphorous pesticides by a novel biosensor based on localized surface plasmon resonance. Biosens Bioelectron 22:513–518

Liu G, Lin Y (2006) Biosensor based on self-assembling acetylcholinesterase on carbon nanotubes for flow injection/amperometric detection of organophosphate pesticides and nerve agents. Anal Chem 78:835–843

Liu T, Su H, Qu X, Ju P, Cui L, Ai S (2011) Acetylcholinesterase biosensor based on 3-carboxyphenylboronic acid/reduced graphene oxide–gold nanocomposites modified electrode for amperometric detection of organophosphorus and carbamate pesticides. Sens Actuators B 160:1255–1261

Liu Y, Wang G, Li C, Wang M, Yang L (2014) A novel acetylcholinesterase biosensor based on carboxylic graphene coated with silver nanoparticles for pesticide detection. Mater Sci Eng C 35:253–258

Liu B, Xiao B, Cui L (2015a) Electrochemical analysis of carbaryl in fruit samples on graphene oxide-ionic liquid composite modified electrode. J Food Compos Anal 40:14–18

Liu Q, Fei A, Huan J, Mao H, Wang K (2015b) Effective amperometric biosensor for carbaryl detection based on covalent immobilization acetylcholinesterase on multiwall carbon nanotubes/graphene oxide nanoribbons nanostructure. J Electroanal Chem 740:8–13

Margolin AL (1996) Novel crystalline catalysts. Trends Biotechnol 14:223–230

Meder F, Hintz H, Koehler Y, Schmidt MM, Treccani L, Dringen R, Rezwan K (2013) Adsorption and orientation of the physiological extracellular peptide glutathione disulfide on surface functionalized colloidal alumina particles. Am Chem Soc 135:6307–6316

Merkoçi A (2013) Nanoparticles based electroanalysis in diagnostics applications. Electroanal 25 (1):15–27

Niemeyer CM (2001) Nanoparticles, proteins, and nucleic acids: biotechnology meets materials science. Angew Chem Int Ed 40:4128–4158

Nunes GS, Jeanty G, Marty JL (2004) Enzyme immobilization procedures on screen-printed electrodes used for the detection of anticholinesterase pesticides: comparative study. Anal Chim Acta 523:107–115

Pohanka M, Kuka K, Jun D (2007) Amperometric biosensor for pesticide methamidorphos assay. Acta Medica (Hradec Kralove) 50:239–241

Qu Y, Sun Q, Xiao F, Shi G, Jin L (2010) Layer-by-layer self-assembled acetylcholinesterase/ pamam-Au on CNTs modified electrode for sensing pesticides. Bioelectrochemistry 77:139–144

Ray SC (2015) Application of Graphene and Graphene-Oxide Based Nanomaterials. Waltham, MA: Elsevier

Ribeiro FWP, Lucas FWDS, Mascaro LH, Morais S, Casciano PN, de Lima-Neto P, Correia AN (2016) Electroanalysis of formetanate hydrochloride by a cobalt phthalocyanine functionalized multiwalled carbon nanotubes modified electrode: characterization and application in fruits. Electrochim Acta 194:187–198

Salam MA, Makki MSI, Abdelaal MYA (2011) Preparation and characterization of multi-walled carbon nanotubes/chitosan nanocomposite and its application for the removal of heavy metals from aqueous solution. J Alloys Compd 509:2582–2587

Sassolas A, Blum LJ, Leca-Bouvier BD (2012) Immobilization strategies to develop enzymatic biosensors. Biotechnol Adv 30:489–511

Schulte-Oehlmann U, Oehlmann J, Keil F (2011) Before the curtain falls: endocrine-active pesticides—a German containation legacy. Rev Environ Contam Toxicol 213:137–159

Shalini J, Sankaran KJ, Lee CY, Tai NH, Lin IN (2014) An amperometric urea bisosensor based on covalent immobilization of urease on N_2 incorporated diamond nanowire electrode. Biosens Bioelectron 56:64–70

Shamagsumova RV, Shurpik DN, Padnya PL, Stoikov II, Evtugyn GA (2015) Acetylcholinesterase Biosensor for Inhibitor Measurements Based on Glassy Carbon Electrode Modified With Carbon Black and Pillar[5]arene. Talanta 144:559–568

Shim J, Kim GY, Moon SH (2011) Covalent co-immobilization of glucose oxidase and ferrocenedicarboxylic acid for an enzymatic biofuel cell. J Electroanal Chem 653:14–20

Sinha R, Ganesana M, Andreescu S, Stanciu L (2010) AChE biosensor based on zinc oxide sol-gel for the detection of pesticides. Anal Chim Acta 661:195–199

Smith AG, Gangolli SD (2002) Organochlorine chemicals in seafood: occurence and health concerns. Food Chem Toxicol 40:767–779

Song Y, Chen J, Sun M, Gong C, Shen Y, Song Y, Wang L (2016) A simple electrochemical biosensor based on AuNPs/MPS/Au electrode sensing layer for monitoring carbamate pesticides in real samples. J Hazard Mater 304:103–109

Song Y, Luo Y, Zhu C, Li H, Du D, Lin Y (2016a) Recent advances in electrochemical biosensors based on graphene two-dimensional nanomaterials. Biosens Bioelectron 76:195–212

Song Y, Chen J, Sun M, Gong C, Shen Y, Song Y, Wang L (2016b) A simple electrochemical biosensor based on AuNPs/MPS/Au electrode sensing layer for monitoring carbamate pesticides in real samples. J Hazard Mater 304:103–109

Suarez-Martinez I, Grobert N, Ewels CP (2012) Encyclopedia of Carbon Nanoforms. Advances in Carbon Nanomaterials: Science and Applications. Pan Stanford Publishing: Singapore

Suprun E, Evtugyn G, Budnikov H, Ricci F, Moscone D, Palleschi G (2003) Acetylcholinesterase sensor based on screen printed carbon electrode modified with prussian blue. Anal Bioanal Chem 383:597–604

Svancara I, Vytras K, Kalcher K, Walcarius A, Wang J (2009) Carbon paste electrodes in facts, numbers, and notes: a review on the occasion of the 50-years jubilee of carbon paste in electrochemistry and electroanalysis. Electroanalysis 21:7–28

Tamayo J, Kosaka P M, Ruz JJ, San Paulo A, Calleja M (2013) Biosensors based on nanomechanical systems. Chem Soc Rev 42:1287–1311

Tegeler T, El Rassi Z (2001) Capillary electrophoresis and electrochromatography of pesticides and metabolites. Electrophoresis 22:4281–4293

Valentini F, Orlanducci S, Terranova ML, Amine A, Palleschi G (2004) Carbon nanotubes as electrode materials for the assembling of new electrochemical biosensors. Sensor Actuat: B Chem. 100(1):117–125

Vandeput M, Parsajoo C, Vanheuverzwijn J, Patris S, Yardim Y, le Jeune A, Sarakbi A, Mertens D, Kauffmann JM (2015) Flow-through enzyme immobilized amperometric detector for the rapid screening of acetylcholinesterase inhibitors by flow injection analysis. J Pharm Biomed Anal 102:267–275

Verma N, Bhardwaj A (2015) Biosensor technology for pesticides—a review. Appl Biochem Biotechnol 175:3093–3119

Vo-Dinh T, Cullum BM, Stokes DL (2001) Nanosensors and biochips: frontiers in biomolecular diagnostics. Sens Actuators B Chem 74:2–11

Waibel M, Schulze H, Huber N, Bachmann TT (2006) Screen-printed bienzymatic sensor based on sol-gel immobilised *Nippostrongylus brasilensis* acetylcholinesterase and a cytocrome P450 BM-3 (CYP102-A1) mutant. Biosens Bioelecton 21:1132–1140

Wang X, Zhao X, Liu X, Li Y, Fu L, Hu J, Huang C (2008) Homogenous liquid-liquid extraction combined with gas chromatgraphy-electron capture detector for the determination of three pesticide residues in soils. Anal Chim Acta 620:162–169

Wong SS, Wong LJC (1992) Chemical crosslinking and the stabilization of proteins and enzymes. Enzyme Microb Technol 14:866–874

Yang SS, Goldsmith AI, Smetena I (1996) Review: recent advances in the residue analysis of n-methylcarbamate pesticides. J Chromatogr A 754:3–16

Yang L, Wang G, Liu Y, Wang M (2013) Development of a biosensor based on immobilization of acetylcholinesterase on NiO nanoparticles–carboxylic graphene–nafion modified electrode for detection of pesticides. Talanta 113:135–141

YanRong L, ZhiYong G, YanFen L, Qian L, JianChun B, ZhiHui D, Min H (2010) Immobilization of acetylcholinesterase on one-dimensional gold nanoparticles for detection of organophosphorous insecticides. Science China Chem 53:820–825

Zejli H, Hidalgo-Hidalgo de Cisneros JL, Naranjo-Rodriguez I, Liu B, Temsamani KR, Marty JL (2008) Alumina sol-gel/sonogel-carbon electrode based on acetylcholinesterase for detection of organophosphorus pesticides. Talanta 77:217–221

Zhai C, Sun X, Zhao W, Gong Z, Wang X (2013) Acetylcholinesterase biosensor based on chitosan/prussian blue/multiwall carbon/hollow gold nanospheres nanocomposite film by one-step electrodeposition. Biosens Bioelectron 42:124–130

Zhao H, Ji X, Wang B, Wang N, Li X, Ni R, Ren J (2015) An ultra-sensitive acetylcholinesterase biosensor based on reduced graphene oxide-Au nanoparticles-ß-cyclodextrin/prussian blue-chitosan nanocomposites for organophosphorus pesticides detection. Biosens Bioelectron 65:23–30

Zhang YY, Arugula MA, Wales M, Wild J, Simonian AL (2015) A Novel Layer by Layer Assembled Multi-Enzyme/CNT Biosensor for Discriminative Detection Between Organophosphorus and Non-Organophosphorus Pesticides. Biosens Bioelectron 67:287–295

Zheng Y, Liu Z, Jing Y, Li J, Zhan H (2015) An acetylcholinesterase biosensor based on ionic liquid functionalized graphene-gelatin-modified electrode for sensitive detection of pesticides. Sens Actuators B 210:389–397

Zhou Q, Yang L, Wang G, Yang Y (2013) Acetylcholinesterase biosensor based on SnO_2 nanoparticles–carboxylic graphene–nafion modified electrode for detection of pesticides. Biosens Bioelectron 49:25–31

Zou MQ, Yang R, Wang DN, Li JF, Jin H (2006) A novel immobilized cholinesterase for on-site screening of organophosphate and carbamate compounds. Pestic Biochem Physiol 86:162–166

Zucca P, Sanjust E (2014) Inorganic materials as supports for covalent enzyme immobilization: methods and mechanisms. Molecules 19:14139–14194

Chapter 19
Nanosensors Based Detection of Foodborne Pathogens

Mohd Hazani Mat Zaid, Jerro Saidykhan, and Jaafar Abdullah

19.1 Introduction

Globalization of the food supply has led to the rapid and widespread foodborne illness by pathogen. The pathogen can easily be transmitted into new and distant environment via travel, migration, and liberalization of trade resulting in an increasingly complex and longer global food chain (Saker et al. 2004). It often occurs when new foods and dietary habits has been introduced into new regions enable spread disease more rapidly (Kearney 2010). While another factor can be due to lack of safe food preparation, poor hygiene and transmitted by other routes: through water, soil, or air; by direct contact between people, or by contact between people and animals (Cabral et al. 2010; Pandey et al. 2017). Foodborne pathogens have been a cause of many diseases worldwide and more in the developing countries. Statistics worldwide indicate 600 million foodborne illnesses with 420,000 deaths in 2010 which the highest burden per population was observed in low income regions such as Africa, and South East Asia (WHO 2015).

Typically, most foodborne contaminants cause due to bacteria, parasites, fungi, and virus. Others less common foodborne illness occurs from accidental chemical poisoning and natural contaminants (WHO 2015). According to Centers for Disease

M. H. M. Zaid
Institute of Advanced Technology, Universiti Putra Malaysia, Serdang, Selangor, Malaysia

J. Saidykhan
Department of Chemistry, Faculty of Science, Universiti Putra Malaysia, Serdang, Selangor, Malaysia

J. Abdullah (✉)
Institute of Advanced Technology, Universiti Putra Malaysia, Serdang, Selangor, Malaysia

Department of Chemistry, Faculty of Science, Universiti Putra Malaysia, Serdang, Selangor, Malaysia
e-mail: jafar@upm.edu.my

S. Siddiquee et al. (eds.), *Nanotechnology: Applications in Energy, Drug and Food*,
https://doi.org/10.1007/978-3-319-99602-8_19

Control and Prevention (CDC), most foodborne illness cases are caused by bacteria consist of *Salmonella, Campylobacter*, *E. coli* and *Listeria*, while foodborne cause by viral consist of Norovirus, Rotavirus and Hepatitis A. Some bacteria also cause intoxication which toxin produced by pathogens such as *Staphylococcus aureus*, *Bacillus cereus* and *Clostridium perfringens* which grow and establish themselves in the human intestinal tract (Sabah Kalyoussef and Feja 2014). Normally, a person got effected by foodborne bacteria would share some common symptom includes diarrhea, severe stomach cramps, vomiting, and nausea. Currently, commercial food may be screened for contamination at the site of production, but it must then pass between packaging facilities, transport vehicles, storage locations, vendor sites and finally homes before its journey ends. At each point in the process, food may come in contact with other products and cross-contamination is a possibility. In fact, foodborne illnesses have recently become more common, despite advances in pathogen screening technology are used. Therefore, initial screening of food products does not ensure that pathogen-free product will be pathogen free when it enters your mouth. Table 19.1 shows a list of common foodborne bacteria, symptoms of disease they cause and food they potentially contaminate.

19.2 Current Methods for Detecting Foodborne Pathogens

Contaminated food cause by pathogens is one of the main reasons incidences of human diseases cases all around the world. Therefore, early detection for food pathogens is very crucial to avoid fatal circumstances. The existing methods for the detection of pathogens are represented a schematic diagram in Fig. 19.1. Traditionally, the characterization and detection of food borne pathogens rely on conventional culturing techniques. It typically performed based on microbiological counts where the bacteria colonies of interest were transferred to nutrient agar slants and allowed to grow. After an appropriate incubation period, the number of colonies that have formed on the culture medium plate could be counted. The best known example which shows high success rate for culture based method is the culture of *E. coli* O157:H7 on Sorbitol MacConkey agar (SMAC) and also another improvement method based agar such as CHROMagar and Cefsulodin Irgasan-Novobiocin (CIN). These cultures normally were examined through gram staining after pre-enrichment and selective enrichment step. However, the biggest drawback in the culture-based method is the slow growth and takes up about 18–24 h to give the exact result (Priyanka et al. 2016).

Enzyme linked immunosorbent assay (ELISA) is one of the most used immunoassays to date for foodborne pathogen detection. From time to time, methodologies in ELISA have been improved to suit the new emerging experiments. The improved version of sandwich ELISA is to detect antibody is well known as double antibody sandwich (DAS)–ELISA. Furthermore, the immunologically based methods for foodborne pathogens detection have been used to explore the specificity of the antibodies (monoclonal or polyclonal) for specific antigens, normally located at

Table 19.1 List of common foodborne bacterial pathogens (From IFT 2004)

Bacteria	Food potentially contaminated	Main symptoms
Campylobacter spp.	Raw chicken, beef, pork, shellfish, raw milk	Watery diarrhea, fever, abdominal, pain, nausea, headache, muscle pain
Bacillus cereus	Meats, milk, vegetables, fish	Watery diarrhea, abdominal pain, nausea and vomiting
Listeria monocytogenes	Raw milk, cheeses, raw vegetables, raw meats, raw and smoked fish, fermented sausages	Nausea, vomiting, diarrhea, meningitis, encephalitis, septicemia in pregnant women and newborns, abortion and stillbirth
Clostridium perfringens	Meat, meat products, gravies	Intense abdominal cramps, diarrhea
Clostridium botulinum	Improperly canned or fermented goods	Weakness, vertigo, double, vision, difficulty swallowing and speaking
Salmonella typhi and *Salmonella paratyphi*	Raw meats, poultry, eggs, milk and dairy products, fish, shrimp, yeast, coconut, sauces, salad dressings	Typhoid-like fever, malaise, headache, abdominal pain, body aches, diarrhea, or constipation
Shigella spp.	Salads (potato, tuna, chicken, macaroni), raw vegetables, bakery products, sandwich fillings, milk and dairy products, poultry	Abdominal pain and cramps, diarrhea, fever, vomiting, blood, pus, or mucus, in stools, tenesmus
Vibrio. cholera Serogroup	Raw or recontaminated oysters, clams, crabs	Diarrhea, abdominal cramps, fever, vomiting, nausea, blood or mucus-containing stools
Vibrio vulnificus	Raw or recontaminated oysters, clams crabs	Fever, chills, nausea, septicemia in individuals with some underlying diseases or taking immunosuppressive drugs or steroids
Staphylococcus aureus	Meat and meat products, poultry, egg products, cream-filled bakery products, milk and dairy products	Nausea, vomiting, retching, abdominal cramps, prostration
Yersinia enterocolitica	Meats, oysters, fish, raw milk	Fever, abdominal pain, diarrhea and/or vomiting

bacteria cellular membrane surface. For this, monoclonal antibodies are preferred over polyclonal antibody as it has monovalency (Zhao et al. 2014).

Beside ELISA, agglutination is another alternative technique have been establish based on the use of antibodies to detect specific antigens related foodborne pathogen. This detection method is based on observation of clumping that occurs when an antigen comes into contact with its corresponding antibody (Miller et al. 2008). In commercial agglutination test, antibody-coated colored latex beads or colloidal gold particles are used for quick confirmation or serological identification of pure culture isolates of bacteria from foods. The main advantage of this technique is ability to obtain semi quantitative results and relatively short time to obtain results. However, similar with existing assay test, this assay techniques need to be carefully interpret

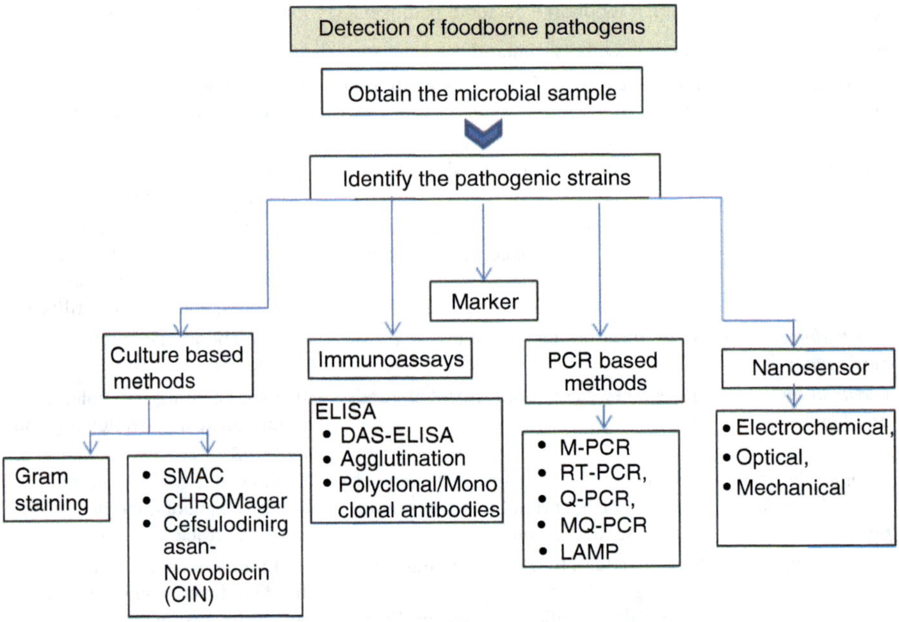

Fig. 19.1 Schematic representation of the existing methods for the detection of pathogens

marginal results and possible risks of interfering substances with others whole-cell antigens (Hajra et al. 2007).

One of the most commonly used molecular-based method for the detection of foodborne bacterial pathogens is polymerase chain reaction (PCR). PCR operates by amplifying a specific target DNA sequence in a cyclic three steps process (Mandal et al. 2011). Currently, multiplex PCR (mPCR) is a variant of simple PCR offers a more rapid detection as compared to conventional PCR through the simultaneous amplification of multiple gene targets. The basic principle of mPCR is like conventional PCR. However, mPCR requires the presence of two or more pair of primers in the reaction, so multiple genes may be amplified in one single reaction used in mPCR assay whereas only one set of specific primers are used in conventional PCR assay (Radhika et al. 2014).

The primary disadvantage of PCR methods in detecting foodborne pathogens are require pre-enrichment times that may vary from 6–8 to 48 h. To overcome this problem, quantitative PCR (qPCR) is suggested to be used in analysis due to no enrichment of the sample procedure. Moreover, it does not require agarose gel electrophoresis and able to monitor the PCR products formation continuously in the entire reaction by measuring the fluorescent signal produced by intercalating dyes. Another approach has been developed for the detection and quantification of multiple pathogens by simultaneous detection like qpCR with more specific detection. This so-called multiplex qPCR assay technique which used Taqman probe with different combinations of fluorophores to detect foodborne pathogen without compromising the sensitivity (Postollec et al. 2011).

Instead of PCR, loop-mediated isothermal amplification (LAMP) is a another nucleic acid amplification method developed by Notomi et al. (2000) which provides a rapid, sensitivity and specific detection of foodborne pathogens. Instead of two primers typically used in PCR, LAMP uses four primers targeting six specific regions and it provides rapid amplification, greater yield of amplification products and lower detection limits. Therefore, LAMP is proven to be more specific and sensitive as compared to PCR assays for the detection of foodborne pathogens (Yamazaki et al. 2008).

In the past decade, many efforts have been made by scientists to develop new methods for rapid detection of foodborne pathogens. As alternatives, various biosensors have been reported with shorter detection time (Hernandez et al. 2014; Poltronieri et al. 2014). A biosensor is an analytical device able to perform chemical or biological analysis theoretically with no considerable sample pre-treatment. In recent years, biosensors have been intensively studied because of their desirable qualities like rapidness, low cost, and easy operation.

In this chapter, we describes the most recent methods (over the last 5 years) applied to the detection of pathogens contaminating food products, with a special focus on nanosensor reported in the literature for detection of several foodborne bacterial pathogens. In addition to this, we summarized and highlighting their principles, advantages, and limitations in terms of simplicity, sensitivity, and multiplexing capability of the developed nanosensor for most common foodborne pathogens responsible for foodborne illness.

19.3 Nanosensors: Definition and Concept

Since foodborne pathogen is classified as epidemic by WHO, nanosensor has become a subject of interest by many researchers worldwide to control an outbreak cause by this pathogen. The need for rapid and sensitive assay methods to detect foodborne pathogens has led to the incorporation of biosensor technology into nanotechnology to develop nanosensors placed in food production, food distribution facilities and in packaging.

The term of nanosensor comes from the word of nano and sensor can be defined as instrumental device for monitoring physical and chemical phenomena which used to convey information related to any biological, chemical, or physical parameters about nanoparticles with at least one of their sensing dimensions being not greater than 100 nm (Malik et al. 2013). The nanosensors basically are based on the nanomaterials incorporated in the development process which these materials whose active elements are employed as catalytic tools, immobilization platforms or as optical or electroactive labels to improve the sensing performance (Inbaraj and Chen 2016). The terminology of nanosensor can also change to nanobiosensor whenever implies the use of biological recognition elements such as enzymes,

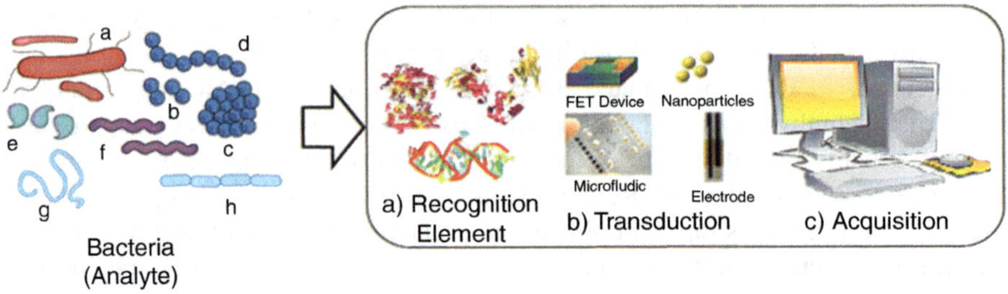

Fig. 19.2 Important prerequisites for designing an efficient of nanosensor (Source: Reference with copyright permission)

antibodies, DNA, aptamers, cell and others in the nanobiosensor fabrication (Chamorro-Garcia and Merkoc 2016; Malik et al. 2013).

Basically, the working scheme for all types of nanosensors involves three step processes (Fig. 19.2), including (a) recognition element, (b) transduction, and (c) signal acquisition. Taking foodborne pathogen as a model, the nanosensor would operate when interaction between the bacteria and recognition element occur and bind together with high affinity and selectively. Furthermore, the unique and tunable physicochemical properties of transducer will convert biochemical response into measurable signal either optical or electrical signal after the loading of biorecognition element. Consequently, the conversion signal ready to be manipulated into digital numeric value measuring an electrical or physical phenomenon such as voltage, current, temperature, pressure, or sound with a computer.

Nowadays, the detection principles in the development of nanosensor have emerged to various types of mechanism and techniques with variety of systems have been designed to demonstrate the advantages of fabricated nanosensor. Therefore, four classes of transducer based nanosensor will be described in this chapter for foodborne pathogen screening which consist of electrochemical, optical, mass/piezoelectric and microfluidic based nanosensor.

19.4 Electrochemical Nanosensor for Foodborne Pathogen Analysis

19.4.1 Concept of Electrochemical Nanosensor

Among all the nanosensor methods that have been grown rapidly over the past fifteen years, nanosensors based on electrochemical technique are intensively studied and well developed for the detection of foodborne pathogens. There are many comprehensive review articles published to illustrate the latest advancement on electrochemical methods for the detection of foodborne pathogen (Cinti et al. 2017).

Basically, electrochemical based nanosensor can be broadly classified to label-dependent and label-free techniques based on their detection strategies. Label-dependent methods apply labels such as enzymes, conductive polymers, metal particles, and more to specifically tag the target analyte (Zhu et al. 2015), whereas the label-free techniques are based on the direct attachment of bacterial cells onto the electrode surface to induce measurable changes in electrical parameters (Ahmed et al. 2014). On the other hand, the nanosensor that use the label-free strategy can finish the detection in much shorter time, making the assay simplest, with fewer variables to control and resources needs. However, these label-free methods usually depend on the electrochemical changes caused by the direct attachment of bacterial cells on the electrode surface. Therefore, they show relatively higher LODs compared with label dependent strategy because their lack of additional amplification (Radhakrishnan and Poltronieri 2017; Priyanka et al. 2016)

Since the mechanism detection is dependent on capturing binders, various biological recognition elements such as enzymes, antibodies, DNA, aptamers, whole cell and more currently lectin have been used in the development of electrochemical nanosensor for foodborne pathogens (Hu et al. 2014; Barreiros dos Santos et al. 2015; Tabrizi and Shamsipur 2015; Li et al. 2015, 2016; Jiang et al. 2017). These powerful biological recognition elements allow avoiding interferences from other microorganisms or molecules that are present in the tested sample with an appropriate selectivity and sensitivity.

Another important key to the successful performance of electrochemical nanobiosensor is the right choice of electrode materials as transducer surface. It is important to choose the right material, so that its functionalization can be achieved with relative simplicity which makes them best suited with biorecognition element (Zhu et al. 2015; Reimhult and Hook 2015). This can be seen in the work presented by Barreiros dos Santos et al. (2015) where *E. coli* antibodies have been immobilized onto ITO surfaces by using covalent attachment and enable to work on electrochemically label-free measurements. In addition to this, the electrode materials for electrochemical biosensor are normally constrained by the requirements for both high electrical conductivity and biocompatible properties to avoid biomolecules denaturation after prolonged exposure to metal surfaces.

19.4.2 The Use of Nanomaterial in the Development of Electrochemical Nanosensor for Foodborne Pathogens

Nanomaterial has aroused great interest in electrochemical studies to achieve a faster electron transfer of biomolecules and higher specificity electrochemical biosensors. Most often function of nanomaterial in electrochemical based nanosensor for foodborne pathogens can be seen in previous study as a selective template for immobilization of target molecules, signal transduction, bio-barcode DNA or aptamer, antibody-functionalized NPs, enzyme-labeled and antibody label (Fig. 19.3).

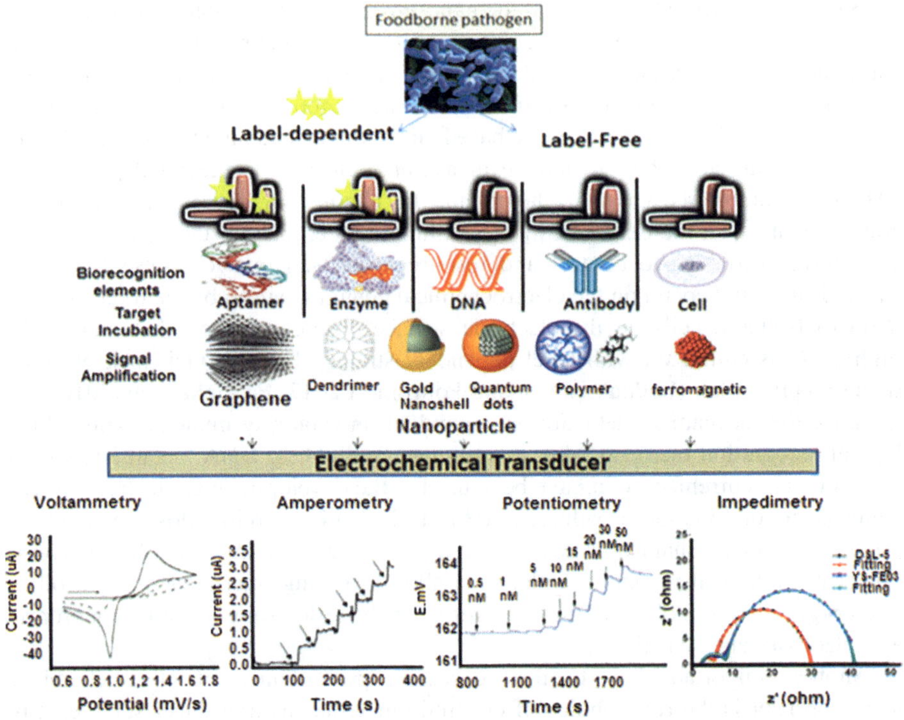

Fig. 19.3 Depiction of the block diagram in fabrication of electrochemical nanobiosensor (Source: Reference with copyright permission)

In recent investigation, graphene oxide (GO)/reduced graphene oxide is in huge demand for the development of label free biosensors. Moreover, the advantages of GO nano-sheets contain abundant surface oxygen groups, including hydroxyl, epoxide, carbonyl, and carboxyl groups (Sharma et al. 2013) provide an array of reaction sites for interaction with others nanoparticles for instance gold NPs and carbon nanotube (Fei et al. 2015; Jia et al. 2015). Meanwhile, for reduced graphene oxide (RGO), implying interaction with another nanomaterial has increases solubility RGO in hydrophilic medium (Muniandy et al. 2017).

In case of signal label approach, gold nanoparticles (AuNPs) seem to be more involved in labels dependent in electrochemical based nanosensor fabrication for foodborne detection and just few implying for label free approach (Hu et al. 2014; Hassan et al. 2015; Xiang et al. 2015; Fei et al. 2015; Li et al. 2016; Das et al. 2014; Jia et al. 2015). Moreover, the advantages of gold with thiol moieties also enable gold to composite with another material, for example Hamidi-Asl et al. (2016) have successfully developed label-free gold-silver core shell NPs for the detection of *Escherichia coli* with the detection limit of approximately 90 CFU mL^{-1}.

Among the nanomaterial, magnetic nanoparticles (MNPs) and magnetic nanobead (Li et al. 2016) possess unique magnetic features for bacterial detection through immunomagnetic separation of the bacteria (Xu et al. 2016). Magnetic properties own by magnetic particles can also be used as a detection label by using a magneto-electrochemical immunosensor for *S. typhimurium* which was described by Brandao et al. (2015) where MNPs can be easily magneto-actuated using permanent magnets. Not only that, their magnetic properties enable them to shares benefits with others nanomaterial to produce new composite material such as magnetic silica nanotube (MSNT).

Besides gold and silver nanoparticles, quantum dots (QDs) semiconductor is one of the most important nanomaterials utilized for multiplex detection of pathogens. More currently, semiconductor based quantum dots such a CdS, PbS and ZnS have been used as a label to link with antibody, DNA, and RNA to differentiate the gene of different foodborne pathogens bacteria by magnetic separation (Fernandes et al. 2014; Silva et al. 2015; Abdalhai, et al. 2015). In the present study, non-protein coding RNA (npcRNA) sequences of each bacterium were used as gene targets (Vijian et al. 2016). Table 19.2 summarizes the collection of some current reported studies on nanomaterial-based electrochemical technique for the detection of bacterial pathogens.

According to type of transducer, electrochemical biosensors can be further divided into sub-categories such as amperometric (measurable of current), impedimetric (ability of a circuit to resist the flow of electrical current, potentiometric (measurable potential or charge accumulation), and conductimetric based on measurable of conductivity change. The fundamental principles of those sensors are very similar except for the units of measurements.

19.5 Electrochemical Technique

19.5.1 Voltammetry

Among electrochemical techniques, voltammetry is the less to prone to noise, it is the most widely used in microbial analysis by biosensors, and was already applied to all types of bioreceptors (Tabrizi and Shamsipur 2015; Das et al. 2014; Xu et al. 2016). In general, voltammetry is a technique typically monitor the changes in current that caused by the oxidation or reduction of the electrochemically active analyte in the electrochemical system, where the electrode potential ramps linearly versus time in cyclical phases (Fig. 19.4a). In typical electrochemical biosensors based voltammetry, cyclic voltammetry (CV) is a simple experiment given useful information about redox reactions in a form which is easily obtained and interpreted (Fig. 19.4b). Other voltammetry method such anodic stripping voltammetry (ASV) (Fig. 19.4c) and Difference pulse voltammetry (DPV) (Fig. 19.4d) was also become popular in sensor research for quantification of resulting current with respect to target analyte's concentration.

This technique generally relies on a three electrodes system: working, reference, and counter electrodes instead of two electrodes in an amperometric technique. Screen printed electrode (SPE) seem to become main choice in developing voltammetry nanobiosensor system for a simple and sensitive DNA biosensor and

Table 19.2 Summaries the use of nanomaterial for foodborne pathogen detection

Nanomaterial	Mode of transducers	Pathogen	Immobilization method	LODs (CFU mL^{-1})	References
NGE	Impedance/ Voltammetry	*Salmonella*	ssDNA/NGE	2.1 pM	Tabrizi and Shamsipur (2015)
AuNPs	Voltammetry	*Salmonella*	HRP-antibody/bio-antibody/ Sreptavidin/gold nanoparticles/4-SPCE	1.95×10^2 CFU/mL^{-1}	Hu et al. (2014)
AuNR	Voltammetry	*E. coli*	Ab–AuNR–FCA/AuNR/SiO_2		Zhang et al. (2015)
AuNPs	Amperometric	*E. coli*	MBs–pECAb/AuNPs–sECAb/SPCEs)	309 CFU mL^{-1}	Hassan et al. (2015)
AuNPs	Voltammetry	*Salmonella*	IHrp-IgG2/IgG/AuNPs-Chitosan	5 CFU mL^{-1}	Xiang et al. (2015)
AuNPs	Voltammetry	*Salmonella*	EC nanoparticle probe/DNA3/DNA1/ MCH/DNA2/AuNP/GE	13 CFU mL^{-1}	Zong et al. (2016)
AuNPs	Voltammetry	*Salmonella*	RCA-DNA/AuNPs/Gold electrode	6 CFU mL^{-1}	Zhu et al. (2014)
AuNPs	Voltammetry	*Salmonella*	Aptamer/AuNPs/Gold electrode	20 CFU mL^{-1}	Li et al. (2016)
AuNPs	Voltammetry	*Salmonella*	ss-DNA/AuNPs-MPTS/SPE	50 pM	Das et al. (2014)
AuNPs	Impedance	*Staphylococcus aureus*	Aptamer modified rGO-ssDNA-AuNPs/ GCE	10 CFU mL^{-1}	Jia et al. (2015)
Graphene	Impedance/ Conducmetric	*E. coli O157:H7*	Graphene/IgG/SiO2-substrates	10 CFU mL^{-1}	Pandey et al. (2017)
rGO	Voltammetry	*Salmonella*	ssDNA/rGO-AP/GCE	101 CFU mL^{-1}	Muniandy et al. (2017)
GO	Potentiometry	*Staphylococcus aureus.*	GO/RGO-Aptamer	8×10^2 CFU/mL^{-1} (cov attachment)	Hernandez et al. (2014)
				$>10^7$ CFU/mL^{-1} (non-cov attachment)	
GO	Impedance	*Pseudomonas aeruginosa*	GO/NaAlg/cell/GE	0.034 μM	Jiang et al. (2017)

AuNPs, GO	Impedance	*Salmonella*	Apt/GO /AuNPs/GCE	3 CFU mL^{-1}	Ma et al. (2014)
AuNPs, rGO	Voltammetry	*Salmonella*	rGO/AuNP/Ab2/S/IMB/4-SPCE	89 CFU mL^{-1}	Fei et al. (2016)
rGO-MWCNT	Impedance	*Salmonella*	rGO-MWCNT/aptamer /GCE	25 CFU·mL^{-1}	Jia et al. (2015)
MWCNT-MB,MNPs	Voltammetry	*Salmonella*	MWCNT-MB/MNPs-IgG/SPCE	17.3 CFU mL^{-1}	Ngoensawat et al. (2017)
MSNT	Impedance	*Salmonella*	Bacteria-MSNT/IDAM	10^3CFU mL^{-1}	Nguyen et al. (2014)
Silica MPs	Amperometric	*Salmonella, Listeria* and *Escherichia coli*	Silica MPs/AntiFlu-HRP/GCE		Liebana et al. (2016)
MNB	Impedimetric	*Escherichia coli* O157:H7 and *Salmonella*	MNBs/Ab/cell/Ab–Gox/SP-IDME	2.05 × 10(3) CFU mL^{-1} (E. coli O157:H7)	Xu et al. (2016)
				1.04 × 10(3) CFU mL^{-1} (S. typhimurium)	
MNB	Impedance	*E. coli* O157:H7	MNB-IgG/SP-IDME	104.45 CFU mL^{-1}	Wang et al. (2015)
MB, AgNPs	Voltammetry	*Staphylococcus aureus*	MB-Apt/bacterium/Apt-AgNPs	1.0 CFU/mL^{-1}	Abbaspour et al. (2014)
QDs, MNP	Potentiometric	*Salmonella*	Ab-MNP/S/ CdS-Ab/Na-ISEs	20 CFU mL^{-1}	Silva et al. (2015)
QDs	Voltammetry	*Vibrio cholerae, Salmonella* sp. and *Shigella* sp.	QDs-RNA/SPCE	34 Am	Vijian et al. (2016)
QDs	Voltammetry	*Escherichia coli* O157: H7and *Salmonella*	QDs-DNA/GE	1 × 10^{-12} mol/L^{-1} (E. coli O157:H7)	Fernandes et al. (2014)
				1 × 10^{-12} mol/L^{-1} (S. aureus)	
QDs	Voltametry	*E. coli* O157:H7	QDs-DNA/(MWCNT/GE	1.97 × 10–14 M	Abdalhai, et al. (2015)

Abbreviation: NPs (Nanomaterial), NGE (Nanoporous glass electrode), GE (Glass electrode), AuNR (Au nanorode), GO/rGO (Graphene oxide/reduce graphene oxide), rGO-MWCNT (Reduce graphene oxide-multi wall carbon nanotube), MNPs (Magnetic nanoparticle), MNB (Magnetic NanoBead), AgNPs (Silver nanoparticle), QDs (Quantumn Dots),MSNT (Magnetic silica nanotube), MIPs (Molecular imprinted polymer), SPCE (Screen printed electrode), GE (Gold electrode)

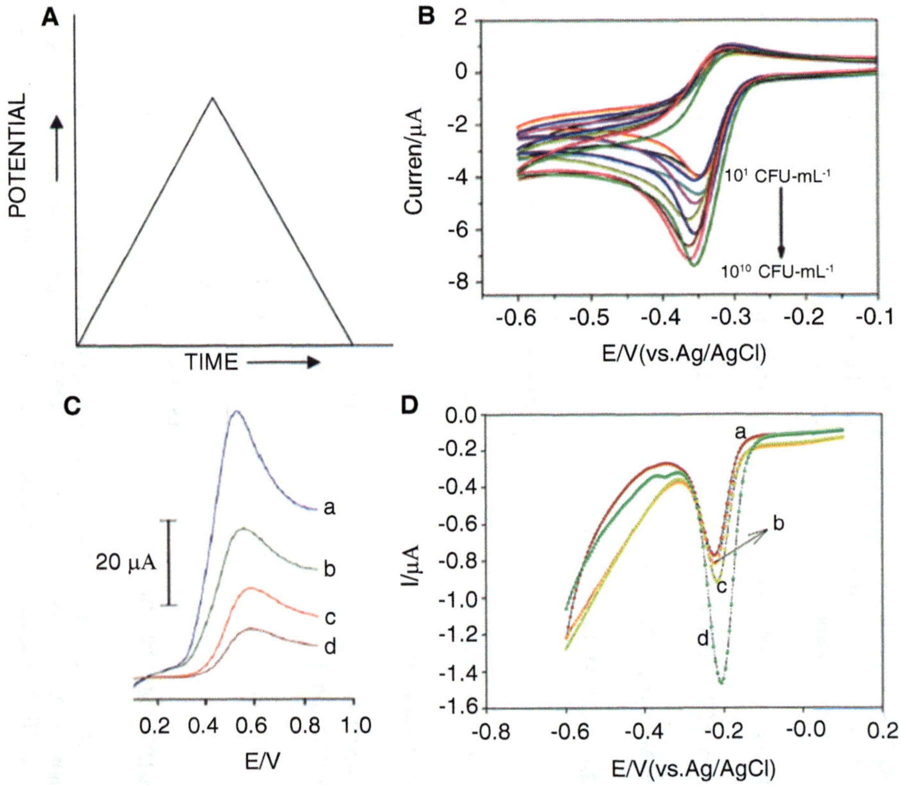

Fig. 19.4 (**a**) One cycle of the potential-excitation signals; (**b**) voltammograms for cyclic voltammetry experiment; (**c**) voltammogram for anodic stripping voltammetry; (**d**) voltammogram for differential pulse voltammetry (Source: Reference with copyright permission)

immunosensor. Das et al. (2014) demonstrated an efficient DNA electrochemical biosensor for rapid detection of Vi gene for *Salmonella typhi* by immobilizing thiol labeled DNA probe on3-mercaptopropyl trimethoxysilane (MPTS) and AuNPs modified SPE (Fig 19.5a). The DNA hybridization was monitored by DPV measurements using methylene blue as an electrochemical indicator. The linearity was found from 1.0×10^{-11} to 0.5×10^{-8} M and the detection limit was 50 (±2.1) pM. In a similar approach, Hu et al. (2014) presented a novel electrochemical enzyme immunosensor based a gold nanoparticles electrodeposited onto the four-channel screen printed electrode (4-SPCE) (Fig 19.5a). In this study, gold nanoparticles acted as an electron transfer agent and bio-molecules absorber which role by streptavidin and bind with HRP-antigen. This sensor capable to detect two different strains of *Salmonella* with the detection limit for *S. pullorum* and *S. gallinarum* determination was 1.95×10^2 CFU/mL (Fig 19.5b).

In recent work, Fei et al. (2015) reported voltammetry immunosensor which used SPCE modified with an ionic liquid, gold nanoparticles and antibody

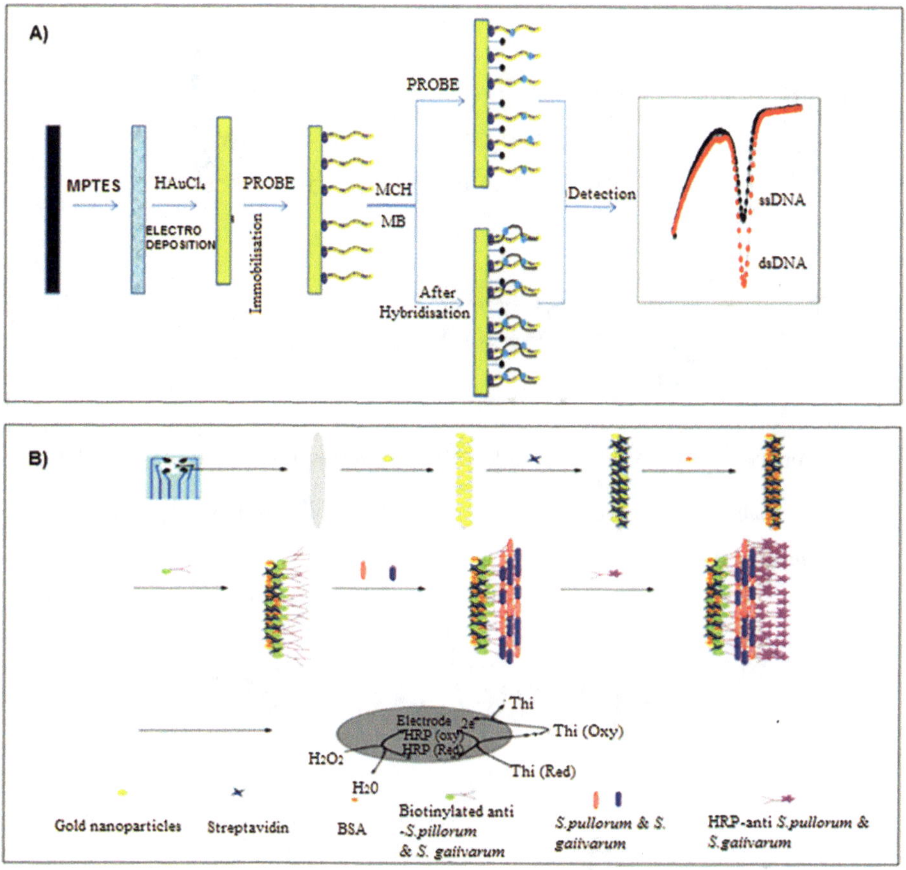

Fig. 19.5 (**a**, **b**) Schematic diagram of the modification process of the electrode and measure mechanism (Hu et al. 2014)

anti-Salmonella pullorum to capture Salmonella from one spiked sample in a sandwich format (Fig. 19.6a). In this work, they incubated the immunocomplex with a secondary antibody labeled with HRP and reaching to a LOD of 3×10^3 CFU mL^{-1}. With the same biosensor designed and electrode (Fig. 19.6b silica modified immunomagnetic beads have been used as capture antibody, while reduced graphene oxide coated with gold nanoparticles have been used as label. As a result, they succeeded to amplifying the electrochemical signal and attained a LOD as low as 89 CFU mL^{-1} (Fei et al. 2015).

Besides SPCE, gold electrode also is the most common electrode used for voltammetry detection. Li et al. (2016) had reported a simple, rapid, and sensitive electrochemical aptasensor based on target-induced strand displacement (Fig. 19.7). In this study, AuNP have been employed as label for electrochemical sensor to enhance the sensitivity and stability for the determination of Salmonella. The principle of target-induced strand displacement is that the aptamer duplex or

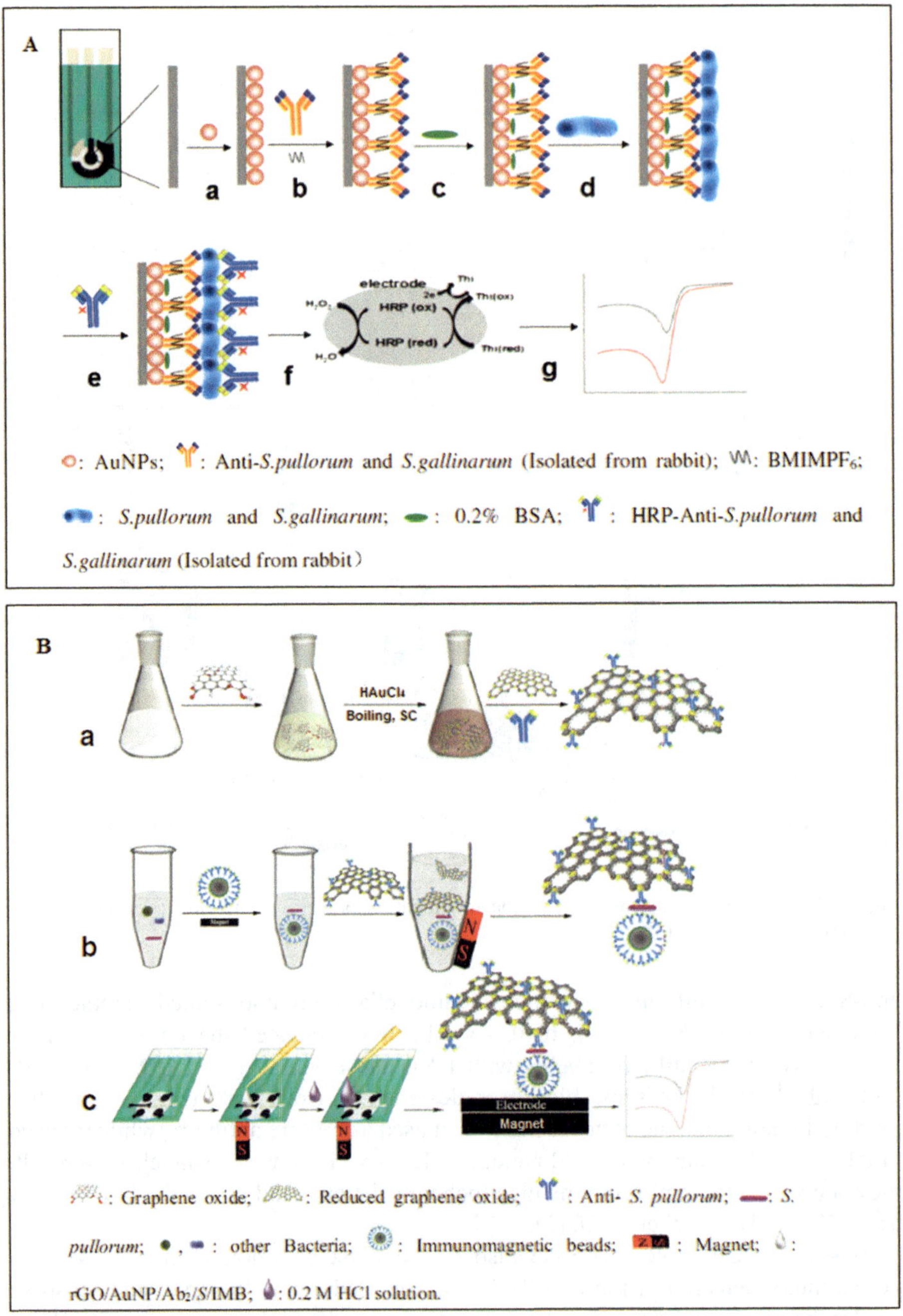

Fig. 19.6 (**a**, **b**) Schematic diagram of the modification process of electrochemical immunosensor and measure mechanism by using SPCE (Fei et al. 2015, 2016)

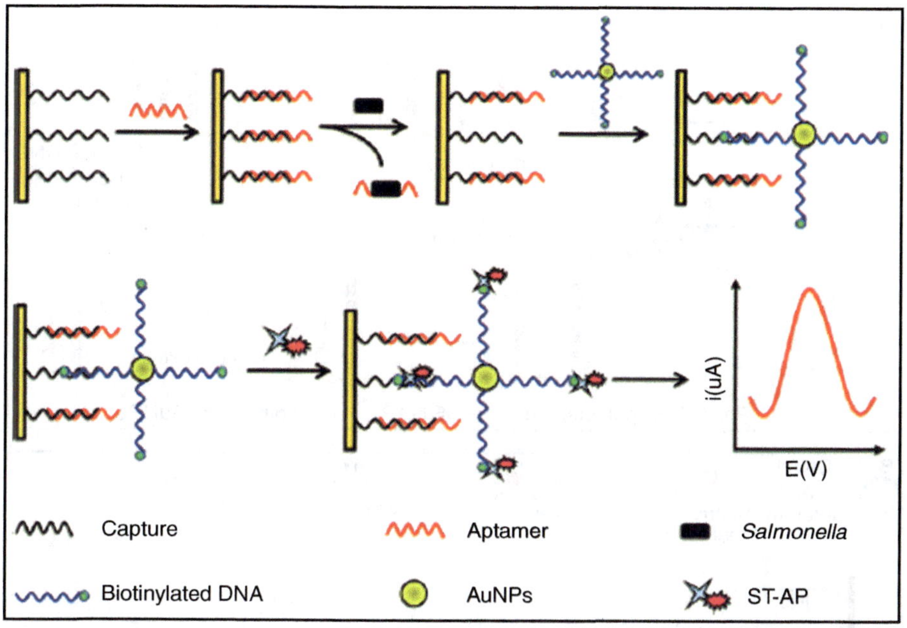

Fig. 19.7 Schematic of the electrochemical aptasensor for *Salmonella* (Li et al. 2016)

complementary DNA is assembled to a supporter. In the presence of the target, the aptamer or its complementary DNA is displaced from the supporter, leading to significant signal changes in the electrochemical signal.

On the other hand, there is several voltammetry nanobiosensor reported work on different types of DNA amplification strategy to avoid the use of PCR amplification process as shown in Fig 19.8a. Zhu et al. (2014) proposed amplification technique based on the detection of nucleic acid sequences of target bacteria by rolling circle amplification (RCA). RCA possesses the advantages of simple and rapid steps, high potential and no requirement for special laboratory conditions, which made it a powerful tool over other amplification technologies. Therefore, an ultrasensitive and rapid biosensor was developed for the detection of Salmonella by combining the dual amplification strategies of RCA and DNA–AuNPs probe which enable for the detection of a complementary target DNA down to 6.76 aM. In addition, the designed assay had been successfully applied to detect PCR amplified products from Salmonella and could detect Salmonella as low as 6 CFU mL^{-1} in real milk samples.

To date, Zong et al. (2016) proposed another nucleic acid sequences amplification strategy using voltammetry DNA biosensor based on entropy-driven molecule switch signal amplification strategy for the ultrasensitive detection of *Salmonella typhimurium* (Fig 19.8b). In this study, the gold electrode modified with gold nanoparticles was used to immobilize hairpin DNA as a capture probe with three different sequences DNA to open the hairpin DNA structure to enable significant

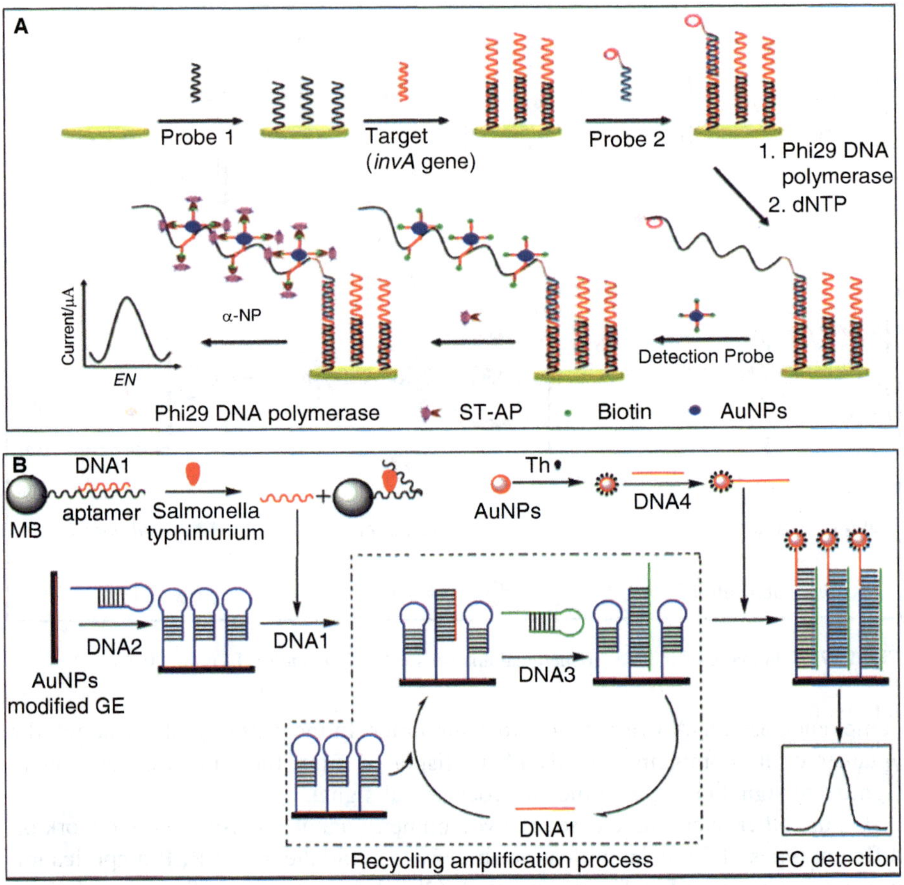

Fig. 19.8 (**a**) Schematic representation of the designed strategy for inv. A detection of *Salmonella* (Zhu et al. 2014); (**b**) Schematic illustration of signal amplified strategy based on entropy-driven molecule switch and EC nanoparticle probe for DNA and Salmonella typhimurium detection (Zong et al. 2016)

amplification of the signal. The reported electrochemical sensor demonstrated excellent sensing performances with ultra-low detection limit of 0.3 fmol L^{-1} for DNA and 13 CFU mL^{-1} for *Salmonella typhimurium*.

19.5.2 *Amperometric*

Amperometric biosensors are extremely explored in the field of electrochemical biosensor nevertheless, it less widely used as transduction technique in foodborne pathogen analysis. Generally, the principle of amperometric biosensor is similar with voltammetric technique where the current was measured from the reaction of

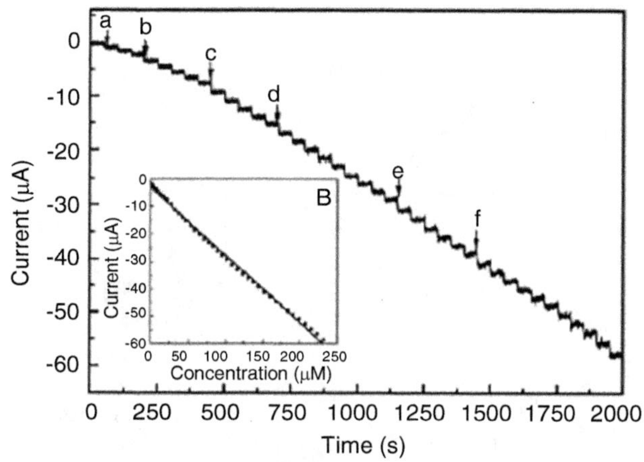

Fig. 19.9 Example of typical chronoamperograms at difference concentration (Source: Reference with copyright permission)

analyzed species at the working electrode where fixed potential over period of times is applied between the working and reference electrodes (Sharma et al. 2013). In the past, this technique is well known with an enzyme biosensor since the pioneering work of Clark (1987) leading to the construction of enzyme electrode and consequently has led to existing glucose biosensor for glucose monitoring. In essence, key advantages of amperometric biosensors are their relative simplicity and ease of miniaturization. They also generally confer excellent sensitivity. However, it limitations include low specificity depending on the applied potential, which if at high applied potential it may allow other redox-active species to interfere the signal and lead to inaccuracies in results (Pohanka and Skladal 2008). Figure 19.9 shows a typical amperometric biosensor based on chronoamperometry measurements obtained against the corresponding concentrations.

Recently, Hassan et al. (2015) had developed immunosensor based on amperometric for the detection of *E. coli* in meat and water samples (Fig. 19.10). In the developed methods, *E. coli* O157:H7 was captured by anti-*E. coli* O157–magnetic beads conjugated (MBs–pECAb) and subsequently labeled with AuNPs modified with secondary antibodies (AuNPs–sECAb) in the presence of control bacteria (Salmonella typhimurium). Taking potential advantages of the electrocatalytic properties of AuNPs on hydrogen formation from hydrogen ions (hydrogen evolution reaction, HER), the developed assay could detect *E. coli* O157:H7 in both beef and water samples without broth enrichment at much lower concentration (457 and 309 CFU/mL, respectively) than other previously reported advanced technologies within shorter time (70–80 min). Additionally, this method showed high specificity rate when tested against *Salmonella typhimurium* and other competing bacteria.

In different study, screen-printed interdigitated microelectrodes (SP-IDMEs) for the firsts time used in amperometric measurement for rapid detection of *E. coli* have been explored by Xu et al. (2016) (Fig. 19.11). In the reported study, Prussian blue was used as the mediator to modify the SP-IDME to monitor the enzymatic reaction of glucose with functional glucose oxidase (GOx)–polydopamine (PDA) based polymeric nanocomposites (PMNCs). For the fabrication of the electrochemical

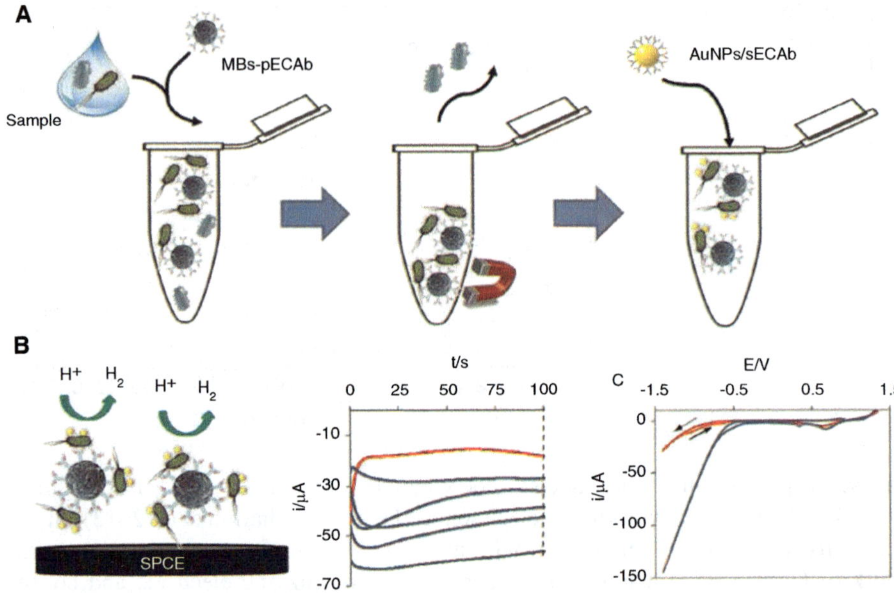

Fig. 19.10 (**a**) Stepwise for preparation magneto sandwich assay using gold nano particle (AuNP) labels, (**b**) detection for labeled *E. coli* O157:H7 through the Hydrogen Evolution Reaction (HER) electrocatalyzed by the AuNP labels, (**c**) Chronoamperograms when holding the working electrode at potential of +1.35 V for 1 min and then applying a negative potential of 1.00 V for 100 s (Hassan et al. 2015)

biosensor, MBs were first bound to GOx through a streptavidin–biotin reaction. After that, a thin layer of polydopamine (PDA) film was synthesized on the MB–GOx conjugates through controlled self-polymerization of polydopamine (PDA) under alkaline conditions. The good biocompatibility of polydopamine (PDA) allowed the immobilized GOx maintain the enzymatic activity to catalyze glucose to produce H_2O_2 which could further reduce $HAuCl_4$ to generate AuNPs. The produced AuNPs was then attached to the surface of the MBs–GOx@PDA PMNCs. With successive adsorption of ABs and additional GOx, the final product ABs/GOxext/AuNPs/MBs–GOx@PDA PMNCs was used to capture the target bacterial cells. By using the filtration technique, the free PMNCs were filtered out and concentrated in the glucose solution for measurement. The filtration treatment helped in the isolation and the concentration of free PMNCs, meanwhile the removal of the bonded PMNCs also reduced the background noise during measurement and the sensitivity of the developed biosensor also improved. The developed biosensor showed a broad detection range from 10^2 to 10^6 CFU mL^{-1} in the pure culture within 1 h. The validation of the developed approach to detect the target bacteria in ground beef demonstrated that it could detect as low as 190 CFU g^{-1} without a pre-enrichment procedure.

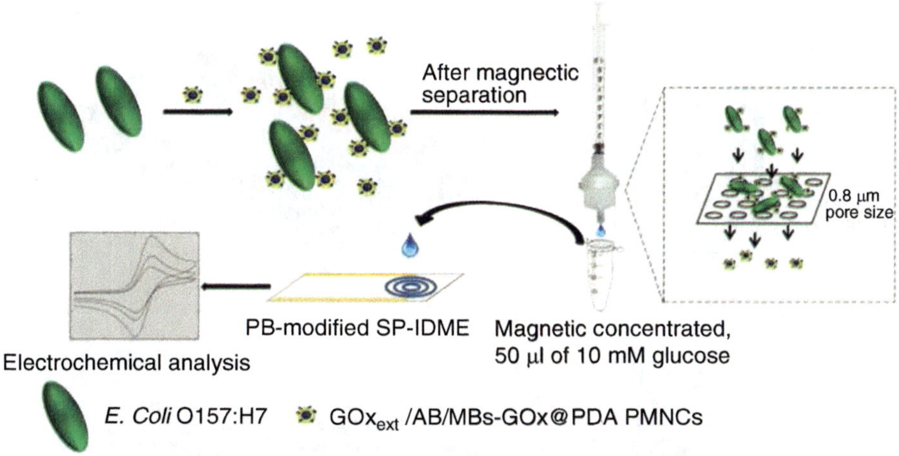

Fig. 19.11 Schematic represents the principle of electrochemical biosensor for rapid detection of *E. coli* O157:H7 with the integration of bifunctional glucose oxidase (GOx)-polydopamine (PDA) based polymeric nanocomposites (PMNCs) and Prussian blue (PB) modified screen-printed interdigitated microelectrodes (SP-IDMEs) (Xu et al. 2016)

19.5.2.1 Potentiometric

Potentiometry (zero current) technique is one of the oldest instrumental methods and well-established position as the analytical techniques for direct use of real sample analysis. Moreover, potentiometric biosensors shows some advantages over other electrochemical transducers due to their recognized capacity of miniaturization plus that in comparison with voltammetric techniques, the signal isn't dependent on the electrode surface area. Generally, this method measures the electrical potential difference between working electrode and reference electrode with a current level almost zero. The reference electrode potential remains invariant during the entire duration of measurement. The working electrode undergoes significant change in its potential even for small changes in the analyte concentration (Ahmed et al. 2014). There are three types of potentiometric electrode commonly used in biosensing. Ion selective electrodes (ISE) are the simplest transducer in the development of potentiometric biosensors. Ion-selective electrodes consist of two electrodes called the indicator electrode (also called the ISE) and the reference electrode as shown in Fig. 19.12a. The indicator electrode differs from the reference electrode by a membrane at the end that can uptake the desired ion species. The most commonly used ISE is the pH probe and these types of electrode can be used to probe detection of ion such as fluoride, bromide, cadmium, and gases in solution such as ammonia, carbon dioxide, and nitrogen oxide.

Consequently, two others important types of potentiometric biosensor namely the ion-sensitive field effect transistors (ISFET) (Fig. 19.12b) and light addressable potentiometric sensor (LAPS) (Fig. 19.12c) basically adopted similar operational

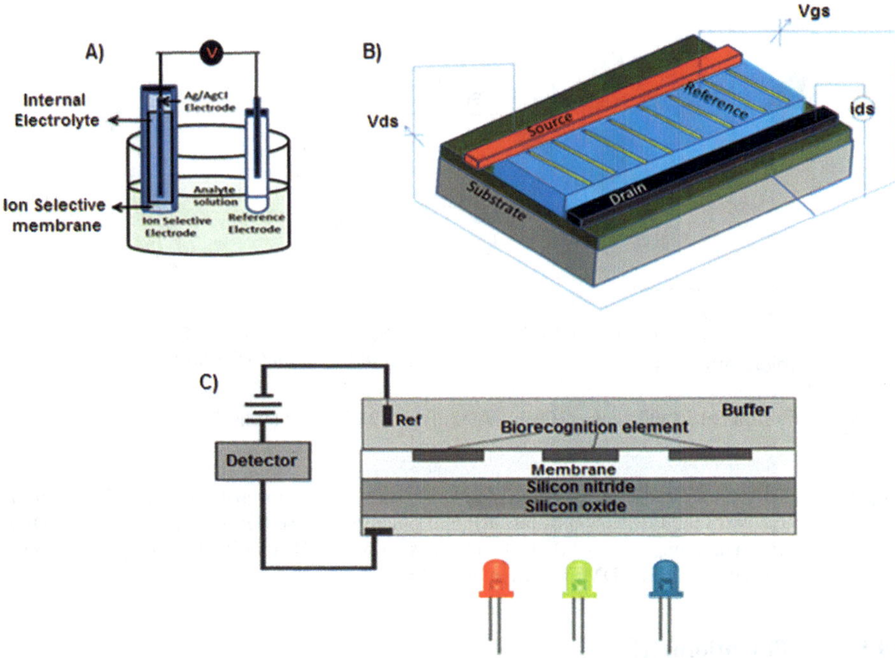

Fig. 19.12 (**a**) Configuration ion selective electrodes (ISE), (**b**) Configuration ion-sensitive field effect transistors (ISFET), (**c**) light addressable potentiometric sensor (LAPS) (Source: Reference with copyright permission)

principle which both system has an electrolyte/insulator/semiconductor structure within its gate region. The difference for both system can be seen where ISFET photocurrent generated by local potential generated by surface ions from a solution. This potential modulates the current flow across a silicon semiconductor. Meanwhile, LAPs is the photocurrent generated based on semiconductor activation by a light-emitting diode (LED). The use of light from LED makes LAPS more sensitive as the location of the light governs the position of the generated photocurrent. The sensor is made from n-type silicon typically coated with 30 nm of silicon oxide, 100 nm of silicon nitride, and indium-tin oxide (Pohanka and Skladal 2008).

In latest update, Silva et al. (2015) have demonstrated the detection of *S. typhimurium* in milk sample by using Ion selective electrodes (ISE) detection based on immunoassay sandwich format (Fig. 19.13). In this work, cadmium and sodium ion selective electrodes were used respectively as indicator and pseudo-reference electrodes were prepared in pipette tips to allow potentiometric measurements in microliter sample volumes. The proposed method was also claimed the first-time utilized coupling potentiometric detection with ion selective electrodes to label sandwich immunoassays using magnetic nanoparticles (MNPs). The limit of detection can be achieved by using this system was 2 cells per 100 μL. The estimated average total time for a complete assay was 75 min.

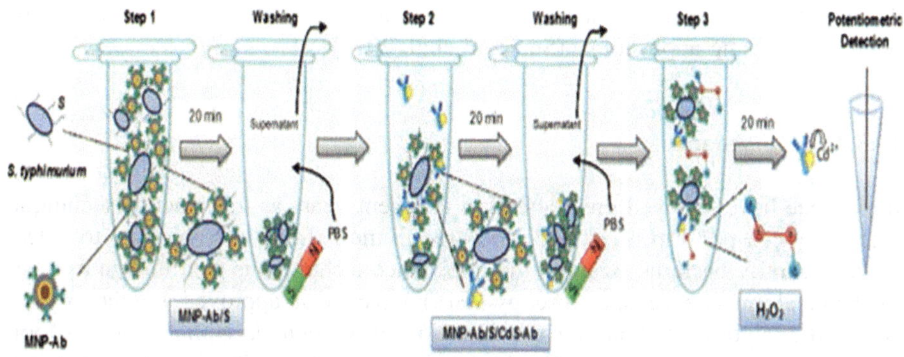

Fig. 19.13 Schematic representation of the potentiometric magnetic immunoassay for S. typhimurium quantification (Silva et al. 2015)

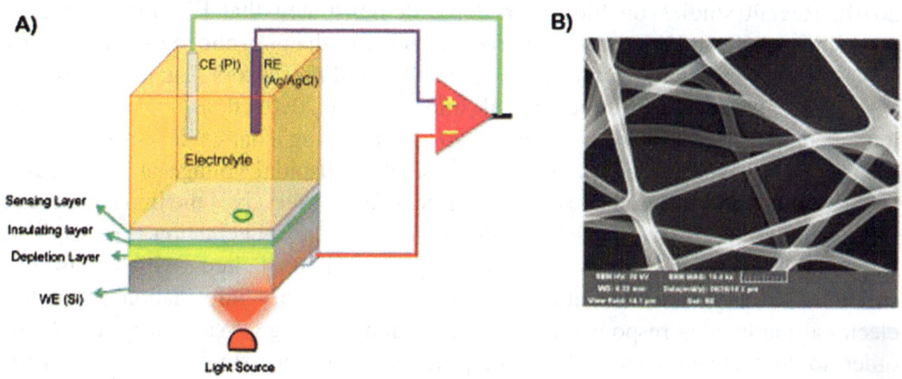

Fig. 19.14 Schematic representation of the NF-LAPS sensor comprising a three-electrode system. (**a**) Si based sensor chip is acting as the WE with pH sensitive PAA/PVA hydrogel NFs, an RE comprising Ag/AgCl and a Pt CE. (**b**) Surface SEM image shows the hydrogel NF sensitive layer under magnification, with the scale bar depicting a 2 μm width (Shaibani et al. 2018)

Another recent development in potentiometric nanosensor demonstrated the portable light Addressable Potentiometric Sensor for rapid *Escherichia coli* detection in orange juice using nanofibers-hydrogel scaffold as a bio-functionalized platform. As shown in Fig. 19.14, this nanofiber are composed of poly (vinyl alcohol)/poly (acrylic acid) (PVA/PAA) has enhanced the sensing properties of LAPS. The sensing mechanism is based on the metabolic activity of *E. coli* toward sugar molecules and the production of acidic products such as lactates and acetates. By functionalizing the surface of the NFs with D-mannose, the need for antibody–antigen complex interactions which reduce sensor reliability over time is eliminated. One of the advantages of the developed sensor was that the overall photocurrent was not interrupted by the opaque samples since the light was illuminated from the backside of the device instead of the front side. Furthermore, the developed photoelectrochemical sensor offers a compact and portable device for on-site biosensing applications (Shaibani et al. 2018).

19.5.2.2 Impedance

Impedance have received great attention in recent years as transduction technique since they can perform direct and label-free in the field of microbiology to detect and/or quantify bacteria. Besides, the impedance technique is well known as electrochemical impedance spectroscopy (EIS) have been approved by the one of accepted analytical methods for the detection of *Salmonella typhimurium* in food products by analytical communities international (AOAC) (Yang and Bashir 2008). Impedance microbiological techniques have also been used to detect many other bacteria species, such as *Enterobacteriaceae, Coliforms, Listeria* spp., and *L. monocytogenes* in various samples (Wawerla et al. 1999; Velusamy et al. 2010). Recent studies on biosensors have demonstrated that EIS measurements produce significantly more sensitive responses than cyclic voltammograms (CV) as far as the fabrication of label-free cell-based biosensors (Liu et al. 2014).

In most literature, electrochemical impedance is defined as the ratio of an incremental change in voltage current to the resulting change in current. Most often this is accomplished by imposing a small sinusoidal voltage at a particular frequency and measuring the resulting current-voltage ratio gives the impedance due to the change in an electric field by the antibody/antigen interaction (Daniels and Pourmand 2007). EIS is often analyzed using an equivalent circuit which is used to curve fit the experimental data and extract the necessary information about the electrical parameters responsible for the impedance change (Wang et al. 2012). In order to understand chemical model process or mechanism involved in every reaction occurs in EIS, equivalent circuit models is often used to fit the experimental data. The simplest, and in fact the most frequently used equivalent circuit for modelling the EIS experimental data is the so-called Randles circuit (Fig. 19.15a), which comprises the uncompensated resistance of the electrolyte (*Rs*), in series with the capacitance of the dielectric layer (Cdl), the charge-transfer resistance (*Rct*) and the Warburg impedance (Zw) (Wang et al. 2012). Typically, a shape of semicircle region will be obtained namely a Nyquist plot which lying on the real axis followed by a straight line. From the Nyquist plot, the values for *Rs* and *Rct* can be easily determined which are the most important electrical parameters in analyzing the impedance signal change for detection of bacteria. In addition, the double layer capacitance also can be calculated from the frequency at the maximum of the semicircle (ù = 2, f = 1/RctCdl).

In recent years, impedimetric biosensors have been designed by immobilizing biorecognition element such as antibodies, nucleic acids, aptamer, and enzyme at the surface of a solid electrode modified with nanomaterial. The used method is considered as a label free technique without complicated and easy to miniaturize, which facilitates their translation to point-of-care systems. Ma et al. (2014) proposed direct detection of *Salmonella. enterica serovar Typhimurium* by EIS technique on glassy

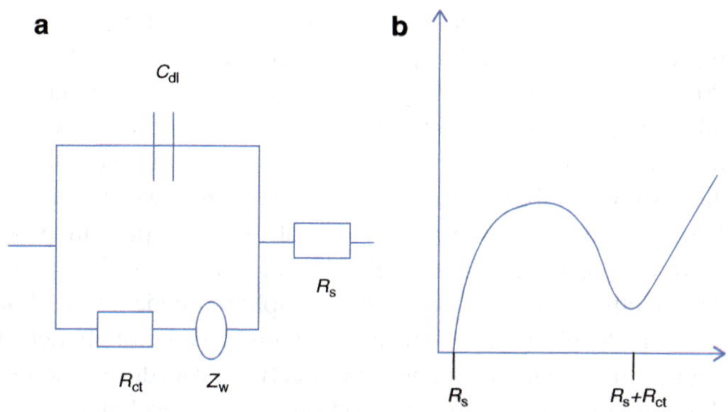

Fig. 19.15 (**a**) Randles equivalent circuit proposed to best fit; (**b**) EIS Nyquist plot (Zimag against Zreal) (Wang et al. 2012)

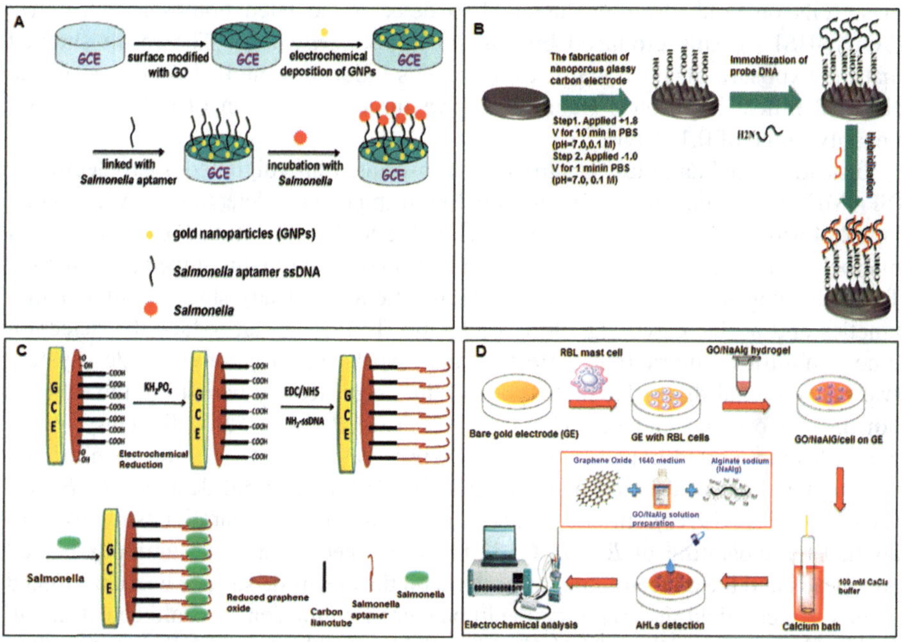

Fig. 19.16 (**a**) Schematic illustration for the aptamer-based electrochemical detection of Salmonella (Ma et al. 2014). (**b**) Fabrication of label-free electrochemical DNA biosensor based on covalent immobilization of salmonella DNA sequences on the nanoporous glassy carbon electrode (Tabrizi and Shamsipur 2015). (**c**) The schematic view of the GCE surface modification and the detection of Salmonella (Jia et al. 2015). (**d**) Schematic illustration for the preparation of RBL-2H3 mast cell sensor and process of AHLs detection (Jiang et al. 2017)

carbon electrode (GCE) which was modified with graphene oxide (GO) and Gold nanoparticle (GNPs) (Fig 19.16a). In the published paper, the DNA aptamers of *Salmonella enterica serovar Typhimurium* were selected and evaluated, exhibited a good linear relationship with impedance, and the detection limit was 3 CFU/mL. In another study, an easy and simple approach has been developed by Tabrizi and Shamsipur (2015). In the reported study, the functionalization step has been used by covalently linked amino modified salmonella ssDNA probe sequence with carboxylic group on the surface of nanoporous glassy carbon electrode to prepare the DNA biosensor. The proposed DNA biosensor in the published work shows the limit of detection of 0.15 pM. Meanwhile, Jia et al. (2015) presented a label-free impedimetric aptasensor for the direct detection of *Salmonella* has been developed by the one-step electro-deposition of reduced graphene oxide and carbon nanotube nanocomposite. The developed aptasensor was claimed can directly detect the whole bacteria without pretreatment steps and an extraction procedure. The detection can be completed within 60 min, even in real food samples. Instead of antibodies, nucleic acids, aptamer, and enzyme, immune cells may act as bioreceptor monitoring pathogen contamination. Jiang et al. (2017) have successful developed Rat Basophilic Leukemia (RBL) mast cells based sensing method by immobilizing living mast cells on gold electrode intended to detect N-acylated homoserine lactones ($3OC_{12}$-HSL) toxins produced by *Pseudomonas aeruginosa*. The result showed that the exposure of $3OC_{12}$-HSL to mast cells induced a marked decreased in the electrochemical impedance signal and gave limit of detection of 0.034 μM with linearity range of 0.1–1 μM.

Instead of gold and glass electrode, screen-printed interdigitated microelectrodes (SP-IME) are increasingly being reported with impedance detection shows promising platform to fabricated nanosensor. This due to the advantages of interdigitated microelectrodes was able to lowering ohmic drop by maximizing impedance change. This advantages will leads to fast establishment of steady-state, rapid reaction kinetics and an increased signal-to-noise ratio. In fact, it can reduce the detection time, minimize interfering effects, and work in a two-electrode system (vs. conventional three-electrode system), thus significantly benefiting the fabrication and performance of electrochemical biosensors (Li et al. 2015). Recently, Wang et al. (2015) reported impedance immunosensor based on the use of screen-printed interdigitated electrodes and magnetic nanobeads for the rapid detection of *E. coli* O157:H7 (Fig. 19.17). In this study, immunomagnetic separation strategy was applied by separating of *E. coli* O157:H7 from media using the antibody-coated nanobeads and the concentration of bacteria on the electrode surface was determined using magnetic field. It is reported the impedance immunosensor could detect *E. coli* O157:H7 at concentration of 104.45 CFU mL^{-1} (~1400 bacterial cells in the applied volume of 25 μL) in less than 1 h without pre-enrichment.

Indeed, impedance biosensor not restricted with label free strategy. Other researchers have used label dependent strategy to amplify the detection signal and achieve lower detection limits. Recently, Xu et al. (2015) demonstrated label-dependent methods for *E. coli* O157:H7 and *S. typhimurium* detection in foods by using glucose oxidase (GOx) as the labeling enzyme. For the fabrication of the immunosensor, the magnetic beads (MBs) with the nanosize of 130 nm coated with streptavidin were first functionalized with the appropriate biotinylated antibody. *E. coli* O157:H7or *S. typhimurium* was then captured by the MBs–antibody

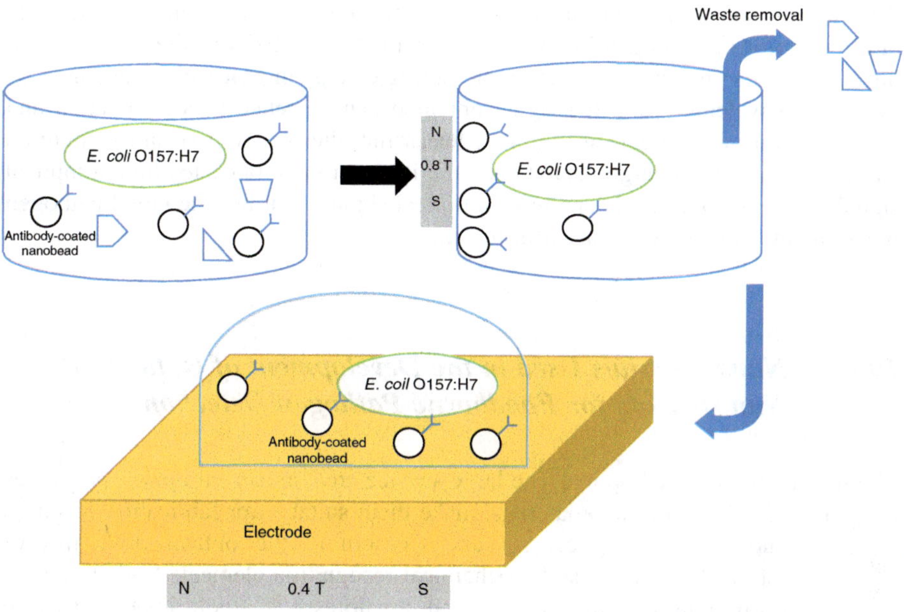

Fig. 19.17 Impedance immunosensor based on the use of screen-printed interdigitated electrodes and magnetic nanobeads for the rapid detection of *E. coli* O157:H7 (Wang et al. 2015)

conjugates. Once the target bacteria were captured, corresponding antibody–GOx conjugates were used to label the bacteria by forming MBs antibody–cell–antibody–GOx sandwich biomass. The final biomass was transferred to the glucose solution with low ion strength. Through the enzymatic reaction, the ion strength of the aqueous samples increased whereas the impedance of the samples decreased.

19.6 Optical Nanosensors for the Detection of Foodborne Pathogens

19.6.1 Concept of Optical Nanosensor

Optical nanosensors are sensors with dimensions in nanometers that use optical transduction to detect chemical or biological analytes. In other words, they are nanosensors that transforms the interaction with analyte optically into a detectable and or measurable analytical signal (Borisov and Klimant 2008). Optical nanosensors used for foodborne pathogen detection are mainly biological or chemical based sensors. Optical biosensors use biological elements e.g. an enzyme, antibody or DNA as a recognition unit while chemical sensors use functional groups like amine, carboxylic acid, thiol etc. as recognition unit. Optical nanosensors for the detection of pathogens are usually composed of biological or chemical recognition unit for capturing pathogen in combination with nanomaterial as the optical transducer. This type of sensors works by measuring change in property of light such as intensity, wavelength, and the signal obtained can be relate to identification and quantification of pathogen in sample. Meanwhile, the recognition unit recognizes and interacts with the target analyte and the interaction is translated into an optical signal by the nanomaterial. The type of optical signal identifies the target pathogen while the magnitude of signal quantifies it.

19.6.2 Nanomaterials Used in the Development of Optical Nanosensors for Foodborne Pathogen Detection

The nanomaterial based optical The large surface area, enhanced optical properties as well as small size of nanomaterials, make them suitable for fabrication of small, portable, and sensitive optical sensors. Combining the optical properties of nanomaterials with chemical and biochemical recognition elements leads to development of different optical sensors such as florescence, colorimetric, surface plasmon resonance (SPR), and surface enhanced Raman scattering (SERS) sensors most of which are being used to detect food borne pathogens (Rowland et al. 2016). Nanomaterials mainly used in design and fabrication of optical sensors include noble metal nanoparticles, magnetic nanoparticles, silica based nanoparticles, metal-based quantum dots and carbon based quantum dots (Shtykov and Rusanova 2008).

Nanoparticles of noble metals like Au and Ag and to some extend Cu and Al, possess intensive evanescence properties, which enhances absorption and scattering of light and accounts for their bright and colorful appearances. As a result, these materials are used as optical transduction agents in many sensors. Absorption and emission of light by these materials is several orders of magnitude higher than conventional dyes thereby significantly improving the sensitivity and detection limits of sensors. In addition, they can be easily functionalized or conjugated with receptors such as enzymes, antibodies, DNA and aptamers for capturing pathogens, their DNA, proteins, and toxins (Shtykov and Rusanova 2008).

Metal based quantum dots such a CdS, CdTe, PbS, CdSe, ZnSe, and PbTe have also been used in fabrication of optical nanosensors. They are semiconductor nanocrystals of about 2–8 nm in diameter with strong photoluminescence properties, making them suitable for optical sensors. They can also be easily linked with recognition and capturing units for pathogens and toxins in different media types. Silica nanoparticles are also used in optical sensors. They are often doped with fluorescence dyes to enhances their photoluminescence properties. Magnetic nanoparticles conjugated with capturing units such as DNA, antibodies. Aptamers etc. have been employed in optical sensors mainly for separation and concentration of analytes in samples. Magnetic nanoparticles include Fe_3O_4, Fe_2O_3, Fe_3S_4, ZnO, and MgO (Shtykov and Rusanova 2008).

Carbon based quantum dots are newly discovered nanomaterials with photoluminescence properties that found applications in optical sensing. They include carbon nanodots (CNDs) and graphene quantum dots (GQDs). CNDs are spherically shaped crystalline materials with diameters below 10 nm while GQDs, composed of few layer of graphene lattice structure are flat with lateral dimensions within 3–20 nm range (Sutarlie et al. 2017). Both have excellent Photoluminescent properties and biocompatibility and can be easily functionalized or bioconjugated with capture receptors for pathogens. They are mainly used as fluorescent probes and FRET donors in chemical and biological sensors.

19.7 Optical Techniques

There are four main types of optical techniques usually explored by researchers in designing and fabrication of nanosensors and these include fluorescence, calorimetry, surface plasmon resonance, and surface enhanced Raman spectroscopy. Each of these techniques are discussed in detail in this section.

19.7.1 Fluorescence

Fluorescence based sensors measure change in frequency or intensity of an emitted radiation stimulated by previous illumination of a fluorescence material. Fluorescence sensing is achieve in three main ways: direct sensing, indirect sensing, and fluorescence resonance energy transfer (FRET) (Touhami 2015). Direct sensing as depicted in Fig. 19.18 is done with target analytes that possess fluorescence

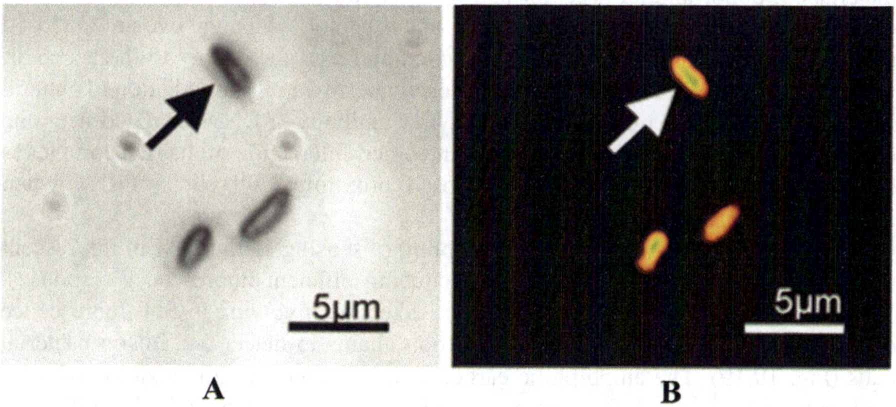

Fig. 19.18 Label free imaging i.e. direct sensing of bacterial cells on gypsum surface, (**a**) visible image of putative bacteria illuminated with white-light, (**b**) fluorescence image of the putative bacteria illuminated with deep UV-light. Adopted from Bhartia et al. (2010)

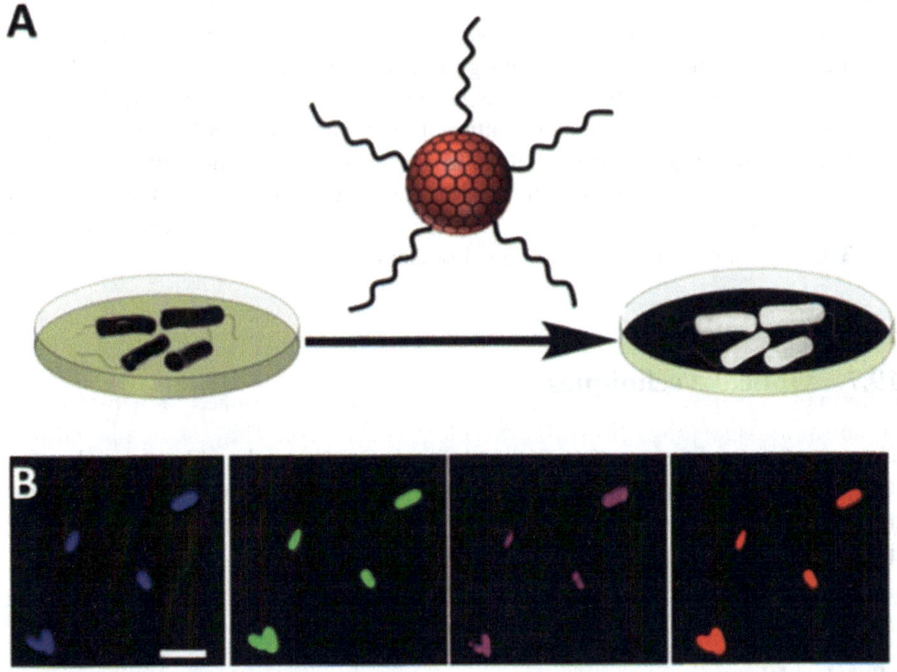

Fig. 19.19 (**a**) Labeling bacteria with amphiphilic carbon dots, (**b**) fluorescence images of *E coli* at different excitation wavelength (Nandi et al. 2015)

properties. However, they are illuminated at a certain wavelength and detected at different wavelength. For analytes that are incapable of fluorescence, indirect fluorescence sensing is used. In such a case, a fluorescent material is added to transduce the presence of the target analyte. Meanwhile, FRET requires a pair of fluorescent materials, a donor and an acceptor and that the emission wavelength of the donor overlaps with the excitation wavelength of the acceptor. When the two are in closed proximity and illuminated, donor gets excited and pass the energy to the acceptor, which in turn becomes excited and then deactivates by emitting radiation (Touhami 2015). In FRET based biosensors, the enzyme, antibody or DNA is tagged with one of the pair while the target is tagged with the other. Interaction of the two complexes brings the donor and acceptor into closed proximity and elicits FRET when illuminated.

Particular attention is given to the coupling of sensing elements with fluorescent reporter dyes and to the methods for producing efficient fluorescence responses. Nandi et al. (2015) has studied florescence detection involving use of fluorescence carbon dots functionalized with hydrocarbon chains to detect and image bacterial cells (Fig. 19.19). The amphiphilic carbon dots prepared by carbonization of acylated fatty acid ester of D-glucose is confirms to possess both hydrophilic and hydrophobic groups and capable of strongly binding to surface membranes of different bacterial species. The method is demonstrated to detect *E. coli, P. aeruginosa, S. typhimurium*, and *B. cereus*. This nanosensor is not only capable of detecting bacterial cells using fluorescence spectroscopy, but able to distinguish different species in a mixture and hence can be used for multiplex bacterial detection. This is possible because the intensity and spectral position of the fluorescence emitted by the carbon dots depend on bacterial species. The method is versatile and can be used in healthcare application, basic research and food safety.

19.7.2 Colorimetric

Colorimetric nanosensors are nanosensors that detect presence of a pathogenic organism through color change observable by naked eye. They are composed of a capturing unit that binds to pathogen and a nanomaterial, which produces a certain color. The binding of many sensor units causes aggregation of the nanoparticles, which then become visible to the eye by producing a color. Colorimetric nanosensors are simple and easy to use devices that can detect microbes without use of any analytical instrument, making them very attractive. Their main disadvantages are that, they are less sensitive and have less potential for multiplex analysis (Yoo and Lee 2016). Colorimetric nanosensors are of two types: flat substrate based and solution based.

Flat substrate based colorimetric nanosensors, a typical example of which is Lateral Flow Assay (LFA), use a paper or glass as a substrate. LFA-based colorimetric sensors utilize capillary flow of the sample over the substrate and color change by aggregation of AuNPs. Yoo and Lee (2016) reported that the AuNPs are functionalized with a pathogen specific receptor such as an antibody and the sensor complex is allowed to capture the target pathogens (Fig. 19.20). The pathogen–sensor systems are then allowed to flow over a membrane, which also has pathogen receptors attached to it. As the samples flows over the membrane, the pathogen–sensor complex interact with the receptor on the membrane, forming a sandwich structure and aggregating the AuNPs which then produce a red line (Fig. 19.20a). An example of such sensor is the sensor produced by DuPont Co which uses AuNPs and antibodies to detect *Salmonella, Listeria* and *E. coli* 0157 in 10 min. Another example is the one produce by BioMerieus, which can detect *Streptococcus* and *legionella pneumophila* in 15 min (Yoo and Lee 2016).

The main drawbacks of LFA-based colorimetric sensing are low sensitivity and limited potential for multiplex detection of microbes. Its sensitivity can be improved by using magnetic beads, which could be attached to the receptors and used to concentrate pathogenic cells before analysis. The use of magnetic beads can improve the sensitivity to about 5 CFU/mL but it also increases the analysis time to about 30 min. Multiplex detection might be achieved by immobilization of different receptors specific for different pathogens on the nanoparticles and different membranes on a single substrate. This also means additional fabrication processes.

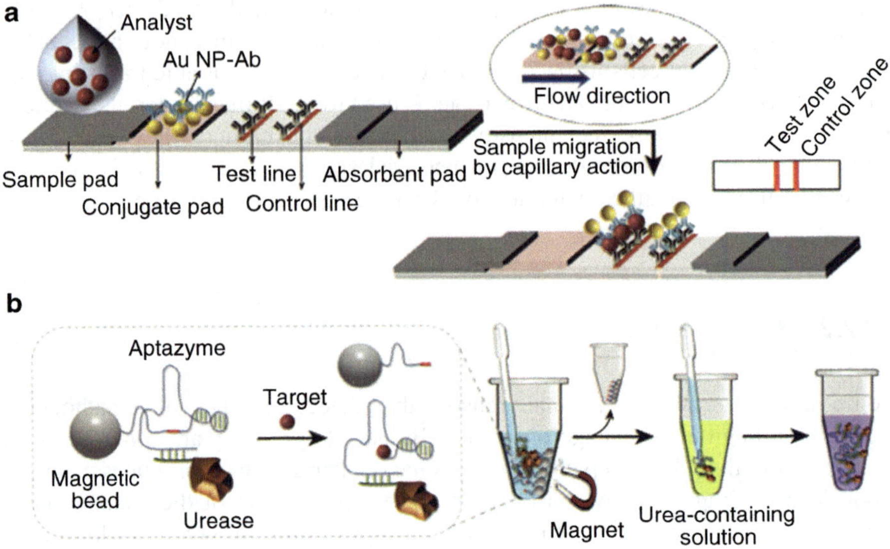

Fig. 19.20 (**a**) LFT-based colorimetric biosensor, (**b**) Solution-based colorimetric biosensor (Yoo and Lee 2016)

In solution based colorimetric biosensors, receptor immobilized on colloidal AuNPs bind to target pathogens and aggregate in solution to produce color. The binding of many sensor units onto a pathogen leads to aggregation of the nanoparticles and hence produce visible color change. A nanoparticle aggregation based solution biosensor linked to a smartphone camera has recently been developed. This approach not only detects *E. coli* in a sample solution but also quantifies it using smart phone camera and an image analysis software on a computer. Another solution based colorimetric sensor using an RNA containing aptazyme linked to nanomagnetic bead at one end and urease tagged with phenol red (a pH sensitive dye) at the other end (Fig. 19.20b) (Yoo and Lee 2016) has also been developed recently. When the aptazyme capture the target pathogen, RNA portion of the aptazyme breaks, separating the part linked to nanomagnetic bead and the part bearing urease. The nanomagnet part is separated from the solution using a magnet and the urease potion in the sample is placed into a urea solution. Urease then hydrolyses urea, increasing pH of the solution, which causes the pH sensitive dye to visibly change color. This method is simple and cheap but its sensitivity is low with a detection limit of 500 CFU/assay (Yoo and Lee 2016).

19.7.3 *Surface Plasmon Resonance (SPR)*

SPR sensors use change in refractive index as an analytical signal for detecting and quantifying presence/binding of analytes at sensor surface. The refractive index of the sample at the sensor surface is determined by measuring the angle of an incident light, which excites surface plasmons (oscillating electrons) at the surface of the sensor (Yoo and Lee 2016). SPR sensors fall into two categories: propagating SPR (PSPR) sensors and localized SPR (LSPR) sensors. PSPR sensors use continuous film of a metal on a glass or prism surface to generate PSPs, which can cover hundred or hundreds of micrometers along the metal/dielectric surface. LSPR sensors, on the other hand, use metal nanoparticles to excite LSPs, which can be turned, based on size, shape and composition of the nanomaterial. SPR nanosensors use for the detection of pathogenic organism and toxins generally belong to LSPR sensors and the nanomaterials use in such sensors include metal nanoparticles (e.g., Au, Ag), magnetic nanoparticles (e.g., Fe_3O_4), liposome nanoparticles, carbon nanostructures, and latex nanoparticles (Zeng et al. 2014) (Fig. 19.21). They are composed of a thin film of nanoparticles functionalized with capturing receptors on a prism or glass surface. The binding or presence of target analytes causes change in the refractive at the sensor–sample interface which, leads to change in the resonance angle of the incident light.

The main advantages this type of sensors compared to ELISA include real time monitoring and detection of label free and broad range of analytes. The selectivity depends on the efficiency of the capturing unit e.g. DNA, antibody or aptamer (Bolduc and Masson 2011). The downsides of SPR nanosensors are low sensitivity due to short penetration of SPR into sensor surface and refractive index similarity between the samples medium and cytoplasm of target pathogen, which sometimes hinders detection of larger analytes such as whole microbial cell. These limitations are being improved by use of gold coated magnetic nanoparticles and use of long range SPRs (LR-SPRs). Gold coated magnetic nanoparticles could improve sensitivity to about 10 × 10^3 CFU/mL by permitting magnetic separation and

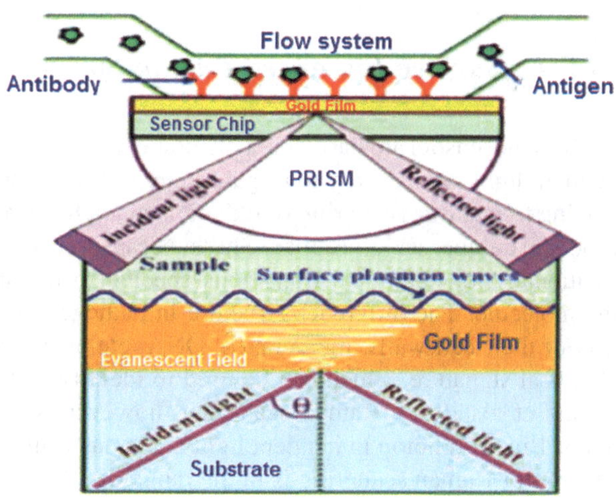

Fig. 19.21 Antibody based SPR nanosensor (Sharma and Mutharasan 2013)

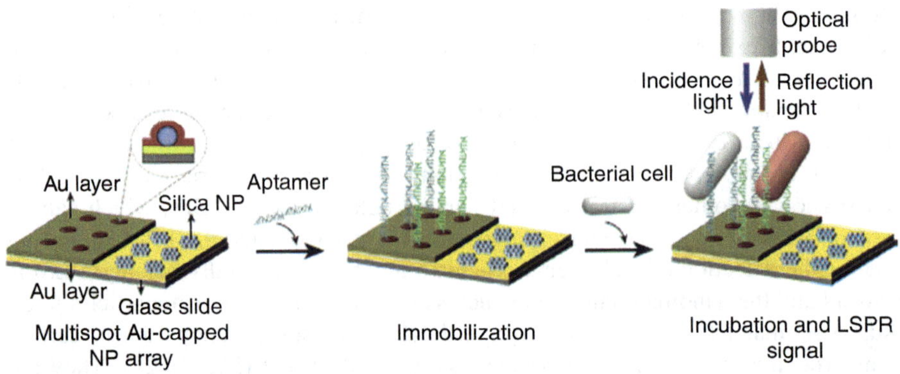

Fig. 19.22 A localized surface plasmon resonance (LSPR) sensor (Yoo and Lee 2016)

concentration of the target analytes. LR-SPR sensors are composing of a film of metal embedded between two dielectric materials that have similar refractive indices. This arranges improves the electric field intensity and penetration distance to over 1 μM compared to conventional SPRs whose penetration depth cannot go beyond 200 nm. Thus, LR-SPR increases sensitivity at longer distance from the sensor surface thereby permitting detection of whole bacterial cells such as *E. coli* (0.7–1.0 μm) (Tangkawsakul et al. 2016).

AuNP based LSPR sensor has been successfully demonstrated to simultaneously detect at 30 CFU/assay level, three different bacterial cells (Fig. 19.22). An LSPR sensor based on Cu coated SiNPs arranged on Au surface that permits ultrasensitive with limit of detection at 10 fM has also been developed (Yoo and Lee 2016).

19.7.4 Surface Enhanced Raman Spectroscopy (SERS)

SERS nanosensors are based on a phenomenon called Raman scattering in which an incident light loses energy to or gains energy from vibrating and rotating molecules leading to inelastic scattering of radiation called Raman spectra. Different molecules produce unique set of Raman Shifts because each type of molecule undergoes rotational and vibrational transitions specific to the structure and composition of the molecule. The difference structure in pathogens involving molecular composition of their cell walls, membranes, DNA, etc. enabling Raman spectra to provide chemical signature, which can be used to identify different pathogens (Table 19.3) (Bantz et al. 2011). Raman scattering, however is very insensitive because only about 1 in 10^8 photon in incident light are inelastically scattered and a typical cross-section for Raman scattering is in the range of 10^{-31}–10^{-29}/molecule, compared to fluorescence with a cross-sections of approximately 10^{-15}/molecule, hence a need to improve or enhance sensitivity of Raman spectroscopy.

Table 19.3 Examples of SERS nanosensors for food borne pathogen detection

Nanomaterial	Recognition unit	Target pathogen	Reference
Graphene nanosheet, Au@Ag NPs	DNA	Bacterial DNA (multiplex)	Duan et al. (2015)
Gold-encapsulated silica NPs	Antibody	*E. coli*, *Salmonella*, *Listeria*	
AgNPs on PVA particles	Direct SERS response	*E. coli*, *Listeria*, *Salmonella*	

NPs = Nanoparticles, PVA = Polyvinyl alcohol

It was discovered that metal films with nanostructured rough surfaces increase signal intensity of Raman scattering to between four and eight orders of magnitude. Rough metal surfaces, sharp metal tips, and metal nanoparticles enable excitation of localized surface plasmons, which increases the intensity of Raman scattering by enhancing the intensity of the incident electric field. The new technique, called Surface Enhanced Raman Scattering drastically increased the molecular cross-section and made possible detection of numerous biomolecules (Bantz et al. 2011).

Quantitative detection of pathogens is one of the challenges of SERS based nanosensors. This is due to the fact that SERS signal intensity depends on the structural features of the surface at nanoscale level. This is being improved upon in a number of ways:

1. Creating uniform gap distances among NPs by addition of cucurbit[n]urils to Au colloids.
2. Use of smooth nanowires with roughness at Å level.
3. Use of substrates with Au coated silicon arranged in regular inverse pyramidal pattern.
4. Use of SER active substrate composed of array of Ag nanorodes.

Coupling SERS with active nanowires for instance, significantly improves SERS sensitivity and lowers the detection limit to 10 pM. Combing SERS incorporation of exonuclease III-aided target-recycling to SERS could enhance detection of multiple fungal DNA by reducing detection limit to 100 fM levels. Detection of whole bacterial cells is also attracting attention of researchers. In newly developed SERS bacterial detection methods, electrostatic attraction between metal cations and negatively charged cell walls of gram positive bacteria and lipopolysaccharides in outer membranes of gram-negative bacteria is used to deposit NPs on bacterial cells. The enables homogeneous SERS signal generation from cells in the sample solution from which different bacterial species are detected and identified based on the fact that cell wall of different bacterial species differ in structure and composition and so produce different SERS spectra. The wide use of such methods is however hindered by the requirements for software analysis tool and standard SERS spectra database for pathogens (Yoo and Lee 2016).

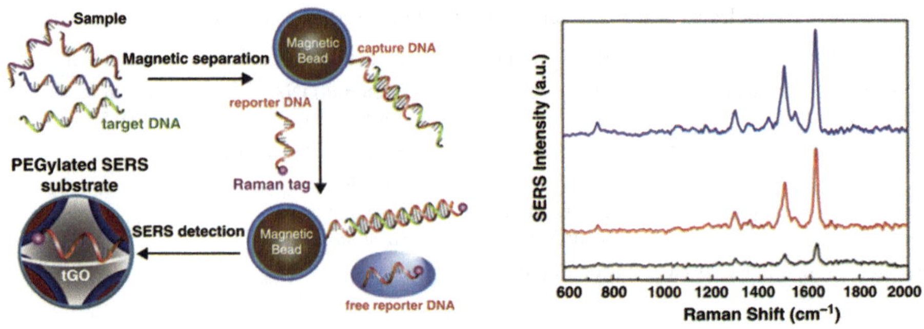

Fig. 19.23 Illustration of the SERS detection of DNA based on the trilayered GO-gapped plasmonic arrays of Au@Ag core–shell nanoparticles (Wang et al. 2017)

An example of SERS nanosensor for pathogen detection is the graphene–nanosheet-gapped plasmonic nanoparticle array SERS sensor for multiplex DNA detection Fig. 19.23.

This sensor make use of a trilayered substrate composed of thiolated graphene oxide nanosheet sandwiched between two layers of closely packed plasmonic nanomaterials and magnetic separation to detect DNA sequence of pathogenic bacteria based on signal turn off. Two oligonucleotides both complementary to the target DNA are tagged with magnetic beads (captured probe) and a Raman (reporter probe) active dye, respectively. The capture probe, hybridizes with the target, separates and concentrates it. The hanging portion of the target DNA hybridizes with the reporter forming a complex structure. In absence of the target DNA, hybridized capture and reporter DNA can permeate the substrate deeper and exhibit strong SERS signal. In presence of the DNA however, the formation of a bulkier sandwich structure do not permit the reporter dye to reach deeper into the substrate, so turn of the signal (Duan et al. 2015).

19.8 Other Types of Nanosensors Used for Pathogens Detection

19.8.1 Piezoelectric/Mass Nanosensors

Piezoelectric sensors are devices that convert mechanical energy or force into electrical potential. They are based on piezoelectric effect, the capacity of certain materials like quartz and tourmaline to change force or displacement into electric potential. Application of pressure to the piezoelectric material causes displacement of charges and the magnitude of charge displacement is proportional to the applied pressure (Priyanka et al. 2013). Quartz Crystal Microbalance (QCM) is one of the most popular piezoelectric sensors used in pathogen detection. It is composed of a thin quartz (SiO_4) plate with a layer of metal electrodes deposited on opposite sides, and the overlapping portion of the electrodes form the active sensing surface. It operates by measuring the change in resonance frequency of the sensor as a function of change in mass i.e. attachment of analytes to sensor surface. The receptors such as antibodies, DNA, aptamers etc. are immobilized on the sensing surface and the interaction of the receptor with the pathogenic analyte leads to change in mass which in turn elicits change resonance frequency.

Like most biosensors, capture receptors are immobilized on the QCM sensor surface by either covalent or noncovalent means. Noncovalent immobilization methods are based on physical adsorption like anchoring the receptor on hydrophobic surface by adsorption. Covalent immobilization involves functionalizing the sensor surface with for example carboxylic acid groups by making a self-assembled monolayer (SAM) and activation the functional groups with 1-ethyl-3-(3-(dimethylamino)propyl)carbodiimide (EDC) and N-hydroxysuccinimide (NHS). The receptors like antibodies, antigens, and DNA are then covalently attracted to NHS (Wu et al. 2017).

Sauerbery equation shows the relationship between change in frequency and change in mass per unit area of QCM (Priyanka et al. 2013).

$$\Delta f = -C_f \times \Delta m, \quad C_f = \frac{2n \times f^2}{\sqrt{\rho_q \times \mu_q}}$$

Solving for change in mass (Δm) gives:

$$\Delta m = \frac{-\Delta f}{C_f} = \frac{\left(f_q - f\right)\sqrt{\rho_q \times \mu_q}}{2n \times f^2}$$

Where Δf = change in frequency (Hz), C_f = sensitivity factor in (Hz/cm^2, Δm = change in mass per unit area (g/cm^2), f = resonance frequency of loaded crystal (Hz), f_q = resonance frequency of unloaded crystal, n = number of harmonic at which the crystal is driven, ρ_q = density of quartz (2.648 g.cm^2), and μ_q = shear modulus of quartz (2.947×10^{11} g cm^{-1} s^{-2}).

This equation is applicable only for thin, even and purely elastic layers. Depositions that lead to complex mass–frequency correlations, such as vacuum and gas depositions might not meet these requirements and so may require some form of calibration to ensure satisfactory measurements (Priyanka et al. 2013).

Figure 19.24 shows an illustration of flow through QCM biosensor system. In this study, the sample is placed into a buffer, which will pass through and introduce it on the sensor surface immobilized with receptors. Frequency shift in real time is monitored as a function of increase mass on sensor surface due to mass of analyte captured on the sensor surface.

One of the main disadvantages of QCM biosensors is their relatively high limit of detection (LOD) which is being improved upon via preconcentration and signal amplification using magnetic nanoparticles which reduces the LOD for *S. typhimurium* to 100 CFU/mL and *E. coli* 0157:H7 to 220 CFU/mL (Farka et al. 2015). Signal amplification is achieved with the use of streptavidin-coated Au nanoparticles as illustrated in Fig. 19.25 for detecting *S. epidermidis*.

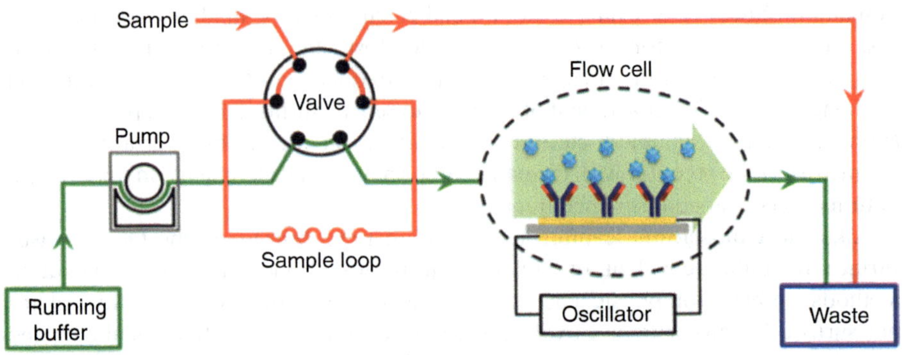

Fig. 19.24 Diagram of a flow through QCM biosensor system (Wu et al. 2017)

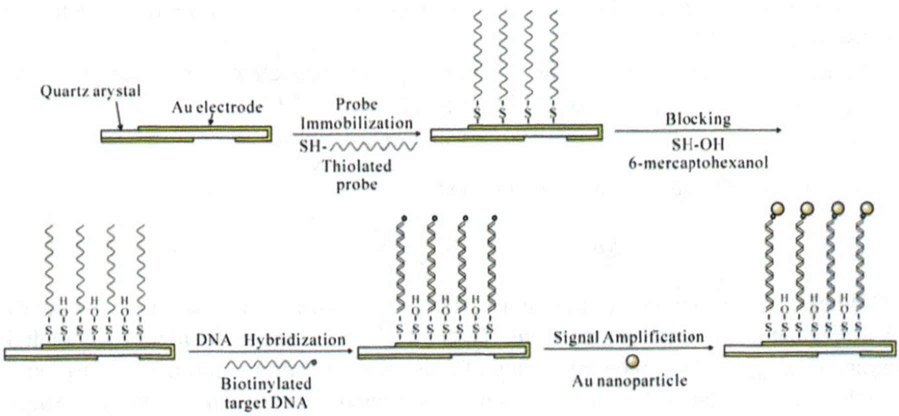

Fig. 19.25 Assembly and signal amplification procedure for QCM nucleic biosensor for *S. epidermidis* detection (Xia et al. 2008)

19.8.1.1 Microfluidic Nanosensors

Microfluidics is the study of control and manipulation of small volumes of fluids within microliter and picoliter range in microchannels of 10s to 100 s of micrometers in diameter. It involves manipulation and transportation of micro to pico volumes of fluid within and through microchannels with the use of pressure or electro-kinetics (Ansari et al. 2016). Microfluidic nanosensors are analytical devices derived from an intersection of nanotechnology, microfluidics, and sensors technology, and utilize nanomaterials and sensing in manipulation and analysis of small volumes of samples. They comprise a network of microchannels for transporting sample/reagents; chambers for preparing samples (to release/separate/concentrate/amplify analytes); and a sensor (s) for detecting and measuring analytes. Microfluidic sensors are products of the concept of miniaturizing all analytical procedures: sample preparation, separation, concentration, analysis and data processing and putting all this on a small portable chip called Lab-On-Chip (LOC) or a Micro Total Analysis System (μTAS). Sensors tend to have sensitivity and selectivity problems due to lack of sample preparation and separation steps, while bench-top instruments lack portability and real-time analysis and the idea was to integrate all analytical steps into a device that combines the small size and portability of sensors with the selectivity and performance of bench-top instruments (Hugh Fan 2013).

Since small volumes of fluids are transmitted in microchannels, laminar flow rather than turbulent flow is favored in microfluidic systems. Forces such as surface tension, viscosity, and friction are significant for small volume of fluids and must be overcome to allow fluids to flow through the channels. Pressure and electro-kinetic are the main mechanisms used to induce motion in micro volumes of fluids and since narrowness of the microchannels limits the effectiveness of pressure, electro-kinetic flow is more favored. Chemical modification of channel surfaces and electrically controlled surface tension are utilized to move fluids in the microchannels. For pressure driven flow, regulating pressure difference between the injection point and sink for the fluid and syringe pumps are employed (Ansari et al. 2016).

Glass, silicon and different types of polymers such as Teflon, poly (methylmethacrylate) (PMM) and poly (dimethylsiloxane) (PDMS) are used as substrates in fabrication of microfluidic devices. As illustrated in Fig. 19.26, typical fabrication of microchannels involves three steps: production of a master mold from a silicon substrate and photoresist using photolithography, generation of PDMS replica with microchannel network on its surface and then attaching the PDMS replica on to a glass substrate to seal the channels.

There are three main types of microfluidic sensor based on sample actuation and manipulation in the microchannels: continuous flow, droplet and digital microfluidic (DMF) as depicted in Fig. 19.27. The continuous flow based is the conventional type that induce fluid flow in microchannels with use of syringe pumps, pneumatic pressure, or electrokinetic. This type of microfluidic is easy to fabricate and operate and compatible with most sensing techniques. It however consumes relatively large volumes of samples and reagents comparted to the other types. Droplet-based microfluidic system creates isolated droplets and moves them through the microchannels using continuous streams of two or more immiscible fluids that intersect at a junction. This type, in addition to being easy to fabricate and operate, it is applicable to analytical procedures that require isolated reaction sites. Difficulty in implementing stable gas-liquid systems and lack of control over individual droplets are its main disadvantages. DMF-based systems are the last to be develops with the further reducing reagent consumption as one of its main objectives. In DMF-based systems, droplet is created on an array of electrodes that are electrostatically operated. Instead of a continuous flow of droplets in microchannels, the droplets are moved forward over the electrodes with use of electorwetting-on-dielectric (EWOD) brought about by employing electric field to change interfacial properties of the liquid. DMF-based systems reduce both power and volume consumption, but they are difficult to fabricate (Luka et al. 2015).

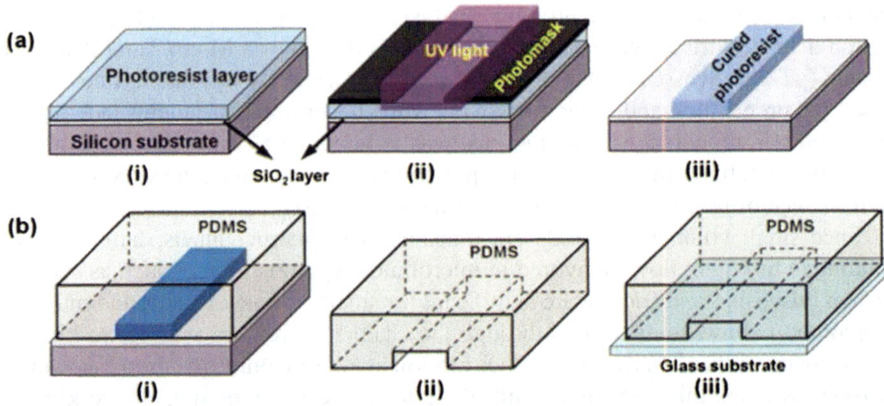

Fig. 19.26 Fabrication methodology of microfluidic channel: (**a**) Generation of master mold using photolithography: (1) deposition of photoresist on a silicon substrate, (2) exposure of photoresist to UV radiation through the photomask, (3) prepared master mold. (**b**) Preparation of the PDMS replica and attachment to glass substrate: (1) PDMS and cross-linker mixture is poured over the master mold, (2) PDMS layer thermally cured and detached from the silicon substrate, (3) PDMS replica bonded to the glass surface, generating sealed channels. Adopted from Ansari et al. (2016)

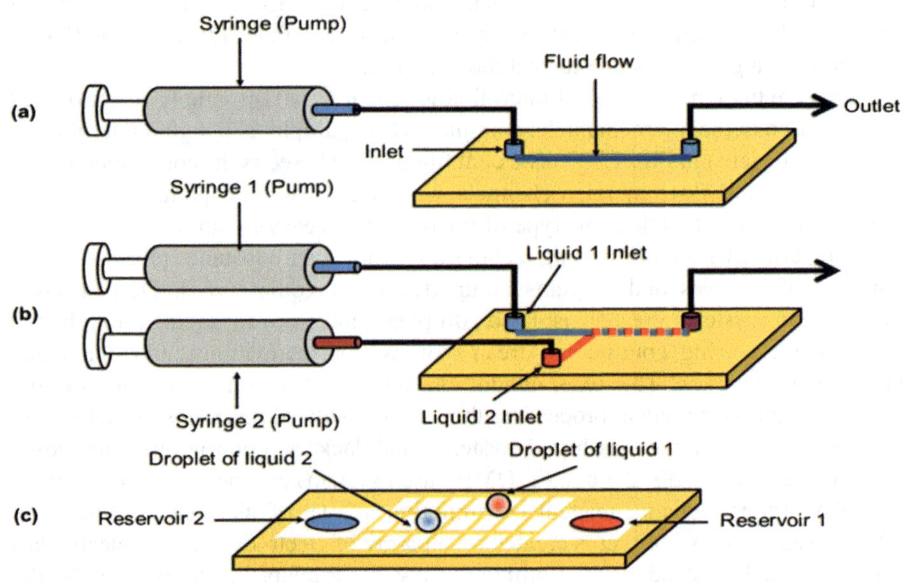

Fig. 19.27 Illustration of the three microfluidic systems (**a**) continuous flow-based, (**b**) drop-based, and (**c**) digital microfluidic. Adopted from Luka et al. (2015)

Microfluidic biosensors in particular, show great potentials for detection pathogenic microbes. They improve pathogen sensing by reducing analysis time, introducing portability and disposability, decreasing reagent consumption, minimizing hazardous chemical handling, and enabling parallel or multiplex analysis. The type of target analyte (DNA, protein, cells), determines method of microfluidic based pathogen detection technique to use which include microfluidic immunoassay for food borne pathogen detection and microfluidic nucleic acid analysis (Safavieh et al. 2015).

Microfluidic integrated plasmon platform for pathogen detection by Onur Tokel's group (Tokel et al. 2015); Microfluidic, urease catalysis, and electrochemical impedance analysis of foodborne pathogens by Chen et al. (2016); and nanostructured microfluidic immunoassay platform infectious pathogen detection by Yu et al. (2017) are clear examples of successful application of microfluidic nanosensors foodborne pathogen detection.

In the work of Tokel et al. (2015), a cheap, portable, multiplex microfluidic-integrated surface plasmon resonance sensor that detects *E. coli* and *S. aureus* have been demonstrated. The sensor can detect and quantify *E. coli* in concentration range of 3.2×10^5 to 3.2×10^7 CFU/mL with high selectivity and sensitivity and multiplex capability. The device (Fig. 19.28) composed of a microfluidic chip, a light emitting diode (LED), a cylindrical lens, a prism and a COSM sensor. The chip, with an inlet port, outlet port and a microchannel, is made of two layers of polymethyl methacrylate (PMMA) and a gold-coated glass substrate. In operation,

Fig. 19.28 Portable, multiplex microfluidic-integrated surface plasmon resonance sensor. (**a**) Microfluidic chip mounted on the top side of the device, (**b**) electronic setup, LED, prism and the CMOS sensor, (**c**) schematic of integrated microfluidic—SPR sensor platform, gold surface with immobilized lipopolysaccharide (LPS) for capturing E. coli. Adopted from Tokel et al. (2015)

the sample is introduced into the inlet port, which flows through the microchannel, over the sensing surface, into the outlet post. The Lipopolysaccharide (LPS) antibodies in the sensing surface capture the pathogen, leading to change in refractive index on the surface, which sifts the resonance angle of light for the LED. The light from the LED passes through the prism to illuminate and be a reflected by the microchip onto CMOS sensor, which is connected to a computer. A software captures the image from the CMOS sensor and calculate the resonance angel in real-time (Tokel et al. 2015).

The microfluidic electrochemical impedance pathogen sensor (Fig. 19.29) can capture and detect Listeria with a capture efficiency of 93% and detection limit of 1.6×10^2 CFU/mL. It is composed of a syringe pump, multiport valve, separation chip, detection chip, and the electrochemical workstation (Chen et al. 2016).

The sample, MNP modified monoclonal antibodies, and AuNP modified polyclonal antibodies are incubated in the separation chip, leading to the formation of urease-AuNP-Listeria-MNP complexes. The complexes are captured on the separation chip with high gradient magnetic field and wash with TBS and DI water to remove unbound species like AuNPs and ions. Catalyzed by urease, the complex is hydrolyzed and suspended by injecting a solution of urea, producing carbonate and ammonium ions, which increases the ionic strength of the solution. The solution is then moved to the detection chip bearing interdigitated array of microelectrodes

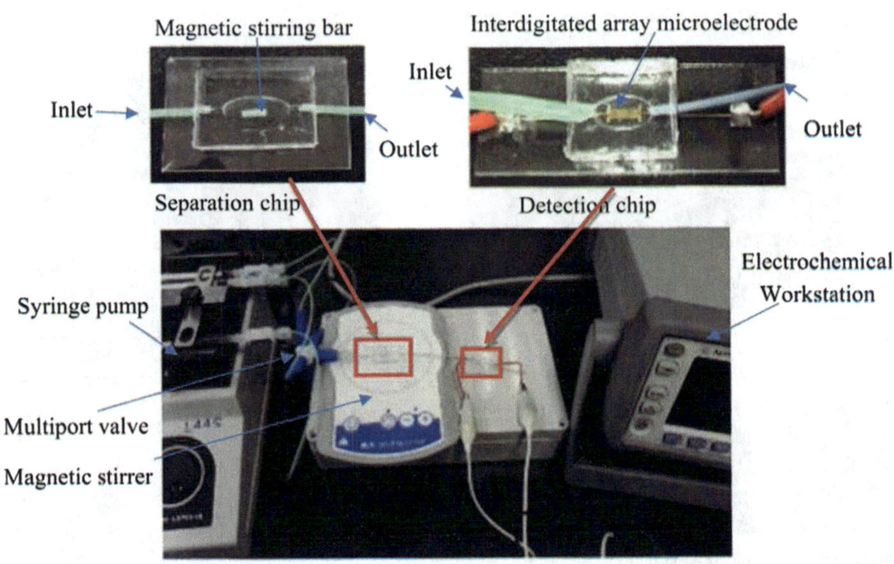

Fig. 19.29 Microfluidic electrochemical impedance pathogen sensor. Adopted from Chen et al. (2016)

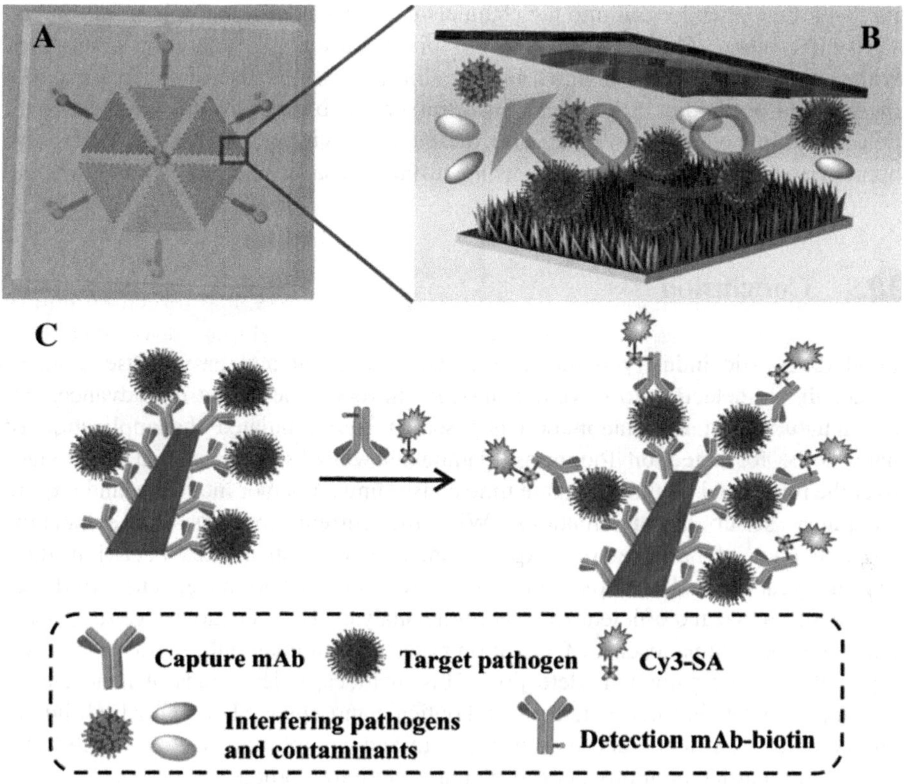

Fig. 19.30 (**a**) Layout of the microfluidic chip with six microchannels, (**b**) capturing of virus on monoclonal antibody (mAB) modified ZnO nanorodes, (**c**) Sandwich assay detection of viral pathogen on surface of ZnO nanorode. Adopted from Yu et al. (2017)

whose impedance change is measured by IM6 electrochemical workstation to quantitatively determine cells of pathogen (Chen et al. 2016).

The nanostructured microfluidic immunofluorescence platform has proven to detect avian influencer virus (AIV), H5N2, with detection limit of 3.6×10^3 at 50% embryo infectious doses per mL (EID_{50}/mL), i.e. 22 times more sensitive than the conventional immunosorbent assay. The microfluidic immunofluorescence chip (Fig. 19.30), which can be loaded with six different samples at a time, contains six branched microchannels arranged in a hexagon (Yu et al. 2017).

Each microchannel contains ZnO nanorodes which can be easily functionalized with monoclonal antibody (mAB) for capturing the viral pathogens. In an analysis process, the microchannels are first loaded with a solution containing the capture mAB and conjugation reagent to functionalize the ZnO nanorodes and form ZnO-mAB. The sample is then loaded to capture the virus to form ZnO-mAB-virus complex, which is then washed with a different reagents to remove interfering pathogens and other contaminants. Biotin-labelled monoclonal antibody, mAB-biotin is then loaded into the channels to for a sandwich complex ZnO-mAB-virus-mAb-biotin. Fluorescence signal is produced using streptavidin conjugated with cyanine dye (Cy3-SA) which binds to biotin. With the use of spatial encoding of captured antibodies, this microfluidic-integrated biosensor can simultaneously detect multiple viruses. Furthermore, the captured viruses can be released using appropriate reagents, making the sensing surface reusable (Yu et al. 2017).

19.9 Conclusion

Food diagnostic industry demands for rapid, accurate and easy-to-use devices, especially for detection foodborne pathogens in food. The increasing advancement of miniaturization and nanomaterials research has stimulated the application of nanosensor for detection foodborne pathogens which show essential advantages over the traditional methods that are time consuming and labor intensive, and require laboratory set-up instrumentations. With the current progress and exhaustive research pace of nanomaterial exploration, new mechanism has been introduce over the years to make nanosensor better in terms of performance, sensitive detection, unobtrusive and efficient for foodborne analysis. In this chapter, we review the latest literature on nanosensor fabrication based on nanomaterial favorable in varies types of forborne pathogens detection. This includes review types of nanosensors developed to date in electrochemical and optical nanosensor, also giving highlight to another important area such as piezoelectric and microfluidic which has recently moved into particular approaches to be maturing into a product.

References

Ahmed A, Rushworth JV, Hirst NA, Millne PA (2014) Biosensors for whole-cell bacterial detection. Clin Microbiol Rev 27(3):631–646

Ansari MIH, Hassan S, Qurashi A, Khanday FA (2016) Microfluidic-integrated DNA nanobiosensors. Biosens Bioelectron 85:247–260

Bantz KC, Meyer AF, Wittenberg NJ, Im H, Kurtuluş O, Lee SH et al (2011) Recent progress in SERS biosensing. Phys Chem Chem Phys 13(24):11551. https://doi.org/10.1039/c0cp01841d

Bhartia R, Salas EC, Hug WF, Reid RD, Lane AL, Edwards KJ, Nealson KH (2010) Label-free bacterial imaging with deep-UV-laser-induced native fluorescence. Appl Environ Microbiol 76(21):7231–7237

Brandao D, Liebana S, M.I. Pividori (2015) Multiplexed detection of foodborne pathogens based on magnetic particles New Biotechnol. 32(5):511–520

Bolduc OR, Masson JF (2011) Advances in surface plasmon resonance sensing with nanoparticles and thin films: nanomaterials, surface chemistry, and hybrid plasmonic techniques. Anal Chem 83(21):8057–8062

Borisov SM, Klimant I (2008) Optical nanosensors—smart tools in bioanalytics. Analyst 133 (10):1302

Cabral, JPS (2010) Water Microbiology. Bacterial Pathogens and Water. Int J Environ Res Public Health 7(10): 3657–3703

Chamorro-Garcia A, Merkoc A (2016) Nanobiosensors in diagnostics. Nanobiomedicine 3:1–26

Chen Q, Wang D, Cai G, Xiong Y, Li Y, Wang M, Lin J (2016) Fast and sensitive detection of foodborne pathogen using electrochemical impedance analysis, urease catalysis and microfluidics. Biosens Bioelectron 86:770–776

Cinti S, Volpe G, Piermarini S, Delibato E, Palleschi G (2017) Electrochemical biosensors for rapid detection of foodborne Salmonella: a critical overview. Sensors (Basel) 17(8):pii:E1910

Clark LC, Lyons C (1987) Electrode systems for continous monitoring in cardiovascular surgery. Annals of the New York Academy of Sciences 102 (1):29–45

Das R, Sharma MK, Rao VK, Bhattacharya BK, Garga I, Venkateshb V, Upadhyay S (2014) An electrochemical genosensor for Salmonella typhi on goldnanoparticles-mercaptosilane modified screen printed electrode. J Biotechnol 188:9–16

Daniels JS, Pourmand N (2007) Label-free impedance biosensors: opportunities and challenges. Electroanalysis 19(12):1239–1257

Duan B, Zhou J, Fang Z, Wang C, Wang X, Hemond HF, Duan H (2015) Surface enhanced Raman scattering by graphene-nanosheet-gapped plasmonic nanoparticle arrays for multiplexed DNA detection. Nanoscale 7(29):12606–12613

Farka Z, Kovar D, Skladal P (2015) Rapid detection of microorganisms based on active and passive modes of QCM. Sensors (Basel) 15(1):79–92

Fei JF, Dou W, Zhao G, (2016) Amperometric immunoassay for the detection of Salmonella pullorum using a screen - printed carbon electrode modified with gold nanoparticle-coated reduced graphene oxide and immunomagnetic beads. Microchimica Acta 183(2):757–764

Fei JF, Dou W, Zhao G (2015) A sandwich electrochemical immunosensor for Salmonella pullorum and Salmonella gallinarum based on a screen-printed carbon electrode modified with an ionic liquid and electrodeposited gold nanoparticles. Microchim Acta 182(13–14):2267–2275

Fernandes AM, Zhang F, Sun X (2014) A multiplex nanoparticles-based DNA electrochemical biosensor for the simultaneous detection of Escherichia coli O157:H7 and Staphylococcus aureus. Int J Curr Microbiol Appl Sci 3(4):750–759

Hamidi-Asl E, Dardenne F, Pilehvar S, Blust R, De Wael K (2016) Unique properties of core shell Ag@ Au nanoparticles for the aptasensing of bacterial cells. Chemosensors 4(3):16

Hassan AR, Muniz AE, Merkoçi A (2015) Highly sensitive and rapid determination of Escherichia coli O157:H7 in minced beef and water using electrocatalytic gold nanoparticle tags. Biosens Bioelectron 67:511–515

Hajra TK, Bag PK, Das CS, Mukherjee S, Khan A, Ramamurthy T (2007) Development of a simple latex agglutination assay for detection of shiga toxin-producing Escherichia coli (STEC) by using polyclonal antibody against STEC. Clin Vaccine Immunol 14(5):600–604

Hu C, Dou W, Zhao G (2014) Enzyme immunosensor based on gold nanoparticles electroposition and Streptavidin-biotin system for detection of S. pullorum & S. gallinarum. Electrochim Acta 117:239–245

Hugh Fan Z (2013) Chemical sensors and microfluidics. J Biosens Bioelectron 4(1):1–2

Inbaraj BS, Chen BH (2016) Nanomaterial-based sensors for detection of foodborne bacterial pathogens and toxins as well as pork adulteration in meat products. J Food Drug Anal 24 (1):15–28

IFT (2004) Bacteria associated with foodborne diseases, 7(August):1–25

Jia F, Duan N, Wu S, Dai R, Wang Z, Li X (2015) Impedimetric Salmonella aptasensor using a glassy carbon electrode modified with an electrodeposited composite consisting of reduced graphene oxide and carbon nanotubes. Microchim Acta 183(1):337–344

Jiang D, Feng D, Jiang H, Yuan L, Yongqi Y, Xu X, Fang W (2017) Preliminary study on an innovative, simple mast cell-based electrochemical method for detecting foodborne pathogenic bacterial quorum signaling molecules (N-acyl-homoserine-lactones). Biosens Bioelectron 90:436–442

Sabah Kalyoussef DO, Feja KN (2014) Foodborne illnesses. Adv Pediatr 61(1):287–312

Kearney J (2010) Food consumption trends and drivers. Philos Trans R Soc Lond B Biol Sci 365 (1554):2793–2807

Liebana S, Brand D, Cortes P, Campoy S, Alegret S, Pividori MA (2016) Electrochemical genosensing of Salmonella, Listeria and Escherichia coli on silica magnetic particles. Anal Chim Acta 904:1–9

Li BM, Yu QL, Duan YX (2015) Fluorescent labels in biosensors for pathogen detection. Crit Rev Biotechnol 35:82–93

Li X, Fu W, He Y, Zhai Q, Guo J, Qing K, Yi G (2016) Electrochemical aptasensor for rapid and sensitive determination of Salmonella based on target-induced strand displacement and gold nanoparticle amplification. J Anal Lett 49:15

Liu Q, Wu C, Cai H, Hu N, Zhou J, Wang P, (2014) Cell-Based Biosensors and Their Application in Biomedicine. Chemical Reviews 114 (12):6423–6461

Luka G, Ahmadi A, Najjaran H, Alocilja E, Derosa M, Wolthers K, Hoorfar M (2015) Microfluidics integrated biosensors: a leading technology towards lab-on-A-chip and sensing applications. Sensors (Basel) 15(12):30011–30031

Ma X, Jiang Y, Jia F, Yu Y, Chen J, Wang Z (2014) An aptamer-based electrochemical biosensor for the detection of Salmonella. J Microbiol Methods 98:94–98

Malik P, Katyal V, Malik V, Asatkar A, Inwati G, Mukherjee TK (2013) Nanobiosensors: concepts and variations. ISRN Nanomater 2013:1–9. https://doi.org/10.1155/2013/327435

Mandal PK, Biswas AK, Choi K, Pal UK (2011) Methods for rapid detection of foodborne pathogens: an overview. Am J Food Technol 6:87–102

Miller RS, Speegle L, Oyarzabal O A, Lastovica A J, (2008) Evaluation of Three Commercial Latex Agglutination Tests for Identification of Campylobacter spp. Journal of Clinical Microbiology 46 (10):3546–3547

Muniandy S, Dinshaw IJ, Teh SJ, Lai CW, Ibrahim F, Thong KL, Leo BF (2017) Graphene-based label-free electrochemical aptasensor for rapid and sensitive detection of foodborne pathogen. Anal Bioanal Chem 409(29):6893–6905

Silva NFD, Magalhães JMCS, Teresa Oliva-Teles M, Delerue-Matos C (2015) A potentiometric magnetic immunoassay for rapid detection of Salmonella typhimurium. Anal Methods 7:4008

Nandi S, Ritenberg M, Jelinek R (2015) Bacterial detection with amphiphilic carbon dots. Analyst 140(12):4232–4237

Ngoensawat U, Rijiravanich P, Surareungchai W, Somasundrum M (2017) Electrochemical immunoassay for Salmonella typhimurium based on an immuno-magnetic redox label. Electroanalysis 29:1–9

Nguyen PD, Trana TB, Nguyen DTX, Min J (2014) Magnetic silica nanotube-assisted impedimetric immunosensor for the separation and label-free detection of Salmonella typhimurium. Sens Actuators B 197:314–320

Notomi T, Okayama H, Masubuchi H, Yonekawa T, Watanabe K, Amino N, Hase T (2000) Loop-mediated isothermal amplification of DNA. Nucleic Acids Res 28:e63

Pandey A, Gurbuz Y, Ozguz V, Niazi JH, Qureshi A (2017) Graphene-interfaced electrical biosensor for label-free and sensitive detection of foodborne pathogenic E. coli O157:H7. Biosens Bioelectron 91:225–231

Poltronieri P, Mezzolla V, Primiceri E, Maruccio G (2014) Biosensors for the detection of food pathogens. Foods 3:511–526

Priyanka B, Patil RK, Dwarakanath S (2016) A review on detection methods used for foodborne pathogens. Indian J Med Res 144(3):327–338

Postollec F, Falentin H, Pavan S, Combrisson J, Sohier D (2011) Recent advances in quantitative PCR (qPCR) applications in food microbiology. Food Microbiol 28:848–861

Pohanka M, Skladal P (2008) Electrochemical biosensors—principles and applications. J Appl Biomed 6:57–64

Priyanka S, Shashank P, Aslam MKM, Prashant S, Pal SK (2013) Nanobiosensors: diagnostic tool for pathogen detection. Int Res J Biol Sci 2(10):76–84

Radhakrishnan R, Poltronieri P (2017) Review label free biosensor methods in detection of food pathogens and Listeria monocytogenes. Biosensors (Basel) 7(4):63

Radhika M, Saugata M, Murali HS, Batra HV (2014) A novel multiplex PCR for the simultaneous detection of Salmonella enterica and Shigella species. Braz J Microbiol 45:667–7663

Reimhult E, Hook F (2015) Design of surface modifications for nanoscale sensor applications (Review). Sensors (Basel) 15:1635–1675

Rowland CE, Brown CW, Delehanty JB, Medintz IL (2016) Nanomaterial-based sensors for the detection of biological threat agents. Mater Today 19(8):464–477

Safavieh M, Nahar S, Zourob M, Ahmed MU (2015) Microfluidic biosensors for high-throughput screening of pathogens in food. In: Bhunia AK, Taitt CR, Kim MS (eds) High throughput screening for food safety assessment: biosensor technologies, hyperspectral imaging and practical applications. Woodhead, Cambridge, pp 327–357

Sharma H, Mutharasan R (2013) Review of biosensors for foodborne pathogens and toxins. Sens Actuators B 183:535–549

Shtykov SN, Rusanova TY (2008) Nanomaterials and nanotechnologies in chemical and biochemical sensors: capabilities and applications. Russ J Gen Chem 78(12):2521–2531

Saker L, Lee K, Cannito B, Gilmore A, Campbell-Lendrum B (2004) Globalization and infectious diseases: a review of the linkages. Special topics in social, economic and behavioural research. World Health Organization, Geneva, No. 3.

Sharma H, Agarwal M, Goswami M, Murugan MS (2013) Biosensors: tool for food borne pathogen detection. Vet World 6(12):968–973

Sutarlie L, Ow SY, Su X (2017) Nanomaterials-based biosensors for detection of microorganisms and microbial toxins. Biotechnol J 12(4):1–26

Shaibani PM, Etayash H, Jiang K, Sohrabi A, Hassanpourfard M, Naicker S, Sadrzadeh M, Thundat T (2018) Portable nanofiber-light addressable potentiometric sensor for rapid Escherichia coli detection in orange juice. ACS Sens 3(4):815–822

Barreiros dos Santos M, Azevedo S, Agusil JP, Prieto-Simón B, Sporer C, Torrents E, Juárez A, Teixeira V, Samitier J (2015) Label-free ITO-based immunosensor for the detection of very low concentrations of pathogenic bacteria. Bioelectrochemistry 101:146–152

Tabrizi MA, Shamsipur M (2015) A label free electrochemical DNA biosensor basedon covalent immobilization of Salmonella DNA sequences on the nanoporous glassy carbon electrode. Biosens Bioelectron 69:100–105

Tangkawsakul W, Srikhirin T, Shinbo K, Kato K, Kaneko F, Baba A (2016) Application of long-range surface plasmon resonance for ABO blood typing. Int J Anal Chem 2016:1–8. https://doi.org/10.1155/2016/1432781

Touhami A (2015) Biosensors and nanobiosensors: design and applications. In: Seifalian A, de Mel A, Kalaskar DM (eds) Nanomedicine. One Central Press, Cheshire, pp 374–400

Tokel O, Yildiz UH, Inci F, Durmus NG, Ekiz OO, Turker B, Demirci U (2015) Portable microfluidic integrated plasmonic platform for pathogen detection. Sci Rep 5:1–9

Vijian D, Chinni SV, Yin LS (2016) Non-protein coding RNA-based genosensor with quantum dots as electrochemical labels for attomolar detection of multiple pathogens. Biosens Bioelectron 77:805–811

Velusamy V, Arshak K, Korostynska O, Oliwa K, Adley C (2010) An overview of foodborne pathogen detection: in the perspective of biosensors. Biotechnol Adv 28(2):232–254

Wang Y, Ye Z, Ying Y (2012) New trends in impedimetric biosensors for the detection of foodborne pathogenic bacteria. Sensors (Basel) 12(3):3449–3471

Wang R, Lum J, Callaway Z, Lin J, Bottje W, Li Y (2015) A label-free impedance immunosensor using screen-printed interdigitated electrodes and magnetic nanobeads for the detection of E. coli O157:H7. Biosensors (Basel) 5:791–803

Wawerla M, Stolle A, Schalch B, Eisgruber H (1999) Review impedance microbiology: applications in food hygiene. J Food Prot 62(12):1488–1496

Wang C, Madiyar F, Yu C, Li J (2017) Detection of extremely low concentration waterborne pathogen using a multiplexing self-referencing SERS microfluidic biosensor. J Biol Eng 11:9

WHO (2015) Global burden of foodborne diseases. WHO Press, Geneva

Wu C, Li X, Song S, Pei Y, Guo L, Pei Z (2017) QCM biosensor based on polydopamine surface for real-time analysis of the binding kinetics of protein-protein interactions. Polymers 9(10):482

Xia H, Wang F, Huang Q, Huang J, Chen M, Wang J et al (2008) Detection of Staphylococcus epidermidis by a quartz crystal microbalance nucleic acid biosensor array using Au nanoparticle signal amplification. Sensors 8(10):6453–6470

Xiang C, Li R, Adhikari B, She Z, Li Y, Kraatz HB (2015) Sensitive electrochemical detection of Salmonella with chitosan-gold nanoparticles composite film. Talanta 1(140):122–127. https://doi.org/10.1016/j.talanta.2015.03.033

Xu M, Wang R, Li YB (2016) Rapid detection of Escherichia coli O157:H7 and Salmonella typhimurium in foods using an electrochemical immunosensor based on screen-printed interdigitated microelectrode and immunomagnetic separation. Talanta 148:200–208

Xu M, Wang R, Li Y (2015) An electrochemical biosensor for rapid detection of E. coli O157:H7 with highly efficient bi-functional glucose oxidase-polydopamine nanocomposites and Prussian blue modified screen-printed interdigitated electrodes, the Royal Society of Chemistry journal Analyst. 21;141(18):5441–5449

Yang L, Bashir R (2008) Research review paper electrical/electrochemical impedance for rapid detection of foodborne pathogenic bacteria. Biotechnol Adv 26:135–150

Yamazaki W, Ishibashi M, Kawahara R, Inoue K (2008) Development of a loop-mediated isothermal amplification assay for sensitive and rapid detection of Vibrio parahaemolyticus. BMC Microbiol. 8:163. https://doi.org/10.1186/1471-2180-8-163

Yoo SM, Lee SY (2016) Optical biosensors for the detection of pathogenic microorganisms. Trends Biotechnol 34(1):7–25

Yu X, Xia Y, Tang Y, Zhang WL, Yeh YT, Lu H, Zheng SY (2017) A nanostructured microfluidic immunoassay platform for highly sensitive infectious pathogen detection. Small *13*(24):1–12

Zhang X, Zhang F, Zhang H, Shen J , Han E , Dong X (2015) Functionalized gold nanorod-based labels for amplified electrochemical immunoassay of E. coli as indicator bacteria relevant to the quality of dairy product. Talanta 132:600–605

Zhao X, Lin CW, Wang J, Oh DH (2014) Advances in rapid detection methods for foodborne pathogens. J Microbiol Biotechnol 24:297–312

Zong Y, Liu F, Zhang Y, Zhan T, He Y, Hun X (2016) Signal amplification technology based on entropy-driven molecular switch for ultrasensitive electrochemical determination of DNA and Salmonella typhimurium. Sens Actuators B 225:420–427

Zhu D, Yan Y, Lei P, Shen B, Cheng W, Ju H, Ding S (2014) A novel electrochemical sensing strategy for rapid and ultrasensitive detection of Salmonella by rolling circle amplification and DNA–AuNPs probe. Anal Chim Acta 846:44–50

Zhu C, Yang G, Li H, Dan D, Lin Y (2015) Electrochemical sensors and biosensors based on nanomaterials and nanostructures. Anal Chem 87(1):230–249

Zeng S, Baillargeat D, Ho H-P, Yong K-T (2014) Nanomaterials enhanced surface plasmon resonance for biological and chemical sensing applications. Chem Soc Rev 43(10):3426

Chapter 20
Electrochemical Methods to Characterize Nanomaterial-Based Transducers for the Development of Noninvasive Glucose Sensors

Nur Alya Batrisya Ismail, Firdaus Abd-Wahab, Nurul Izzati Ramli, Mamoun M. Bader, and Wan Wardatul Amani Wan Salim

20.1 Introduction

Electrochemical biosensors are integrated devices that include an electrochemical transducer, which is a physically or chemically modified electrode with a biological recognition layer (Thevenot et al. 1999). Electrochemical biosensors have emerged in the past two decades as useful analytical tools with wide applications in many fields, one of which is biomedicine, with glucose meters being the most popular (Lee et al. 2016; Bruen et al. 2017; Arduini et al. 2016; Vashist 2012; Wisitsoraat et al. 2013).

Ever since the invention of the Clark oxygen electrode by Professor Leland Charles Clark in 1956, the most successful commercial biosensors on today's market have been for blood glucose; glucose biosensors have been the historical market driver of biosensor technology. Although there are many manufacturers of glucose biosensors, these biosensors have not changed in operational principle over the years (Chambers et al. 2002). The traditional GOx-based glucose biosensors relied on anodic H_2O_2 detection; glucose biosensors are now based predominantly on detection principles of either optical reflectance or electrochemical reaction (Clarke and Foster 2012). The layers in a conventional glucose meter include the chemistry component, the circuitry, and the sample chamber, with the chemistry component

N. A. B. Ismail · F. Abd-Wahab · N. I. Ramli · W. W. A. Wan Salim (✉)
Department of Biotechnology Engineering, Faculty of Engineering, International Islamic University Malaysia, Selangor, Malaysia
e-mail: asalim@iium.edu.my

M. M. Bader
Department of Chemistry, College of Science and General Studies, Alfaisal University, Riyadh, KSA

S. Siddiquee et al. (eds.), *Nanotechnology: Applications in Energy, Drug and Food*,
https://doi.org/10.1007/978-3-319-99602-8_20

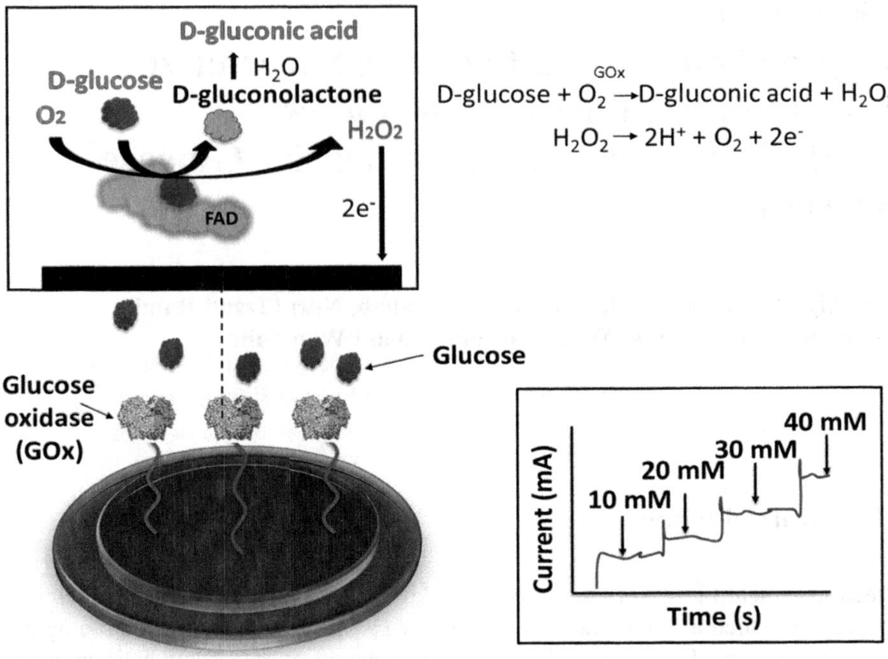

Fig. 20.1 Basic principle of glucose-meter operation: glucose analyte reacts with glucose oxidase enzyme that catalyzes the oxidation of glucose to D-gluconolactone, which can then hydrolyze itself to gluconic acid. The glucose oxidase is a dimer with two protein subunits (monomer). Each monomer has two domains; one binds to the glucose substrate while the other binds to a cofactor and an oxidizing agent, flavin adenine dinucleotide (FAD). In the reaction between glucose and glucose oxidase, FAD is reduced (accepts electrons) to $FADH_2$, which is subsequently oxidized and the oxygen reduced to hydrogen peroxide (H_2O_2), which then produces two electrons and two protons. The electrons flow to a conductive electrode of a glucose meter strip (Source: Reference with copyright permission)

consisting of a chemical mixture of mainly enzyme and mediators that turns glucose into electricity. Enzyme-based amperometric detection is the most developed and frequently used detection principle in the development of glucose meters (Mongra 2012). The principle is based on the measurement of current changes as a function of analyte/substrate addition. Current is produced from the oxidation and reduction of an electroactive product generated from the catalytic reaction at working-electrode surfaces. The basic principle of current glucose-meter operation is illustrated in Fig. 20.1.

Since the mid-1970s, glucose monitoring for diabetics has shifted from tests conducted in a doctor's office to self-monitoring by patients, resulting in the increased number of smaller commercial laboratory analyzers; pioneer companies are Ames, MediSense, and Boehringer Mannheim (Zhang et al. 2015). The home blood-glucose monitor was among the first biosensors on the market and now features various brands, including ReliOn™, FreeStyle Lite™, Precision Xtra™,

OneTouch Ultra™, among others. The monitoring system determines approximate concentration of glucose in the blood and is used mainly by or for people who have diabetes or hypoglycemia.

Current development in glucose meters includes the move towards noninvasive glucose monitoring, and research and development focuses on either enhancing the electrode transduction layer to achieve low detection limit of glucose from non-blood bodily fluids (e.g., saliva and tears), or developing wearable devices with auto-insulin delivery (Arduini et al. 2016; Vashist 2012; Lin 2017). Current glucose meters involve the act of finger pricking ~8–12 times a day, and can create skin irritation and pain, which warrants the need for non-invasive monitoring. Furthermore, the push towards noninvasive glucose monitoring is also a result of many clinical studies that show a strong correlation between glucose levels in blood and those of other bodily fluids such as saliva and tears (Zhang et al. 2015; Satish et al. 2014).

Saliva has become an attractive choice for non-invasive glucose monitoring owing to its accessibility, availability, and its unique attributes as a mirror to human health. According to one study, the mean blood glucose level in a diabetic group is 136.30 ± 28.58 mg/dL, while the mean salivary glucose level in the group is 8.47 ± 4.20 mg/dL (Mongra 2012). Even though the value of mean salivary glucose is much lower in comparison to the mean blood glucose, research shows a strong positive correlation between blood glucose and salivary glucose with r-value close to 88%. In addition, the higher mean salivary glucose level in diabetics in comparison to non-diabetics observed in the aforementioned research was acknowledged in previous studies (Gupta et al. 2015; Sener et al. 2009; Vasconcelos et al. 2010; Carda et al. 2006; Aydin 2007; Zhao et al. 2011). However, studies pertaining to the correlation between blood glucose and salivary glucose have registered varying values of mean salivary glucose level; several factors, such as different methods of saliva collection and analysis, age groups, and medical conditions of the participants, may contribute to this finding. Hence, establishment of a standardized clinical procedure and proper design of reliable clinical studies may enable the use of saliva as a body fluid replacing blood in diabetes diagnostics.

Improvements on glucose meters to achieve detection of such low glucose levels in non-blood body fluid can be achieved only through incorporation of nanomaterials in the electrode transduction layer. Common nanomaterials used for these purposes are carbon nanomaterials such as graphene (Lee et al. 2016; Zhao et al. 2011; Zhu et al. 2015) and carbon nanotubes (Mani et al. 2013; Shi et al. 2011), noble-metal nanoparticles such as gold [Pingarrón et al. 2008; Kesik 2014; Nirala et al. 2015; Liang and Zhuobin 2003) and palladium (Santhosh et al. 2009; Huang et al. 2008), or conductive polymers and ion-exchange resins that can facilitate ion transfer from redox reactions at the electrode-solution interface. Contribution of these nanomaterials towards enhancing electrochemical transduction can be understood through electrochemical voltammetry techniques such as cyclic voltammetry (CV), linear sweep voltammetry (LSV), and electrical impedance spectroscopy (EIS).

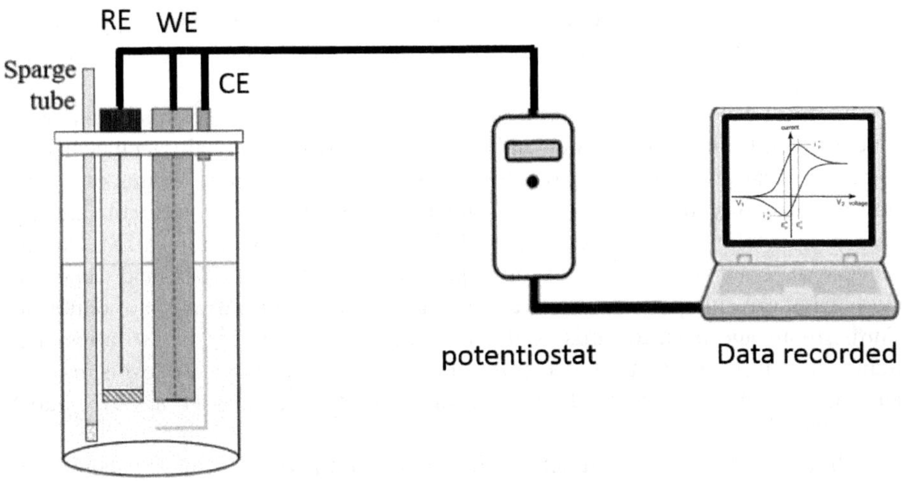

Fig. 20.2 A three electrode system consisting of a noble-metal working electrode, a Ag/AgCl reference electrode, and a noble metal counter electrode immersed in potassium ferricyanide as the electrolyte species. The system is connected to a potentiostat that amplifies the signal at the working electrode, and the amplified signal is measured with a computer (Source: Reference with copyright permission)

Voltammetry techniques are usually performed using a three-electrode system set-up connected to a potentiostat, as shown in Fig. 20.2. A three-electrode system consists of a working electrode, often made from inert material such as glassy carbon or a noble metal like gold or platinum, a reference electrode that maintains constant potential with respect to the working electrode without passing current through it and is usually silver/silver chloride (Ag/AgCl), and a counter electrode or an auxiliary electrode, which allows current to pass and is usually made of the same material as the working electrode (Bard and Faulkner 2001). If oxidation occurs at the working electrode, reduction with the same magnitude of current occurs at the counter electrode, maintaining no current flow between the working and the reference electrode. The high input impedance between the working and the reference electrode ensures that measurements at the working electrode are influenced only by the addition of analyte (Bruen et al. 2017).

This chapter aims to introduce readers to the electrochemical voltammetry techniques used to characterize nanomaterial-modified electrodes. The shape and size of the peaks reveal the relative kinetic rates and diffusion coefficient of the electrochemical system. Characterization of the electrode, functioning as transducer in a biosensor system, allows for better understanding of nanomaterial contribution to improvement of device performance in terms of sensitivity, detection limit, response time, and stability. In this chapter, we will focus on characterization of screen-printed glassy carbon electrodes modified with reduced graphene–conductive polymer composite. To explain the concepts effectively, examples and analysis of voltammetry results from current research will be provided.

20.2 Nanomaterials Enhance Electrochemical Transduction

Electrodes are modified with nanomaterials that can enhance electrochemical transduction in the following ways: improve surface-to-volume ratio that allows for enhanced biochemical interaction between the analyte and the recognition layer, and increase electron transfer. Carbon-based materials such as graphene are popular owing to the materials being chemically inert; any resulting interaction is strictly from the biochemical reactions that result between the recognition layer and the analyte. In our research, we focus on graphene, owing to the 2D honeycomb lattice structure that allows for effective biomolecule immobilization and enhanced electron transfer. Graphene was obtained in an oxidized form called graphene oxide (GO), which is reduced via an electrochemical method into reduced graphene oxide (rGO). To provide room for biomolecule immobilization, GO is partially reduced to retain its functional groups while still being conductive for electron transfer (Wang 2006; Nirala et al. 2015; Abraham et al. 2015).

Similar to graphene, conducting polymers have advantages in terms of inherent biocompatibility, ease of surface modification, and large surface area. Conducting polymers have a π electron spine with alternating single and double bonds that contributes to their unique properties such as relatively high electrical conductivity, high electron affinity, low-energy optical transitions, and low ionization potential (Park et al. 2016). One state-of-the-art conducting polymer is poly (3,4-ethylenedioxythiophene) (PEDOT); owing to the insolubility of most conducting polymers, PEDOT was initially produced to enhance the absence of α,β- and β,β-coupling within the polymer structure and is formed by a standard oxidative or electrochemical polymerization method. The only limitation of PEDOT is its insolubility, which can be overcome by mixing with a water-soluble polyelectrolyte to assist the charge-balancing dopant during polymerization: poly(styrene sulfonic acid) (PSS). The result is a water-soluble PEDOT:PSS with high conductivity, high visible-light transmittance, and high stability (Chambers et al. 2002; Park et al. 2016).

By incorporating both PEDOT:PSS and rGO as nanocomposite, an improved and attractive material with promising properties to enhance electrochemical transduction for biosensing purposes is fabricated for our work. The nanocomposite material is produced by mixing GO and PEDOT:PSS homogeneously and drop casting onto a screen-printed glassy carbon working electrode, with subsequent reduction using a potentiostat to result in a rGO-PEDOT:PSS composite for electrochemical voltammetry characterization. The modified electrode is illustrated in Fig. 20.3.

Fig. 20.3 A screen-printed glassy carbon electrode with working electrode modified with rGO-PEDOT:PSS composite for enhancement of electron transfer at the electrode-solution interface (Source: Reference with copyright permission)

20.3 Electrochemical Techniques to Characterize Materials for Transduction

The voltammetry technique works by regulating a voltage waveform that creates a potential difference between the working electrode and the reference electrode. Current consequently flows through the electrolyte between the working electrode and the counter electrode, while no current flows between the working and the reference electrode. The magnitude of the electrolytic current can then be plotted against its corresponding voltage. In linear sweep voltammetry (LSV), a single waveform of a linear voltage profile is relayed through the electrochemical system and the corresponding output current is measured; a current-voltage curve is typically plotted. Cyclic voltammetry (CV) runs on the same principle as in LSV, but with multiple runs that cycle between two specified voltages. The output current–voltage plot will then have more than one cyclic plot.

Electrical impedance spectroscopy is a technique that is commonly used in complement to cyclic voltammetry tests. While in both linear sweep voltammetry and cyclic voltammetry the voltages that are passed through the electrochemical system are in dc mode, electrical impedance spectroscopy tests are conducted by applying ac potentials.

20.3.1 Cyclic Voltammetry

In CV, the voltage sweep that is generated by the potentiostat is a two-way step that linearly changes back and forth between two specified voltage levels. The forward scan sweep can be likened to a linear sweep voltammetry waveform, as the voltage linearly decreases from an initial value to a lower value. Once it reaches the specified end-point voltage, or the switching potential, the direction of the sweep is then reversed and the voltage linearly increased back to its initial value.

The waveform of the voltage applied, when plotted against time, is a triangular-shaped curve, and the slope of this curve gives us information about the scan rate. A scan-rate value tells us how fast the voltage change is applied to the working electrode in both the forward and reverse scans (Parker 1986). A typical output of an electrode reacting with the surrounding electrolytes, as the voltage sweep is applied, is illustrated in the current vs. electrode potential curve plot (I-V curve, Fig. 20.4). We can qualitatively obtain information on our electrochemical systems by analyzing the I-V curve from a CV run. This provides us with initial

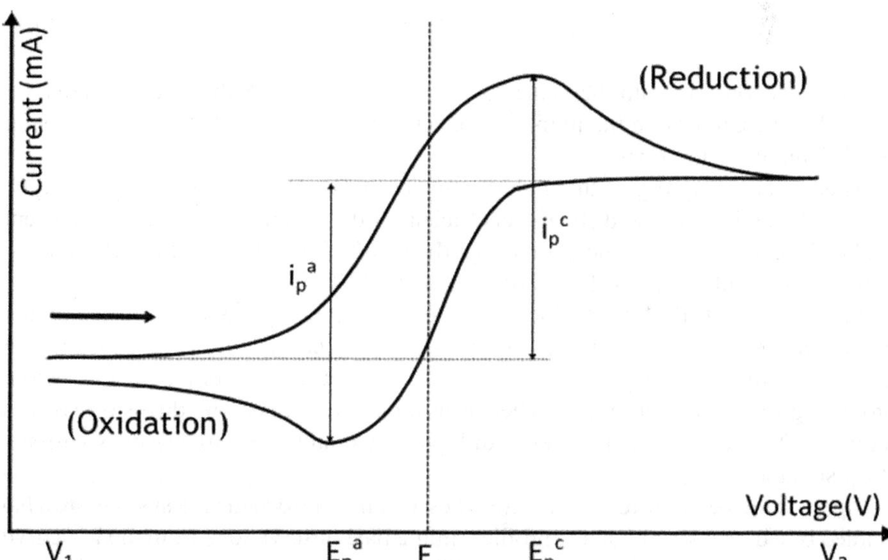

Fig. 20.4 Cyclic voltammogram for a typical reversible redox reaction; the current is continuously monitored as the voltage sweep is run from a high positive potential (V_1) to a low negative potential (V_2) and back to V_1. As it goes from V_1 to V_2, the electrolytic solution (analyte) is reduced at the electrodes and current is produced. At the point when the sweep reaches peak cathodic potential, the reaction produces a peak current (i_p^c). As the voltage is reversed from V_2 to V_1, the analyte is oxidized at the electrode surface. A peak anodic current (i_p^a) is produced upon reaching the peak anodic potential E_p^a. The voltammogram above is displayed according to the US convention, whereby the high voltage is shown on the left-hand side of the axes, and the reduction curve is plotted above the oxidation plot. The IUPAC convention is a 180°-rotated version of this convention (Source: Reference with copyright permission)

characterization of our electrodes, or even comparative analysis of the effect of introducing transducer materials on the surface of the electrode.

As the scan moves in the forward direction, reduction occurs in the electrochemical solution. Charge is transferred from the electrode to the electrolytic species residing at the surface of the electrode (Bard and Faulkner 2001; Elgrishi et al. 2018). This is translated as current in the I-V curve. Upon reaching the switching voltage that has been specified in the potentiostat, the scan direction is then reversed, and the voltage is linearly returned back to its initial value. At this stage, oxidation takes place, and charge is transferred from the electrolyte to the electrode, which can be seen in Fig. 20.4 as a reversed current plot.

Most electrode reactions that occur in the electrochemical solution involve rapid charge transfer, such that the amplitude of the current at each voltage step is mainly dependent on the diffusion rate of the electrolytic species at the surface of the electrode (Brownson and Banks 2014). At the point in time where the diffusion is highest, the current amplitude will reach a peak value, as shown in the voltammogram (I_p^c). The magnitude of the peak current, I_p, is governed by the Randles-Sevcik equation (Elgrishi et al. 2018);

$$I_p = kn^{3/2}AD^{1/2}C^b\upsilon^{1/2} \tag{20.1}$$

where k is a constant with the value 2.69×10^5 (at 25 °C), A the electrode area, C^b the bulk concentration, n the number of electrons per mole species, D the diffusion coefficient, and v the scan rate.

This equation highlights the high dependence of the peak current on electrolytic species. In addition to the diffusion coefficient of the species itself, the peak current is also dependent on its concentration gradient, which is influenced by the concentration at the bulk solution (Bard and Faulkner 2001).

Scan rate affects the rate at which the potential step is applied to the electrode, and this regulates the diffusion layer thickness at the electrode surface. At lower scan rates, the diffusion layer grows beyond the electrode surface to form a thicker region, producing lower current output, whereas at higher scan rates, the diffusion layer is reduced. This permits the occurrence of higher flux, and hence higher peak currents (Elgrishi et al. 2018).

Apart from peak current, characteristics of an electrochemical system are also studied based on a set of voltage profiles. Immediate observations from the I-V curve would provide us with peak potentials for both the anodic (oxidation of species) and cathodic (reduction of species) sides. These peak potentials would correspond to the potential where peak currents are observed. In reversible electrode reactions, where the charge transfer is rapid, the peak potential is independent of the scan rate (Bard and Faulkner 2001; Mirkin 2007). This means that regardless of the scan rate used, the peak current will appear at the same voltage in the I-V curve. An irreversible reaction, on the other hand, will undergo a shift in the peak potential when the scan rate is changed (Elgrishi et al. 2018).

To illustrate further the use of cyclic voltammetry to provide qualitative analysis of electrochemical systems, we shall use the cyclic voltammogram shown in Fig. 20.5. The experimental design of this cyclic voltammetry test uses a composite electrode comprised of reduced graphene oxide and a conductive polymer (PEDOT: PSS). A Ag/AgCl electrode is used as the reference in this system, and the cell is filled with 0.1 M potassium ferricyanide as the electrolyte species. The reaction taking place at the electrode surface is a ferricyanide/ferrocyanide coupled redox reaction. The forward scan (reduction) process is shown below, and the oxidation process is merely the reverse direction.;

$$\mathrm{Fe(CN)_6}^{-3} + \mathrm{e} \rightarrow \mathrm{Fe(CN)_6}^{-4} \tag{20.2}$$

The I-V curve shown here illustrates the cyclic voltammetry test done at different scan rates, ranging between 25 and 300 mV/s. From here qualitative analysis may be deduced. First, the peak currents appear to increase linearly in both directions. In relation to the Randles-Sevcik equation, this can be seen clearly if one were to plot the peak currents against the root-squared scan rates (i_p vs. $\nu^{1/2}$, Fig. 20.5) (Elgrishi et al. 2018; Brownson and Banks 2014; Randviir and Banks 2013). The diffusion coefficient may be obtained from the slope of this plot, given that the electrode area and the bulk concentration are known. Second, the peak potentials appear to be independent of the scan rates. In this case, peak anodic current occurs at 0.07 V, whereas peak cathodic current occurs at 0.25 V, at all scan rates. This signifies that the electrode reaction taking place in our system is a reversible reaction. However, theoretically for reversible reactions, the peak potential difference should correspond to a value of 59 mV/n, where n is the mole number of electrons per mole of the electrolytic species. Since we know that the conversion of ferricyanide to ferrocyanide involves one electron ($n = 1$), the peak potential difference, ΔEp, should then be 59 mV (Brownson and Banks 2014). While detailed analysis of the cause behind this discrepancy is beyond the scope of this text, it is important to highlight here that the theoretical value of 59 mV for peak potential difference is rarely observed in practice. Nonetheless, the independence of peak currents from scan rates remains a clear indicator of reversible reactions.

20.3.2 Linear Sweep Voltammetry

Linear sweep voltammetry (LSV) measures cell current as a function of time. A potentiostat is used to apply the voltage at the working electrode (WE) and reference electrode (RE) in a single scan direction. The direction of the voltage scan is determined by the reaction of interest. For instance, voltage scan from a positive potential to a negative potential will give insight into a reduction reaction, while voltage scan from a negative potential to a positive potential will give insight into an oxidation reaction (Bard and Faulkner 2001). Figure 20.6 shows an example of a

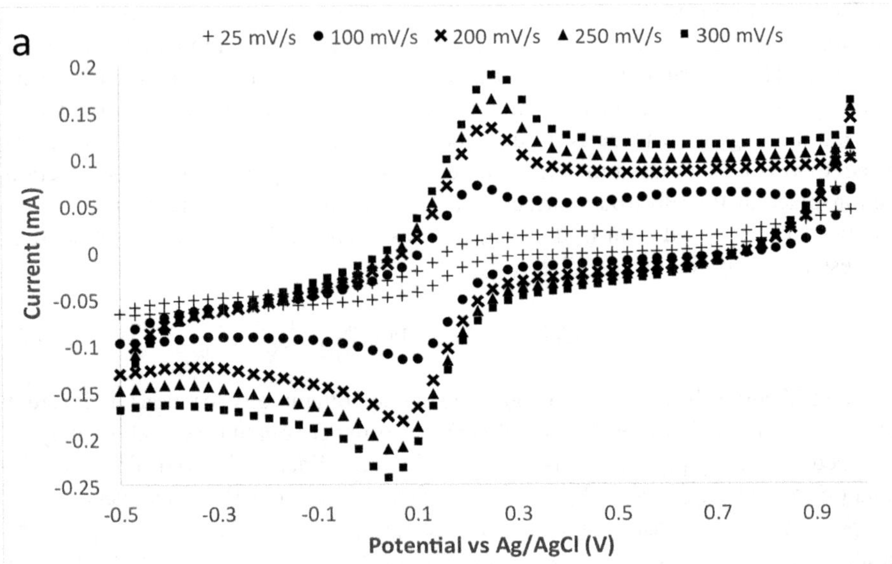

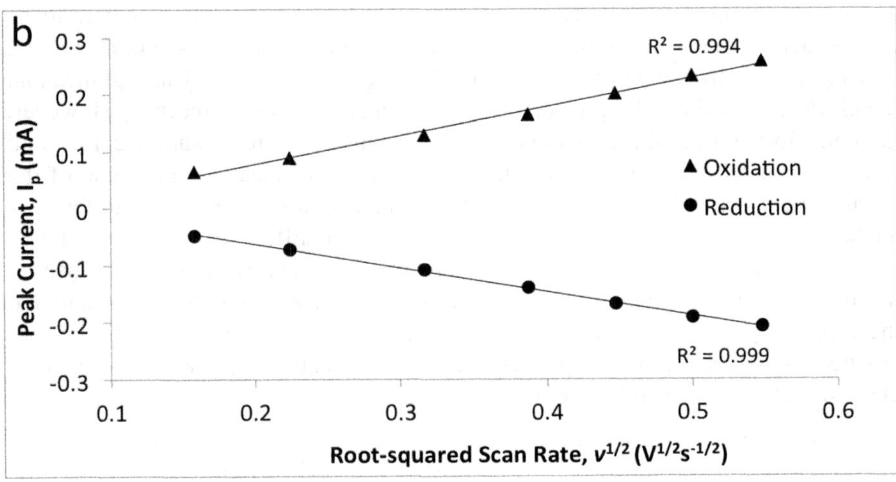

Fig. 20.5 Scan-rate characterization of the electrode reaction; (**a**) cyclic voltammetry at different scan rates for a composite electrode containing reduced graphene oxide and PEODT:PSS in the transducer. The analyte used in this experiment is 0.1 M potassium ferricyanide. The sweep was run at potentials from 1 to −0.5 V and back. The arrow shows the direction of the sweep. The peak currents can be located in the curve at peak potentials of 0.07 and 0.25 V for reduction and oxidation currents, respectively. (**b**) The current function plot of peak currents vs. root squared scan rate. Both plots, each for oxidation and reduction reactions, display a linear relationship. This provides qualitative justification that the electrode reactions are reversible (Source: Reference with copyright permission)

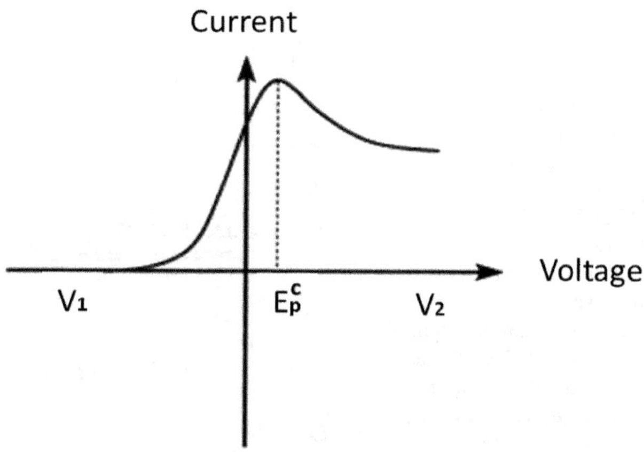

Fig. 20.6 Linear sweep voltammogram showing a one-direction voltage sweep from the right (V1) to the left (V2); a peak current is observed between these two voltages (Source: Reference with copyright permission)

LSV; the direction of the voltage scan starts from the right side (V1) towards the left side (V2) of the plot and a peak current (I_p) can be observed between the fixed voltages. The peak current (I_p) is the largest current measured, denoting the maximum amount of $Fe(CN)_6{}^{3-}$ ion that can be reduced by the electrode before the diffusion layer on the electrode surface slows down the $Fe(CN)_6{}^{3-}$ ion delivery from bulk solution, causing the decline of the current measurement. The peak potential (E_p) is the potential of the peak current (Wang 2006).

LSV was performed on the WE of a screen-printed glassy carbon electrode modified with the rGO:PEDOT:PSS composite, in 0.1 M potassium ferricyanide, $K_3[Fe(CN)_6{}^3]$, redox active solution, as shown in Fig. 20.7. The voltage is swept from 1 to −0.5 V for scan rates of 25, 50, 100, 150, 200, 250, and 300 mV/s. As the $Fe(CN)_6{}^{3-}$ is being reduced to $Fe(CN)_6{}^{4-}$ at the WE, the output current is recorded by a potentiostat. As can be observed from Fig. 20.6, the output current increases with increasing scan rates. The peak current is the largest at the highest scan rate (300 mV/s) at a potential of 0.04 V. Alteration of scan rate will affect the current response owing to the size of the diffusion layer on the electrode surface. When the scan rate is increased, the diffusion layer above the electrode surface is reduced. As a result, higher current can be observed. There are several factors affecting the value of peak current, including the concentration of analyte, the kinetics of electron transfer and kinetics of mass transport of the analyte. The peak current can also be determined using the Randles-Sevcik equation.

20.3.3 Electrical Impedance Spectroscopy

Electrochemical impedance spectroscopy (EIS) is a useful technique to study the interfacial behavior of molecules and adsorption properties on the surface of the electrode and was introduced in the late nineteenth century by mathematician Oliver Heaviside. This technique yields real values of impedance in temporal space. The

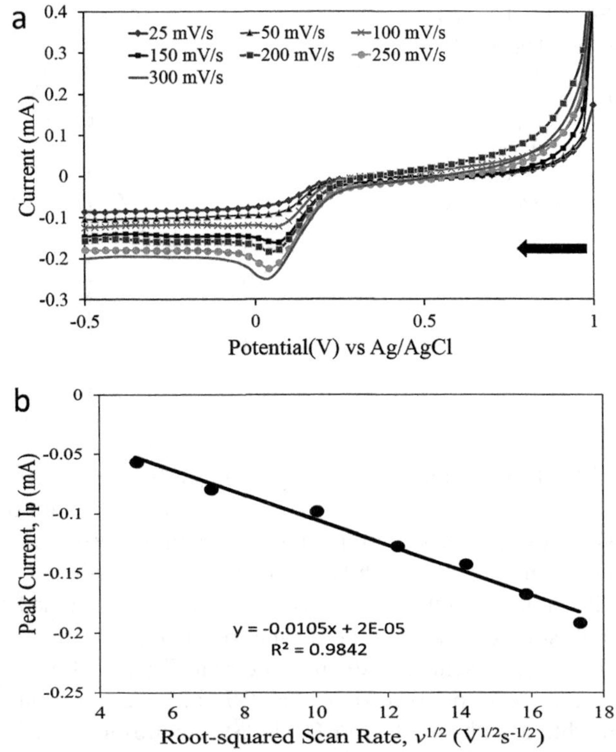

Fig. 20.7 Linear sweep voltammetry (LSV); (**a**) graph showing LSV at different scan rates for a composite electrode containing reduced graphene oxide and PEDOT:PSS in the transducer. The analyte used in this experiment is 0.1 M potassium ferricyanide. The sweep was run at potentials from 1 to −0.5 V. The arrow shows the direction of the sweep. (**b**) The current function plot of peak currents vs. root squared scan rate. Both plots, each for oxidation and reduction reactions, display a linear relationship. This provides qualitative justification that the electrode reactions are reversible (Source: Reference with copyright permission)

relationship between impedance and the diffusional electrochemical system was reported by Warburg in 1899 (Randviir and Banks 2013). Generally, EIS can be utilized as a supplementary tool to other electrochemical characterization methods.

EIS can identify the effects of modifying electrodes with the rGO-PEDOT:PSS composite, including electrochemical contributions from polar, ionic, and dielectric relaxation processes in the electrolyte system, as well as within the electrode, at the electrode surface, and within the double-layer region. Exchange-current densities, charge-transfer resistances, double-layer capacitances, and other key parameters of an electrochemical system under investigation can also be measured. EIS can also be used to estimate physical parameters such as surface roughness and the porosity of a modified electrode (Fernández-Sánchez et al. 2005). An EIS experiment is conducted using a potentiostat by applying a fixed sinusoidal voltage across an electrochemical cell containing the electrolyte and analyte of interest. The data obtained from EIS can be represented by Nyquist plots (Fig. 20.8a), where the real versus imaginary impedance components are plotted, and by Bode plots (Fig. 20.8b), where impedance phase angle is plotted against frequency. A Randles equivalent circuit (Fig. 20.8c) is commonly used to simulate EIS experiments and to understand the different components contributing to the overall impedance of the simplified circuits (Orazem et al. 2006).

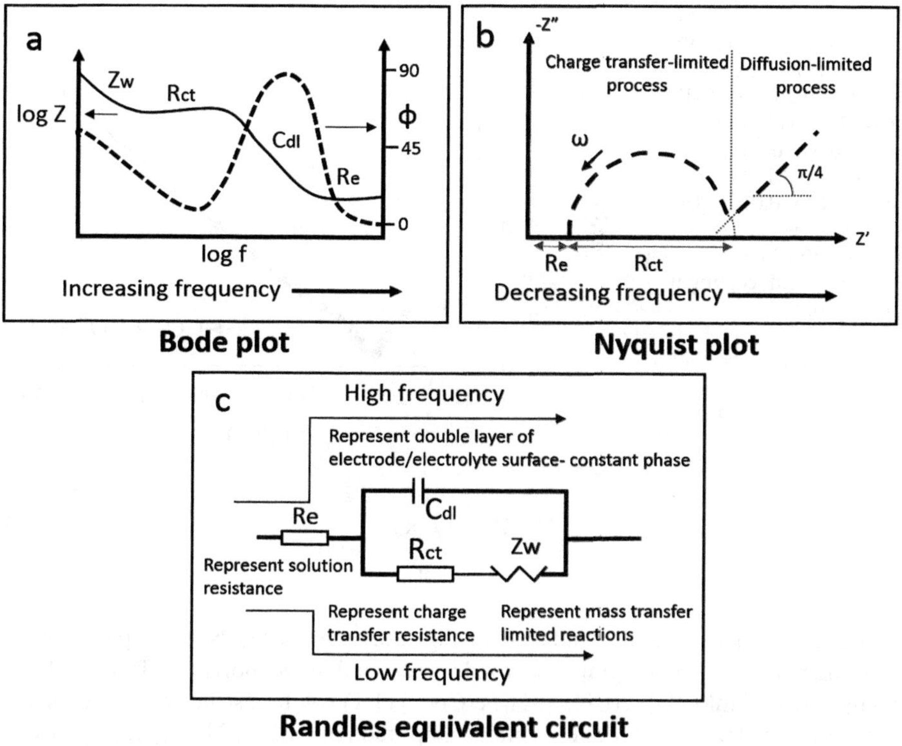

Fig. 20.8 (**a**) Bode plot: log of impedance and phase angle against log of frequency. (**b**) Nyquist plot: imaginary impedance against real impedance. (**c**) Simple Randles equivalent circuit that best describes the impedance behavior of a typical electrochemical cell. The Bode plot shows an increase in frequency from the left-hand side to the right-hand side of the plot, whereas in Nyquist plot, frequency increases in the opposite direction (Source: Reference with copyright permission)

In principle, impedance is simply the force in opposition to electrical current in a circuit and is measured in the same units as resistance, Ω. However, resistance differs from impedance because resistance obeys Ohm's law, and is observed in dc circuits where the resistance is technically impedance at zero phase angles since the current is not alternating (Orazem and Tribollet 2008). The capacitance and inductance are related to current and voltage as seen in Eqs. (20.3) and (20.4):

$$I = CV \tag{20.3}$$

$$I = V/L \tag{20.4}$$

where I is the current, C is the capacitance, V is the voltage, and L is the inductance. Hence, the impedance function Z can be written in terms of capacitance or inductance as seen in Eqs. (20.5) and (20.6), where $\omega = 2\pi f$, and f is frequency:

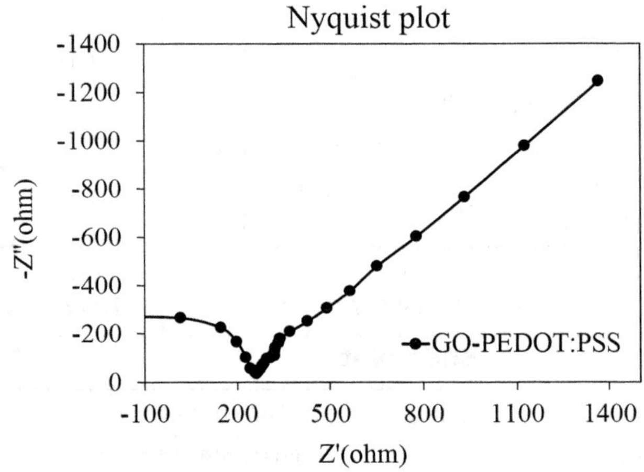

Fig. 20.9 Nyquist plot of modified GO-PEDOT:PSS nanocomposite electrode in 5 mM $K_3[Fe(CN)_6]$:$K_4[Fe(CN)_6]$ (1:1) electrolyte scanned at a frequency of 100 kHz to 0.1 Hz and amplitude of 0.02 V. The frequency increases from right to left (Source: Reference with copyright permission)

$$Z = 1/\omega C \tag{20.5}$$

$$Z = \omega L \tag{20.6}$$

An experimental result displayed in Fig. 20.9 shows the Nyquist plots for an electrode modified with graphene oxide and conductive polymer (PEDOT:PSS) composite in 5 mM $K_3[Fe(CN)_6]$:$K_4[Fe(CN)_6]$ (1:1) solution scanned at a frequency range of 100 kHz to 0.1 Hz and a voltage amplitude of 0.02 V. The Nyquist plot was used to interpret and deduce solution resistance, charge transfer resistance, Warburg impedance, and time constant. The results from the figure suggested that the properties of the composite in the modified electrode are heavily influenced by diffusion-controlled regions. The linear line observed in the low-frequency region can be represented by the Warburg element, as seen in the Randles equivalent circuit. The Warburg element accounts for the diffusion of ions in solution in an electrochemical reaction. For instance, at high frequencies, Warburg impedance is not observed, as migration occurs over much longer time periods than the operational frequency and thus the relatively slow movement of molecules in solution renders impedance contributions negligible and will shift towards the double-layer capacitance elements represented by the small semicircle observed in Fig. 20.9.

20.4 Conclusions

This chapter describes voltammetry techniques commonly used to characterize nanomaterial-modified electrode surfaces for electrochemical biosensing applications. In this chapter, all electrodes were modified with a composite of rGO-PEDOT:PSS; the composite was formed by drop-casting a mixture of GO and PEDOT:PSS onto a screen-printed glassy carbon electrode followed by reduction to a rGO-PEDOT:PSS composite. We believe from the results of our research that the composite can enable the development of glucose meters capable of detecting glucose at the low levels in saliva and can help advance the development of non-invasive glucose meters.

To understand the behavior of screen printed glassy carbon electrodes modified with rGO-PEDOT:PSS and whether rGO-PEDOT:PSS can be used as to enhance electrochemical transduction, we must understand the material properties in terms of the electron-transfer process relative to the voltage scan rate. The shape and size of the peaks in CV and LSV indicate the relative kinetic rates and diffusion coefficient of the electrochemical system. EIS, on the other hand, can provide rich information on the interfacial behavior of molecules and adsorption properties on the surface of the electrode. From the CV, LSV, and EIS graphs, one can say that electrodes modified with rGO-PEDOT:PSS composite are diffusion limited. From the CV graph of the composite-modified electrodes, one can also obtain the diffusion coefficient from the slope of the graph, and the independence of the peak current from the change in scan rates implies a reversible reaction. On the other hand, LSV is used when you are interested in a process that occurs in the direction of the potential applied. The Nyquist plot from the EIS measurements shows that the composite before reduction is heavily influenced by a diffusion-controlled region, supporting the results of the CV and LSV. However, further study is needed to analyze the Nyquist plot of the reduced form (rGO-PEDOT:PSS) to further characterize the reduction process.

The results from CV, LSV and EIS inform us about the effect of the composite on electrochemical transduction, allowing for further improvement and modifications to be made on the electrodes. However, the results are in no way conclusive to determine overall sensor performance, which also depends on biomolecule immobilization on the composite as well as on the electronics. The concentration and activity of immobilized enzyme, in our case glucose oxidase, plays an important role in the performance of biosensors. Therefore, modification of the electrodes and the immobilization techniques for these biomolecules must be carefully considered for further improvement of device performance. Our current work focuses on optimizing the transduction along with glucose oxidase immobilization and design of suitable electronics. Our aim is to develop a glucose meter that can measure glucose in the salivary glucose range, with intention for non-invasive glucose monitoring.

References

Abraham S, Srivastava S, Kumar V et al (2015) Enhanced electrochemical biosensing efficiency of silica particles supported on partially reduced graphene oxide for sensitive detection of cholesterol. J Electroanal Chem 757:65–72. http://sci-hub.tw/10.1016/j.jelechem.2015.09.016

Arduini F, Micheli L, Moscone D et al (2016) Electrochemical biosensors based on nanomodified screen-printed electrodes: recent applications in clinical analysis. Trends Anal Chem 15:1–33. http://sci-hub.tw/10.1016/j.trac.2016.01.032

Aydin S (2007) A comparison of ghrelin, glucose, alpha-amylase and protein levels in saliva from diabetics. J Biochem Mol Biol 40:29–35. http://sci-hub.tw/10.5483/BMBRep.2007.40.1.029

Bard AJ, Faulkner LR (2001) Electrochemical methods fundamentals and applications, 2nd edn. Wiley, New York

Brownson DAC, Banks CE (2014) Interpreting electrochemistry. In: The handbook of graphene electrochemistry. Springer, London, pp 23–77

Bruen D, Delaney C, Florea L, Diamond D (2017) Glucose sensing for diabetes monitoring: recent developments. Sensors (Basel) 17:1–21. http://sci-hub.tw/10.3390/s17081866

Carda C, Mosquera-Lloreda N, Salom L et al (2006) Structural and functional salivary disorders in type 2 diabetic patients. Med Oral Patol Oral Cir Bucal 11:309–314

Chambers JP, Arulanandam BP, Matta LL et al (2002) Biosensor recognition elements. Curr Issues Mol Biol 10:1–12

Clarke SF, Foster JR (2012) A history of blood glucose meters and their role in self-monitoring of diabetes mellitus. Br J Biomed Sci 69:83–93

Elgrishi N, Rountree KJ, McCarthy BD et al (2018) A practical beginner's guide to cyclic voltammetry. J Chem Educ 95:197–206. http://sci-hub.tw/10.1021/acs.jchemed.7b00361

Fernández-Sánchez C, McNeil CJ, Rawson K (2005) Electrochemical impedance spectroscopy studies of polymer degradation: application to biosensor development. Trends Anal Chem 24:37–48. http://sci-hub.tw/10.1016/j.trac.2004.08.010

Gupta S, Sandhu SV, Bansal H, Sharma D (2015) Comparison of salivary and serum glucose levels in diabetic patients. J Diabetes Sci Technol 9:91–96. http://sci-hub.tw/10.1177/1932296814552673

Huang J, Liu Y, Hou H, You T (2008) Simultaneous electrochemical determination of dopamine, uric acid and ascorbic acid using palladium nanoparticle-loaded carbon nanofibers modified electrode. Biosens Bioelectron 24:632–637. http://sci-hub.tw/10.1016/j.bios.2008.06.011

Kesik M (2014) A functional immobilization matrix based on a conducting polymer modified with PMM/Clay nanocomposites and gold nanoparticles: applications to amperometric glucose biosensors. Middle East Technical University, Ankara

Lee J, Kim J, Kim S, Min D-H (2016) Biosensors based on graphene oxide and its biomedical application. Adv Drug Deliv Rev 105:1–13. http://sci-hub.tw/10.1016/j.addr.2016.06.001

Liang W, Zhuobin Y (2003) Direct Electrochemistry of glucose oxidase at a gold electrode modified with single-wall carbon nanotubes. Sensors 3:544–554. http://sci-hub.tw/10.3390/s31200544

Lin T (2017) Non-invasive glucose monitoring: a review of challenges and recent advances. Curr Trends Biomed Eng Biosci 6:1–8. http://sci-hub.tw/10.19080/CTBEB.2017.06.555696

Mani V, Devadas B, Chen SM (2013) Direct electrochemistry of glucose oxidase at electrochemically reduced graphene oxide-multiwalled carbon nanotubes hybrid material modified electrode for glucose biosensor. Biosens Bioelectron 41:309–315. http://sci-hub.tw/10.1016/j.bios.2012.08.045

Mirkin MV (2007) Determination of electrode kinetics. In: Zoski CG (ed) Handbook of electrochemistry, 1st edn. Elsevier B.V, Oxford, pp 639–660

Mongra AC (2012) Commercial biosensors: an outlook. J Acad Ind Res 1:310–312

Nirala NR, Abraham S, Kumar V et al (2015) Partially reduced graphene oxide-gold nanorods composite based bioelectrode of improved sensing performance. Talanta 144:745–754. http://sci-hub.tw/10.1016/j.talanta.2015.05.059

Orazem ME, Tribollet B (2008) Electrochemical impedance spectroscopy. Wiley, Hoboken

Orazem ME, Pébère N, Tribollet B (2006) Enhanced graphical representation of electrochemical impedance data. J Electrochem Soc 153:B129. http://sci-hub.tw/10.1149/1.2168377

Park CS, Lee C, Kwon OS (2016) Conducting polymer based nanobiosensors. Polymers (Basel) 8:1–18. http://sci-hub.tw/10.3390/polym8070249

Parker VD (1986) Linear sweep and cyclic voltammetry. In: Bamford CH, Tipper CFH, Compton RG (eds) Comprehensive chemical kinetics. Elsevier, Amsterdam, pp 145–202

Pingarrón JM, Yáñez-Sedeño P, González-Cortés A (2008) Gold nanoparticle-based electrochemical biosensors. Electrochim Acta 53:5848–5866. http://sci-hub.tw/10.1016/j.electacta.2008.03.005

Randviir EP, Banks CE (2013) Electrochemical impedance spectroscopy: an overview of bioanalytical applications. Anal Methods 5:1098. http://sci-hub.tw/10.1039/c3ay26476a

Santhosh P, Manesh KM, Uthayakumar S et al (2009) Fabrication of enzymatic glucose biosensor based on palladium nanoparticles dispersed onto poly(3,4-ethylenedioxythiophene) nanofibers. Bioelectrochemistry 75:61–66. http://sci-hub.tw/10.1016/j.bioelechem.2008.12.001

Satish BNVS, Srikala P, Maharudrappa B et al (2014) Saliva: a tool in assessing glucose levels in diabetes mellitus. J Int Oral Health 6:114–117

Sener A, Jurysta C, Bulur N et al (2009) Salivary glucose concentration and excretion in normal and diabetic subjects. J Biomed Biotechnol 2009:2018. http://sci-hub.tw/10.1155/2009/430426

Shi J, Claussen JC, McLamore ES et al (2011) A comparative study of enzyme immobilization strategies for multi-walled carbon nanotube glucose biosensors. Nanotechnology 22:1–10. http://sci-hub.tw/10.1088/0957-4484/22/35/355502

Thevenot D, Toth K, Durst R, Wilson G (1999) Electrochemical biosensors: recommended definitions and classification. Pure Appl Chem 71:2333–2348. http://sci-hub.tw/10.1081/AL-100103209

Vasconcelos ACU, Soares MSM, Almeida PC, Soares TC (2010) Comparative study of the concentration of salivary and blood glucose in type 2 diabetic patients. J Oral Sci 52:293–298. http://sci-hub.tw/10.2334/josnusd.52.293

Vashist SK (2012) Non-invasive glucose monitoring technology in diabetes management: a review. Anal Chim Acta 750:16–27. http://sci-hub.tw/10.1016/j.aca.2012.03.043

Wang J (2006) Analytical electrochemistry, 3rd edn. Wiley, Hoboken

Wisitsoraat A, Pakapongpan S, Sriprachuabwong C et al (2013) Graphene-PEDOT:PSS on screen printed carbon electrode for enzymatic biosensing. J Electroanal Chem 704:208–213. http://sci-hub.tw/10.1016/j.jelechem.2013.07.012

Zhang W, Du Y, Wang ML (2015) Noninvasive glucose monitoring using saliva nano-biosensor. Sens Biosens Res 4:23–29. http://sci-hub.tw/10.1016/j.sbsr.2015.02.002

Zhao F, Wang F, Zhao W et al (2011) Voltammetric sensor for caffeine based on a glassy carbon electrode modified with Nafion and graphene oxide. Microchim Acta 174:383–390. http://sci-hub.tw/10.1007/s00604-011-0635-y

Zhu X, Xu J, Duan X et al (2015) Controlled synthesis of partially reduced graphene oxide: enhance electrochemical determination of isoniazid with high sensitivity and stability. J Electroanal Chem 757:183–191. http://sci-hub.tw/10.1016/j.jelechem.2015.09.038

MIX
Papier aus verantwortungsvollen Quellen
Paper from responsible sources
FSC® C105338

Printed by Books on Demand, Germany